Free Radicals and Cancer

Free Radicals and Cancer

edited by ROBERT A. FLOYD

Biomembrane Research Laboratory
Oklahoma Medical Research Foundation
Oklahoma City, Oklahoma

MARCEL DEKKER, INC. New York and Basel

Library of Congress Cataloging in Publication Data
Main entry under title:

Free radicals and cancer.

Includes bibliographical references and indexes.
1. Carcinogenesis. 2. Radicals (Chemistry)
I. Floyd, Robert A. [date]. [DNLM: 1. Free
radicals. 2. Neoplasms--Etiology. QZ 202 F853]
RC268.5.F73 1982 616.99'4071 82-13046
ISBN 0-8247-1551-9

MARCEL DEKKER, INC.
270 Madison Avenue, New York, New York 10016

Current printing (last digit):
10 9 8 7 6 5 4 3 2 1

PRINTED IN THE UNITED STATES OF AMERICA

CONTENTS

CONTRIBUTORS

Larry S. Andrews, Ph.D.,[*] Laboratory of Chemical Pharmacology, National Heart, Lung, and Blood Institute, National Institutes of Health, Bethesda, Maryland

John E. Biaglow, Ph.D., Department of Biochemistry and Radiology, Case Western Reserve University, Cleveland, Ohio

Albert M. Bobst, Ph.D., Department of Chemistry, University of Cincinnati, Cincinnati, Ohio

B. Bösterling, Ph.D.,[†] Department of Spectroscopy, Max-Planck-Institut für Biophysikalische Chemie, Göttingen, Federal Republic of Germany

J. Cadet, Ph.D., Département de Recherche Fondamentale, Laboratoire de Radiobiochimie, Grenoble, France

Ercole Cavalieri, Ph.D., Eppley Institute for Research in Cancer, University of Nebraska Medical Center, Omaha, Nebraska

R. Claus, Ph.D., Department of Spectroscopy, Max-Planck-Institut für Biophysikalische Chemie, Göttingen, Federal Republic of Germany

Patricia A. Craven, Ph.D., Department of Medicine, Veterans Administration Hospital, and University of Pittsburgh, Pittsburgh, Pennsylvania

Edward R. Davis, Ph.D.,[‡] Department of Chemistry, Guelph-Waterloo Centre for Graduate Work in Chemistry, University of Guelph, Guelph, Ontario, Canada

Frederick R. DeRubertis, M.D., Department of Medicine, Veterans Administration Hospital, and University of Pittsburgh, Pittsburgh, Pennsylvania

---

*Present affiliation: ARCO Chemical, New Town Square, Pennsylvania

†Present affiliation: Stanford University, Stanford, California

‡Present affiliation: Department of Chemistry, University of Delaware, Newark, Delaware

N. M. Emanuel, Ph.D., Department of Chemical and Biological Kinetics, Institute of Chemical Physics, Academy of Sciences of the USSR, Moscow, USSR

Robert A. Floyd, Ph.D., Biomembrane Research Laboratory, Oklahoma Medical Research Foundation, Oklahoma City, Oklahoma

James R. Gillette, Ph.D., Laboratory of Chemical Pharmacology, National Heart, Lung, and Blood Institute, National Institutes of Health, Bethesda, Maryland

Jack A. Hinson, Ph.D.,* Laboratory of Chemical Pharmacology, National Heart, Lung, and Blood Institute, National Institutes of Health, Bethesda, Maryland

Yoshikazu Ioki, Ph.D., Biophysics Division, National Cancer Center Research Institute, Tokyo, Japan

Edward G. Janzen, Ph.D., Department of Chemistry, Guelph-Waterloo Centre for Graduate Work in Chemistry, University of Guelph, Guelph, Ontario, Canada

Teruyuki Kimura, Ph.D., Biophysics Division, National Cancer Center Research Institute, Tokyo, Japan

Cameron J. Koch, Ph.D., Cross Cancer Foundation, Edmonton, Alberta, Canada

Masahiko Kodama, M.D., Biophysics Division, National Cancer Center Research Institute, Tokyo, Japan

Chikayoshi Nagata, Ph.D., Biophysics Division, National Cancer Center Research Institute, Tokyo, Japan

I. Reitz, Department of Spectroscopy, Max-Planck-Institut für Biophysikalische Chemie, Göttingen, Federal Republic of Germany

Eleanor Rogan, Ph.D., Eppley Institute for Research in Cancer, University of Nebraska Medical Center, Omaha, Nebraska

Robert Roth, Ph.D.,† Eppley Institute for Research in Cancer, University of Nebraska Medical Center, Omaha, Nebraska

Rajagopalan Sridhar, Ph.D., Biomembrane Research Laboratory, Oklahoma Medical Research Foundation, Oklahoma City, Oklahoma

*Present affiliation: Division of Carcinogenesis Research, National Research Center for Toxicological Research, Jefferson, Arkansas

†Present affiliation: Midwest Research Institute, Kansas City, Missouri

A. Stier, Ph.D., Department of Spectroscopy, Max-Planck-Institut für Biophysikalische Chemie, Göttingen, Federal Republic of Germany

Harold M. Swartz, Ph.D., M.D., College of Medicine, University of Illinois, Urbana, Illinois

R. Téoule, Ph.D., Département de Recherche Fondamentale, Laboratoire de Radiobiochemie, Grenoble, France

Marie E. Varnes, Ph.D., Department of Biochemistry and Radiology, Case Western Reserve University, Cleveland, Ohio

PREFACE

The idea that cancer development and free radicals are in some undefined manner related has been held by a few investigators for at least 30 years. The vagueness of the hypothetical relationship has recently, as a result of experimentation, undergone transformation such that it is now becoming possible to define certain areas where there is strong association and other areas where there is apparently no association between free radicals and cancer. The purpose of this book is to bring together the thoughts and experimental results of some of the best researchers in this field. It has been my goal to collect in one book an overview of how free radicals relate to cancer from as many different aspects as possible. We have sought to clarify what is really known about the subject and thus aid the oncologist's focus on problems which will most likely yield useful results.

In attempting to summarize the field, I present in schematic form how free radical processes and cancer development appear to be related. Cancer development is viewed as progressing through a series of sequential events. There may be two or more paths from one particular carcinogen to the same end result, namely tumor. A particular path may proceed through certain steps which are associated with free radical events, and therefore appear to belong to the free radical domain.

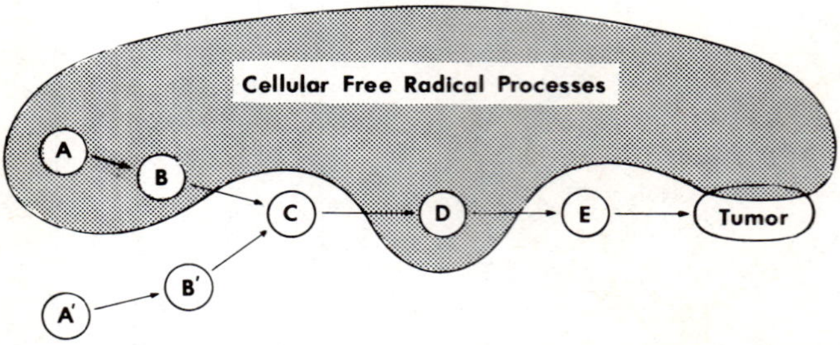

If this scheme represents a close approximation to the true events that occur in tumor development, then it is understandable how various investigators may reach different conclusions depending upon the phase of tumor development with which they are experimenting. The most important question—Does tumor development proceed through at least one step involving free radical events?—cannot be answered unequivocally at the present time. Neither has it been unequivocally proven that for any particular case of tumor development there is a free radical event that controls the overall rate of the process. On the other hand, as attested by the collection of data presented here by many workers in the field, there is much circumstantial evidence that strongly indicates that free radicals do play a significant role in many particular cases of tumor development. It is certain however, concerning x-irradiation treatment of cancer, that free radical processes are very important, particularly in the initial primary events.

Robert A. Floyd

Free Radicals and Cancer

1
FREE RADICALS PRODUCED FROM CHEMICAL CARCINOGENS AND THEIR SIGNIFICANCE IN CARCINOGENESIS

Chikayoshi Nagata, Masahiko Kodama, Yoshikazu Ioki, and Teruyuki Kimura

National Cancer Center Research Institute
Tokyo, Japan

I. INTRODUCTION

Biochemical and biophysical studies on the metabolism of chemical carcinogens during past decades have opened new prospects in the field of chemical carcinogenesis. Thus, various kinds of chemical carcinogens were found to act as triggers to initiate malignant transformation after being metabolized to the active form which binds to biopolymers. At present, it is accepted by many that the active electrophiles produced from esterified N-hydroxyamino compounds or from hydrocarbon diol-epoxides bind to nucleophiles in the cells [1,2]. However, this does not necessarily exclude other metabolic activation pathways. In fact, involvement of another pathway in the genesis of tumor is suggested by the finding that 2-acetylamino-fluorene (2-AAF) can cause cancer in such tissues as mammary gland or sebaceous gland of the external auditory canal which lacks the esterification (sulfonation) activity [3]. Furthermore, the accelerating effect of 1,1,1-trichloropropene-2,3-oxide (TCPO) on benzo-(a)pyrene [B(a)P] carcinogenesis [4-6] strongly suggests that besides diol-epoxides, other active species exist as an ultimate form of B(a)P. TCPO has been shown to almost completely inhibit the formation of the diol-epoxide of B(a)P and decrease the binding of this epoxide to DNA [7-9]. Furthermore, anthanthrene which lacks the "bay region," a necessity in order to form the postulated ultimate carcinogenic diol-epoxide, was found to be moderately carcinogenic in mice [10].

In view of these apparent contradictions, it is of importance to search for new activation pathways other than those mentioned above. A plausible pathway is free radical formation, and interesting facts suggesting the possible roles of free radicals in carcinogenesis have been found. Szent-Györgyi et al. have already pointed out that carcinogenic aromatic compounds are converted to free radicals by forming a molecular complex with iodine [11]. The nitroxyl radical is formed as an intermediate from N-hydroxyacetylaminofluorene (N-OH-AAF) in its conversion to N-acetoxy-AAF and 2-nitrosofluorene [12-14]. It has also been shown that B(a)P is converted to a free radical during its enzymatic oxygenation in the liver as well as in the lung and skin [15-17]. Similarly, strong hepatocarcinogens such as dimethylaminoazobenzene and its derivatives were shown to be converted to free radicals by incubating them with liver microsomes from methylcholanthrene (MC)-treated rats [18]. The enzymatic formation of free radicals was also observed for aromatic amines [18], 2-AAF [18], and quinone anticancer agents [19-21]. Thus, chemical carcinogens which are stable in their parent forms were found to be converted to free radicals after being metabolized. On the other hand, *direct carcinogens* such as N-methyl-, and N-ethyl-N'-nitro-N-nitrosoguanidine [22] or *proximate forms of chemical carcinogens* such as 4-hydroxyaminoquinoline-1-oxide (4-HAQO) [23], aminonaphthols, and naphthylhydroxylamines [24] were found to be converted to free radicals nonenzymatically.

In this chapter, the formation and behavior of free radicals produced from various kinds of chemical carcinogens are described and their casual significance in chemical carcinogenesis is discussed.

II. AROMATIC HYDROCARBONS

A. Photochemical Formation of Free Radicals

Aromatic hydrocarbons cause photoinactivation of enzymes [25] and exert a photodynamic action on microbes as well as on mammalian cells [26,27]. Abundant data have also accumulated on the enhancing effect of light on carcinogenicity, although in some cases reduction of carcinogenicity was reported [28-32]. Even though the mechanisms underlying the biological action have remained obscure, it has been generally believed that aromatic hydrocarbons manifest their activities after conversion to an active form, such as a free radical.

1. *Free radical produced from irradiated benzo(a)pyrene [B(a)P]*

Among aromatic hydrocarbons, B(a)P was found to be most sensitive to photoirradiation, resulting in the formation of a free radical [33]. Thus, 5-min irradiation of B(a)P in benzene was sufficient to form a large electron spin resonance (ESR) signal. The signal was split into a hyperfine structure when the radical solution was degassed to 10^{-3} torr (Fig. 1). The g value of the signal was 2.004, and the line width was 13.7 G.

The photoinduced free radical is not a peroxy radical, because the g value of this radical is 2.004, which is quite different from the reported g value of 2.015 for the peroxy radical [34]. The hyperfine structure of this radical is clearly different from the signal of a cation or anion radical of B(a)P [15,33], and also different from the azulene-type radical obtained upon heating B(a)P [35]. Further, the ESR signal of this free radical is different from the 3,6-, 1,6-, and 6,12-B(a)P-semiquinone radicals, both in the pattern of its hyperfine structure as well as in line width [15]. Finally, it was found that the hyperfine structure of the signal was entirely the same as that of the 6-oxy-B(a)P radical which was produced by abstracting hydrogen atom from 6-hydroxy-B(a)P [6-OH-B(a)P] with ceric sulfate [15,33] (Fig. 2). Spin densities of the 6-oxy-B(a)P radical calculated by means of the unrestricted self-consistent field molecular orbital (SCF MO) method within π approximation, reflect well the observed coupling constants (Table 1). This is an additional support for the conclusion that the photoinduced radical is the 6-oxy-B(a)P radical.

2. *Conditions for formation of the free radical*

Formation of the photoinduced free radical is dependent upon the wavelength of the irradiation light. Thus, B(a)P in benzene irradi-

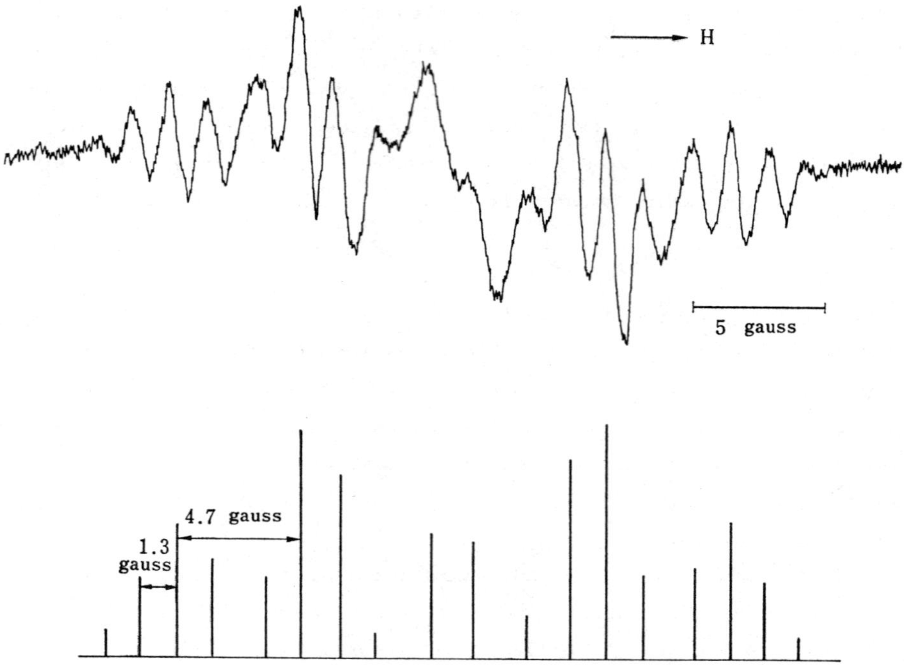

A

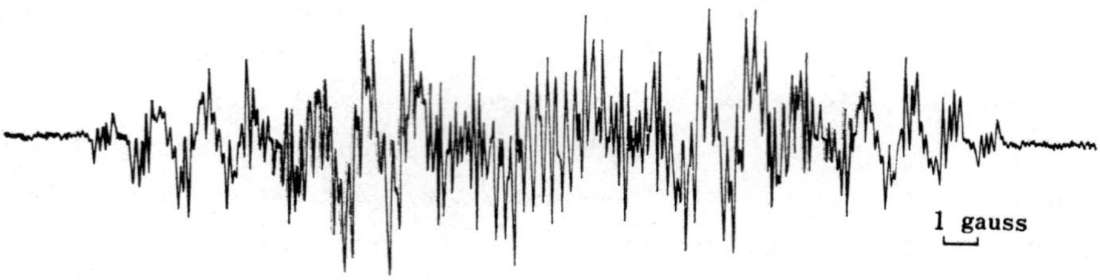

B

Figure 1 Hyperfine structure of the ESR signal of the oxy radical produced from irradiated benzene solution of B(a)P (5.66×10^{-2} M). The light source was a high-pressure mercury lamp of 300W (5.05×10^{2} erg sec^{-1} mm^{-2}). The ESR measurement was made at room temperature with a Model JES-3BX using a 100 kHz field modulation. Hyperfine structure was observed only after degassing the sample tube down to 10^{-3} torr or less. Modulation amplitude: (A) 1 G; (B) 0.08 G. (From Ref. 33.)

Figure 2 Hyperfine structure of the ESR signal of 6-oxy-B(a)P radical produced by oxidizing 6-hydroxybenzo(a)pyrene [6-OH-B(a)P] with ceric sulfate in dimethyl sulfoxide.

ated with a monochromatic light of 255, 271, 288, 303, and 401 nm [wavelengths all in the region of the absorption of B(a)P] produced the free radical, whereas irradiation with wavelengths outside the absorption band of B(a)P, 440 and 452 nm, were inactive in forming the free radical. This indicates that the free radical is produced by direct excitation of the B(a)P molecule [33].

Table 1 Comparison of the Calculated Coupling Constants with Observed Values for 6-Oxy-B(a)P Radical

Position	Coupling constant, A (gauss)	
	Calculated[a]	Observed
1	3.98	4.7
3	4.03	
4	4.39	
12	4.05	
2	1.33	1.3
5	1.69	
11	1.57	
7	0.50	
8	0.79	
9	0.50	
10	0.92	

[a]Spin densities, ρ, were calcuated by means of unrestricted SCF MO method; coupling constant, A, was obtained according to the formula, $A = Q\rho$, where $Q = 22.5$ G.

Formation of the free radical was strongly dependent upon the solvent used. The ESR signal was produced when B(a)P was irradiated in benzene, CS_2, cyclohexane, and CCl_4, but no signal was observed in dioxane and acetone. This may not necessarily mean that the free radical was not produced in dioxane or in acetone, for failure to detect the free radical in such solvents may be due to the instability of the free radical in these solvents. In the presence of air, the amount of free radical increased with irradiation time and reached a plateau after 4- or 5-min irradiation. However, when the sample tube was degassed before irradiation to about 10^{-3} torr, no ESR signal could be detected [33].

3. *Mechanism of formation of the oxy radical and its conversion to quinones*

We have presented a mechanism to account for the formation of the oxy radical from B(a)P. The scheme of Cusachs and Steele [36] was taken into consideration.

$$\mathrm{BP} \longrightarrow {}^1\mathrm{BP}^* \qquad (1)$$

$$^1\mathrm{BP}^* \longrightarrow {}^3\mathrm{BP}^* \qquad (2)$$

$$^3\mathrm{BP}^* + {}^3\mathrm{O}_2 \longrightarrow \mathrm{BP} + {}^1\mathrm{O}_2^* \qquad (3)$$

$$^1\mathrm{BP}^* + {}^3\mathrm{O}_2 \longrightarrow {}^3\mathrm{BP}^* + {}^1\mathrm{O}_2^* \qquad (3')$$

$$\mathrm{BP} + {}^1\mathrm{O}_2^* \longrightarrow \mathrm{BPOOH} \qquad (4)$$

$$2\mathrm{BPOOH} \longrightarrow 2\mathrm{BPO}\cdot + \mathrm{H_2O} + \tfrac{1}{2}\,{}^3\mathrm{O}_2 \qquad (5)$$

BPOOH and BPO• denote B(a)P-6-hydroperoxide and 6-oxy-B(a)P radicals, respectively. The left shoulder suffix represents the spin multiplicity, and the asterisk represents the excited state. The importance of the reactions shown in Eq. (3') was stressed by Steele et al. in connection with the carcinogenesis mechanism of aromatic hydrocarbons [37], as well as the fact that the energy difference between the lowest excited singlet and triplet level of B(a)P, i.e., 1.58 eV, is enough to excite the ground state of oxygen to its lowest singlet state. It is reasonable to consider that in Eq. (4) a singlet molecular oxygen attacks position 6 of B(a)P, since it is known that this position is the most reactive in the B(a)P molecule.

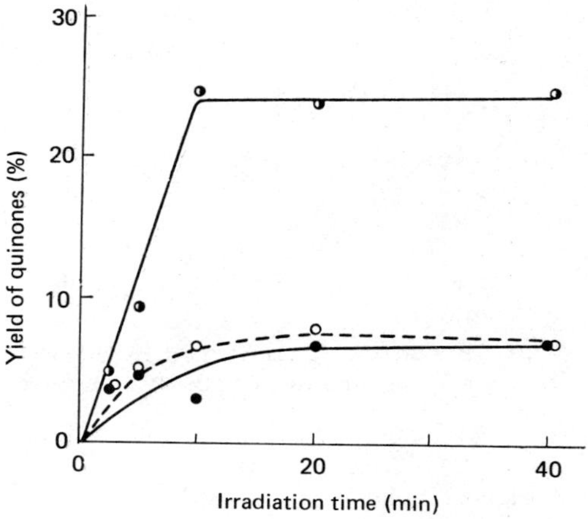

Figure 3 Amount of B(a)P-quinones produced from photoirradiated 6-OH-B(a)P in benzene (7.1 x 10^{-4} M). (From Ref. 33.)

It has been shown that the main photoproducts of B(a)P in organic solvent are the B(a)P-quinones [38,39]. This was confirmed in experiments using thin-layer chromatography [33]. Photoirradiation of 6-OH-B(a)P resulted in the conversion of this compound into the 6-oxy-B(a)P radical, which was further oxidized to B(a)P-quinones with a yield amounting to about 40% of 6-OH-B(a)P (Fig. 3). This result indicates that the oxy radical is the intermediate between B(a)P and B(a)P-quinones.

A molecular orbital study on the electron distribution of 6-oxy-B(a)P radical gave additional support to the above supposition. Thus, as indicated in Figure 4, the spin densities at positions 1, 3, and 12 are extremely large compared with other positions, and this seems to be the reason why the corresponding 1,6-, 3,6-, and 6,12-B(a)P-quinones are formed. The spin density at position 4 is comparable to that at the 1, 3, and 12 positions. Nevertheless, 4,6-B(a)P-quinone has never been detected as a photoproduct.

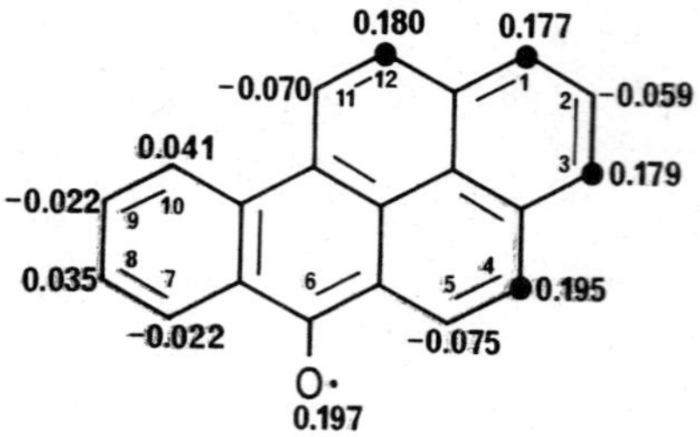

Figure 4 Spin densities of 6-oxy-B(a)P radical calculated by means of the unrestricted SCF MO method. Bond distances for C—C and C—O bonds were taken as 1.4 Å.

4. *Free radicals produced from 6-substituted B(a)P and other aromatic hydrocarbons*

When irradiated in benzene, 6-substituted B(a)P derivatives, such as 6-acetoxy, 6-amino, 6-bromo-, and 6-nitro-B(a)P yield free radicals, the structures of which were identified as the 6-oxy-B(a)P radical. The order of efficiency of radical production was as follows: 6-nitro- > 6-amino- > 6-acetoxy-B(a)P. When a benzene solution of aromatic hydrocarbons such as anthracene, 2,3,7,8-dibenzophenanthrene, benzo(a)anthracene, 3,4,9,10-dibenzo(a)pyrene, and 1,2,4,5-dibenzo(a)pyrene were irradiated, only the latter two compounds yielded ESR signals. These two compounds are the only ones of the above which have the B(a)P skeleton. The g values of these signals were the same as that of the 6-oxy-B(a)P radical, i.e., 2.004. This suggests that the free radicals produced were the oxy radicals of these hydrocarbons.

B. Enzymatic Formation of 6-Oxy-B(a)P Radicals

The metabolism of B(a)P has been of major concern in hydrocarbon carcinogenesis. More than twenty metabolites have been identified [40-43]. However, experimental detection of the hydroxylated compound at the 6 position has long been unsuccessful, although in a few cases it was detected indirectly [40,44]. This has been an enigma in B(a)P metabolism because Berenblum et al. identified the main metabolite of B(a)P as 3-OH-B(a)P [45], although the 6 position is the most reactive toward all kinds of reagents, i.e., electrophilic, nucleophilic, and radical [46]. As was described in the previous section, it was found that B(a)P was converted to the 6-oxy-B(a)P radical by photoirradiation. The same free radical was also found to be generated in an enzyme system.

Figure 5 shows the ESR signal of the free radical obtained after incubating B(a)P with liver microsomes (or homogenates). Degassing the sample tube gave a hyperfine structure (Fig. 5C) which is equivalent to that of the 6-oxy-B(a)P radical. This clearly shows that the 6-oxy-B(a)P radical is formed as a metabolite of B(a)P.

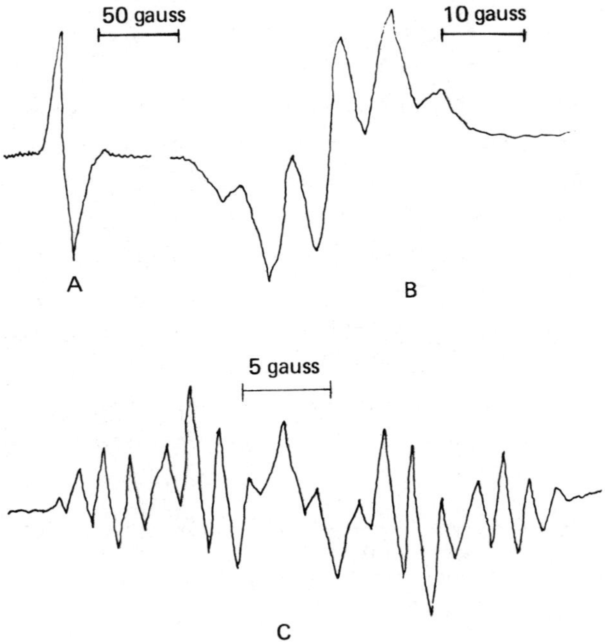

Figure 5 The ESR signal of the enzymatically formed 6-oxy-B(a)P radical. The reaction mixture, in a total volume of 15.75 ml contained 5-9 mg microsomal protein, 50 mM of Tris-chloride buffer, 0.15 M KCl, pH 7.5, 0.8 mM of NADPH, and 10^{-4} M B(a)P (added in 0.6 ml methanol just prior to incubation). After incubation (37°, 10 min), the reaction mixture was extracted with cold benzene. Following evaporation of the solvent, the residue was dissolved in 0.3 ml of benzene for the ESR measurement: (A) before degassing; (B and C) after degassing to 10^{-3} torr. Modulation amplitudes: (A and B) 4 G; (C) 1 G. (From Refs. 15 and 16.)

1. Pathways of 6-oxy-B(a)P radical formation

The ESR signal of the oxy radical increased gradually after benzene was added to the evaporated residues of the benzene extract from the incubation mixture, reaching a maximum after some 2-3 hr. Thereafter, the ESR signal began to decay with a half-life of about 50 hr. This is consistent with the case of the 6-oxy-B(a)P radical formed by photoirradiation [33]. The most reasonable explanation for the increase of the ESR signal in benzene with time is that the 6-oxy-B(a)P radical is produced from a precursor by autoxidation [15,47,48]. Although the chemical structure of the precursor of the radical has not yet been identified directly, 6-OH-B(a)P most likely is the precursor. In fact, 6-OH-B(a)P was found to be labile and easily converted to the 6-oxy-B(a)P radical under physiological conditions [15,33,47,48]. Thus, 6-OH-B(a)P dissolved in benzene showed an ESR signal due to the 6-oxy-B(a)P radical formed by autoxidation of 6-OH-B(a)P. This signal was increased severalfold after 6-OH-B(a)P was incubated with heat-treated (60°, 15 min) microsomes at 37° for 10 min. This indicates that the conversion of 6-OH-B(a)P to the 6-oxy-B(a)P radical occurs nonenzymatically. Of course, this does not exclude the possibility that 6-OH-B(a)P is metabolized enzymatically. In fact, Wiebel showed that 6-OH-B(a)P was oxidized enzymatically during incubation with liver microsomes [49]. From these results, together with the previous finding demonstrating that the 6-oxy-B(a)P radical was detected in the enzyme system, the following scheme is proposed:

B(a)P —(enzymatic)→ 6-OH-B(a)P —(enzymatic and nonenzymatic)→ 6-oxy-B(a)P radical

6-oxy-B(a)P radical —(enzymatic and nonenzymatic)→ B(a)P-quinones and other products

In connection with this metabolic pathway, the findings of Marnett et al. are of particular interest. They demonstrate that during prostaglandin biosynthesis B(a)P undergoes significant cooxidation, resulting in the formation of B(a)P-quinones exclusively [50].

2. *Participation of cytochrome P_{450} in the formation of 6-oxy-B(a)P radical precursor*

In the experiment using B(a)P and liver microsomes, no ESR signal could be detected when heat-treated microsomes (60°, 15 min) were used or when microsomes were omitted. Without NADPH, the amount of free radical was negligible. The ESR signal was small when NADH was used instead of NADPH, but when NADH was added together with NADPH the effect was synergistic.

The formation of 6-oxy-B(a)P radical was inhibited by addition of CO or SKF 525-A, but it was scarcely affected by addition of CN^-. This indicates that the enzyme responsible belongs to the microsomal mixed function oxidase, i.e., cytochrome P_{450} (Table 2). Rispin et al. found that $^{17}O_2$ was incorporated into the 6-oxy-B(a)P radical under a $^{17}O_2$ atmosphere [51]. This is further confirmation of the participation of cytochrome P_{450} in formation of the precursor of the 6-oxy-B(a)P radical.

The enzymatic hydroxylation of aromatic rings is an important metabolic pathway of endogenous as well as xenobiotic substances. It is generally believed that phenolic metabolites are formed by nonenzymatic isomerization via initial formation of arene oxides. However, from analysis of the deuterium isotope effect, occurrence of a direct hydroxylation of the aromatic ring has recently been

Table 2 Effect of Carbon Monoxide, SKF 525-A, and CN^- on Formation of 6-Oxy-B(a)P Radical and 3-OH-B(a)P in Hepatic Microsomes from Noninduced Rats

Condition		6-Oxy-B(a)P Radical	3-OH-B(a)P
In air	—	100	100
N_2-O_2	9:1	71.3	91.9
CO-O_2	9:1	45.1	31.9
Concentration	0	100	100
of SKF 525-A	10^{-7} M	52.9	36.1
	10^{-6} M	36.8	14.4
	10^{-5} M	0	2.1
Concentration	0	100	100
of CN^-	10^{-4} M	94.6	106.3
	10^{-3} M	71.0	83.5

suggested in some cases [52]. Detection of 6-oxy-B(a)P radical strongly supports the idea that hydroxylation of B(a)P can occur not only via epoxidation but also directly, since epoxidation—including that at the 6 position—is considered to be a difficult process. The epoxide at the 6 position has never been detected in an enzyme reaction or in model systems, although the epoxide at this position has been synthesized [53].

3. *Formation of 6-oxy-B(a)P radicals in skin and lung*

After discovery of the formation of 6-oxy-B(a)P radicals in liver microsomes (or homogenates), the study was extended to lung and skin, which are known to be the target tissues for B(a)P [16,17]. In Table 3, the amount of 6-oxy-B(a)P radical formed by incubating B(a)P with skin homogenates is compared with the amount of 3-OH-B(a)P formed, which is regarded as the main phenolic metabolite of B(a)P [54].

As seen in Table 3, the amount of 6-oxy-B(a)P radical formed is only one-eighth that of 3-OH-B(a)P after 10-min incubation. However, the amount of the oxy radical increased with incubation time, and after 45- and 60-min incubation, it became larger than the amount of 3-OH-B(a)P formed. After 75-min incubation, the amount of radical decreased rapidly—as was observed with liver homogenates [16,48].

Incubating B(a)P with microsomes or homogenates of lung from MC-treated rats yielded a large amount of ESR signal [17]. After degassing the sample tube, the hyperfine structure of the signal was obtained. It was equivalent to that of the 6-oxy-B(a)P radical. The magnitude of the signals obtained by incubating B(a)P with homogenates or microsomes from control and phenobarbital-treated rats was much smaller than that of the signal obtained with MC-treated rats, and in fact identification of the free radical was unsuccessful in these cases.

Table 3 Amounts of 6-Oxy-B(a)P Radical and 3-OH-B(a)P with Various Incubation Times when Incubating B(a)P with Homogenates of Mouse Skin Epidermis or Skin Corium (Values for Rat Liver Microsomes are Given for Comparison)

	Incubation time (min)	6-Oxy-B(a)P radical[a]		3-OH-B(a)P (pmole/mg protein)
		$Signal/Mn^{2+}$/R.M.[b]	pmole/mg protein	
Homogenates of mouse skin epidermis[c]	10	14.5	2.9	23.7
	20	62.8	12.6	34.0
	30	162.4	32.6	41.0
	45	540.5	108.6	66.0
	60	820.9	165.0	89.0
	75	153.7	30.9	113.8
Homogenates of mouse skin corium[c]	30	280.4	51.7	31.6
	60	851.0	156.4	45.8
Rat liver microsomes[d]	10	17.1	20.7	612.9
	30	27.9	33.8	1336.6
	60	28.3	34.3	1467.1

[a]Identification of the free radical was made by comparing the ESR signals. The amount of the oxy radical was estimated by comparing the area of the ESR signal with that given by known concentration of diphenylpicrylhydrazil. The amount of 3-OH-B(a)P was derived from the fluorescence spectrum.

[b]$Signal/Mn^{2+} = 1$ corresponds to 0.1033×10^{-8} mole; R.M. = reaction mixture.

[c]Protein in the reaction mixture (100 ml) of skin epidermis and corium was 51.4 and 56.2 mg respectively.

[d]Protein in the reaction mixture (13.75 ml) was 8.52 mg.

C. Formation of 6-Oxy-B(a)P Radical by Stirring B(a)P with Proteins or Skin Homogenates

A mixture of albumin and B(a)P in distilled water was stirred at 4° in complete darkness. The subsequently lyophilized sample yielded a large ESR signal. The free radical was extracted with benzene and from the hyperfine structure, the radical was identified as the 6-oxy-B(a)P radical [15,55,56]. Stirring for 3 days or more was necessary to obtain sufficient signal. No ESR signal could be observed in control samples lacking B(a)P. Other carcinogenic hydrocarbons such as methylcholanthrene and 7,12-dimethylbenz(a)anthracene were effective in generating a free radical; but noncarcinogenic compounds such as dibenzo(a,c)anthracene, pyrene, tetra cene, anthracene, phenanthrene, and naphthalene were ineffective. In this sense, it appears that free radical formation and carcinogenicity are correlated. However, this is not necessarily true in all cases, because the noncarcinogenic perylene gave an ESR signal by stirring with albumin.

Globulin was also effective in producing the 6-oxy-B(a)P radical, but no signal was observed after stirring B(a)P with such biopolymers as DNase, chymotrypsin, insulin, DNA, and RNA. Since heat-treated albumin (60°, 30 min) was effective in producing the free radical, formation of 6-oxy-B(a)P radical does not occur via enzyme reaction. Photoproduction of the radical was also excluded by complete shielding of light during stirring as well as in the course of preparing the sample. Most likely, the free radical was formed via autoxidation of B(a)P. Without stirring, no ESR signal was observed. This indicates that stirring is necessary in order to effectively incorporate oxygen into the reaction mixture. As in the case of albumin, stirring of B(a)P with skin homogenates resulted in the formation of a free radical, identification of which was unsuccessful because of the difficulty in resolution of the ESR signal [55,56].

Okano et al. found that stirring of methyl-substituted benz(c)-acridines with albumin in water resulted in the production of a free radical [57]. Globulin, histone, and DNase were also effective in producing the free radical, although albumin was the most effective.

Interestingly, carcinogenic benz(c)acridines were more effective in producing the free radical than were noncarcinogenic benzacridines. Although the structure of the free radical has not yet been identified, the oxy or semiquinone radical was presumably formed as in the case of B(a)P [57].

D. Formation of 6-Oxy-B(a)P Radical in Model Systems

Oxygenation of biologically important chemical carcinogens using model systems had never been achieved until Ioki et al. found that B(a)P could be oxygenated using Fenton's reagent and trifluoroperoxyacetic acid [58,59]. In model systems, the relative ratio of phenolic products depends on the solvent used. Thus, when B(a)P dissolved in benzene was mixed with Fenton's reagent, fluorescence of 6-OH-B(a)P was observed (Fig. 6A). On the other hand, when trifluoroperoxyacetic acid was used, the greater part of the fluorescence was ascribed to 3-OH-B(a)P (Fig. 6B). In both systems, however, production of 6-oxy-B(a)P radical was proven by analysis of the hyperfine structure of the ESR signal. Trifluoroperoxyacetic acid was far more efficient in producing 6-oxy-B(a)P radical than Fenton's reagent, and this was considered to be the reason why 6-OH-B(a)P is oxidized to the 6-oxy-B(a)P radical, which is then further oxidized to quinones. On the other hand, 3-OH-B(a)P is more stable than 6-OH-B(a)P and, therefore, easily detectable.

In the enzymatic reaction, the excitation spectrum of the oxygenated products is similar to that of 3-OH-B(a)P. The fluorescence spectrum of 6-OH-B(a)P has never been detected [54]. Comparing the results of the enzymatic reaction with those of model systems suggests that the enzymatic reaction of B(a)P corresponds to a more efficient and stronger oxygenation reaction than in a model system such as the one with trifluoroperoxyacetic acid [59].

When B(a)P was treated with air or with singlet oxygen generated from 9,10-diphenylanthracene peroxide, no derivatives of B(a)P were detected using ESR, fluorometry, and thin layer chromatography. This observation suggests that molecular oxygen, both in the ground triplet ($^3\Sigma_g$) and in the singlet ($^1\Delta_g$) states, is not involved in the enzymatic reaction, although it does not exclude participation of singlet oxygen in the photoproduction of the oxy radical [33,59] as was discussed previously.

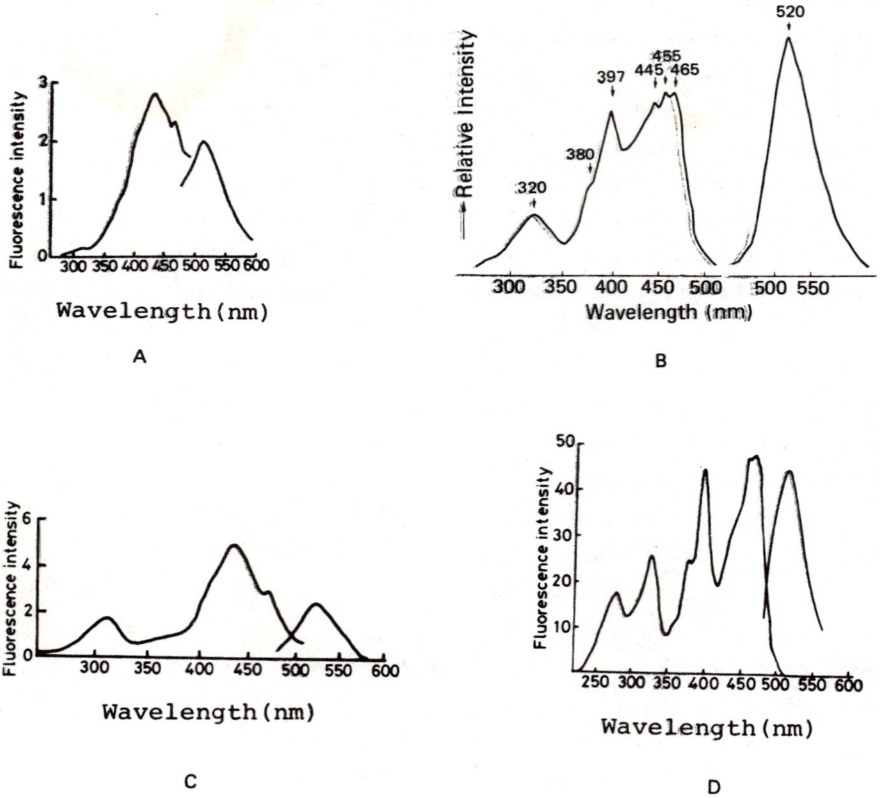

Figure 6 Excitation and fluorescence spectra (in alkaline solution) of products obtained from B(a)P in benzene with (A) Fenton's reagent or (B) trifluoroperoxyacetic acid. Spectra (C) and (D) are for the authentic samples of 6-OH-B(a)P and 3-OH-B(a)P, respectively. (Part B from Ref. 58; parts A, C and D from Ref. 59).

E. Binding of Free Radicals with DNA and Polynucleotides

One of the criteria for determining whether a metabolite is involved in the genesis of tumor is the covalent binding of the metabolite to nucleic acids. It has been found that 6-OH-B(a)P, a precursor of 6-oxy-B(a)P radical, binds covalently to DNA in vitro and that the binding increases parallel to the radical concentration [16,47]. This strongly suggests a participation of the 6-oxy-B(a)P radical in the binding with DNA. Direct proof was obtained when the ESR signal of the bound complex was detected [60]. The ESR signal obtained by incubating 6-OH-B(a)P with poly(G), shown in Figure 7B, is compared with that of the ESR signal obtained with a shorter incubation when $K_3Fe(CN)_6$ or chloranil was added, thereby accelerating oxy radical formation.

Since the strong ESR signal was observed even after extensive washing with organic solvents, the binding is considered to be covalent. The covalent binding was further confirmed by the enzymatic hydrolysis of the bound complex with RNase T_1 (Fig. 7C). A remarkable fact is that the free radical of the bound complex was extremely stable. In aqueous solution it showed no decay after standing for 2 days at room temperature and only decayed 10% even after 10 days (Fig. 8). Among homopolynucleotides, poly(G) was especially effective in producing the bound complex. Poly(A), poly(C), and DNA yielded only small signals, but poly(U) yielded no signal.

The characteristics of the signal of the bound complex (Fig. 7B) is somewhat different from that of the free 6-oxy-B(a)P radical (Fig. 7A) when they are compared with respect to line width and g value (Table 4). This suggests two possibilities: the first is that the signal pattern of the free 6-oxy-B(a)P radical was altered by binding with the guanine moiety; the second is that the oxy radical was further converted to the more active free radical such as the semiquinone radical during oxidation.

The first possibility may be excluded because the line width of ESR signal of a molecule is generally increased upon formation of a complex with a polymer. In order to test the second possibility,

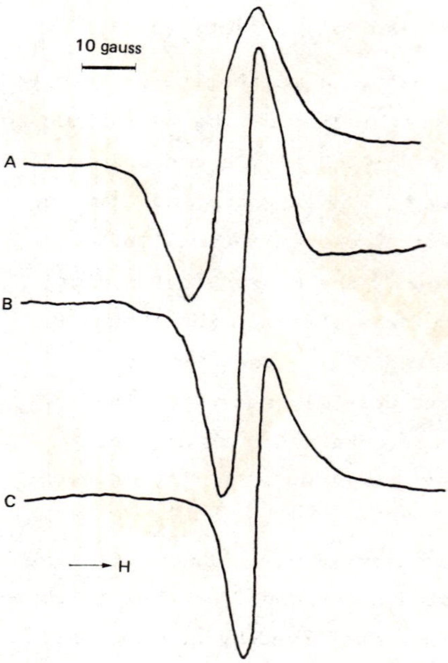

Figure 7 The ESR signal of the covalently-bound complex formed by incubating 6-OH-B(a)P with poly(G) in comparison with that of 6-oxy-B(a)P radical. 6-OH-B(a)P (1.5 mg/2 ml in ethanol) was mixed with an aqueous solution of poly(G) (10 mg/5 ml) and incubated at 37° for 5 hr or more. After incubation, the mixture was extracted with chloroform, phenol, and ether, followed by repeated precipitation from ethanol. The dried sample was measured with the ESR apparatus using a modulation amplitude of 4 G. Curves: (A) 6-oxy-B(a)P radical in benzene; (B) the covalently-bound complex; (C) sample B after treatment with RNase T_1 followed by washing with organic solvents. (From Ref. 60.)

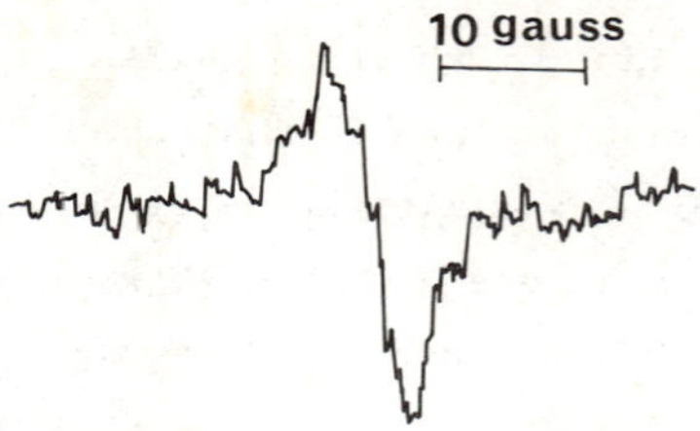

Figure 8 The ESR signal of the bound complex between 6-OH-B(a)P and poly(G) in water. Modulation amplitude: 4 G.

Table 4 Line Widths and g Values for the Free State of the 6-Oxy-B(a)P Radical and B(a)P-semiquinone Radicals in Comparison with Those of Bound States

Compound	Line width (gauss)	g Value
6-Oxy-B(a)P radical in ethanol	13.7	2.004
6-Oxy-B(a)P radical (solid state)	12.3	2.004
6-Oxy-B(a)P radical + poly(G) (solid state)	6.5	2.005
6-Oxy-B(a)P radical + poly(G) in water	5.2	2.005
3,6-B(a)P-semiquinone anion radical in DMSO[a]	5.3	2.005
1,6-B(a)P-semiquinone anion radical in DMSO[a]	5.3	2.005
6,12-B(a)P-semiquinone anion radical in DMSO[a]	6.3	2.005
3,6-B(a)P-semiquinone anion radical + poly(G) (solid state)	7.2	2.004
1,6-B(a)P-semiquinone anion radical + poly(G) (solid state)	7.2	2.004
3,6-B(a)P-semiquinone anion radical + DNA (solid state)	5.3	2.004
1,6-B(a)P-semiquinone anion radical + DNA (solid state)	5.3	2.004

[a]DMSO: dimethylsulfoxide.

a binding experiment was carried out between the B(a)P-semiquinone radical and DNA or polynucleotide [61,62]. As may be seen in Figure 9, the ESR signal due to the bound complex 3,6-semiquinone radical was far larger than that of 1,6-semiquinone radical. It is remarkable that no binding could be detected when the 6,12-semiquinone radical was interacted with DNA. This result is difficult to explain even if we take into account the fact that the half-life of the 6,12-B(a)P-semiquinone radical is far shorter than those of the 3,6- and 1,6-semiquinone radicals. Another explanation is that the 12 position is the site of binding of the free radical to the base moiety. In connection with this proposition, the finding of Hecht et al. [63] is significant. They found that substitution at the 12 position in 5-methylchrysene with a fluoro or methoxy group diminished the carcinogenicity considerably. Among homopolynucleotides the ESR signal due to the bound complex was extremely large for poly(G), whereas other polynucleotides yielded only a small signal. This

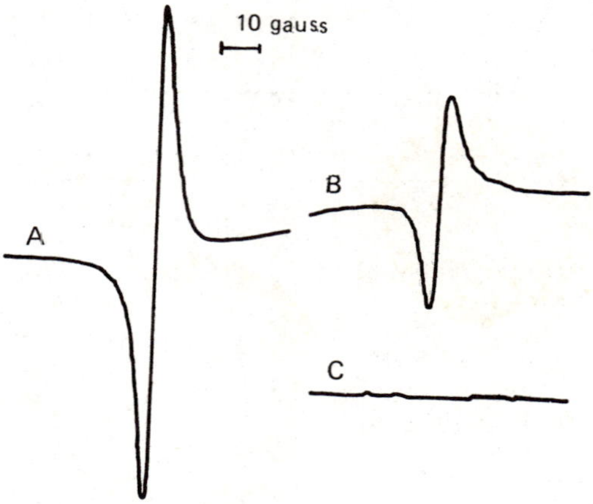

Figure 9 The ESR signal of the bound complex between B(a)P-semiquinone anion radicals and DNA. 3,6-, 1,6-, and 6,12-B(a)P-quinones (1 mg/0.5 ml in dimethylsulfoxide) were reduced by adding dropwise an aqueous solution of sodium borohydride (1 mg/0.5 ml). An aqueous solution of calf thymus DNA (10 mg/10 ml) was added to the solution and incubated at 37° for 18 hr. After extensive extraction with 80% phenol and ether, the sample was dried for ESR measurement. Curves: (A) 3,6-B(a)P-quinone; (B) 1,6-B(a)P-quinone; (C) 6,12-B(a)P-quinone. (From Ref. 61.)

shows that guanine is the site of binding as in the case of the 6-oxy-B(a)P radical.

Some differences were observed between the binding of 6-oxy-B(a)P radical and those of B(a)P-semiquinone radicals. For 6-oxy-B(a)P radical, the signal due to the bound complex with poly(G) was far larger than the signal with DNA, whereas the reverse relation was seen for B(a)P-semiquinone radicals. As may be seen in Table 4, the g values and line widths are not the same for the two types of free radicals, although the differences are not very large [62]. Concerning the semiquinone radicals, the anion radical was used in the binding experiment. Under neutral reaction conditions, the ESR signal of the bound complex between the B(a)P-semiquinone radical and DNA was too small to permit us to analyze and identify the chemical structure. The g value and line width of the neutral semiquinone radicals are considered to be somewhat different from those of semiquinone anion radicals. Thus, at the present state of knowledge, it is difficult to identify the bound complex derived from 6-oxy-B(a)P radical as being of a semiquinone type ([X] in Fig. 10).

F. Anthanthrene and Other Aromatic Hydrocarbons

Anthanthrene has been known to be noncarcinogenic for a long time, and this seems to lend support to the aforementioned "bay region" theory [64-66] because this compound lacks the bay region. However, recently Cavalieri et al. showed that anthanthrene induced malignant epidermal neoplasma in mice after repeated skin application, and from this result anthanthrene was classified as a relatively strong carcinogen [10]. This new finding is of special importance in view of the structural condition for carcinogenicity of aromatic hydrocarbons. Thus, the existence of a bay region is not necessarily required for the carcinogenicity of aromatic hydrocarbons.

An activation pathway other than epoxidation in the bay region therefore must be involved in the metabolism of anthanthrene. As a possible candidate for the activation pathway, oxy radical formation due to hydroxylation at the meso position was proposed—as in the case of B(a)P.

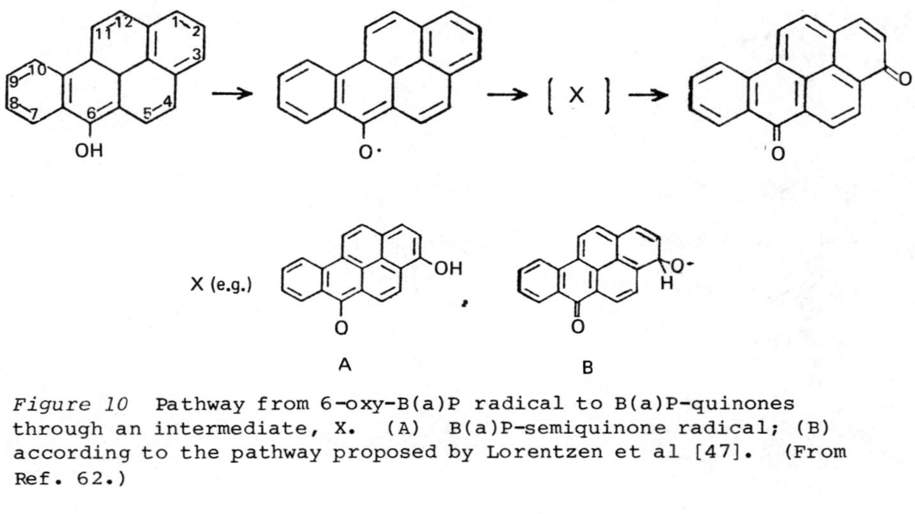

Figure 10 Pathway from 6-oxy-B(a)P radical to B(a)P-quinones through an intermediate, X. (A) B(a)P-semiquinone radical; (B) according to the pathway proposed by Lorentzen et al [47]. (From Ref. 62.)

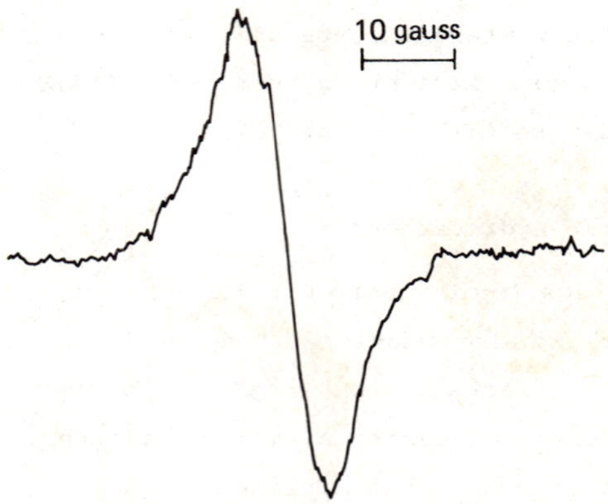

Figure 11 ESR signal of free radical produced by incubating anthanthrene with liver microsomes. The condition of the enzymatic reaction and the preparation of the sample are the same as in the case of B(a)P in Figure 5.

After incubation of anthanthrene with liver microsomes, the sample for ESR measurement was prepared according to the method used for detection of 6-oxy-B(a)P radical. As shown in Figure 11, an ESR signal with a g value of 2.005 was observed. No ESR signal could be detected when NADPH was omitted from the reaction mixture. This indicates that the mixed function oxidase in the microsomes is involved in the formation of the free radical or its precursor. Contrary to the case of B(a)P, degassing of the sample tube did not bring about resolution of the hyperfine structure.

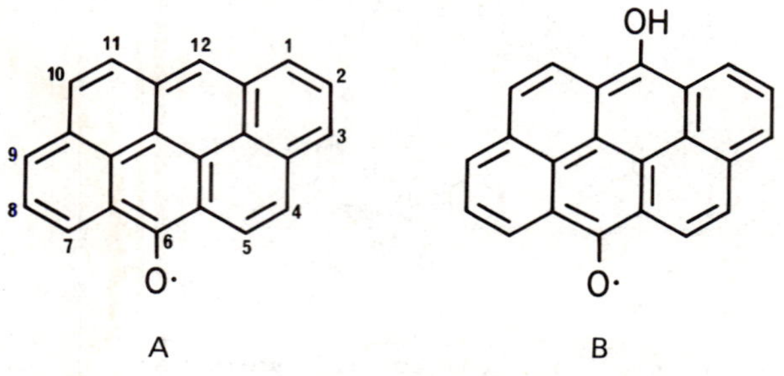

Figure 12 (A) 6-Oxy-anthanthrene radical; (B) 6,12-anthanthrene semiquinone radical.

The line width of the signal is 9.8 G, which is smaller than that of the 6-oxy-B(a)P radical but is rather similar to the value of the B(a)P-semiquinone radicals [15,33]. This suggests that the free radical is a mixture of oxy radical and semiquinone radical in which two equivalent positions (6 and 12) are oxygenated (Fig. 12).

Aromatic hydrocarbons such as 3-methylcholanthrene, 7,12-dimethylbenzo(a)anthracene (DMBA), pyrene, and anthracene yielded no free radical after incubation with liver microsomes.

As indicated for 6-OH-B(a)P and 3-OH-B(a)P, some of the hydroxy compounds of the aromatic hydrocarbons were shown to be converted to free radicals which subsequently became bound to DNA or to polynucleotides with a retention of their free radical form [16,60,62]. In this connection, it is of value to test the convertibility of the hydroxylated aromatic compounds to the free radical and subsequent binding ability of these compounds. Six hydroxy derivatives of DMBA were tested, i.e., 2-OH-, 3-OH-, 4-OH-, 5-OH-, 9-OH-, and 10-OH-DMBA, which were generously supplied by Dr. Melvin S. Newman (Department of Chemistry, The Ohio State University). Among these compounds, 2-OH-DMBA gave the largest ESR signal of the covalently bound complex with poly(G) (Fig. 13), whereas 4-OH- and 5-OH-DMBA yielded covalently bound complexes in smaller amounts. The other three hydroxy derivatives yielded no ESR signal. Carcinogenicity testing has not been conducted on these compounds, but it is of interest to know whether these compounds, particularly 2-OH-DMBA, is carcinogenic.

III. 4-NITROQUINOLINE-1-OXIDE (4-NQO) AND RELATED COMPOUNDS

Two decades have passed since Nakahara et al. discovered that 4-nitroquinoline-1-oxide (4-NQO) had potent carcinogenic activity on the skin of mice [67]. Since then a large amount of data has accumulated on the biological properties of this compound. Thus, 4-NQO has been characterized as having a wide spectrum of biological activities such as phage induction [68], mutagenicity [69-72], high toxicity to a single cell system [73], malignant transformation [74,75], chromosome aberration in mammalian cells in tissue [76,77], and photodynamic activity [78].

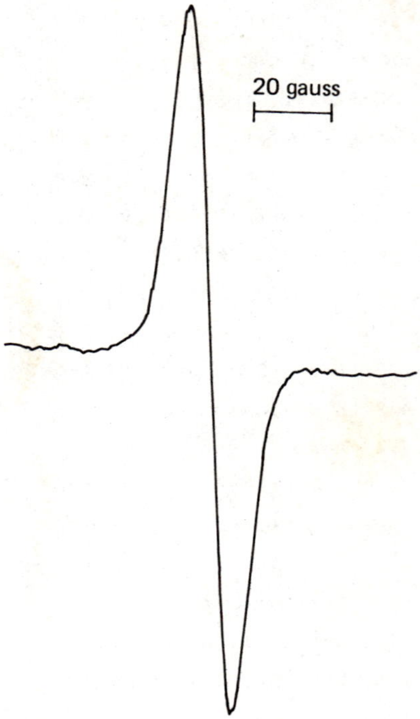

Figure 13 ESR signal of the covalently-bound complex between 2-OH-DMBA and poly(G). Reaction conditions and sample preparation were the same as in Figure 7, except that the incubation time was 48 hr. Diluted H_2O_2 (0.5 ml H_2O_2/10 ml H_2O), 0.1 ml, was added to accelerate free radical formation. Modulation amplitude: 4 G.

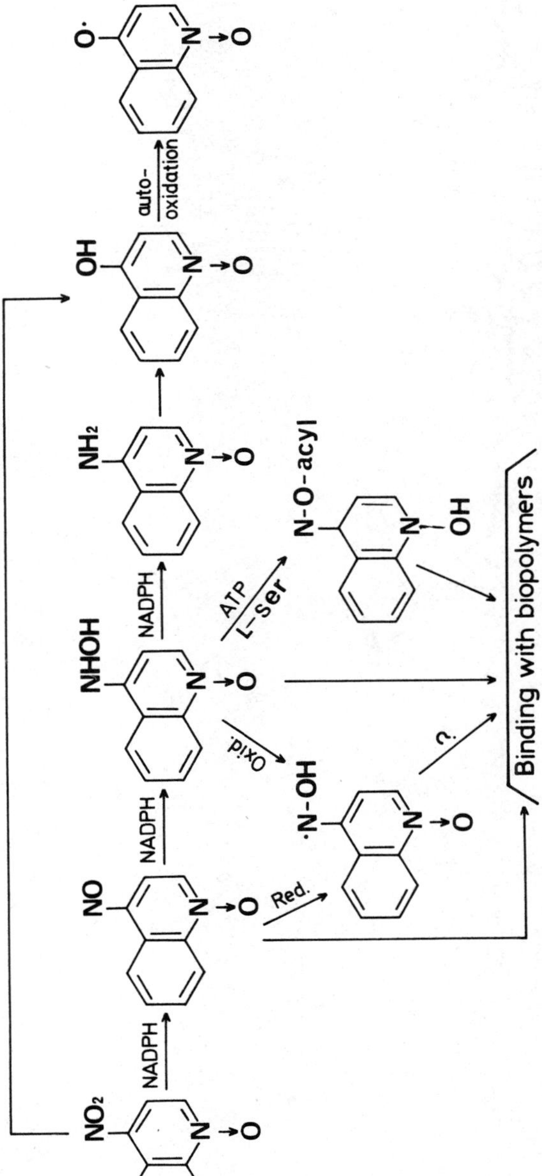

Figure 14 Metabolic pathways of 4-nitroquinoline-1-oxide and free radicals produced from the metabolites.

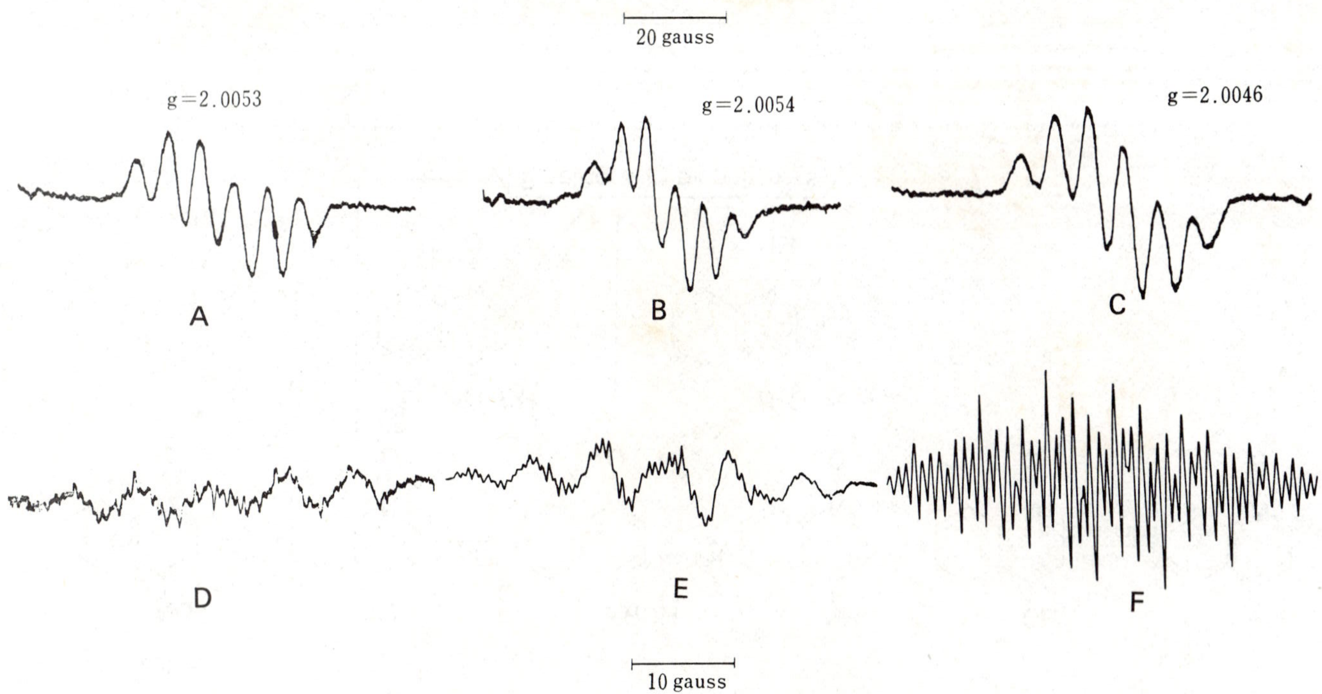

Figure 15 ESR signals of 4-HAQO in water (A and D), methanol (B and E), and dioxane (C and F). Modulation amplitude: 4 G for A, B, and C; 0.7 G for D, E, and F. (From Ref. 23.)

Special effort has been devoted to clarifying the carcinogenic mechanism of 4-NQO, and a scheme has been generally accepted whereby 4-NQO exerts its activity after being reduced to 4-hydroxyamino-quinoline-1-oxide (4-HAQO), the seryl form of which has been proven to bind to biopolymers [79-81] (Fig. 14). In connection with this scheme, it is to be emphasized that a free radical is generated from 4-HAQO without activation [23] and that the same free radical is generated from 4-nitrosoquinoline-1-oxide [82].

A. Free Radical Produced from 4-Hydroxyaminoquinoline-1-oxide (4-HAQO) and 4-Nitrosoquinoline-1-oxide

4-HAQO, a proximate form of 4-NQO, is unstable and was found to be converted to a free radical in organic as well as in aqueous solution [23] (Fig. 15). This is in sharp contrast to 4-NQO, where an ESR signal was observed only after photoirradiation in organic solvents [83,84].

1. Determination of the chemical structure of the free radical

In order to identify the free radical, three ^{15}N-labeled derivatives (the ring nitrogen, the hydroxyamino nitrogen, and the compound with both of these nitrogens labeled) and derivatives of 4-HAQO having the hydrogens replaced by deuterium were synthesized. ESR spectra of free radicals of these derivatives were analyzed by comparison with each other and with the ESR spectra of the free radical of the parent compound [85]. As a result, the coupling constants of nitrogen and protons were determined as indicated in Table 5. Bearing these values in mind and also taking into account the fact that the free radical was oxidatively formed from 4-HAQO [23], three candidate forms were considered for the structure of the free radical from 4-HAQO (Fig. 16). Spin densities for these candidate structures were calculated by means of the unrestricted SCF MO method (Table 5). Comparing the calculated coupling constants with calculated ones, structure A seems to be the most plausible one of the free radicals produced from 4-HAQO. Thus, only structure A conforms with the observed result that the coupling constant of H-3 is larger than that of H-2. Furthermore, the relative magnitude of the coupling constant of H-5 in structure A best fits the observed value (Table 5).

Table 5 Comparison of the Observed Hyperfine Coupling Constants for Structures A, B, and C in Figure 16 with Those Calculated by Means of Unrestricted SCF MO Method[a]

Atom	Hyperfine coupling constant, A_H — Calculated values for structure A	B	C	Observed
N_{ring}	4.71	6.15	1.85	5.9
N_{NHOH}	11.03	2.00	8.53	5.9
H-3	4.68	5.83	5.15	5.9
H-2	4.21	8.28	6.95	3.0
H-5	1.69	3.87	2.07	1.5
H-6	1.13	2.50	1.51	0.74
H-7	1.17	2.61	1.55	0.74
H-8	1.13	2.45	1.69	0.74

[a]The coupling constant for proton, A_H, was estimated by the equation $A_H = Q\rho_C$ where ρ_C is the spin density at the carbon atom to which the proton is attached and Q is as usual taken as 22.5 G. For the nitrogen atom, a Q value of 25.3 G was used.

Source: From Ref. 85.

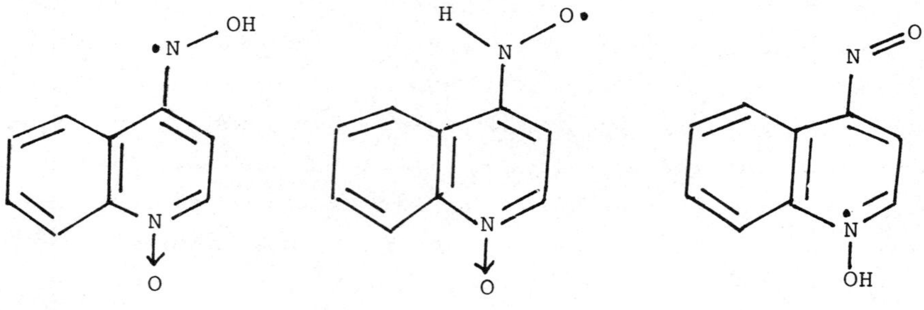

A B C

Figure 16 Three possible structures for the free radical produced from 4-HAQO.

2. *Conditions for the formation of the free radical*

Whereas a large percentage of 4-HAQO was converted to the free radical in alkaline aqueous or alcohol solution (>pH 8) or in organic solvents (dioxane and benzene), no free radical was detected in acidic solutions [23]. Free radical formation was accelerated by addition of H_2O_2, benzoyl peroxide, or potassium permanganate [86]. On the other hand, addition of benzaldehyde, catalase, or ascorbic acid prevented formation of the free radical or quenched the produced radical [23].

3. *Free radical formed from 4-nitrosoquinoline-1-oxide*

As seen in the scheme of Figure 14, 4-nitrosoquinoline-1-oxide has been anticipated as an intermediate between 4-NQO and 4-HAQO. It was thought for a long time that 4-nitrosoquinoline-1-oxide was too unstable to be synthetically obtained, but recently it has been successfully synthesized [87]. 4-Nitrosoquinoline-1-oxide yielded a larger ESR signal in organic solvents such as dioxane, ethanol, and dimethyl sulfoxide—and in water, although the signal was considerably smaller than that of organic solvents [82]. After degassing a dioxane solution of 4-nitrosoquinoline-1-oxide down to 10^{-3} torr, a hyperfine structure of 52 lines was obtained. This agrees entirely with the signal obtained from 4-HAQO and clearly shows that 4-nitrosoquinoline-1-oxide was converted to the same free radical as that formed from 4-HAQO. The free radical thus formed did not decay even under anaerobic conditions in the degassed ESR tube. On the other hand, the ESR signal produced from 4-HAQO decayed progressively in the degassed tube, which means that the free radical is formed reductively from 4-nitrosoquinoline-1-oxide in contrast to the case of 4-HAQO, where the free radical was produced oxidatively (Fig. 14). All this is consistent with the finding that addition of a reducing agent such as sodium borohydride to the dioxane solution of 4-nitrosoquinoline-1-oxide results in a marked increase of the ESR signal whereas the signal of 4-HAQO is decreased [82].

It is not certain whether this particular free radical is involved in carcinogenesis of 4-NQO. However, it is significant that the precursors of this radical, 4-nitrosoquinoline-1-oxide and 4-HAQO, bind to poly(G) to a considerable extent [82]. Thus, both 4-nitrosoquinoline-1-oxide and 4-HAQO were found to actively bind to biopolymers, with the former compound being more active than the latter one. On the other hand, 4-NQO did not react to form a complex (Fig. 17). Such results suggest that in addition to the activation pathway to seryl-4-HAQO, 4-nitrosoquinoline-1-oxide and 4-HAQO themselves could be ultimate carcinogens. Recent experiments have shown that 4-nitrosoquinoline-1-oxide is highly carcinogenic in mice [88], which may substantiate the above suggestion.

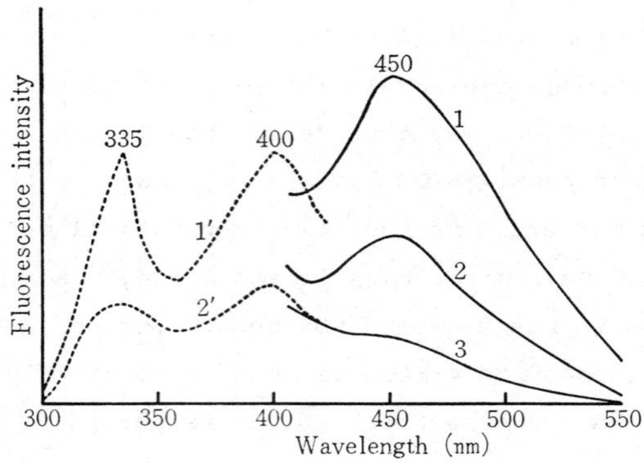

Figure 17 Emission spectra (solid lines) with excitation at 359 nm and excitation spectra (dotted lines) with emission at 460 nm of poly(G)-bound 4-nitrosoquinoline-1-oxide (1 and 1'), 4-HAQO (2 and 2'), and 4-NQO (3). (From Ref. 82.)

4. Enzymatic formation of free radical from 4-NQO

Figure 18 shows the ESR signal obtained by incubating 4-NQO with liver homogenate (9,000 x *g*) in the presence of NADPH. The microsomal fraction was inactive in forming the free radical in contrast to the liver homogenate. The triplet structure with equal coupling constant as shown in Figure 18A indicates that the unpaired electron is localized largely on the nitrogen atom. By degassing the sample tube to 10^{-3} torr, a hyperfine structure was obtained (Fig. 18B). This signal pattern was quite different from that of the free radical formed from 4-HAQO shown in Figure 15. Various forms were considered as candidates for the free radical, among which the 4-oxy radical was thought to be the most likely in view of the formation

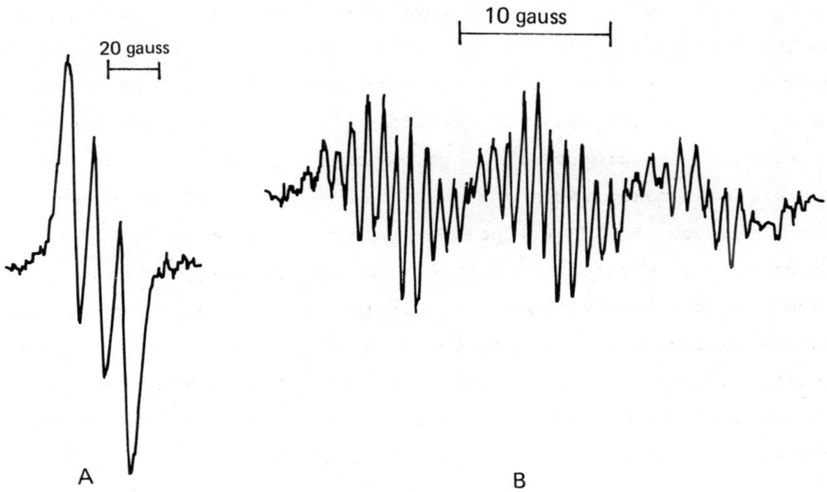

Figure 18 ESR signal obtained in an enzymatic reaction by incubating 4-NQO with supernatant of liver homogenates (9,000 x g) in 0.1 M phosphate buffer (pH 7.4) containing NADPH (0.8 mM). Prior to incubation 4-NQO (10^{-4} M) was added. Incubation was at 37° for 10 min. After incubation, 4-NQO and its metabolites were extracted with secondary butanol. After evaporation to dryness, 0.3 ml of dioxane was added for ESR measurement. (A) Before degassing; modulation amplitude: 4 G. (B) Degassed to 10^{-3} torr; modulation amplitude: 0.8 G.

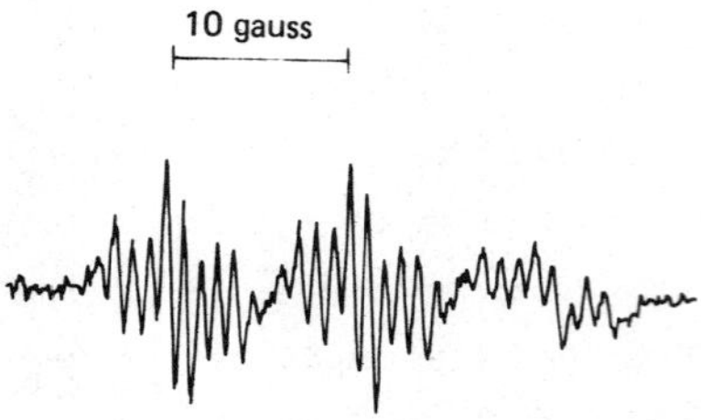

Figure 19 Hyperfine structure of the free radical produced from 4-hydroxyquinoline-1-oxide by abstracting a hydrogen atom with ceric sulfate in dimethylsulfoxide. Degassed to 10^{-3} torr. Modulation amplitude: 0.8 G.

of 6-oxy radical from B(a)P and 6-nitro-B(a)P. 4-Hydroxyquinoline-1-oxide was synthesized, and the 4-oxy radical was formed by abstracting a hydrogen atom from the hydroxyl group with ceric sulfate (Fig. 19). The hyperfine structure thus obtained was entirely the same as that of the free radical obtained in the enzymatic reaction (Figs. 18B and 19). In this case, it is difficult to determine whether the oxy radical or its precursor, 4-hydroxyaminoquinoline-1-oxide, is formed via four metabolic steps (as shown in Figure 14) or directly formed. However, the direct pathway is considered more likely to occur in view of the following facts: the oxy radical can be obtained in pure form without contamination of the N-hydroxy radical which was formed from 4-HAQO or 4-nitrosoquinoline-1-oxide. Further, even after a short incubation time (10 min), a large amount of the oxy radical was formed, which suggests that the oxy radical is not formed via several metabolic steps but directly from 4-NQO.

IV. N-NITROSO COMPOUNDS

N-Alkyl-N-nitroso compounds have many biological activities such as inactivation of transforming activity [89], mutagenicity [90-92], carcinogenicity [93-95], carcinostatic activity [96,97] and prophage induction [90]. In connection with these activities, methylation and deamination of nucleic acid bases, reaction with proteins, and metabolism in animal cells have been studied [98-103]. Another characteristic reaction of N-alkyl-N-nitroso compounds, especially of N-methyl-N'-nitro-N-nitrosoguanidine (MNNG), is the denitrosation reaction, the main product of which is N-methyl-N'-nitroguanidine. Although this product was proven to be noncarcinogenic, one cannot exclude participation of the denitrosation reaction in MNNG carcinogenesis. Actually, the nitroxide radical was proven to be formed during the course of the denitrosation reaction [104,105]. The released NO radical was shown to be sufficiently reactive to form a complex with nonheme protein in liver and other tissues [22].

A. Photochemical Formation of Nitroxide Radical from MNNG and Its Related Compounds

MNNG, N-ethyl-N'-nitro-N-nitrosoguanidine (ENNG), N-propyl-N'-nitro-N-nitrosoguanidine (PNNG), N-butyl-N'-nitro-N-nitrosoguanidine (BNNG), and methylnitrosourea (MNU) were irradiated in oxygen-saturated organic solvents such as benzene, tetrahydrofuran, and methanol and found to be converted to free radicals [105]. As seen in Figure 20, MNU yielded a signal composed of six peaks, as did MNNG, whereas ENNG, PNNG, and BNNG produced a signal with five peaks. Replacement of a methyl group by ethyl, propyl, or butyl is responsible for the change in the ESR signal from six to five peaks, and this indicates that the proton of the methyl or methylene group adjacent to the nitrogen atom interacts with the unpaired electron through hyperconjugation. Photoirradiation of degassed solution gave no signal, indicating that oxygen is essential for production of the free radical.

For identification of the free radical, ^{15}N-methyl-N'-^{15}N-nitrosoguanidine (<u>1</u>), N-methyl-N'-nitro-N-nitroso(^{15}NO)-guanidine (<u>2</u>), and deuterated MNNG (<u>3</u>) were synthesized. ESR spectra of the synthesized compounds were obtained [105]. After illumination of compound 1 in benzene, an ESR signal composed of five peaks was obtained (Fig. 21, <u>1</u>). This shows that the nuclear spin of the nitrogen atom of the N-methyl group interacts appreciably with the spin of the unpaired electron. On the other hand, the signal of compound <u>2</u> is exactly the same as that of MNNG (Fig. 21, <u>2</u>), indicating that the nitroso group of MNNG is not included as a part of the free radical. In fact, denitrosation by photoirradiation proved this point [104].

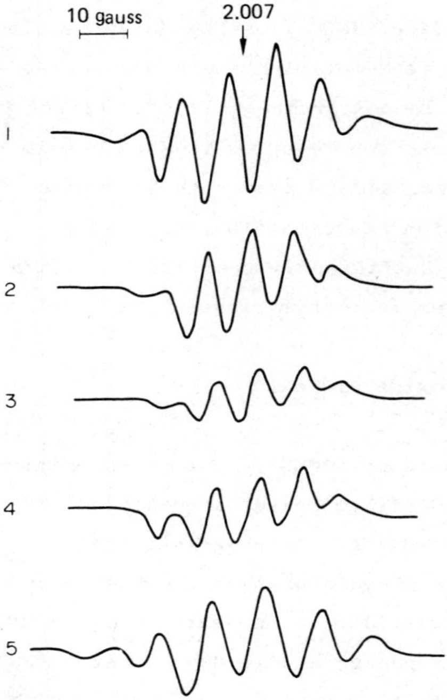

Figure 20 Hyperfine structures of ESR signals produced by photo-irradiation in benzene solution of (1) MNNG, (2) ENNG, (3) PNNG, (4) BNNG, and (5) MNU. Irradiation times for those compounds were 3 min, 15 min, 30 sec, 10 sec, and 3 min, respectively. (Reprinted with permission from Y. Ioki, A. Imamura, and C. Nagata, *Photochem. Photobiol. 21*. Copyright 1975, Pergamon Press, Ltd.)

The signal of compound 3 is exactly the same as that of MNNG (Fig. 21, 3), which shows that the imino group does not effectively interact with the unparied electron. Bearing these facts in mind, the structures shown in Figure 22 are proposed as candidates for the free radical. The coupling constant and total energy were calculated by means of the INDO (Intermediate Neglect of Differential Overlap) MO method (Table 6). After comparison between the calculated and observed coupling constants, structure A was rejected. Further, taking into account the fact that the free radical with the same ESR signal as that from MNNG was produced from N-methyl-N,N'-dinitroguanidine in a degassed tube, structure C was also dismissed,

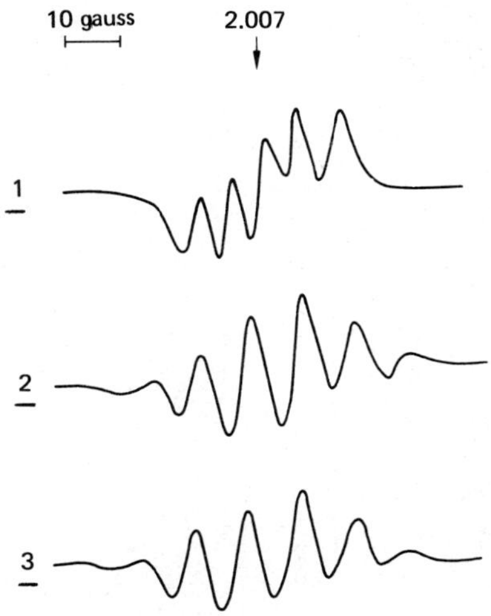

Figure 21 Hyperfine structures of ESR signals produced by 3-min illumination of isotopically replaced MNNG in benzene solution. Curves: (1) ^{15}N-methyl-N'-nitro-^{15}N-nitrosoguanidine; (2) N-methyl-N'-nitro-N-nitroso(^{15}NO)-guanidine; (3) D-replaced MNNG. (Reprinted with permission from Y. Ioki, A. Imamura, and C. Nagata, *Photochem. Photobiol. 21*. Copyright 1975, Pergamon Press, Ltd.)

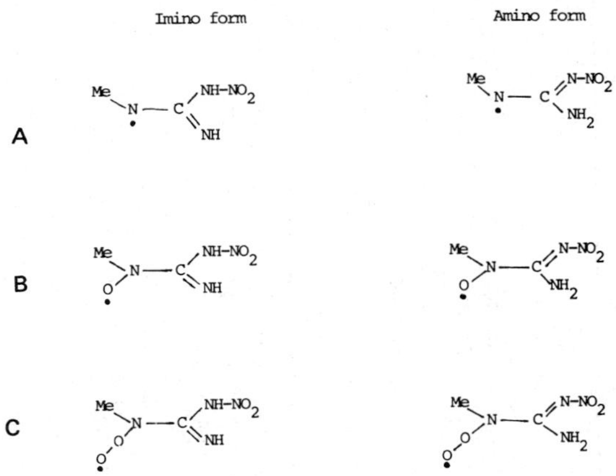

Figure 22 Possible chemical structures of the free radical. (Reprinted with permission from Y. Ioki, A. Imamura, and C. Nagata, *Photochem. Photobiol. 21*. Copyright 1975, Pergamon Press, Ltd.)

Table 6 Comparison of Calculated Coupling Constants and Total Energies for Various Possible Structures of MNNG with those Observed

Structure		Calculated coupling constants (gauss)				Total energy (a.u.)
		$a_{H(CH_3)}$	$a_{N(N-CH_3)}$	$a_{H(imino\ or\ amino)}$	$a_{N(imino\ or\ amino)}$	
Structure A	Imino	1.7	-2.4	23.3	33.8	-97.1506
	Amino	20.7	13.6	-1.4	0.6	-97.2163
Structure B	Imino	5.0	5.0	-3.4	1.9	-114.9170
	Amino	4.4	4.3	-0.7	0.4	-114.9414
Structure C	Imino	7.5	8.7	-4.9	2.3	-132.4031
	Amino	6.1	6.7	-0.9	0.7	-132.3909
Observed values for MNNG[a]	—	9.6	8.5	<~0.5	<~0.5	—

[a]The observed coupling constants for nitrogen and proton were calculated from the hyperfine structures of the parent and isotope-labeled MNNG.

Source: From Ref. 105.

because it is known that under anaerobic conditions the peroxy radical is not easily formed from N-methyl-N,N'-dinitroguanidine. Calculated coupling constants from the amino form but not the imino form of structure B agree well with the observed values. Furthermore, the calculated total energy predicts that the amino form is more stable than the imino form. Based on these results, the amino form of structure B was adopted as the most probable structure of the free radical [105].

B. Free Radical Produced by Stirring MNNG in Buffer Solution

Exactly the same free radical as the photoinduced nitroxide radical was produced by stirring MNNG in buffer solution [104]. A mixture of 10 mg of MNNG and 20 mg of albumin in 10 ml of distilled water (or buffer solution) was stirred by a magnetic stirrer in a light-shielded glass flask at 4°. After stirring, the samples were lyophilized in vacuum. Stirring for 1 hr was sufficient to produce the nitroxide radical, although the signal was small. Production of the nitroxide radical increased with increased stirring time, but after some 2-3 days, it reached a plateau. Stirring for more than 3 days did not increase the signal further.

Lysine, oxidized and reduced glutathione, histone, and DNA produced the same level of signal as albumin. This indicates that the presence of a specific type of biomolecule is not necessary for the formation of the nitroxide radical. In fact, the same level of ESR signal was measured after MNNG was stirred in buffer solution and then lyophilized in the presence of albumin added to prevent the escape of MNNG as was obtained when albumin was present during stirring. Of course, this does not entirely exclude the participation of albumin or other biomolecules in the formation or in the stabilization of the nitroxide radical. An especially large ESR signal was observed when cysteine or cystine was included during the stirring of MNNG [104].

The amount of nitroxide radical produced depends on the pH value. The optimum pH was in a range between pH 3 and 6. The signal at pH 7 was much smaller than that at pH 6, and only a trace was seen at pH 8. No signal was observed in ranges above pH 9 and below pH 2.

The distinct signal seen in the case of MNNG was not observed when N-ethyl-, N-propyl-, and N-butyl-N'-nitro-N-nitrosoguanidines and N-methyl-N-nitrosourea were stirred. This is probably due to the poor solubility of these compounds in water.

C. Formation of Paramagnetic Complex by Mixing MNNG with Liver Supernatant

Following incubation of a mixture of MNNG, liver supernatant (105,000) x *g*) and NADH, and subsequent drying of the sample yielded a characteristic ESR signal with g value of 2.039, 2.015, and 2.007 (Fig. 23, curve a) [22]. Mitochondria, microsomes, and the nuclear fraction were ineffective in producing the paramagnetic complex. Liver supernatant and NADH without MNNG did not yield the characteristic signal although a symmetrical signal with g value of 2.007 was observed (Fig. 23, curve b). Paramagnetic complex formation required NADH or NADPH. When NAD was substituted for NADH or NADPH, the signal was considerably smaller, and when cofactors were omitted the signal was almost negligible (Fig. 23, curve c). Supernatant from perfused and nonperfused liver yielded the same amount of signal, indicating that blood has little effect on the formation of this paramagnetic complex. Interestingly, the same ESR signal was observed after the liver was perfused with an aqueous solution of MNNG in vivo and lyophilized. Perfusion of a KCl solution without MNNG was ineffective in producing this specific ESR signal. However, the symmetrical signal due to the liver itself was present.

1. *Identification of the paramagnetic complex*

When the lyophilized sample was heated in a water bath at 100°, the signal with g = 2.007 increased with time of heating whereas the others scarcely changed. Because of this different behavior toward

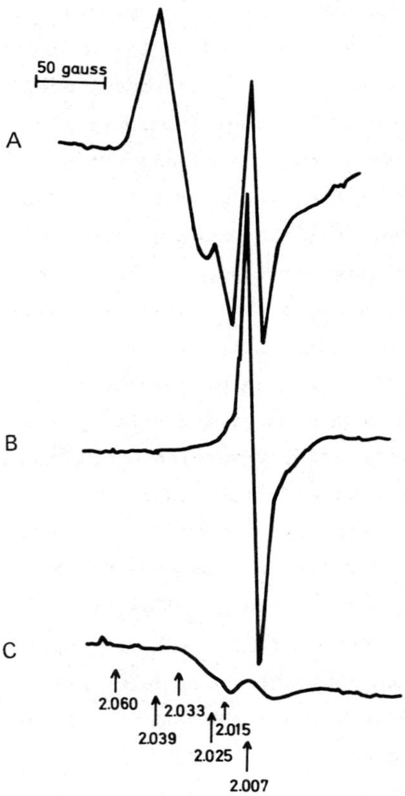

Figure 23 ESR signals of lyophilized samples of the following mixtures: (A) MNNG, liver supernatant, and NADH; (B) liver supernatant and NADH; and (C) liver supernatant and MNNG. Five milliliters of the supernatant were mixed with 5 ml of 0.01 M phosphate buffer containing 25 mg of NADH, to which 5 mg of MNNG in 0.05 ml acetone was added. The reaction mixture was incubated at 37° or maintained at room temperature for 5 min and then lyophilized. (From Ref. 22.)

heating, the signal with g = 2.007 is considered to be a species different from the components with g = 2.039 and 2.015. Based on the similarity of g value, the signal with g = 2.007 was considered to be the same as that obtained in the case of liver supernatant without MNNG (Fig. 23B). When $FeCl_2$ was added, the signal increased considerably, especially in the region of g = 2.039 and 2.015. The paramagnetic-complex formation was completely inhibited when the liver supernatant was dialyzed against ethylenediaminetetraacetate (EDTA) or citric acid beforehand. No change of the signal was found when the liver supernatant was dialyzed against KCl (0.15 M). This suggests that some nonheme iron is involved in the paramagnetic components. Furthermore, sulfhydryl compounds such as L-glutathione (reduced and oxidized), 2-mercaptobenzoic acid, 2-mercaptoethylamine, L-cysteine ethyl ester, and L-cysteine were effective in producing the same ESR signal upon mixing these sulfhydryl compounds with MNNG in aqueous $FeCl_2$ solution. From these results, it was assumed that the characteristic ESR signal originated from some paramagnetic complex between MNNG and an iron-thiol component in the liver supernatant. In this connection, it is worth noticing that the signal was very similar to that obtained in vivo by Commoner and associates from the liver of rats fed hepatic carcinogens such as AAF and dimethylaminoazobenzene (DAB) [106-110]. In light of the study of McDonald and coworkers [111], the signal obtained in vivo was identified by Commoner and associates as an iron-nitric oxide-thiol protein complex [108-111]:

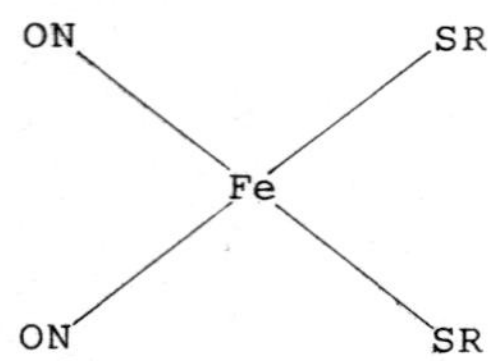

(I)

Although the structure of the paramagnetic complex formed by mixing MNNG with liver supernatant has not yet been completely identified, structure (I) is the most probable candidate.

2. *Conditions for the formation of the paramagnetic complex*

When the heat-treated (100° for 20 min) or pronase-treated liver supernatant was used, the ESR signal was somewhat larger than in the case of nontreated liver supernatant. Therefore, formation of the paramagnetic complex is not due to an enzymatic reaction. The enhancing effect of heat or pronase treatment may be explained by an unmasking of SH groups in the proteins. Formation of the paramagnetic complex was dependent upon the pH value of the reaction mixture; optimum values were obtained in a range between 5.3 and 7.4.

3. *Formation of the paramagnetic complex by mixing MNNG with supernatant of stomach mucosa or other tissues*

It is notable that the same type of paramagnetic complex was detected by mixing MNNG with supernatant of stomach which is a target organ of MNNG. Glandular stomach was just as effective in producing the specific paramagnetic complex as liver, but no complex was produced by the forestomach. Other organs such as intestines, lung, and brain were far less effective in producing the paramagnetic complex (Table 7).

Table 7 Formation of Paramagnetic Complex by Mixing MNNG with Various Kinds of Normal Tissues

Supernatant (105,000 x *g*) from tissue of:	Magnitude of ESR signal
Liver	50
Kidney	50
Glandular stomach	50
Large intestine	25
Small intestine	12
Lung	10
Brain	5
Duodenum	5
Forestomach	0

Table 8 The Paramagnetic Complex Produced by Mixing MNNG with Normal Liver or Cancerous Tissues

Supernatant (105,000 x *g*) from tissue of:	Magnitude of ESR signal
Normal liver	50
Regenerating liver	30
Morris hepatoma 7316A	20
Morris hepatoma 7794A	20
Liver from tumor-bearing rats	15
Ascites hepatoma (AH-130)	0
Yoshida sarcoma	0

4. *Formation of the paramagnetic complex in normal liver and cancerous tissues*

No ESR signal could be observed by mixing MNNG with supernatants of ascites hepatoma (AH-130) or Yoshida sarcoma. AH-130 was transplanted to the back of rats. The liver from these tumor-bearing rats was much less effective in its ability to produce the specific paramagnetic complex than was the liver of normal animals; on the other hand, regenerating liver and Morris hepatoma were intermediate between normal liver and hepatoma or sarcoma (Table 8).

Although no ESR signal could be detected for the cancerous tissue, the signal appeared upon addition of ferrous ion as indicated in Figure 24. The signal increased gradually as the ferrous ion content was increased, reached a maximum, and then began to decrease. On the other hand, for normal liver the signal increased sharply as the ferrous ion content was increased.

5. *The component in the liver responsible for the formation of the paramagnetic complex*

In order to identify the component responsible for the formation of the paramagnetic complex, liver components were fractionated with Bio-gel P-200. The ability of each fraction to form the paramagnetic complex was tested using the ESR method. As can be seen in Figure 25, the components responsible for the formation of the paramagnetic complex were included in the rapidly eluted fraction, the molecular weight of which is around 6-7 x 10^4.

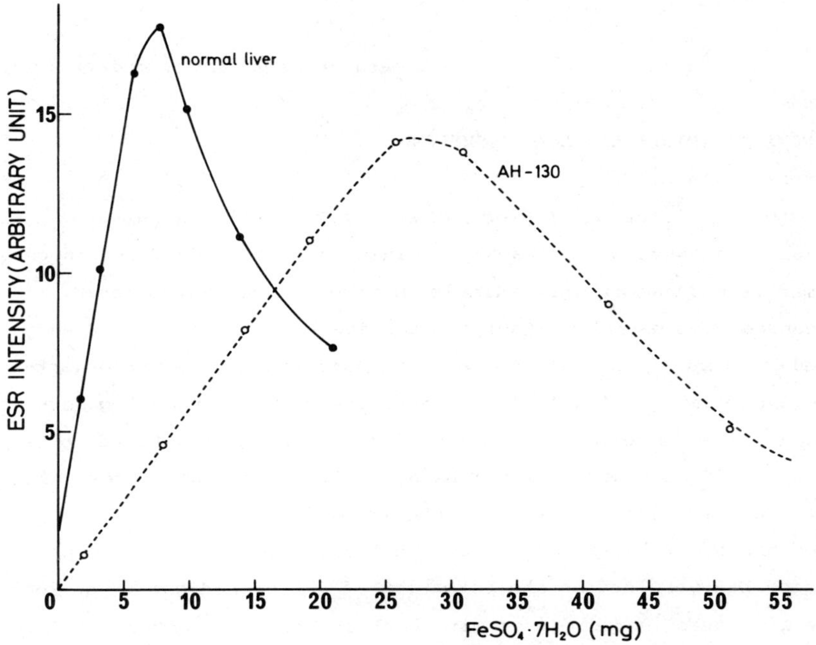

Figure 24 Effect of ferrous ion on the formation of the paramagnetic complex. The experimental conditions are the same as in Figure 23.

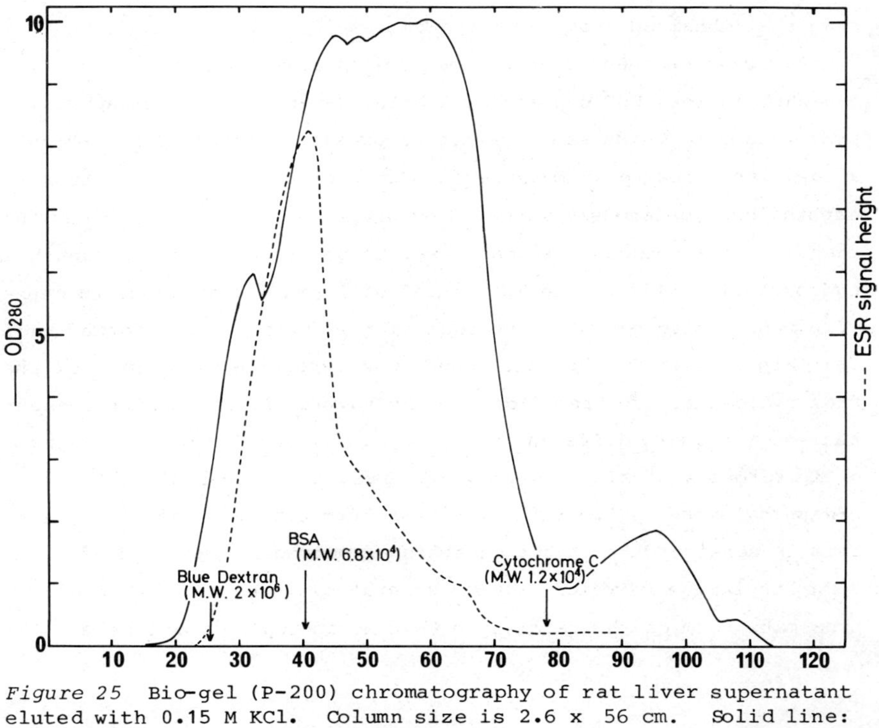

Figure 25 Bio-gel (P-200) chromatography of rat liver supernatant eluted with 0.15 M KCl. Column size is 2.6 x 56 cm. Solid line: OD_{280}. Dotted line: magnitude of the ESR signal (arbitrary unit).

V. AROMATIC AMINES AND AZO COMPOUNDS

A. Aromatic Amines

1. *Enzymatic formation of free radical from naphthylamines*

Activation pathways of the naphthylamines, strong bladder carcinogens in human as well as in test animals, have been well established. Thus nonreactive naphthylamines are considered to be activated enzymatically to such proximate forms as the aminonaphthols and N-naphthyl-hydroxylamines [112-117]. As described in the preceding sections, a large number of chemical carcinogens have been found to be converted to free radicals after being activated to their proximate forms. This is also the case for the aromatic amines.

Incubation of 2-naphthylamine with liver microsomes in the presence of NADPH resulted in the formation of a free radical (Fig. 26A). In the absence of cofactors or with heat-treated microsomes, no ESR signal could be detected. This indicates that the free radical is formed enzymatically. 1-Naphthylamine also gave an ESR signal after incubating with liver microsomes, but the signal was much less strong than that obtained from 2-naphthylamine.

In order to identify the free radical formed enzymatically from 2-naphthylamine, ESR signals of 2-amino-1-naphthol and 2-naphthylhydroxylamine, which are regarded as proximate forms of 2-naphthylamine, were measured. Measurements were also taken for 2-nitrosonaphthalene (generously supplied by Dr. M. Okada, Tokyo Biochem. Res. Inst.). This compound is considered to be produced from 2-naphthylhydroxylamine [117]. The ESR signal of 2-amino-1-naphthol in benzene (Fig. 26B) is very similar to that of the enzymatically formed radical (Fig. 26A). On the other hand, the hyperfine structures of the free radical formed from 2-naphthylhydroxylamine and 2-nitrosonaphthalene are quite different from the signal pattern of the enzymatically formed radical (Fig. 26C,D). Thus, the free radical from 2-naphthylamine is identified as that from 2-amino-1-naphthol, and this is consistent with the established metabolic pathways of 2-naphthylamine in which 2-amino-1-naphthol is regarded as a more proximate form of the carcinogen than 2-naphthylhydroxylamine [117].

2. *Ability to form free radicals and carcinogenicity of aminonaphthols and naphthylhydroxylamines*

Whereas 2-naphthylamine did not yield an ESR signal in organic solvents or water, its proximate forms, 2-amino-1-naphthol and 2-naphthylhydroxylamine, yielded large ESR signals in organic solvents such as dioxane or benzene [24]. An ESR signal from these two compounds was also observed in alkaline or in acidic aqueous solutions. However, in neutral aqueous solution it was difficult to detect the signal because of the insolubility of these compounds. The same was true for 1-amino-2-naphthol and 1-naphthylhydroxylamine. On the other hand, the noncarcinogenic aminonaphthols such as 5-amino-1-naphthol, 5-amino-2-naphthol, and 8-amino-2-naphthol did not yield an ESR signal under the same conditions (Table 9). 4-Amino-1-naphthol is known to be noncarcinogenic, but this compound yielded an ESR signal in acidic (<pH 4.5) or alkaline (>pH 12) ethanol as well as in alkaline aqueous solution, though no signal could be detected in dioxane and benzene. Except for this compound, the relationship between the ability to form free radicals and the carcinogenicity of aminonaphthols and naphthylhydroxylamines is fairly well established [24].

B. Acetylaminofluorene

It is now accepted by many that the ultimate reactive form of acetlyaminofluorene (AAF) is produced by esterification of N-OH-AAF, which is itself formed after N-hydroxylation of AAF. As described in Section I, it is necessary to consider alternative routes of activation, because AAF causes tumors of mammary gland and ear duct gland of the rat even though these glands lack the esterification enzyme [3]. The most plausible route in this regard is the free radical mechanism, substantiated by the finding that nitroxide radical was formed from N-OH-AAF. Indeed, Bartsch and Hecker [13] and Floyd et al. [118] have found that in the presence of horseradish peroxidase and H_2O_2, N-OH-AAF is converted to the nitroxide radical which dismutates to form N-acetoxy-AAF and 2-nitrosofluorene (Fig. 27). Following this finding, Floyd et al. demonstrated that in the

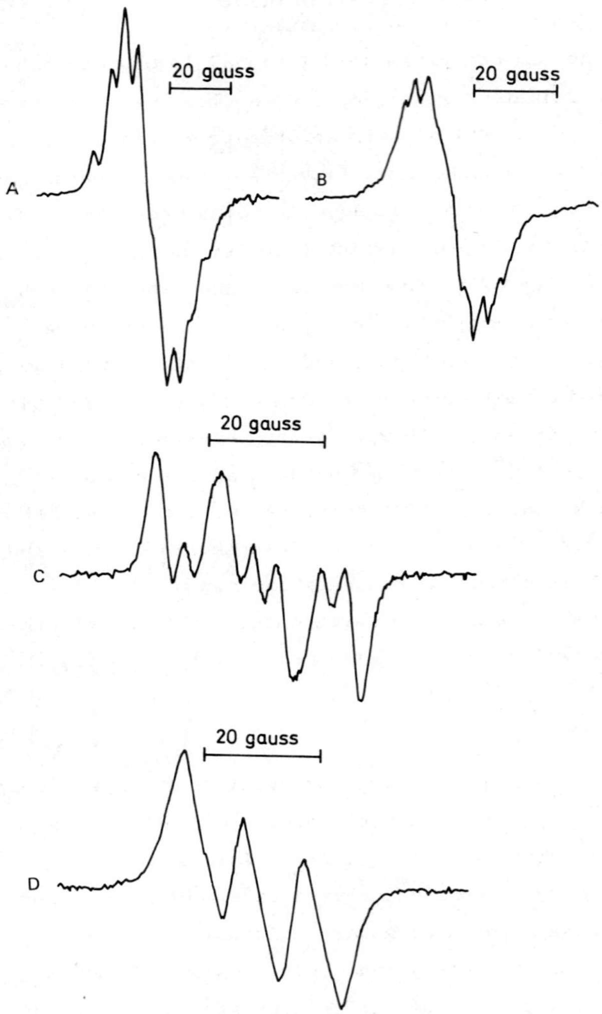

Figure 26 (A) The ESR signal of the free radical produced by incubating 2-naphthylamine with liver microsomes fortified with NADPH. The reaction condition and the sample preparation were the same as in Figure 5. (B, C, and D) ESR signals of 2-amino-1-naphthol, 2-naphthylhydroxylamine, and 2-nitrosonaphthalene in benzene (1 mg/ml), respectively. Modulation amplitude: 4 G.

Table 9 Relation Between Free Radical Formation Capacity and Carcinogenicity of Aminonaphthols, Naphthylhydroxylamines, and Naphthols

Compound	Free radical formation in Dioxane	Free radical formation in Ethanol	Free radical formation in Acidic or alkaline Aqueous Solution	Carcinogenicity
2-Amino-1-naphthol	+	+	+	+
1-Amino-2-naphthol	+	+	+	+
2-Naphthylhydroxylamine	+	+	+	+
1-Naphthylhydroxylamine	+	+	+	+
4-Amino-1-naphthol	−	+	+	−
5-Amino-1-naphthol	−	−	−	−
5-Amino-2-naphthol	−	−	−	−
8-Amino-2-naphthol	−	−	−	−
1-Naphthol	−	−	−	−
2-Naphthol	−	−	−	−

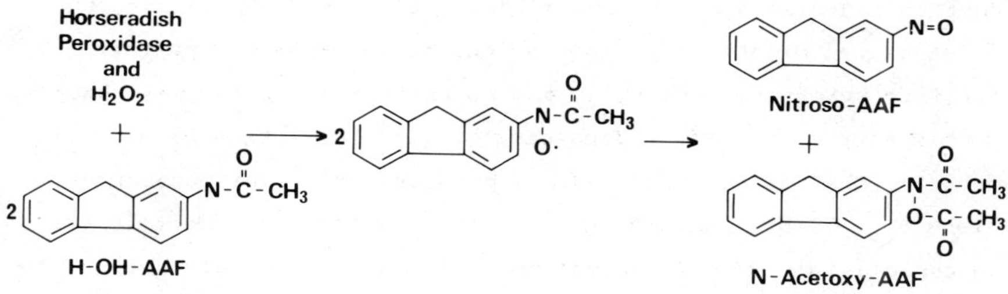

Figure 27 Reaction scheme for the activation of N-OH-AAF into N-acetoxy-AAF and 2-nitroso-AAF. (Proposed by Bartsch and Hecker [13] and Floyd et al. [118].)

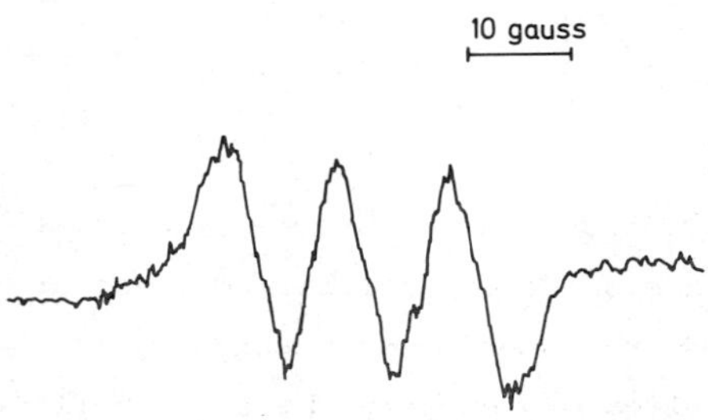

Figure 28 The ESR signal of the free radical produced by incubating AAF with liver microsomes fortified with NADPH. The reaction condition and the sample preparation were the same as in Figure 5. Modulation amplitude: 4 G.

presence of methemoglobin or hematin, N-OH-AAF was activated with lipid hydroperoxide in the same way as with peroxidase and H_2O_2 [12]. Scribner and Naimy proposed that the radical cation is formed upon decomposition of 2-acetoxy-AAF [119]. From these results, it is naturally anticipated that the free radical could be formed during the activation process of AAF in the liver microsome system.

In fact, a triplet ESR signal was detected after incubating AAF with liver microsomes (Fig. 28). The signal pattern of the radical is very similar to that of the nitroxide radical obtained from N-OH-AAF [12,13], but it is difficult to identify it as nitroxide radical for the following reasons: (a) The coupling constant of the free radical is 11 G, which is larger than that of the nitroxide radical of N-OH-AAF, which is ≃7 G [12,13,120]; this coupling constant is nearly the same as that of the ESR spectrum of the 2-nitrosofluorene adduct to 2,3-dimethyl-2-butene [121]. (b) On the other hand, the triplet signal of the nitroxide radical of N-OH-AAF was further split into hyperfine structure, and no further splitting was observed for this signal even after degassing at 10^{-3} torr; this behavior is similar to that of the free radical produced from 2-nitrosonaphthalene (Fig. 26D). Accordingly, it is reasonable to identify the free radical in Figure 28 as due to 2-nitroso-AAF rather than to the nitroxide radical of N-OH-AAF. Probably, in the microsomal system, the nitroxide radical of N-OH-AAF would instantaneously dismutate to form nitrosofluorene and 2-acetoxy-AAF in accord with the proposed reaction scheme of N-OH-AAF [12,13].

C. Aminoazo Compounds

From the accumulated experimental evidence, it has become clear that the metabolic pathways of aminoazo compounds are similar to those of AAF and other aromatic amines. Thus, N-oxidation of secondary and tertiary aminoazo compounds is an obligatory step in their metabolic conversion to the proximate forms, whereas ring-hydroxylation and azo reduction are deemed to be detoxification reactions. The N-hydroxy compound is further metabolized to the esterified form, which is considered to be an ultimate form of aminoazo carcinogens. The ester is easily converted to a reactive electrophile and binds covalently with nucleic acids and proteins [1,2,117,122]. In view of the convertibility of the proximate forms of carcinogens to free radicals, it is expected that the free radical is involved in the activation pathway of aminoazo compounds. Actually the generation of free radicals from 3'-methyldimethylaminoazobenzene (3'-methyl-DAB) and other aminoazo compounds has been demonstrated in an enzyme system in vitro as well as in vivo [18].

1. Free radical formation in an enzyme system

The potent hepatic carcinogen, 3'-methyl-DAB, was shown to be converted to a free radical upon incubation with liver microsomes from MC-treated rats (Fig. 29). The supernatant fraction was ineffective in producing the free radical. No signal was observed when the assay was carried out in the absence of NADPH or with heat-treated microsomes (60°, 10 min). Also the signal was considerably smaller when microsomes from phenobarbital-treated rats or nontreated rats were used. The half-life of the free radical in benzene was about 2 hr in an ESR tube open to air. When 3'-methylmonomethylaminoazobenzene (3'-methyl-MAB) was used instead of 3'-methyl-DAB, a much greater ESR signal having the same hyperfine structure was observed (Fig. 29B). This indicates that the same species of free radical is formed from 3'-methyl substituted DAB and MAB.

In order to identify this free radical, the hyperfine structure was compared with the nitroxide radicals generated from authentic N-OH-MAB and N-hydroxyaminoazobenzene (N-OH-AB) (Fig. 29C,D). As can clearly be seen in the figure, the hyperfine structure of the free radicals generated enzymatically from 3'-methyl-MAB and 3'-methyl-DAB are the same in their signal patterns as the one obtained from N-OH-MAB. This type of ESR signal is characteristic of a nitroxide radical with a methyl group as in the case of MNNG [105], where the coupling of the unpaired electron was restricted to the nitrogen atom and three methyl group protons without spreading to the other conjugated parts in the molecule [105]. Such is the case for

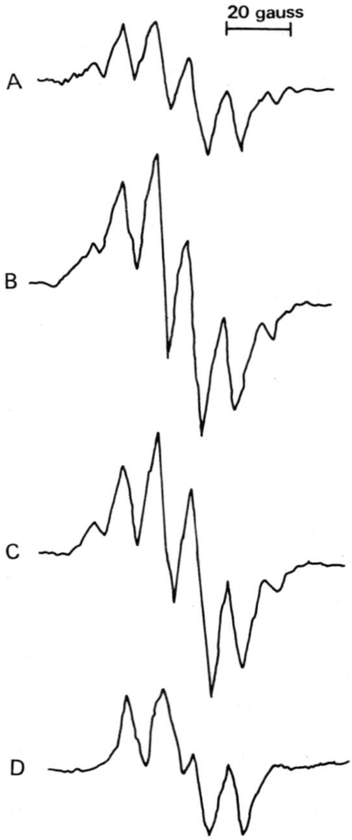

Figure 29 The ESR signals obtained by incubating 3'-methyl-DAB or 3'-methyl-MAB with liver microsomes from MC-treated rats. Incubation conditions and sample preparation are the same as in Figure 5. Curves: (A) 3'-Methyl-DAB; (B) 3'-methyl-MAB; (C and D) authentic N-OH-MAB and N-OH-AB in benzene (1 mg/ml). Modulation amplitude: 4 G. (From Ref. 18.)

the nitroxide radical of N-OH-MAB. Accordingly, the methyl group at the 3' position has no effect on the signal pattern of the nitroxide radical. As a result, the enzymically formed free radical from 3'-methyl-DAB and 3'-methyl-MAB was unambiguously identified as the nitroxide radical of N-OH-MAB, which is of particular interest in regard to the metabolic pathways of MAB proposed by Kadlubar et al. [122] (Fig. 30). Thus, N-hydroxylation of MAB has been proven to be the metabolic activation route resulting in the formation of unstable N-OH-MAB (Fig. 30, path A) even though N-OH-MAB is labile and has never been isolated and detected, except by thin-layer chromatography/mass spectroscopy [122]. All this is consistent with the fact that a large amount of the nitroxide radical is formed—presumably generated from N-OH-MAB.

It is generally believed that N-OH-MAB is a proximate form of MAB and that this metabolite is further activated to the esterified compound [1,2,122]. However, the finding that N-OH-MAB is easily converted to the nitroxide radical suggests a causal significance

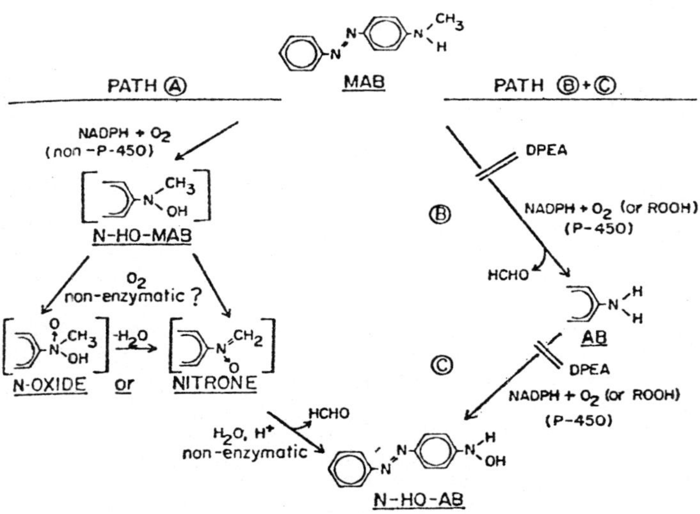

Figure 30 Metabolic oxidation of MAB by rat liver microsomes as proposed by Kadlubar et al. (From Ref. 122.)

of this free radical in carcinogenesis by aminoazo compounds. In fact, it was shown that DAB binds covalently with DNA or protein in the microsomal enzyme system without participation of the esterification enzyme [123]. In this regard, it is also of interest that the amount of free radical produced from several aminoazo compounds parallels the carcinogenicity of these compounds. The order in the amounts of free radical formed is 3'-methyl-MAB > 3'-methyl-DAB > 3'-Cl-DAB > MAB > DAB > 2'-Cl-DAB > 4'-Cl-DAB. The noncarcinogenic 4'-methyl-DAB yielded an extremely small signal. The ESR signal of the free radical produced by incubating AB with liver microsomes was the same as that given by the nitroxide radical obtained from N-OH-AB (Fig. 29D). This presumably corresponds to pathway C shown in Figure 30, which is not directly related to the carcinogenic process.

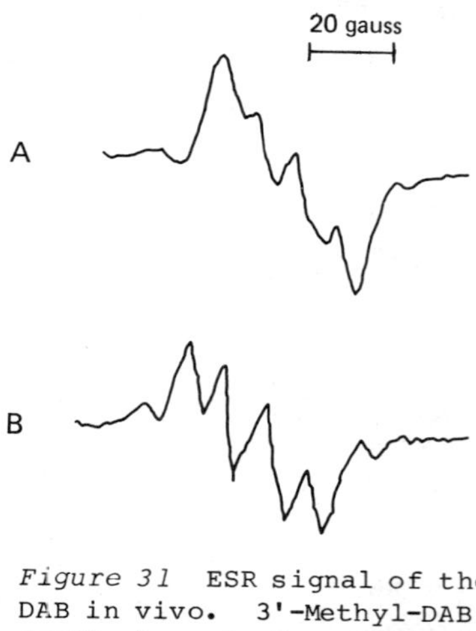

Figure 31 ESR signal of the free radical produced from 3'-methyl-DAB in vivo. 3'-Methyl-DAB (40 mg/2 ml olive oil) was administered orally to male Sprague-Dawley rats and after intervals (6, 12, 24, and 48 hr), the rats were killed and livers were homogenized with 8 vol KCl-Tris buffer in polytron for 1 min. After extraction with an equal volume of benzene, the extract was evaporated to dryness and dissolved in 0.3 ml of benzene for ESR measurement. Curves: (A) 6 hr; (B) 12 hr after oral administration of 3'-methyl-DAB. Modulation amplitude: 4 G.

2. *Free radical formation in vivo*

After oral administration of 3'-methyl-DAB to rats, the liver, containing the dye and its metabolites, was homogenized and extracted with cold benzene. The ESR signal obtained was very similar to that observed in the in vitro experiment, indicating that the same type of free radical was metabolically formed in vivo (Fig. 31). The signal could be detected 6 hr after oral administration, although the hyperfine structure was rather obscure. After 12 hr, a more clearly resolved hyperfine structure was obtained, but after 24 hr and 48 hr, the ESR signals were too small to be resolved into hyperfine structure.

VI. CARCINOGENIC AND CARCINOSTATIC ANTIBIOTICS

Haddow et al. [124] first called attention to a substantial correlation between carcinogenicity and growth-inhibitory power of polycyclic hydrocarbons. Since then abundant data have been accumulated showing that this paradoxical relation holds for a wide variety of chemical carcinogens. This relation is especially important for the practical use of carcinostatic compounds, and considerable efforts have been devoted to clarifying the mechanism underlying the carcinostatic and carcinogenic activities. Subsequently, it has been found that some carcinostatic antibiotics were converted to free radicals during their activation processes.

Biological activity of mitomycin C (MMC) is considered to be due to a cross-linking of the complementary DNA strand [125,126] and also to a single point alkylation of nucleic acids [127]. Chemical reduction was shown to be a requisite for the activation of MMC, and Iyer and Szybalski [125] have found that in order for cross-linking to take place DNA must be added within seconds after completion of the reduction reaction. Probably because of the extremely short lifetime, an attempt to detect the intermediate was unsuccessful. By means of the rapid scan spectrometer and using a flow method in combination with ESR, Nagata and Matsuyama have succeeded in detecting the short-lived intermediate of MMC and have shown that this intermediate exists as the semiquinone radical [19].

Figure 32 shows the ESR signal obtained by mixing MMC with sodium borohydride. No signal could be observed when MMC or sodium borohydride was measured separately. The half-life of this radical was several seconds, and this was considered to correspond to a new absorption band appearing at 518 nm in the rapid scan spectrometer, the half-life of which was several seconds. Since MMC contains the quinone structure in the molecule, it is reasonable to expect the formation of semiquinone radical upon reduction of MMC. Based on the similarity in g value and line width between this short-lived radical ($g = 2.0034$, 9.7 G) and the p-benzosemiquinone radical ($g = 2.0047$, 9.48 G), it was indeed identified as the semiquinone radical of MMC. This result supports the proposition put forth by Murakami that MMC would be converted to the semiquinone radical during its activation process [128]. A scheme showing the activation process of MMC is presented in Figure 33.

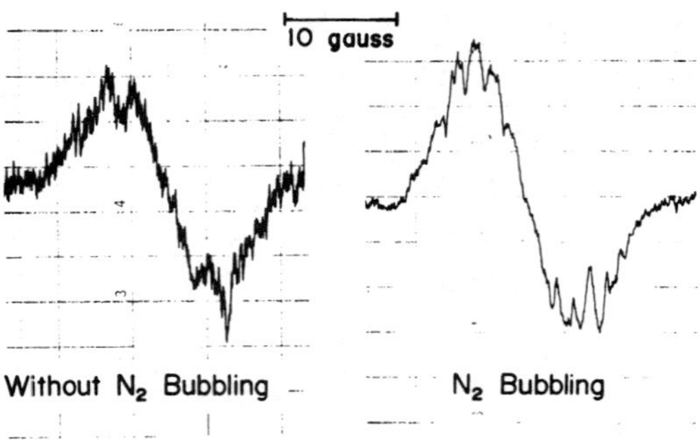

Figure 32 ESR signal of the free radical produced by mixing MMC (100 mg/100 ml water). Measurement was carried out using the flow system of the ESR apparatus. Flow rate was 0.1 ml/sec. (From Ref. 19.)

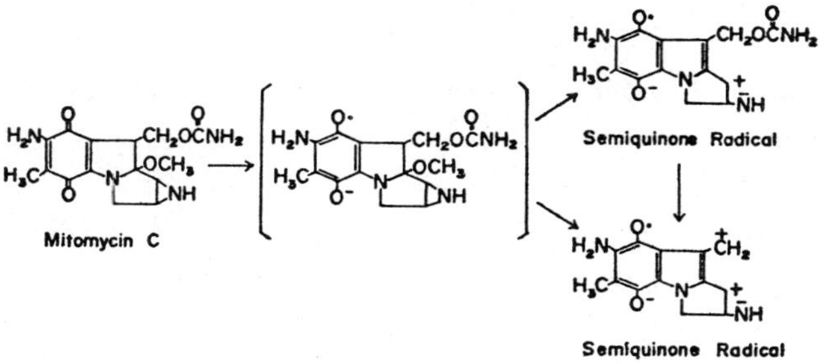

Figure 33 Proposed scheme of the formation of semiquinone radical from MMC. (From Ref. 19.)

Along this same line, the finding of Handa and Sato [20] is of interest. They showed that the quinone-containing antibiotics MMC, daunomycin, and carbazilquinone were able to initiate sulfate oxidation in the presence of microsomes and NADPH, implying that these chemicals were converted to semiquinone radicals. Heat-denatured (80°, 10 min) microsomes were completely ineffective, indicating that the reaction is enzymatic. Anticancer chemicals which do not contain a quinone group, such as cyclophosphoamide, toyomycin, and thioTEPA were totally ineffective. Free radicals, generated from quinone anticancer agents, however, were directly detected [21,129]. Thus, adriamycin, aclacinomycin A, and streptonigrin gave the characteristic ESR signal after mixing with rat liver microsomes in the presence of NADPH. No signal could be detected from MMC due to the short lifetime of the free radical [19].

Ishida and Takahashi found that DNA chain breakage by bleomycin was strongly inhibited by agents such as 2-mercaptoethanol, dithiothreitol, ascorbic acid, and hydrogen peroxide. They suggested that the superoxide radical is involved in this intensification process and postulated that bleomycin is activated by the superoxide radical [130]. Further experiments to detect the active form, probably a free radical, are awaited.

VII. SUMMARY AND CONCLUDING REMARKS

Potent chemical carcinogens such as aromatic hydrocarbons, 4-NQO, aromatic amines, aminoazo compounds, and carcinogenic antibiotics have been shown to form free radicals after enzymatic activation to the proximate forms or during photoirradiation. Direct-acting nitroso compounds were found to be converted to free radicals without enzymatic activation:

1. The 6-oxy-B(a)P radical was detected either by photoirradiation of B(a)P or by incubating it with liver microsomes (or homogenates), lung microsomes (or homogenates), and skin homogenates. In the enzymatic reaction, the precursor of the radical was presumably 6-OH-B(a)P. This oxy radical was proven to be an obligatory intermediate in the conversion of B(a)P to B(a)P-quinones and it was thought to be one of the proximate acting forms of B(a)P. Thus, mixing with 6-OH-B(a)P, a precursor of the 6-oxy-B(a)P radical, resulted in the formation of a covalently bound complex with DNA or polynucleotides, while retaining the free radical structure. Anthanthrene which lacks the bay region was also found to be converted to a free radical in an in vitro enzymatic reaction.

2. 4-HAQO and 4-nitrosoquinoline-1-oxide, which are known as proximate forms of 4-NQO, were easily converted to a free radical. Free radical formation occurred in the absence of either photoirradiation or enzyme activation. The structure of the radical was identified as the N-hydroxy radical. 4-NQO was enzymically hydroxylated to 4-hydroxyquinoline-1-oxide, and this finding seems to add a new metabolic pathway in the activation of 4-NQO.

3. N-Alkyl-N-nitroso compounds, which are reported to be direct-acting carcinogens, were found to be converted to free radicals in the absence of enzymatic activation. The free radical was identified as the nitroxide radical. The

same radical was shown to be formed by photoirradiation. Incubation of N-alkyl-N-nitroso compounds with liver homogenates of liver perfusion with a buffer solution containing these compounds resulted in the release of the NO group. The released NO group formed a complex with SH-containing nonheme proteins. Liver and glandular stomach tissues were very active in forming this complex. However, no such complex was found when the supernatant of cancerous tissues such as ascites hepatoma (AH-130) and Yoshida sarcoma were used. Whereas the ability to form the complex was considerably decreased in liver from tumor-bearing (on the back) rats, regenerating liver was effective. Fractionation of tissue homogenate with Bio-gel P-200 revealed that the component responsible for complex formation was found to have a molecular weight of about 6×10^4.

4. The bladder carcinogen 2-naphthylamine was converted to a free radical upon incubation with liver microsomes. The ESR signal of this radical was similar to that yielded by 2-amino-1-naphthol in benzene. This indicates that the enzymatically formed metabolite is 2-amino-1-naphthol, which is easily converted to a free radical. On the other hand, no ESR signal could be detected for the noncarcinogenic aminonaphthols.

5. Incubation of 3'-methyl-DAB with liver microsomes resulted in the appearance of an ESR signal with six peaks. The same signal was observed in the in vivo experiment in which the dye was administered orally. By comparing its ESR signal with that of N-OH-MAB, this free radical was identified as the nitroxide radical formed from N-OH-MAB. The same ESR signal was observed for 3'-methyl-MAB, 3'-Cl-DAB, MAB, DAB, 2'-Cl-DAB and 4'-Cl-DAB, but was negligibly small for 4'-methyl-DAB.

6. Carcinostatic and carcinogenic quinone antibiotics such as MMC, daunomycin, carbazilquinone, adriamycin, aclacinomycin A, and streptonigrin were found to generate semiquinone radicals during reductive activation processes.

All of the observations outlined above have shown that almost all classes of chemical carcinogens are able to generate free radicals by enzymatic reactions. On the other hand, noncarcinogenic compounds are rather stable and difficult to convert to free radicals under the same conditions. These results suggest a causal relationship of free radicals in chemical carcinogenesis and indicate that further investigation of free radical reactions during carcinogenesis is needed in addition to the extensively studied ionic reaction mechanisms.

REFERENCES

1. J. A. Miller, *Cancer Res. 30*: 559 (1970).
2. J. A. Miller and E. C. Miller, in *Screening Tests in Chemical Carcinogenesis* (R. Montesano, H. Bartsch, and L. Tomatis, Eds.). IARC Publn. No. 12 (1976).
3. C. C. Irving, D. H. Janss, and L. T. Russell, *Cancer Res. 31*: 387 (1971).
4. D. L. Berry, T. J. Slaga, A. Viaje, N. M. Wilson, J. D. Giovanni, M. R. Juchau, and J. K. Selkirk, *J. Nat. Cancer Inst. 58*: 1051 (1977).
5. M. Watanabe, K. Watanabe, and H. Sato, *Gann 70*:83 (1979).
6. M. Kodama, Y. Ioki, and C. Nagata, *J. Cancer Res. Clin. Oncol 98*:105 (1980).
7. J. K. Selkirk, R. G. Croy, P. P. Roller, and H. V. Gelboin, *Cancer Res. 34*: 3474 (1974).
8. M. D. Burke, H. Vadi, B. Jernstrom, and S. Orrenius, *J. Biol. Chem. 252*: 6424 (1977).
9. K. Alexandrov and M. H. Thompson, *Cancer Res. 37*: 1443 (1977).
10. E. Cavalieri, P. Mailander, and A. Pelfrene, *Z. Krebsforsch. 89*: 113 (1977).
11. A. Szent-Györgyi, I. Isenberg, and S. L. Baird, Jr., *Proc. Nat. Acad. Sci. U.S. 46*:1444 (1960).
12. R. A. Floyd, L. M. Soong, R. N. Walker, and M. Stuart, *Cancer Res. 36*: 2761 (1976).
13. H. Bartsch and E. Hecker, *Biochim. Biophys. Acta 237*: 567 (1971).
14. H. Bartsch, M. Traut, and E. Hecker, *Biochim. Biophys. Acta 237* 556 (1971).

15. C. Nagata, M. Inomata, M. Kodama, and Y. Tagashira, *Gann 59* 289 (1968).
16. C. Nagata, Y. Tagashira, and M. Kodama, in *Chemical Carcinogenesis* (P. O. P. Ts'o and J. A. DiPaolo, Eds.). Dekker, New York, 1974, p. 87.
17. T. Kimura, M. Kodama, and C. Nagata, *Biochem. Pharmacol. 26:*671 (1977).
18. T. Kimura, M. Kodama, and C. Nagata, *Biochem. Pharmacol. 28.* 557 (1979).
19. C. Nagata and A. Matsuyama, *Proceedings of the 6th International Congress of Chemotherapy.* Univ. of Tokyo Press, Tokyo, 1970, p. 423.
20. K. Handa and S. Sato, *Gann 66.* 43 (1975).
21. N. R. Bachur, S. L. Gordon, and M. V. Gee, *Cancer Res. 38.* 1745 (1978).
22. C. Nagata, Y. Ioki, M. Kodama, Y. Tagashira, and M. Nakadate, *Ann. N.Y. Acad. Sci. 222:* 1031 (1973).
23. C. Nagata, N. Kataoka, A. Imamura, Y. Kawazoe, and G. Chihara, *Gann 57:* 323 (1966).
24. C. Nagata, Y. Ioki, M. Inomata, and A. Imamura, *Gann 60:* 509 (1969).
25. G. C. Mills and J. L. Wood, *Cancer Res. 13:*69 (1953).
26. S. Epstein, M. Burroughs, and M. Small, *Cancer Res. 23:*35 (1963).
27. Y. Morimura, P. Kotin, and H. L. Falk, *Cancer Res 24:* 1249 (1964).
28. J. J. Morton, E. M. Luce-Clausen, and E. B. Mahoney, *Cancer Res. 2:*256 (1942).
29. J. C. Mottram and I. Doniach, *Am. J. Cancer 39.* 234 (1940).
30. L. Santamaria, G. G. Giordano, M. Alfisi, and F. Cascione, *Nature 210:* 824 (1966).
31. A. U. Khan and M. Kasha, *Ann. N.Y. Acad. Sci. 171:* 24 (1970).
32. I. R. Politzer, G. W. Griffin, and J. L. Laseter, *Chem.-Biol. Interactions 3.* 73 (1971).
33. M. Inomata and C. Nagata, *Gann 63:* 119 (1972).
34. R. W. Fessenden and R. H. Schuler, *J. Chem. Phys. 39* 2147 (1963).
35. W. F. Forbes and J. C. Robinson, *Nature 217* 550 (1968).
36. L. C. Cusachs and R. H. Steele, in *The Jerusalem Symposium on Quantitative Chemistry and Biochemistry* (E. D. Bergman and B. Pullman, Eds.). Israel Acad. Sci. and Humanities, Jerusalem, 1966; *Int. J. Quantum Chem. 1S:*175 (1967).
37. R. H. Steele, L. C. Cusachs, and S. P. McGlynn, *Int. J. Quantum Chem. 1S.* 179 (1967).
38. Y. Masuda and M. Kuratsune, *Int. J. Air Water Pollut. 10* 805 (1966).
39. W. J. P. Neish, *Acta Unio Int. Contra Cancrum 7* 137 (1950).
40. H. L. Falk, P. Kotin, S. Suklee, and A. Nathan, *J. Nat. Cancer Inst. 28* 699 (1962).
41. P. Sims, *Biochem. Pharmacol. 19.* 795 (1970).
42. J. K. Selkirk, R. G. Croy, and H. V. Gelboin, *Cancer Res. 36.* 922 (1976).
43. G. Holder, H. Yagi, P. Dansette, D. M. Jerina, W. Levin, A. Y. H. Lu, and A. H. Conney, *Proc. Nat. Acad. Sci. U.S. 71:* 4356 (1974).

44. O. Pihar and J. Spaleny, *Chem. Listy 50* 296 (1956).
45. I. Berenblum and R. Schoental, *Cancer Res. 6* 699 (1946).
46. L. F. Fieser and E. B. Hershberg, *J. Am. Chem. Soc. 60:* 2542 (1938); *61:* 1565 (1939).
47. R. J. Lorentzen, W. J. Caspary, S. Lesko, and P. O. P. Ts'o, *Biochemistry 14:* 3970 (1975).
48. S. Lesko, W. Caspary, R. Lorentzen, and P. O. P. Ts'o, *Biochemistry 14* 3978 (1975).
49. F. J. Wiebl, *Arch. Biochem. Biophys. 168* 609 (1975).
50. L. J. Marnett, G. A. Reed, and J. T. Johnson, *Biochem. Biophys Res. Commun. 79* 5691 (1977).
51. A. S. Rispin, H. Kon, and D. W. Nebert, *Mol. Pharmacol. 12:* 476 (1976).
52. J. E. Tomaszewski, D. M. Jerina, and J. W. Daly, *Biochemistry 14.* 2024 (1975).
53. M. S. Newman and L.-F. Lee, *J. Org. Chem. 30:* 1446 (1974).
54. D. W. Nebert and H. V. Gelboin, *J. Mol. Biol. 243:* 6242 (1968).
55. C. Nagata, Y. Tagashira, M. Kodama, and A. Imamura, *Gann 57:* 437 (1966).
56. C. Nagata, M. Kodama, and Y. Tagashira, *Gann 58* 493 (1967).
57. T. Okano, T. Horie, M. Imaizumi, and N. Motohashi, *Gann 69* 109 (1978).
58. Y. Ioki, M. Kodama, Y. Tagashira, and C. Nagata, *Gann 65* 379 (1974).
59. Y. Ioki and C. Nagata, *J. Chem. Soc. (Perkin II)* p. 1172 (1977).
60. M. Kodama and C. Nagata, *Gann 68.* 125 (1977).
61. M. Kodama, Y. Ioki, and C. Nagata, *Gann 68.* 253 (1977).
62. C. Nagata, M. Kodama, and Y. Ioki, in *Polycyclic Hydrocarbons and Cancer* (P. O. P. Ts'o and H. V. Gelboin, Eds.), Vol. 1. Academic Press, New York, 1978, p. 243.
63. S. S. Hecht, N. Hirota, M. Lay, and D. Hoffman, *Cancer Res. 38.* 1694 (1978).
64. D. M. Jerina, R. E. Lehr, H. Yagi, O. Hernandez, P. M. Dansette, P. G. Wislocki, A. W. Wood, R. L. Chang, W. Levin, and A. H. Conney, in *In Vitro Metabolic Activation in Mutagenesis Testing* (F. J. de Serres, J. R. Fouts, J. R. Rend, and R. M. Philpot, Eds.). Elsevier—North-Holland Biochem. Press, Amsterdam, 1976, p. 159.
65. P. G. Wislocki, J. Kapitulnik, W. Levin, R. Lehr, M. S. Riddev, J. M. Karle, D. M. Jerina, and A. H. Conney, *Cancer Res. 38:* 693 (1978).
66. A. H. Conney, A. W. Wood, W. Levin, A. Y. H. Lu, R. M. L. Chang, P. G. Wislocki, G. M. Holder, P. M. Dansette, H. Yagi, and D. M. Jerina, in *Reactive Intermediates. Formation, Toxicity and Inactivation* (D. Jallow, J. Kocsis, R. Snyder, and H. Vanio, Eds.). Plenum Press, New York, 1977, p. 335.
67. W. Nakahara, F. Fukuoka, and T. Sugimura, *Gann 48:* 129 (1957).
68. H. Endo, M. Ishizawa, and T. Kamiya, *Nature 198* 195 (1963).
69. T. Okabayashi, M. Endo, A. Yoshimoto, and M. Otsubo, *Chem. Pharmacol. Bull. 13:* 610 (1965).
70. M. Ishizawa and H. Endo, *Mutation Res. 9*:134 (1970); *12*:1 (1971); *Gann 63*:511 (1972).

71. L. Prakash, J. Stewart, and F. Sherman, *J. Mol. Biol. 85:*51 (1974).
72. B. N. Ames, F. D. Lee, and W. E. Durston, *Proc. Nat. Acad. Sci U.S. 70:*78 (1973).
73. A. Zahalsky, M. M. Keane, S. H. Hutner, K. J. Lubart, M. Kittrell, and D. Amsterdam, *J. Protozool. 10:*421 (1963).
74. H. Sato and T. Kuroki, *Proc. Japan. Acad. 42:*1211 (1966).
75. J. Kamahora and T. Kakunaga, *Proc. Japan. Acad. 42:*1079 (1966).
76. T. Yoshida, K. Kurita, and K. Morikawa, *Gann 56*, 523 (1965).
77. H. F. Stich and R. H. C. San, *Mutat. Res. 10*. 389 (1970).
78. C. Nagata, K. Fujii, and S. S. Epstein, *Nature 215*. 972 (1967).
79. T. Matsushima and T. Sugimura, in *Chemistry and Biological Actions of 4-Nitroquinoline-1-Oxide* (H. Endo, T. Ono, and T. Sugimura, Eds.). Springer, New York, 1971, p. 53.
80. M. Tada and M. Tada, *Nature 255*. 510 (1975).
81. M. Tada and M. Tada, in *Fundamentals in Cancer Prevention* (P. N. Magee, T. Sugimura, S. Takayama, and T. Matsushima, Eds.). Univ. of Tokyo Press, Tokyo, and Univ. Park Press, Baltimore, 1976, p. 217.
82. M. Kodama, C. Nagata, M. Mochizuki, and M. Okada, *Gann 69* 285 (1978).
83. N. Kataoka, A. Imamura, Y. Kawazoe, G. Chihara, and C. Nagata, *Chem. Pharm. Bull. (Tokyo) 14*:897, 1171 (1966).
84. C. Nagata, in *Chemistry and Biological Actions of 4-Nitrosoquinoline-1-Oxide* (H. Endo, T. Ono, and T. Sugimura, Eds.). Springer, New York, 1971, p. 17.
85. N. Kataoka, A. Imamura, Y. Kawazoe, G. Chihara, and C. Nagata, *Bull. Chem. Soc. Japan 40*. 62 (1967).
86. N. Kataoka, S. Shibata, A. Imamura, Y. Kawazoe, G. Chihara, and C. Nagata, *Chem. Pharm. Bull. 15*. 220 (1967).
87. R. A. Abramovitch and E. M. Smith, *J. Heterocycl. Chem. 12:*969 (1975).
88. K. Takahashi, Y. Kawazoe, M. Tada, M. Tada, N. Ito, and M. Okada, *Gann 69*:499 (1978).
89. A. Terawaki and J. Greenberg, *Biochim. Biophys. Acta 95*. 170 (1965).
90. S. Iwahara, K. Yanagimachi, S. Kamiya, M. Nakadate, and I. Suzuki, *Chem. Pharm. Bull. (Tokyo) 19*:1914 (1971).
91. J. D. Mandell and J. Greenberg, *Biochem. Biophys. Res. Commun. 3:*575 (1960.
92. A. J. Muller and T. Gichner, *Nature 201*. 1149 (1964).
93. H. Druckrey, R. Preussmann, S. Ivankovic, B. T. So, C. H. Schmidt, and J. Bucheler, *Z. Krebsforsch 68* 87 (1966).
94. R. Schoental, *Nature 209*. 726 (1966).
95. T. Sugimura, M. Nagao, and Y. Okado, *Nature 210*, 962 (1966).
96. M. O. Greene and J. Greenberg, *Cancer Res. 20*. 1166 (1966).
97. W. A. Skinner, H. F. Gram, M. O. Greene, J. Greenberg and B. R. Baker, *J. Med. Pharm. Chem. 2:*299 (1960).
98. P. D. Lawley, *Nature 218*. 580 (1968).
99. D. R. McCalla, *Biochim. Biophys. Acta 155* 114 (1968).

100. M. Nagao, T. Yokoshima, H. Hoso, and T. Sugimura, *Biochim. Biophys. Acta 192*: 191 (1969).
101. U. Schulz and D. R. McCalla, *Can. J. Chem. 47*: 2021 (1969).
102. T. Kawachi, K. Kogure, Y. Kamijo, and T. Sugimura, *Biochim. Biophys. Acta 222*: 409 (1970).
103. A. Tanaka and T. Sano, *Experientia 27*: 1007 (1971).
104. C. Nagata, M. Nakadate, Y. Ioki, and A. Imamura, *Gann 63*: 471 (1972).
105. Y. Ioki, A. Imamura and C. Nagata, *Photochem. Photobiol. 21*: 387 (1975).
106. B. Commoner, J. C. Woolum, B. H. Senturia, Jr., and J. L. Ternberg, *Cancer Res. 30*: 2091 (1970).
107. A. J. Vithayathil, J. L. Ternberg, and B. Commoner, *Nature 207* 1246 (1965).
108. J. C. Woolum, E. Tiezzi, and B. Commoner, *Biochim. Biophys. Acta 160*: 311 (1968).
109. J. C. Woolum and B. Commoner, *Biochem. Biophys. Acta 201*: 131 (1970).
110. R. W. Chang, J. C. Woolum, and B. Commoner, *Biochim. Biophys. Acta 257*:452 (1972).
111. C. C. McDonald, W. D. Phillips, and H. F. Mower, *J. Am. Chem. Soc. 87*: 3319 (1965).
112. G. M. Bonser, L. Bradshaw, D. B. Clayson, and J. W. Jull, *Br. J. Cancer 10*: 539 (1956).
113. D. B. Clayson, *Br. J. Cancer 7*:460 (1953).
114. J. A. Miller and E. C. Miller, *Lab. Invest. 15* 217 (1966).
115. W. Troll and N. Nelson, *Fed. Proc. 20*: 41 (1961).
116. E. Boyland, E. R. Busby, C. E. Dukes, P. L. Grover, and D. Manson, *Br. J. Cancer 18*: 575 (1964).
117. J. C. Arcos and M. F. Argus, *Chemical Induction of Cancer*, Vol. II, Pt. B., Academic Press, New York, 1974.
118. R. A. Floyd, L. M. Soong, and P. L. Culver, *Cancer Res. 36*: 1510 (1976).
119. J. D. Scribner and N. K. Naimy, *Cancer Res. 33*: 1159 (1973).
120. A. R. Forrester, M. M. Ogilvy, and R. H. Thomson, *J. Chem. Soc. (C)* p. 1081 (1970).
121. R. A. Floyd, L. M. Soong, M. A. Stuart, and D. L. Reigh, *Arch. Biochem. Biophys. 185*: 450 (1978).
122. F. F. Kadlubar, J. A. Miller, and E. C. Miller, *Cancer Res. 36*: 1196 (1976).
123. M. Meunier and J. Chauveau, *Int. J. Cancer 6* 463 (1970).
124. A. Haddow, R. J. C. Harris, G. A. R. Kon, and E. M. F. Roe, *Phil. Trans. R. Soc. London, Ser. A 241*: 147 (1948).
125. V. N. Iyer and W. Szybalski, *Science 145*: 55 (1964).
126. W. Szybalski, in *Mechanism of Action and Biosynthesis of Antibiotics* (D. Gottlieb and F. D. Shaw, Eds.). Springer, New York, 1967, p. 211.
127. A. Weissbach and A. Lisio, *Biochemistry 4*: 196 (1965).
128. H. Murakami, *J. Theoret. Biol. 10*: 236 (1966).
129. J. R. White and H. H. Dearman, *Proc. Nat. Acad. Sci. U.S. 54* 887 (1965).
130. R. Ishida and T. Takahashi, *Biochem. Biophys. Res. Commun. 66*: 1432 (1975).

2

NITROXIDE METABOLISM IN LIVER MICROSOMES: SPIN LABELS AND AROMATIC NITROXIDES

A. Stier, R. Claus, B. Bösterling,* and I. Reitz

Max-Planck-Institut für Biophysikalische Chemie Göttingen, Federal Republic of Germany

I. INTRODUCTION

Two types of nitroxide radicals appearing sequentially in rabbit liver microsomal incubations have been detected by electron spin resonance (ESR) spectroscopy. These partially published results [1,2,2a] will be described in this chapter. We will discuss the tentative importance of these findings, proffering a working hypothesis which mainly emphasizes the hazardous oxidation-reduction couple of aromatic hydroxylamine nitroxides and the possible role of endogenous carriers of ultimate or proximate carcinogens for selective interaction with targets. In this provocative hypothesis, aromatic N-hydroxy esters are considered preferentially as those carriers; their established position as ultimate carcinogens has been weakened [3,4]. Since this is not a critical review that aims to compare alternative candidates for ultimate carcinogens among the aromatic amines, the literature chosen is highly selective. The discussion will be based on the somatic mutation theory of chemical carcinogenesis, although we are keenly aware that the lack of a comparably defined theory of tumor promotion does not mean that the chemical processes involved in tumor promotion and manifestation are of secondary importance.

*Present affiliation: Stanford University, Stanford, California

Table 1 Radical formation and carcinogenicity of different aromatic N-compounds[a]

Substance incubated with rabbit liver microsomes (O_2, 37°, NADPH)	Radical detected by ESR	Carcinogenic activity	Experimental tumors formed	
			In tissue[b]	In species[b]
2-Aminofluorene	+	+	b	m,d,h,mon
2-Acetylaminofluorene	+	+	li,b,i,s,mg,eg	r,mo,ra,h,d
N-Hydroxy-2-acetylaminofluorene	+	+	st,mg,li,i,eg,s	r,ra,mi,h,gp
9-Aminofluorene	0	0		
2-Aminonaphthalene	+	+	b	m,mon,ra,d,mo
1-Aminonaphthalene	0	0[c]	l	mo
3-Nitro-2-aminonaphthalene	0	+	mg,st,eg	r,mo
1,2-Dichloro-3-nitronaphthalene	0	+	mg	
2-Aminoanthrancene	+	+	li,b,mg,s	mi,h
6-Aminochrysene	0	(0)[c]	li,l	mi
4-Nitroquinoline-1-oxide	+	+	l,s	mi,r,ra
4-Aminostilbene	+	+	li,l,mg,i	r,mi
4,4-Diaminostilbene	+	+		
Azobenzene	0	0		
4-Aminoazobenzene	+	+	li	r
N-Methyl-p-(phenylazo)aniline	+	+	li	r,mo
4-Diethylaminoazobenzene	0	0		
Benzidine	+	+	li,b,eg	r,d,mo
4-N-Dimethylaminodiphenyl	+	+	mg,eg,i	
Aniline	+	0		
N-Methylaniline	+	0	k	r
Acetanilide	0	0		
N,N-Dimethyl-p-phenylenediamine	0	0		
Phenacetin	+	(0)[c]	k	

[a]Data on carcinogenicity have been collected from Refs. 8 and 9.

[b]Abbreviations used: b, bladder; li, liver; i, intestine; s, skin; mg, mammary gland; eg, earduct gland; st, stomach; l, lung; k, kidney; ag, adrenal gland; m, man; d, dog; h, hamster; mon, monkey; r, rat; mo, mouse; ra, rabbit; mi, mice; gp, guinea pig.

[c]Marginal carcinogens; see text.

II. CARCINOGENICITY AND NITROXIDE RADICALS: A COMPILATION RATHER THAN A CORRELATION

Nitroxide radicals as possible metabolites of amines were first discussed in relation to the toxic effects of these compounds [5]. Bartsch and Hecker [6] found nitroxide radicals utilizing ESR spectroscopy when they oxidized N-hydroxy-2-acetylaminofluorene by a peroxidase but did not consider them as ultimate carcinogens. In rabbit liver microsomal incubations of various carcinogenic aromatic amines, nitroxide radicals formed under conditions of mixed function oxidation were detected with the same technique [7]. In the meantime, many carcinogenic and noncarcinogenic nitrogen compounds, mainly amines, have been investigated [1,2a] (see Table 1).

There appears to be a relationship between the detectability of radicals and their ability to produce cancer; that is, of the nonaromatic amines investigated, not one produced a radical. 9-Aminofluorene, amino-tert-butane, diaminoethane, and piperidine have been tested. The special case of tetramethylpiperidine, which was found to yield a radical, is discussed in Section IV.A. Of the 14 carcinogenic aromatic compounds investigated, 12 produced radicals; of the 10 noncarcinogenic or only very weakly carcinogenic aromatic compounds, four showed radicals after in vitro incubation. The exceptions in the group of carcinogens which did not yield radicals were the nitrocompounds. 3-Nitro-2-aminonaphthalene is known to produce tumors of mammary gland, earduct, and intestines when administered by stomach tube to rats. Compared to other 2-aminonaphthalene analogs, it was found to be even more active with regard to intestinal tumors but less effective toward other tissues [8]. The other exception, 2,2-dichloro-3-nitronaphthalene, is known to induce mammary tumors only in rats [8]. Perhaps in the latter case the nitro group is not reduced as effectively as the nitro group of 4-nitroquinoline-N-oxide, which is known to be a carcinogen and yielded a radical in our incubations. On the other hand, the neighboring nitro group of 3-nitro-2-aminonaphthalene may impede the oxidation of the amino group as well as the formation of the stabilized secondary radical (see Section III). The exceptions in the group of the noncarcinogenic aromatic amines are the monocyclic amines aniline and N-methylaniline, which are believed to be noncarcinogenic. We have reasons to assume that the primary nitroxides of these compounds, as well as the ones formed from some amines which are known to be only marginally carcinogenic, are less reactive than nitroxides formed from the other, more carcinogenic amines (see Section III).

Some of the other substances included here in the groups of noncarcinogenic compounds may be considered as marginally carcinogenic. 2-Aminonapthalene clearly produces bladder tumors in dogs [8], but 1-aminonapthalene does not; on the other hand, in mice the 1-isomer has been found to produce lung tumors, but much less than 2-aminonaphthalene [8]. 6-Aminochrysene was found to produce tumors only in newborn, predominantly male mice [8], whereas in other animals it inhibited even hydrocarbon-induced or spontaneous tumors [8]. Phenacetin is shown in Table 1 as a noncarcinogen since it was not found to produce tumors in animal carcinogenicity testing [8,9]. These substances demonstrate that the assessment of carcinogenicity is not a qualitative but a quantitative problem. Under certain experimental conditions in certain species in a certain organ, compounds in addition to the ones with established carcinogenicity may be found to induce tumors. General reasons for not considering this compilation of carcinogen radical data as a correlation demonstrating a causal relationship are obvious: carcinogenic effects in a variety of species and on a great diversity of tissues are compared nonquantitatively with the detectability of radicals produced in incubations prepared from liver microsomes of one species—the rabbit. The only substance we have investigated with microsomes from male Wistar and Sprague-Dawley rats, 2-N-acetylaminofluorene, gave a similar radical spectrum as that obtained with rabbit liver microsomes, but only after a prolonged period of incubation and only in phenobarbital pretreated animals. The detection of radicals is limited by the sensitivity of the experimental technique. On the other hand, biotransformation of aromatic amines in liver microsomal membranes is

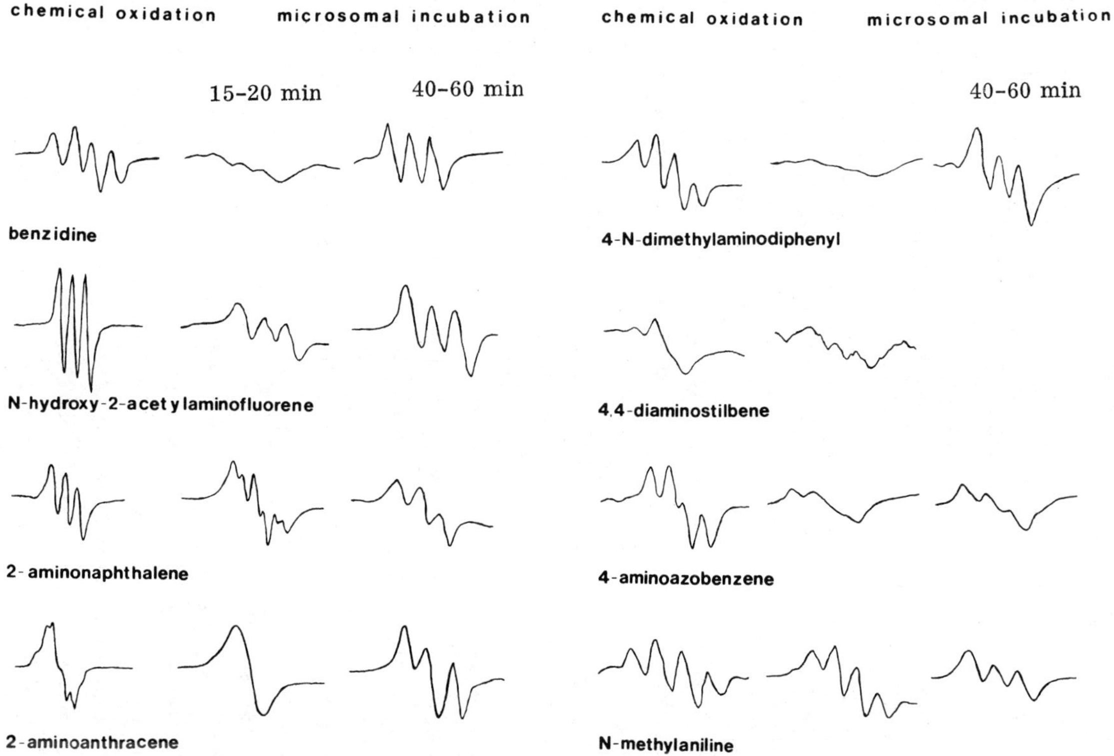

Figure 1 ESR spectra of methanol:chloroform (2:1, vol:vol) extracts of mixed function oxygenation incubations of different amines with liver microsomes from phenobarbital-pretreated rabbits and of solutions of these amines to which 3-chloroperbenzoic acid had been added. Modulation amplitude: 6.25 G. (From Ref. 1.)

a fairly good model for the first step of tumor initiation, bioactivation, since the liver manifests nearly all enzymatic activities involved in bioactivation which may also occur in other organs. Liver may play a central role in the production of transport forms of proximate carcinogens from where they are distributed throughout the organism to be finally activated into ultimate carcinogens in their target organs (see Section IV.B.).

III. TRANSFORMATION OF REACTIVE PRIMARY RADICALS TO LONG-LIVED SECONDARY RADICALS

ESR spectra had to be recorded utilizing a high modulation amplitude in order to obtain a sufficient signal-to-noise ratio. Hence, the resolution is usually too low to give a detailed fine structure of the spectra. To obtain a preliminary identification we compared the spectra of solvent extracts from microsomal incubations with spectra obtained by chemical oxidation of the corresponding amines under the same instrumental conditions. From Figure 1 it can be seen that only in the case of 2-aminoanthracene, 2-aminonaphthalene, 4-aminoazobenzene, and N-methylaniline did the spectral shape correspond in the two oxidizing systems and only during the earlier incubation times. The spectra were converted as incubation continued into broad-band triplet spectra. This transformation can best be observed in the case of 2-aminonaphthalene incubation. At early incubation times, two types of nitroxide radical spectra appear superimposed—one triplet spectrum with a larger hyperfine splitting of about 14 G, the other spectrum with a splitting of ≃10 G. It can be seen from Figure 2 that the spectrum with the smaller hyperfine splitting (characteristic for the nitrogen coupling of an aromatic nitroxide), which is similar in shape to the spectrum produced by chemical oxidation of 2-aminonaphthalene, disappears. Simultaneously the other radical with the larger splitting increases. Apparently the primarily produced aromatic nitroxide is converted to another nitroxide radical. The spectral shapes observed after prolonged incubation are similar in the case of the aforementioned substances as well as

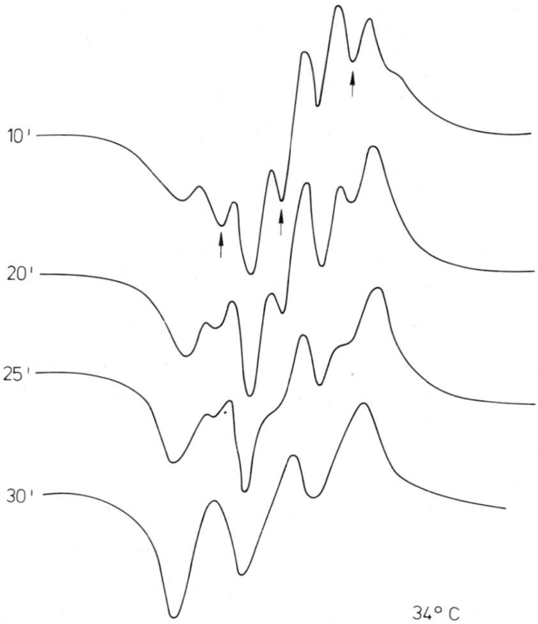

Figure 2 ESR spectra of microsomal solvent extracts after various incubation times of 2-aminonaphthalene. For other conditions, see Figure 1. The bands of the spectrum corresponding to the spectrum in the case of chemical oxidation are marked. (From Ref. 1.)

in all cases in which we could find a radical from aromatic amines. In the case of 4-nitroquinoline-N-oxide, we found only one broad radical band. These secondarily formed radicals are very stable and can be detected in the microsomal suspension after many hours of incubation. The fact that after prolonged incubation all radicals formed show a similar spectral shape with nitrogen splittings between 14 and 17 G may be explained by covalent binding of the nitrogen to the same endogenous chemical structures. The longevity of the secondarily formed radicals is consistent with a spin-label-like structure since the latter compounds are well known to be quite stable nitroxides. These results suggest that the aromatic amines are oxidized via the N-hydroxy compounds to short-lived nitroxides which by reaction with an endogenous microsomal constituent are transformed into a relatively stable spin-label-like nitroxide preserving the radical group. This transformation from one radical form to the other takes place during incubation at a rate that depends on the particular compound. In most cases the interconversion is too fast for the primarily formed aromatic nitroxide to be detectable. The reason for the slower interconversion of the aromatic nitroxides of 2-naphthylamine, N-methylaniline, 2-aminoanthracene, and 4-aminoazobenzene may be because of their lower reactivity compared to the other nitroxides. Therefore the primary nitroxides of these substances accumulate. Interestingly most of these substances are noncarcinogenic or only marginally carcinogenic (see Section II and Table 1).

The structure of the secondary radicals is difficult to elucidate in a direct way, due to the small amount formed. We used two indirect methods: First the high stability of the observed radicals allowed us to test their chromatographic behavior. We had observed that the amount of secondary radicals formed is higher in incubations of N-hydroxy compounds than in incubations of amines. This can be explained by the assumption that the oxidation of the amine to the N-hydroxy compound is rate limiting. The formation of the radicals then proceeds at a relatively higher rate starting from the N-hydroxy compound. Therefore we incubated phenylhydroxylamine with phenobarbital-induced rabbit liver microsomes. By chromatography of the chloroform-methanol extract of these incubations on Sephadex LH 20 we could separate two fractions in which the secondary radical spectrum was found. The first fraction, which accounted for about one-half of the total amount of radicals formed, was eluted together with the microsomal phospholipids lecithin and sphingomyelin. The smaller second fraction could not be assigned to any membrane constituent. In silica gel thin-layer chromatography of the extract of microsomal incubations, however, the stable radical behaved like a fatty acid or cholesterol but different from all phospholipids. This could mean that two different secondary radicals are formed which behave differently in column chromatography but similarly in thin-layer chromatography. Another possibility is that only one

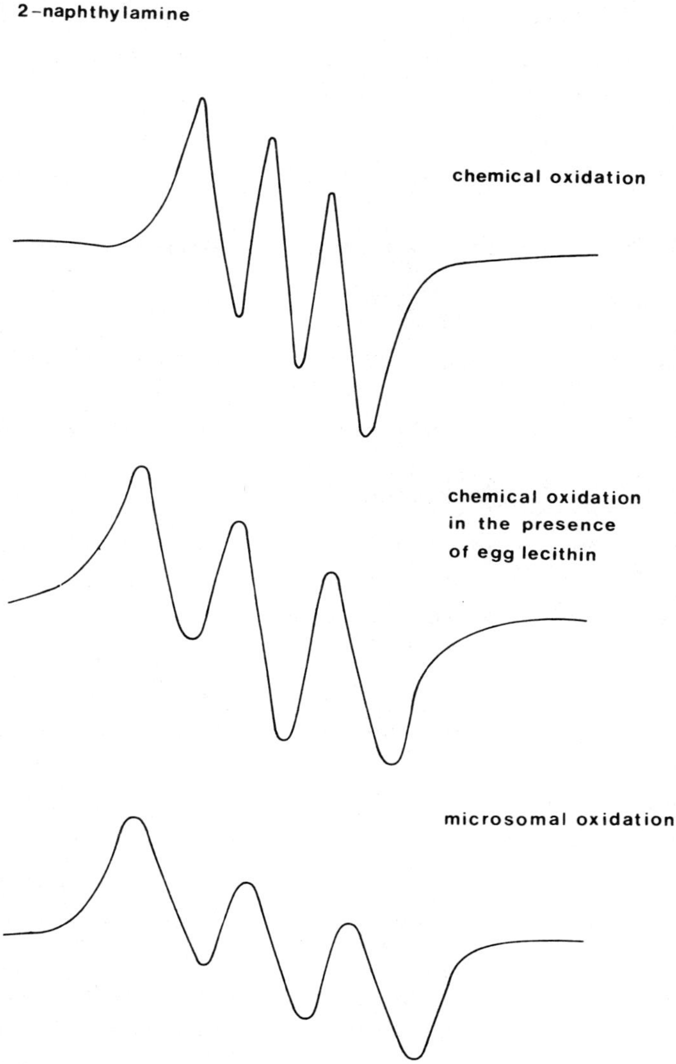

Figure 3 ESR spectra of 2-aminonaphthalene oxidized in different systems. Modulation amplitude: 2.5 G. For other conditions, see Figure 1. (From Ref. 1.)

secondary type of radical has been formed which is carried on partially in column chromatography with the phospholipids.

In the second approach we investigated the formation of the secondary radicals characterized by their typical ESR spectrum in chemical model reactions of the primary nitroxides with several endogenous compounds. This approach was based on the observation that styrene, vinyl monomers, and higher aromatic hydrocarbons are known to act as radical traps by adding reactive radicals to their double bonds. This results in the formation of stable radicals [10,11]. Figure 3 shows that 2-naphthylamine oxidized by perbenzoic acid in the presence of egg lecithin forms a radical which has the same spectral shape as the secondary radical formed in microsomal incubations and which differs from the ESR spectrum of the radicals formed by the same oxidant in the absence of egg lecithin. In the presence of dipalmitoyl lecithin, the spectral shape was the same as in the case of oxidation without addition of any lipid. Another example is shown in Figure 4: Phenylhydroxylamine oxidized in the presence of oleic acid gives a different spectrum when compared to the same compound oxidized without addition of oleic acid. The former spectrum appears identical with a spectrum observed in an extract of microsomal incubation.

The secondary radical formed must be highly immobilized within the microsomal membrane. It can be sedimented with microsomes, and it is in most cases only detectable in solvent extracts of microsomes, not in the microsomal incubations themselves. This can be explained by strong line broadening due to immobilization. This explanation is directly supported by an experiment in which under favorable conditions—the substrate was N-hydroxy-2-acetylaminofluorene—a radical spectrum can be detected directly in the microsomal incubation which is characteristic for a strongly immobilized nitroxide (see Fig. 5). The large increase of the signal due to fast isotropic mobility is clearly demonstrated in the spectrum of the lipid extract. A similar property was shown for the radical formed in microsomal incubations of nitrosofluorene [12].

The transformation of the primary to the secondary nitroxide may be formulated as a radical addition to a double bond (see Fig. 6)

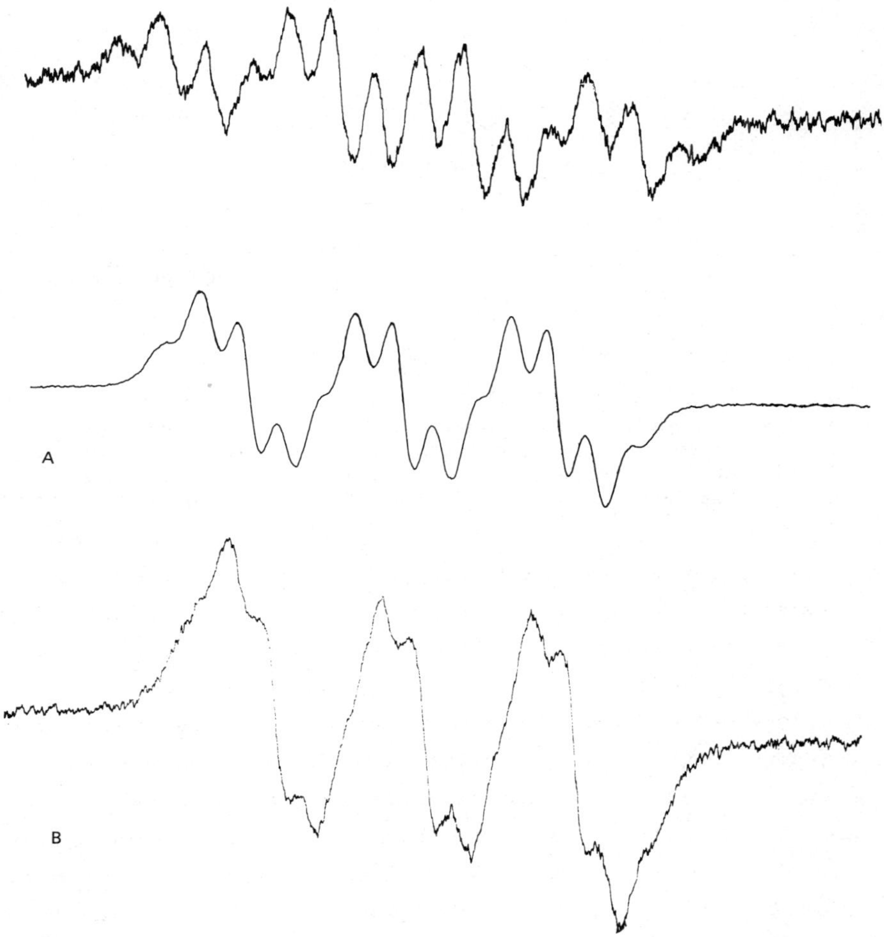

Figure 4 The two upper ESR spectra were taken from oxidized solutions of phenylhydroxylamine. The second spectrum was taken from a solution which contained oleic acid. The third spectrum was obtained from solvent extracts of microsomal incubations of phenylhydroxylamine. NADPH was not added. For other conditions, see Figure 1. (From Ref. 1.)

analogous to the reaction of a bistrifluoromethyl nitroxide with alkenes [13]. At present it is unknown whether aromatic nitroxides would be as reactive as the latter compound. For the production of a similar type of nitroxide from 2-nitrosofluorene with rat liver microsomal membranes, a reaction analogous to a pseudo-Diels-Alder reaction [12] has been assumed. In a subsequent oxidation, the N-hydroxy compound would then be oxidized to the nitroxide. In

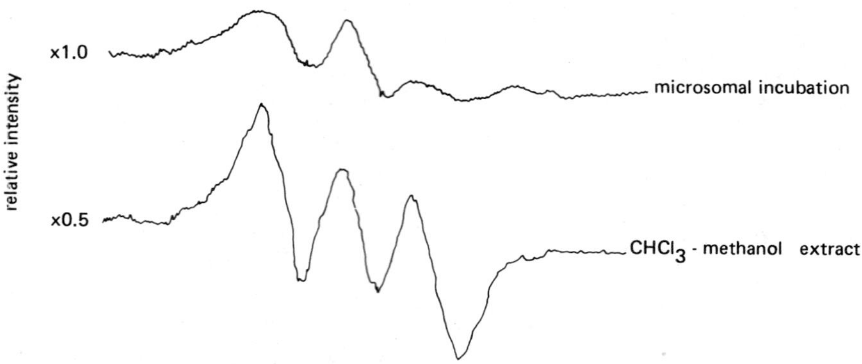

Figure 5 ESR spectra of microsomal incubations of N-hydroxy-2-acetylaminofluorene (upper spectrum) and of the solvent extract of this incubation (lower spectrum). Note the different vertical scales of the spectra. For other conditions, see Figure 1. (From Ref. 1.)

principle, the same mechanism could operate in the conversion of the primary nitroxide to the secondary nitroxide. This would involve an intermediary oxidation step of the primary nitroxide to the corresponding nitrosofluorene. Our observation of an apparent direct interconversion of the nitroxide of 2-aminonaphthalene to the secondary nitroxide, however, would favor a different mechanism, which would involve direct interaction of the nitroxide with the endogenous reaction partner. The involvement of the radical chain of lipid peroxidation in the formation of the secondary radical may be excluded on the basis of the finding that propyl gallate did not inhibit the formation of the secondary radical in incubations with N-hydroxy-2-acetylaminofluorene—but even accelerated this reaction. Propyl gallate is known to be a strong inhibitor of the microsomal lipid peroxidation initiated by CCl_4 [14]; on the other hand, it was

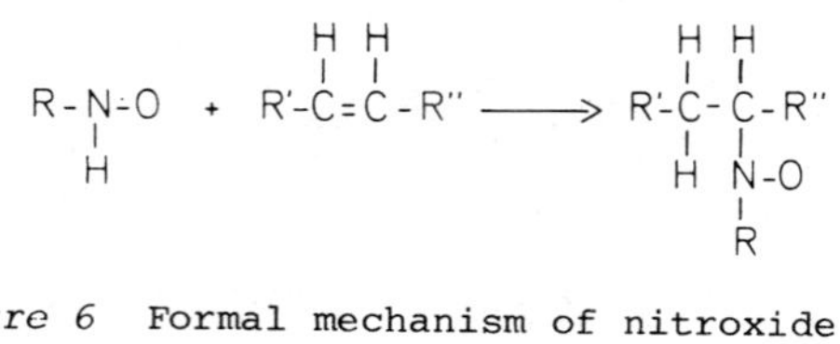

Figure 6 Formal mechanism of nitroxide radical addition to the double bond.

shown recently that propyl gallate only partially, if at all, inhibits NADPH- or NADPH-plus-CCl_4-induced formation of a radical which can be trapped by tert-butylnitrone [15]. Perhaps propyl gallate does not inhibit the formation of radicals initiating the radical chain of lipid peroxidation but only the propagation of this chain.

IV. NITROXIDE RADICALS IN CARCINOGENESIS: A TENTATIVE VIEW

To evaluate possible biological consequences of the formation of nitroxide radicals, we must differentiate between the biological consequences of mechanisms in which the primary aromatic nitroxide radicals are involved and those in which the spin-label-like secondary nitroxide radicals are involved. Since the role of chemicals in tumor initiation is much better defined—on the basis of the theory of somatic mutation by covalent binding to DNA—than the processes of tumor promotion, the discussion here will be restricted to the first topic. The reaction of the aromatic nitroxide radical with membrane constituents may play a crucial role in chemical carcinogenesis. The importance of this reaction in tumor promotion has yet to be fully assessed.

A. Aromatic Hydroxylamines and Nitroxides: Dangerous Redox Couples?

So far we do not know whether aromatic nitroxides can react chemically with DNA. However, the reactivity with microsomal membrane constituents points to their electrophilic reactivity. Furthermore, it has been shown that they react with other radicals in a radical recombination reaction, thereby inhibiting radical chains [16]. Even a compound as stable as the spin-label 2,2,6,6-tetramethyl-4-piperidine-N-oxyl was found to react with •OH-induced DNA radicals produced by pulse radiolysis [17]. This explains the known radiosensitizing ability of the compound [18]. From our own experiments on microsomal oxygenation of N-hydroxy-2-acetylaminofluorene (N-OH-AAF) and of 2,2,6,6-tetramethylpiperidine (TMP)—the latter being transformed to the stable nitroxide TEMPO, well known as a spin-label—

we have indications that: (a) in the course of their oxidation, reactive oxygen species such as superoxide anions and •OH radicals might be formed; and, (b) these rather unspecific reactions are not confined to liver nor to the endoplasmic reticulum. N-OH-AAF was found, among the amines we investigated, to be the substance most effectively transformed to the stable nitroxide. This transformation is highly dependent on the presence of oxygen and microsomes but not on NADPH addition; in fact, NADPH even retards the formation of the radical [19]. These results are supported by experiments on the oxidation of TMP and the hydoxylamine of TMP (TMP-OH) to the TEMPO radical [20,21]. In the overall reaction of TMP via TMP-OH to TEMPO, the oxidation of TMP to TMP-OH is the rate-limiting step whereas oxidation of TMP-OH is a rapid process which occurs at oxygen saturation even in the absence of microsomes. Even at oxygen concentrations as low as 10^{-5} M TMP-OH is readily reoxidized, as can be seen from experiments of various spin-labels. When NADPH is added to microsomes containing spin-label, the nitroxide is enzymatically reduced. At oxygen saturation, reoxidation of the reduced spin-label proceeds at a higher rate than the reduction. Therefore, an initial lag phase occurs in which the oxygen is consumed, and then the reduction becomes zero order in which all spin-label is reduced. The lag period could be abolished only when the oxygen concentration was kept below 10^{-5} M. Interestingly, this reduction was inhibited by CO in microsomes and even more inhibited in a bilayer vesicle system [reconstituted from purified rabbit cytochrome P_{450} (LM_2) and purified cytochrome P_{450} reductase with a mixture of egg lecithin and egg phosphatidyl ethanolamine]. This NADPH-dependent enzymatic reduction is therefore predominantly mediated by cytochrome P_{450}. The reoxidation of spin-labels is also mediated by cytochrome P_{450}, as was shown in experiments in which the reoxidation could be greatly augmented by the addition of purified cytochrome P_{450} (LM_2) reconstituted into phospholipid vesicles. This oxidation process is not cytochrome P_{450} specific because the addition of myoglobin also had some effect.

Since the oxidation of the aromatic hydroxylamine compounds by oxygen (as well as TMP-OH) must be a one-electron transfer, the assumption of the formation of superoxide anions is quite reasonable [22]. This oxygen species is known to disproportionate into oxygen and •OH radicals. This means that, in an unspecific reaction which can occur in many tissues, aromatic N-hydroxy compounds may be oxidized to the corresponding nitroxides concomitant with production of very reactive radicals known to transform DNA to a radical form which may subsequently recombine with the nitroxide. On the other hand, the nitroxides are good electron acceptors. They may initiate a redox cycle which may become highly effective in production of active oxygen species. A similar cycle has been assumed to be the cause of methemoglobin formation by aromatic amines [23]. These speculations would make unnecessary the assumption of the occurrence of specific activating enzymes in target tissues.

B. Carriers for Selective Interaction

Selectivity of interaction of reactive compounds with their targets resulting in a covalent binding means low competition of other cellular components with these reactive compounds. The proposed electrophilicity of ultimate carcinogens is a property not so specific as to allow for a high degree of selectivity of interaction with the targets.

Some of the N-hydroxy esters assumed to be ultimate carcinogens may have specific functions for tumor initiation different from the ones allotted to them hitherto. For example, they may serve as carriers distributing the ultimate carcinogen to target tissues: Glucuronides transported into kidney and intestines where the hydroxylamine compound is released by hydrolysis could start the hazardous redox cycle.

The long lifetime of the secondary radical may allow its transport within the cell and perhaps even out of the cell. On the other hand, our spin-label experiments indicate that these radicals may also be able to interact covalently with DNA and cooperate in the hazardous redox cycle described earlier. The endogenous part of the secondary radical, whose nature is still unknown, might act as a relatively specific carrier which is able to transport the radical by lateral diffusion in certain areas of the endoplasmic reticulum membrane where a covalent binding with the specific target can occur. Diffusion within the cell could be restricted by the specific properties of the carrier in such a way that only a limited number of targets in or at the membrane can be reached. Generally the two-dimensionality of a membrane might favor a selective interaction. It is obvious that in the case of targets apposed to the membrane, such as mRNA, encounters of the ultimate carcinogens—formed within the membrane and approaching their targets by lateral diffusion—with competing nontargets might be fewer than corresponding encounters of ultimate carcinogens formed in the plasma space of the cell and reaching their targets by diffusion within the three-dimensional space of the cell. The nitroxide group might be in a distinct steric orientation at the surface of the endoplasmic reticulum membrane. This specific binding would promote an interaction with target molecules like mRNA or certain proteins which are themselves apposed on the surface of the membrane. These speculations imply that the covalent binding of the ultimate carcinogen to the biological target occurs at or in the membrane of the endoplasmic reticulum. Thus one should be able to explain how a somatic mutation can be induced by this binding if one wants to conform to the current working hypothesis of chemical carcinogenesis. Several possibilities may be considered: (a) A covalent binding to mRNA could be retranslated by the reverse transcription to DNA. (b) A covalent binding to mRNA could induce a modified structure of DNA polymerase; this modified polymerase might have a higher mutation rate during replication of DNA. (c) A structural defect of the DNA polymerase could be produced by direct covalent binding of the ultimate carcinogen to this protein during its biosynthesis on the membrane of the endoplasmic reticulum. All these difficulties of interpretation could be circumvented if one assumes that the endogenous part of the nitroxide serves as a specific carrier which transports the appending radical to its target within or out of the cell.

REFERENCES

1. A. Stier, R. Clauss, B. Bösterling, A. Lücke, and I. Reitz, *Xenobiotica 10*:661-673 (1980).
2. R. Clauss, *Inaugural-Dissertation*, Göttingen (1977).
2a. A. Stier, R. Clauss, A. Lücke, R. Rösen, and P. Rosenberg, *Panminerva Med. 18*:449-453 (1976).
3. H. Bartsch, C. Malaveille, H. F. Stich, E. C. Miller, and J. A. Miller, *Cancer Res. 37*:1461-1467 (1977).
4. C. E. Weeks, W. T. Allaben, S. C. Louie, E. J. Lazear, and C. M. King, *Cancer Res. 38*:613-618 (1978).
5. W. Heubner, *Arch. Exp. Pathol. Pharmakol. 205*:310 (1948).
6. H. Bartsch and E. Hecker, *Biochim. Biophys. Acta 237*:567-578 (1971).
7. A. Stier, I. Reitz, and E. Sackmann, *Naunyn-Schmiedebergs Arch. Pharmacol. 274*:189-191 (1972).
8. J. C. Arcos and M. F. Argus (Eds.), *Chemical Induction of Cancer*. Academic Press, New York, 1974.
9. D. Schmähl (Ed.), *Entstehung, Wachstum und Chemotherapie Maligner Tumoren*. Cantor, Aulendorf in Württemberg, Germany, 1970.
10. J. C. Bevington, *Trans. Faraday Soc. 51*:1392-1397 (1955).
11. M. Levy and M. Szwarc, *J. Am. Chem. Soc. 77*:1949-1955 (1955).
12. R. A. Floyd, L. M. Soong, M. A. Stuart, and D. L. Reigh, *Arch. Biochem. Biophys. 185*:450-457 (1978).
13. R. E. Banks, R. N. Haszeldine, and B. Justin, *J. Chem. Soc. (C)* pp. 2777-2785 (1971).
14. T. F. Slater and B. C. Sawyer, *Biochem. J. 123*:823-828 (1971).
15. A. Ingall, K. A. K. Lott, T. F. Slater, S. A. E. Finch, and A. Stier, *Biochem. Soc. Trans. 6*:962-964 (1978).
16. S. Nigam, K.-D. Asmus, and R. L. Willson, *Trans. Faraday Soc. 72*:2324-2340 (1976).
17. K. F. Nakken, T. Sikkeland, and T. Brustad, *FEBS Lett. 8*:33-36 (1970).
18. P. T. Emmerson, *Radiat. Res. 30*:841-849 (1967).
19. A. Stier and I. Reitz, unpublished data.
20. A. Stier and I. Reitz, *Xenobiotica 1*:499-500 (1971).
21. B. Bösterling and A. Stier, in preparation.
22. L. L. Poulsen, B. S. S. Masters, and D. M. Ziegler, *Xenobiotica 6*:481-498 (1976).
23. M. Kiese, *Pharmacol. Rev. 18*:1091-1095 (1966).

3
ELECTRON SPIN RESONANCE STUDIES OF CANCER USING COMPLEX BIOLOGICAL SYSTEMS

Harold M. Swartz

College of Medicine
University of Illinois
Urbana, Illinois

I. INTRODUCTION

A. Scope of This Chapter

This chapter attempts to provide a critical overview of ESR studies of cancer in complex biological samples such as cells and tissues. This is a difficult task because the experiments have been carried out to test several related but different theoretical concepts and the resulting data often provide apparent empirical correlations unrelated to previously existing theories. I shall attempt to separate out these various theoretical and empirical concepts, provide a résumé of experimental results, relate these results to the various theories, and finally indicate likely directions of research in the future. The treatment of all topics will necessarily be brief, and the reader is referred to more detailed reviews for a comprehensive description of experimental results [1,2]. I would like to emphasize that this chapter is based on my interpretations of the data and their application to the various theories; others in the field may seriously disagree with my conclusions.

Electron spin resonance (ESR) is a very sensitive technique that detects most molecules that have unpaired electrons. In biological samples these are principally free radicals and certain valance states of transition elements (e.g., Fe^{2+}, Fe^{3+}, Cu^{2+}, Mn^{2+},

Mo^{5+}). Because of the sensitivity of ESR (concentrations of 10^{-9} M can be routinely detected), sample preparation and spectroscopic techniques can greatly affect the observed spectra. This has led to considerable confusion in this field, especially because it involves sophisticated biology, chemistry, and physics. Many of the investigators have lacked background in one or more of these areas. Some of these problems are discussed in more detail, along with descriptions of the ESR technique itself in some recent reports [2-5].

B. Summary of Theories Related to ESR Studies of Cancer

ESR studies of cancer, although usually described and considered in terms of one particular theory or viewpoint, actually involve several quite different concepts. These will be discussed in detail in Section III of this chapter, but I briefly summarize them here to provide perspectives for the review of experimental findings.

The first theory historically and also the most general is that cancer is caused by uncontrolled free radical reactions. Several other concepts have developed from this but are really somewhat different and will be considered separately.

The second major theory is that the active forms of chemical carcinogens are free radicals. This subdivides into the following postulates: (a) all carcinogens act via a free radical mechanism, or (b) certain classes of carcinogens involve free radical intermediates.

A third set of theories state that uncontrolled free radical reactions are generally deleterious to cells and that one of the manifestations of such reactions is an increased incidence of malignant transformation.

A fourth set of concepts which are somewhat related but really distinctly different involve a specific role of free radicals in the rates and regulation of several physiological processes, especially cell division. These theories relate rates of cell division to the level of free radicals and substances that modify these levels.

In the course of investigations designed to test the aforementioned theories, experimental data have been obtained that have led to empirical correlations unrelated to the theories that generated the experiments. Such empirically based correlations with cancer include not only considerations of free radicals existing in living cells but also paramagnetic trace elements and free radicals induced in the processing of samples (i.e., lyophilization-induced free radicals). In addition to potentially providing useful tools by means of which we may be able to detect or follow tumors, empirical correlations may provide fundamental knowledge if they reflect underlying biochemical differences that lead to the observed paramagnetic changes. Consider, for instance, that changes in redox potential of cells may shift the distribution of trace metals between valence states including some paramagnetic states and/or affect the levels of detectable free radicals.

II. EXPERIMENTAL FINDINGS IN ESR STUDIES OF CANCER

A. Concentration of Free Radicals in Tumors

1. *Nature of free radicals in cells*

For the purposes of this chapter we define free radicals as species having ESR signals in the $g = 2.00$ region with line shapes and power saturation characteristics typical of organic free radicals. While paramagnetic trace elements may have spectral components in the $g = 2.00$ area, they usually do not readily power saturate. Most free radicals that occur in biological systems do power saturate readily, often at microwave powers of about 100 μW at 77°K [5]. This is an important differentiation to make because there are a number of reports in the literature in which signals in the $g = 2.00$ area are termed free radicals but their line shapes and power saturation characteristics indicate that they are more likely due to paramagnetic metals. For the purposes of this chapter, complexes involving paramagnetic metals are considered separately, as are free radicals induced by the processing of the samples.

The nature of the free radical signal usually observed in cells and tissues remains unclear [5,6a,12]. Although a number of authors refer to these as semiquinone free radicals [5,7-10], the evidence for such an assignment is not rigorous. Most tissues have detectable free radicals, and the levels of free radicals appear to be related to the metabolic activity of the tissues [5,6,11,12,16]. Most of the free radicals are in the mitochondria [5,13,16]. The free radical signal seen in tissues or cells is probably due to a number of different free radicals formed as intermediates in oxidation-reduction reactions. The level of these free radicals is remarkably stable, persisting for hours or days in isolated tissues even under relatively unphysiological conditions [14-17]. This stability is somewhat puzzling and suggests that free radicals in tissues may include components that are not enzymatic intermediates. It is very difficult experimentally to differentiate overlapping free radicals. There is some experimental evidence that the kinds of free radicals in a tissue may change with time even though the shape of the ESR signal and the level of the free radicals may appear to be constant [5]. An understanding of the complexity of naturally occurring free radical signals in tissue and the nonspecificity of the information contained in the usual studies of the signals seems essential for adequate interpretations of changes that may occur in the concentration of free radicals in tumors.

2. *Concentration of free radicals in mature tumors*

The amount of free radicals is usually lower in mature tumors as compared to normal tissue. Commoner and Ternberg's original report comparing normal liver with a hepatoma indicated that the level of free radicals was decreased in the hepatoma [16]. Subsequent experimental results have generally yielded data consistent with this original observation [18-20]. Figure 1 illustrates the decreased level of free radicals in the spleen of a mouse with experimental leukemia. Note that in the ESR studies at high microwave power levels (200 mW) in the same region (g = 2.00) in which one ordinarily observes free radicals the signal apparently increased as the leukemia developed,

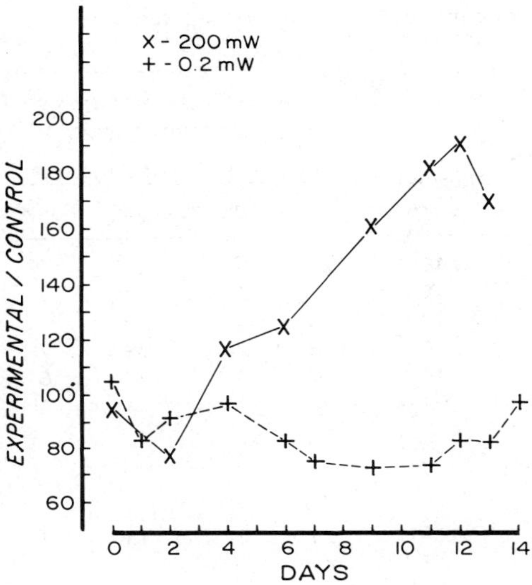

Figure 1 Relative ESR intensity in g = 2.00 region of samples of spleen from mice injected with AKR leukemia cells. The line shapes and power saturation data indicate that at 200 mW the spectra were primarily due to unpaired electrons associated with a paramagnetic metal and at 0.2 mW were primarily due to organic free radicals. (From Ref. 61; see Ref. 20 for further details of the experiment.)

but a study of the same samples at low microwave power (0.2 mW) indicated that the concentration of free radicals actually decreased. Increased amounts of paramagnetic trace elements caused the apparent increase at 200 mW, whereas the data at 0.2 mW reflect the actual decrease in free radicals. Almost all experiments indicate low concentrations of free radicals in experimental tumors, especially if one omits studies in which the spectroscopic conditions were not specific for free radicals or in which the free radicals were produced by the experimental procedure.

Data are limited on free radical concentrations in human tumors, but in general these are the same as or less than those observed in comparable normal tissues (H.M. Swartz, unpublished data; see also Ref. 21).

Table 1 Free Radical and Mn^{2+} Concentrations for DMBA-Induced Tumors and Normal Breast Tissue

Tissue	Time between tumor appearance and biopsy (days)	Number of samples	Free radicals $\bar{x}$	Free radicals SEM	Manganese $\bar{x}$	Manganese SEM
Tumor	0-7	18	29	3	22	3
Tumor	8-14	16	32	3	24	3
Tumor	15-21	10	35	3	28	4
Tumor	22-28	10	38	2	39	7
Tumor	>28	10	37	5	42	6
Normal (apparently uninvolved breast from an animal whose tumor was biopsied)	—	61	16	1	14	1
Normal (experimental animal, prior to appearance of any tumor)	—	17	21	2	13	1
Normal (control)	—	13	16	1	15	2

Column group heading: Average signal intensity[a]

[a]Average peak height (power set 30 dB), 35 GHz spectra.
Source: From Ref. 22.

A few instances have been reported in which there appear to be real increases in concentrations of free radical in tumors. The principal experimental system in which this is noted is the dimethylbenzanthracene (DMBA)-induced breast carcinoma in Sprague-Dawley rats [22,23] (Table 1). In this system the breast tumors have free radical levels twice as high as normal breast tissue. This increase is observed as soon as the tumors are detectable and persists for at least several weeks. It does not seem to be attributable to a particular type of cell. As far as I know this is the only report of increased concentrations of free radicals in experimental cancers except for melanin-containing tumors [24-26] and lyophilized samples. In melanomas the increased levels of free radicals are due to melanin and simply reflect an increased amount of melanin in some of the tumor cells. The free radical in melanin is relatively inert and probably has a different significance than other free radical signals in tissues [5,27]. There is one report of an increased level of free radicals in patients having carcinoma of the cervix [28]; but other studies indicate that this may be unusual even for this type of cancer (H.M. Swartz, unpublished data; also Ref. 29).

3. *Occurence of unique free radicals in tumors*

One of the goals of ESR studies of cancer has been to find free radicals characteristic of tumors. Several such reports have appeared [30-32], but none of the signals found has proven to be unique to cancer. To date no characteristic qualitative changes in free radicals have been found in either experimental or human tumors.

4. *Significance of experimental findings on concentrations of free radicals in tumors*

First we consider whether a reduction of free radicals in tumors has actually been demonstrated. This requires a proper tissue with which to compare the tumor. Most experimental tumors have properties so different from the normal tissue from which they are derived that it is reasonable to question the validity of comparing them with their parent tissues. Perhaps the best approach is to examine chemically induced tumors as they develop. Several such studies have been carried out for hepatic tumors, and the resulting tumors had lower concentrations of free radicals than the normal liver [18,32-33]. In such cases a reduction of free radicals in the tumor accurately describes the results. Comparisons may be less valid for tumors such as the Walker carcinoma that have been transplanted over a long period of time and are compared with muscle or subcutaneous tissue in which they are grown [34-36]. The study of cells in culture that undergo transformation from normal to malignancy in one cell cycle may be the best biological material with which we can compare free radical levels between the normal and malignant state. To date no such studies have been reported. In fact in the only published study of tissue culture cells studied by ESR [37], free radicals may not have been observed. This study was carried out at relatively high microwave power levels, and consequently paramagnetic trace elements probably were the main source of the observed signal and the contribution of free radicals may have been minor [5].

Accepting the fact that at least in some tumor systems there appears to be a genuine reduction in free radicals, the next question is "What does this reduction mean?" One simple interpretation is that it reflects the fact that tumor cells have fewer mitochrondria than do normal tissues [16]. Most of the free radicals appear to be present in the mitochrondria [5,13,16]. In preliminary studies in our laboratory we were unable to account for changes in intensity of ESR signals in breast tissues on the basis of the number of mitochondria [22].

If differences in the number of mitochondria do not account for the observed decrease in free radicals in tumors, there are several possible alternate explanations. The most likely explanations involve changes in redox equilibria. Presumably many free radicals in tissues are intermediates in oxidation-reduction equilibria; for example, consider the following:

$$\underset{\text{(reduced form)}}{\text{hydroquinone}} \rightleftharpoons \underset{\text{(free radical)}}{\text{semiquinone}} \rightleftharpoons \underset{\text{(oxidized form)}}{\text{quinone}} \qquad (1)$$

The relationship between redox levels and intensities of free radicals is complex because of the many factors that may effect such an equilibrium, including the nature of the products and reactants, pH, oxygen, and the effect of other redox biochemicals such as ascorbic acid, glutathione, and vitamin E. The maximum level of free radical intermediates should occur at intermediate redox potentials at which both oxidized and reduced forms are present. The observable free radical concentration can also be modified by changes in metal concentrations because the free radical may form complexes with metals, thereby shifting the redox equilibrium of Eq. (1) toward the free radical [38]. The semiquinone metal complex would yield an ESR signal indicative of a free radical:

$$\text{semiquinone} \rightleftharpoons \text{semiquinone-metal complex} \qquad (2)$$

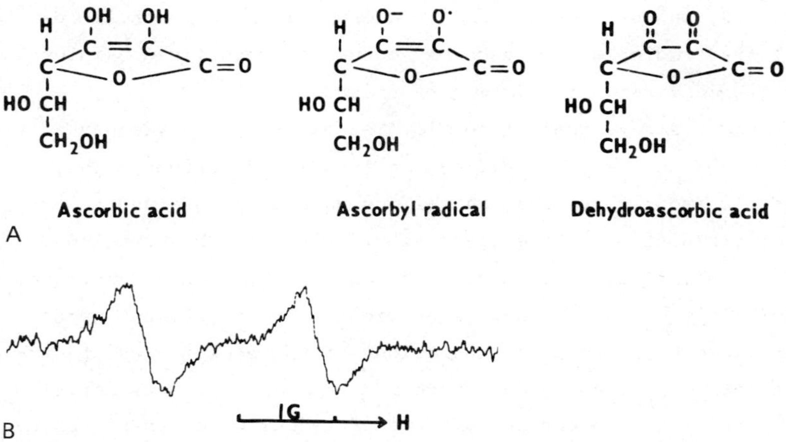

Figure 2 Ascorbyl free radicals. (A) Chemical formulas of the radical and the two major nonradical species of the equilibrium. (B) ESR spectrum of the ascorbyl free radical in tissue at room temperature. (From Ref. 41.)

B. Ascorbyl Radical

1. Nature of the ascorbyl radical

The ascorbyl radical is considered separately because it can be specifically identified in tissues and because its significance may differ from that of the other free radicals that have been noted in tumors and normal tissues. Figure 2 shows the chemical structure of the ascorbyl free radical and its ESR spectrum. There appears to be no other commonly occurring free radical in complex cellular systems that has an ESR spectra similar to the ascorbyl radical. Therefore, the presence of two narrow lines ($\simeq$0.35 G wide) 1.8 G apart and centered at $g = 2.0054$ provides adequate evidence for the presence of the ascorbyl radical.

Because of the narrowness of the lines, rather low concentrations of ascorbyl radicals can be detected; but it is quite easy to fail to detect it if suboptimal spectroscopic conditions are used. The two-line spectrum can be observed only in the liquid state and only by employing low modulation amplitudes. Immobilizing the free radical (e.g., by freezing) broadens the spectrum and decreases its detectability by at least two orders of magnitude. As indicated in

Section II.F, one can detect large concentrations of immobilized ascorbyl free radicals and these may occur in lyophilized samples.

2. Occurrence during irradiation of tissues

Commoner and his co-workers (see Floyd et al. [39] and Vaughan et al. [40]) detected the ascorbyl radical in irradiated tissues. They produced the ascorbyl radical in various normal tissues and tumors by direct irradiation in the ESR cavity. Although they concluded that radiosensitivity correlated with concentration of ascorbyl radicals, this is probably not so. The original "correlations" were based on radiotherapeutic considerations of the sensitivity of different tissues to irradiation. Such radiotherapeutic considerations include both acute and chronic effects of radiation on these tissues. The proper parameter with which to correlate molecular changes is acute radiation sensitivity, because chronic radiation sensitivity is not based on the immediate response of the cells but on chronic changes in small blood vessels which critically reduce blood flow. When one considers acute cellular radiation sensitivity, the correlations found between concentration of the ascorbyl radical and radiosensitivity disappear. An additional problem is the high radiation doses required to observe the ascorbyl radical in these experiments. The dose rate was several thousand rads per second, and the tissues were observed for several minutes to an hour under these conditions. At doses of this magnitude, extensive molecular changes occur that are probably not representative of the significant events that occur at doses which kill cells. Lethal doses for cells are about 150 rads.

3. Concentrations in tumors and normal tissue

The ascorbyl radical has also been observed in nonirradiated tissues, especially by Dodd's group, [41-44]. They detected the ascorbyl radical in several tumor tissues. Under certain conditions they could also observe the radical in normal tissues. The concentration of the ascorbyl radical was independent of the total amount of ascorbic acid and appeared to be related to redox conditions in the cell. The ascorbyl radical concentration increased with necrosis and/or rejection of the tumors, indirectly indicating that these conditions cause a change in the redox equilibrium of the cell. Dodd concluded that in vivo increases in ascorbyl radical are associated with increased oxidation of ascorbic acid but not with total ascorbic acid content.

4. *Origin and significance*

Data on the occurrence and intensity of the ascorbyl free radical have not provided significant insights into cancer. Its occurrence depends on complex redox equilibria, including the concentration of other oxidation-reduction substrates and enzymes. This appears to preclude a straightforward interpretation of its changes. As additional knowledge is gained on the relationship of redox equilibria to normal and pathological metabolisms, the level of ascorbyl free radicals may become a more useful parameter. The radical itself is quite unreactive, being much less reactive than nonradical forms of ascorbic acid [45]. Because of this somewhat surprising property and the low concentration of the ascorbyl radical, reactions of the ascorbyl radical itself usually need not be considered in vivo.

C. Nonheme Iron Sulfur Protein Nitric Oxide Signal

1. *Description of the signal*

Commoner's laboratory in 1965 reported an ESR signal in tumors having a g factor of 2.03 [32]. Initially these findings indicated that some of the fondest hopes of investigators using ESR in cancer might be realized, i.e., the development of a specific test that would detect cancer prior to changes detectable by other means. In animals fed a number of different hepatic carcinogens, the g = 2.03 signal was observed before microscopic or biochemical evidence of malignant transformation. It appeared to meet the criteria for an ESR change that preceded other evidence of malignancy.

2. *Origin and significance of the signal*

This same group, however, also reported considerable difficulty in reproducing these experiments. In some experiments the signal occurred in most of the animals fed the carcinogen while in apparently duplicate experiments, few if any of the animals had the same changes. The investigators eventually resolved this problem when they recognized that the amount of nitrate in the water supply caused the observed experimental variation. By adding nitrate (or nitrite) to drinking water, they could reproducibly produce the g = 2.03 signal in the experimental animals. They could also produce a similar signal in animals on an otherwise normal diet by feeding large amounts of nitrate. The origin of the signal was then shown to be a complex of nonheme iron proteins with nitrogen oxides [46-48] (Fig. 3). Either nitrate or nitrite can be the source of the nitrogen oxide because of the reduction of the compounds in vivo.

Commoner's group also studied the biological significance of this ESR signal. They found a negative correlation between its occurrence and the incidence of tumors [49]. That is, under condi-

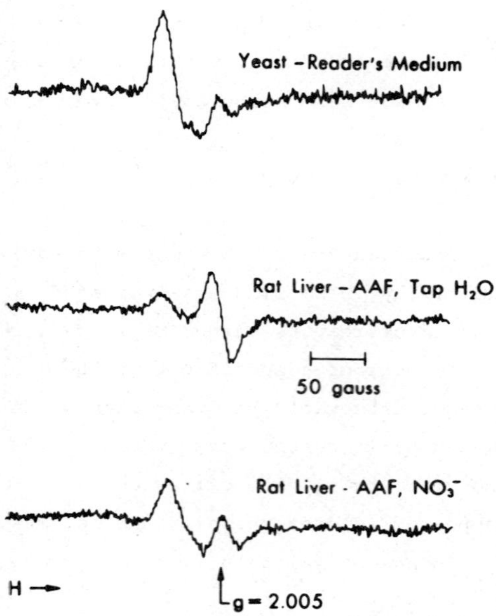

Figure 3 (Upper) ESR signals from baker's yeast cultured anaerobically in Reader's medium for 3 days; (middle) from liver of a rat fed aminoacetylfluorene (AAF) in synthetic diet and provided with tap water; (lower) from liver of a rat on the aminoacetylfluorene synthetic diet but provided with 2.4 mg/liter of NO_3^- nitrogen in drinking water. Computer averages of 15 successive spectra. Modulation amplitude: 16 G. (From Ref. 46.)

tions in which the g = 2.03 signal was most prominent, fewer tumors developed than were observed when the carcinogen was fed without nitrogen oxides. This result, however, is contrary to that observed in J. Mallard's laboratory (personal communication).

D. Heme Iron Protein-Nitric Oxide Signal ("Triplet Signal")

1. *Description of the signal*

Brennan et al. in 1966 reported the occurrence of a "unique" hyperfine pattern associated with tumors [30]. Emanuel and co-workers reported a similar spectrum which they also considered specific for tumors [31]. Figure 4A (ascites hepatoma) illustrates this type of ESR spectrum. The major features are a three-line spectrum, with the lines separated by 16-18 G centered at g = 2.012 and an additional broad line at the lower field.

2. *Origin and significance of the signal*

Maruyama and co-workers [50] and Vanin et al. [51] have clearly demonstrated the molecular basis of this ESR signal. They could induce the signal in any tissue, including normal tissues, if the tissue was kept anoxic for a sufficiently long period in the presence of nitrogen oxides, particularly nitric oxide. They could get essentially identical spectra by adding nitric oxide to purified preparations of heme proteins, as shown in Figure 4. The signal is due to a complex of heme iron with nitric oxide, with the three-line splitting due to interactions of the unpaired electron with the nitrogen atom.

The triplet signal is more frequently observed in tumor tissues than in normal tissues. Its increased frequency in tumors is probably related to the fact that tumors often outgrow their blood supply. The resulting anoxia results in nitrate reduction to nitric oxide, which reacts with hemoglobin or other heme proteins. Thus the occurrence of the triplet indicates a poor blood supply and is not specific for tumors. This signal does cause considerable experimental problems if one attempts to study relatively large tumors because these inevitably have considerable necrosis and therefore a large

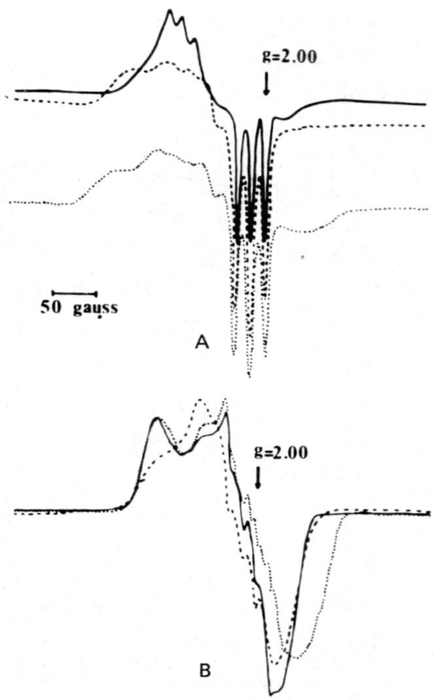

Figure 4 A comparison of the spectra of ascites hepatoma (whole) with those of NO-hemoglobin and NO-cytochrome c. Curves: solid line, ascites hepatomas; dashed line, NO-hemoglobin; dotted line, NO-cytochrome c. (A) Denatured form. (B) Native form. (From Ref. 50).

"triplet" signal. Even at low microwave power this signal may dominate the spectra, interfering with observations of other signals. But because the shape of the triplet signal is known, one can remove it by spectral subtraction.

E. Manganous (2+) Ion Levels in Tissues

Recent studies at 35 GHz have led to the observation of the Mn^{2+} spectra in tissues. Studies at 35 GHz allow detection of the Mn^{2+} signal with much greater sensitivity than do conventional studies at 9 GHz [22,23,36]. Apparently in cells there is less g factor anisotropy in the Mn^{2+} spectra than in spectra of most other paramagnetic metal systems. Therefore, the lines of the latter are broadened at high frequency, leaving the Mn^{2+} as the predominant observable species. In DMBA-induced breast tumors the Mn^{2+} levels gradually rose as the tumors grew, whereas in the Walker carcinoma Mn^{2+} levels were lower than in muscle. These divergent results are not unexpected because several factors such as total Mn concentration and the redox potential of the cell could affect the observable Mn^{2+} signal. It is possible that further understanding of the conditions that affect Mn^{2+} levels will eventually lead to use of the Mn^{2+} signal as a useful monitor of intracellular redox conditions.

F. Concentrations of Free Radicals in Lyophilized Tissues

1. *Original experimental findings*

Lyophilized samples have been used in a large number of studies to study the role of free radicals in cancer [2,52a]. These studies have provided data indicating strong correlations between concentrations of free radicals and cancer, including increases in free radicals very early in tumor development. But there is considerable controversy over the validity and significance of these findings. This is because:

1. Lyophilization can create artifacts in samples prepared for ESR spectroscopy [52a,53-55].
2. Most of these findings are from the Soviet Union, where the style of publication is considerably briefer than in Western journals; therefore, reproducing the experimental conditions is difficult.
3. The results conflict with many reports from Western laboratories.

Consequently the results on lyophilized samples tended to remain unused by most Western scientists.

Emanuel recently summarized the basic findings of the aforementioned studies [52]. As shown in Figure 5, the investigators cited by him found an amazingly consistent pattern of increased concentra-

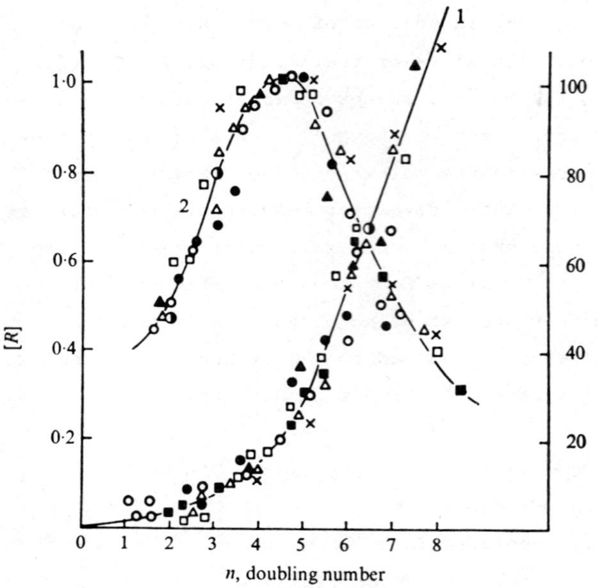

Figure 5 Generalized kinetic curves for relative growth (1) and free radical content (2) of tumor tissue, revised from Emanuel (52): (R) = relative intensity of free radicals as determined by peak-peak measurements of ESR spectra; *n* = relative weight of tumor. The various symbols correspond to seven mouse tumors and one rat tumor (Walker carcinosarcoma, □). (Data from Ref. 52.)

tions of free radicals early in the development of tumors, followed by a decrease to below control values as the tumors became large. Emanuel feels that this pattern accounts for both the increase in free radicals early in cancer development postulated by some theories and the experimentally observed decrease in free radicals in mature tumors. The results obtained by studies of lyophilized samples have frequently been supported by studies in which free radical levels have been studied more directly by measuring so-called antioxidant activity [56-58]. These studies, also from the Soviet Union, are based on inhibition of autoxidation of lipids by tissue extracts. Several other techniques such as chemiluminescence and inhibition of polymer cross-linking have been used similarly [59,60].

Several fundamental questions arise in the interpretation of these data:

1. What is the relationship between the ESR signals seen in the lyophilized tissues and those seen before lyophilization?
2. Is the signal seen in lyophilized tissue reproducible?
3. If this signal is reproducible, does it change characteristically with the development of the tumor?
4. What is the molecular nature of the signal in the lyophilized sample?
5. What factors affect its development?

In our laboratory, we have recently carried out a number of studies [20,22,23,35,36] to help start answering these questions and thereby permit evaluation of the vast amount of data obtained by use of lyophilized samples.

2. Relationship to other studies

In general, the pattern of changes in the intensity of ESR signals that had been observed in lyophilized preparations have not been observed in nonlyophilized samples (either unfrozen or frozen without drying). Until quite recently, however, there had been no studies directly comparing the same tumors by the two methods. Our laboratory recently completed such studies in two tumor systems: Walker carcinosarcoma and DMBA-induced breast tumors in Sprague-Dawley rats. We examined samples of tumor and normal tissues at various intervals after initiation of the tumor. We have utilized procedures which permit examination of the same sample before and after lyophilization under similar geometric conditions at 77°K (Fig. 6).

Emanuel's laboratory extensively studied the Walker carcinosarcoma. The changes observed there are also shown in Figure 6. The results we obtained depended on the sample preparation. Samples examined after freezing but prior to lyophilization had decreased ESR signals (peak-peak heights). These signals remained depressed throughout the experiment. After lyophilization of these same samples, the intensity apparently increased. Close analysis of the

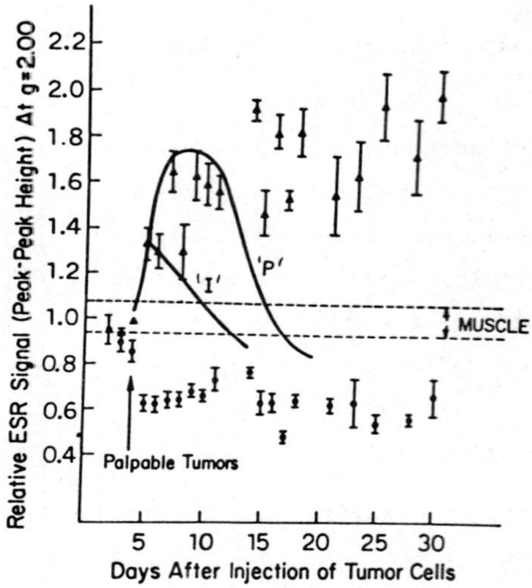

Figure 6 Peak-peak heights of ESR spectra from lyophilized (triangles) and nonlyophilized (circles) Walker 256 tumor and normal muscle samples of Sprague-Dawley rats. The solid lines are data from Saprin et al. [34] on the same system, with samples derived from the periphery (P) and the internal parts (I) of the tumor. We show our results for the internal part of the tumor; the periphery gave similar results if necrotic areas were avoided. Tumors were palpable by day 4. The sample for day 4 contains 10-15% muscle. From day 5 on, the tumors were well defined and samples were 100% tumor. All samples exceeded the cavity's sensitive volume. The peak-peak heights were obtained from 9-GHz spectra at 0.01-mW output power, 8-G modulation amplitude at 77°K. A 2,2-diphenylpicrylhydrazyl (DPPH) standard was used for daily calibration and g value calculations. (From Ref. 36.)

data indicated that the shape of the ESR signals changed after lyophilization. This invalidated peak-peak height measurements and required double integrations to determine if the total number of free radicals changed. We found that the integrated intensity had not changed (Table 2).

These experimental results apparently confirm the findings of Emanuel's group in regard to early increases in peak-peak height during tumor development but indicate that their interpretation of these as representing an increase of free radicals may not have been correct. The increase in peak-peak heights was really a reflection of the change in signal shape rather than a change in the total number of free radicals. In fact, without lyophilization a real

Table 2 Comparison of ESR Peak-Peak Heights and Integrated Signal Intensities for Walker-256 Tumors[a]

Day after injection of 10^4 cells	No. of samples	Line width ΔH (Gauss)	Peak-peak height arbitrary units)	Relative integrated intensity
A. Lyophilized samples:				
Muscle	12	13.1 ± 0.2[b]	14.4 ± 1.0[b]	48 ± 1.6[b]
6	2	10.5 ± 0.4	20.2 ± 0.8	52 ± 10.0
7	3	9.6 ± 0.4	24.4 ± 0.2	42 ± 3.0
9	3	10.3 ± 0.9	28.4 ± 3.8	48 ± 9.0
10	3	10.0 ± 0.3	24.9 ± 1.9	42 ± 0.6
15	4	10.3 ± 0.3	24.9 ± 1.1	41 ± 3.0
17	3	9.8 ± 0.2	24.7 ± 1.0	41 ± 4.0
21	3	9.7 ± 0.3	21.8 ± 1.5	40 ± 1.0
28	2	9.5 ± 0.4	26.9 ± 0.8	58 ± 3.0
30	2	9.0 ± 0.0	33.7 ± 1.4	48 ± 0.5
B. Nonlyophilized samples:				
Muscle	9	13.6 ± 0.2	13.5 ± 0.2	34 ± 4.0
2	3	13.0 ± 0.6	13.9 ± 1.1	42 ± 4.0
6	3	13.3 ± 0.3	8.6 ± 0.6	25 ± 7.0
15	3	13.3 ± 0.3	8.4 ± 0.6	18 ± 6.0
17	3	13.3 ± 0.3	6.6 ± 0.5	12 ± 1.0
28	3	14.0 ± 0.6	8.3 ± 0.4	35 ± 4.0

[a]Integrations were performed by the method of moments.
[b]Standard error of the mean.
Source: From Ref. 36.

decrease in free radicals occurred in Walker tumors compared to normal muscle. The results do seem to indicate, however, that a qualitative change in the type of free radical occurred in lyophilized samples, comparing the tumor tissue with normal muscle.

We obtained quite different results in an analogous experiment utilizing DMBA-induced breast tumors. The carcinogen was administered to the animals, and samples were taken from tumors at different stages of development, studied first in the frozen state and then again after lyophilization. The studies were conducted at 35 GHz in order to deal with the small samples available from early tumors and normal breasts. Prior to lyophilization free radical intensity increased in tumors compared to normal breasts. This was an abrupt increase, observable as soon as the tumors were detectable, and it did not change over the duration of the experiment. The intensity and shape of the ESR signals after lyophilization were similar to those seen before lyophilization. Thus these results neither fit the pattern reported by the Soviet scientists nor the pattern observed in our laboratory with the Walker carcinoma. These different results cannot be due to the use of a higher frequency ESR spectrometer because repetition of the experiment with the Walker carcinoma at 35 GHz gave results that were similar to those found in the original study at 9 GHz [23].

3. *Origin and significance*

The line shape of the lyophilized Walker tumor differed from that of normal muscle or breast (Table 2 and Fig. 7). These differences suggest a possible additional use of the ESR data: namely, the identification of biochemical differences between tumor and normal cells.

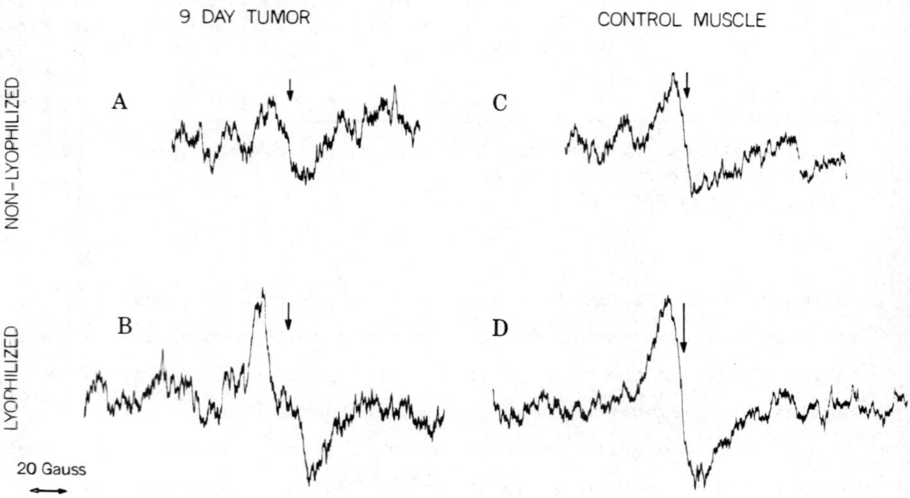

Figure 7 Typical 35-GHz ESR spectra from lyophilized and nonlyophilized muscle and tumor tissues. The tumor tissues are from a 9-day tumor; controls are healthy muscle tissue. The difference in line shape between tumor and control in lyophilized samples can be seen better at this frequency than at 9.1 GHz. The receiver gain is the same for all the spectra shown. Other settings are 30 dB microwave power attenuation from 60 mW (~ 0.05 mW), 8-G modulation amplitude at 123°K. The unmarked arrows point to g ≃ 2.004 + 0.001. (From Ref. 36.)

The differences are of interest even if the ESR spectra that reflect them occur only as a result of an artifact of the sample preparation technique and occur only in some types of tumors.

The spectra of Walker tumors resembled those seen after exposure of lyophilized tissues to oxygen (Fig. 8B), even though the Walker tumor had not been exposed to oxygen. We do not know whether the spectra of lyophilized Walker tumors are typical of other tumors or represent an exception. The only other tumor studied by comparable technique, the DMBA-induced breast tumors, had no qualitative differences from normal breast. There are no other reports in the literature of studies of lyophilized samples at 35 GHz.

Some evidence indicates that the ESR signal seen in most lyophilized tissues after exposure to oxygen may be related to the ascorbyl free radicals [61]. In view of the similarity of spectra of Walker tumors to such spectra, study of oxygen-exposed lyophilized samples provide some insights into the origin of the ESR signals in lyophilized Walker tumors. Most tissues show an increase in intensity and a narrowing of the 9-GHz spectra after exposure to air. Integration of the spectra indicates that the increase in intensity is real. Studies at 35 GHz resolved an additional peak in the samples. The shape of the spectrum at 35 GHz is typical of π-system radicals with axial symmetry, including semiquinones. The intensity of the ESR signal after exposure to oxygen and the increased amount of detail in the ESR spectra indicate that we may be able to determine the molecular nature of this signal. Such studies are underway in our laboratory. The preliminary results suggest that this may be an immobilized ascorbyl free radical (Fig. 8). The evidence includes the following observations:

1. A similar spectrum is observed in lyophilized ascorbic acid irradiated by ultraviolet (UV) light.
2. A similar ESR spectrum is observed upon lyophilization and air exposure of a solution of albumin and ascorbic acid.
3. An even closer fit to the ESR spectrum is observed upon lyophilization and air exposure of a mixture of ascorbic

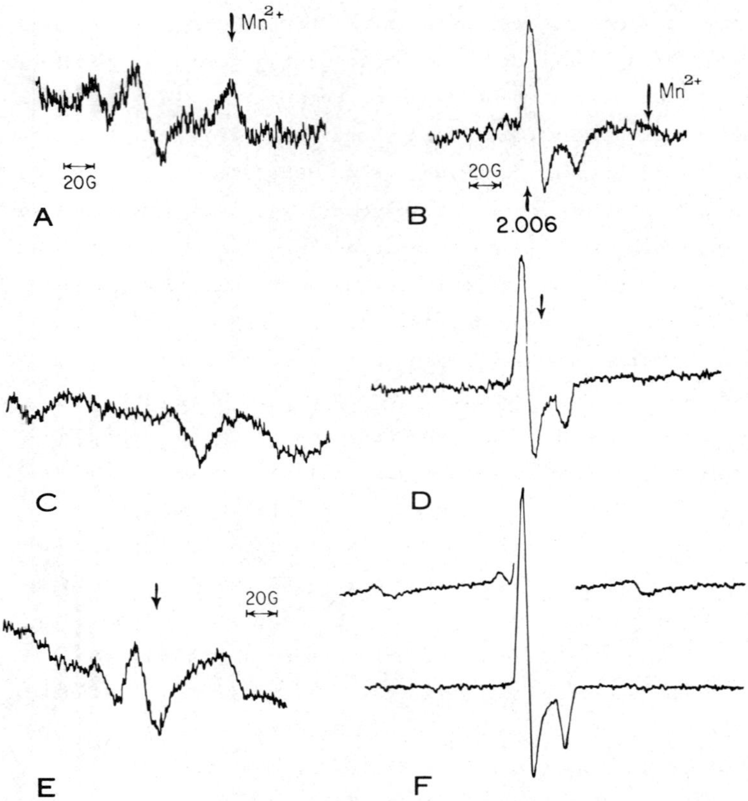

Figure 8 ESR spectra of lyophilized samples at 35 GHz after exposure to air for 45 min. (A) DMBA-induced breast tumor (before exposure to air). (B) The same tumor after exposure to air. (C) Bovine serum albumin. (D) Bovine serum albumin plus 10^{-4} M ascorbic acid. (E) Rat liver microsomes (1 ml contained microsomes from 0.5 g liver). (F) Rat liver microsomes plus 10^{-4} M ascorbic acid. ESR conditions were similar for all spectra except for amplification; temperature, -150°C; incident microwave power, 0.06 mW; modulation amplitude, 8 G. In part F, a spectrum obtained at 60 mW is also shown to indicate the position of the second, third, and fourth peaks of Mn^{2+}. The position of the fourth Mn^{2+} peak is also indicated in parts A and B. (From Ref. 61.)

acid with a more complex biological system such as microsomes.

4. Significant amounts of ascorbyl radicals have been identified in tissues [41-44].

The argument that the signal observed in lyophilized tissues exposed to air is due to the ascorbyl radical remains incomplete. The spectroscopic information, while consistent, is not specific. The ESR spectra of UV-irradiated ascorbic acid and other model systems differ from the spectra observed in tissues; the splitting between g-parallel and g-perpendicular components is 31 G in tissue and some 1 to 5 G less in the model systems. Differences of this magnitude could be due to differences in the immediate environment of the ascorbyl free radical. Additional studies underway include (a) the determination of ENDOR spectra of ascorbyl radicals and the tissue radicals, (b) experiments with ascorbic acid enriched with carbon 13 at C-1, and (c) experiments and computer simulations to determine the relationships between ESR spectra of ascorbyl radicals in liquids and frozen solutions.

G. Paramagnetic Changes in Blood

1. *Introduction*

The examination of blood for paramagnetic changes is particularly attractive because of the ready accessibility of this type of sample. Were characteristic changes found in blood, these would have greater practical significance than changes in tissues, at least for diagnosis and screening [62]. Accordingly, there have been a number of studies of peripheral blood, looking for changes in either free radicals or paramagnetic trace elements. Both frozen and lyophilized samples have been utilized. The usual studies have examined the blood of patients with relatively advanced cancer and compared them to normal controls. A few experiments have also studied patients with other types of disease. Whole blood, plasma, serum, and red cells from both patients and experimental animals have been examined. I will limit the discussion to studies of human blood.

2. Free radicals in frozen blood

Frozen blood or frozen plasma of normal volunteers or cancer patients examined under conventional conditions does not have a free radical ESR signal [63]. A small free radical signal in blood of normal volunteers has been reported by D. C. Borg and K. Schaich (personal communication), using a much larger than ordinary sample at the X-band.

3. Free radicals in lyophilized blood

There have been a number of interesting reports from the Soviet Union on free radical changes in lyophilized blood in cancer patients, especially patients with leukemia [46,64,65]. The investigators reported a correlation between free radical levels and the status of leukemia in a large number of patients. They found that patients with leukemia had higher levels of free radicals, which decreased upon treatment and which rose again before the usual signs of relapse. These spectacular results have not been duplicated in other laboratories, and the Soviet scientists are no longer pursuing these studies. Studies at other laboratories, including ours, indicate a high degree of variability in the ESR signal observed after lyophilizing blood (H. M. Swartz, unpublished data; B. M. Baysal, personal communication; J. Paxton, personal communication; see also Ref. 66). Thus the interpretation and even the reproducibility of studies using lyophilized blood remains unsettled. Because of the potential usefulness of studies of this type, it is likely that work will continue on such systems.

4. Paramagnetic trace metals in blood

In 1972 we reported an apparent correlation of the intensity of a signal at $g = 2.05$ in frozen plasma with the presence of cancer in patients [63]. We had initiated the studies to test whether some of the abnormal paramagnetic trace elements reported to occur in tumors [2,66] might be detected in blood. A second possible source of differences was a systematic reaction to the presence of tumors which might be reflected by a change in paramagnetic elements in blood either directly or as a result of redox changes occurring secondary

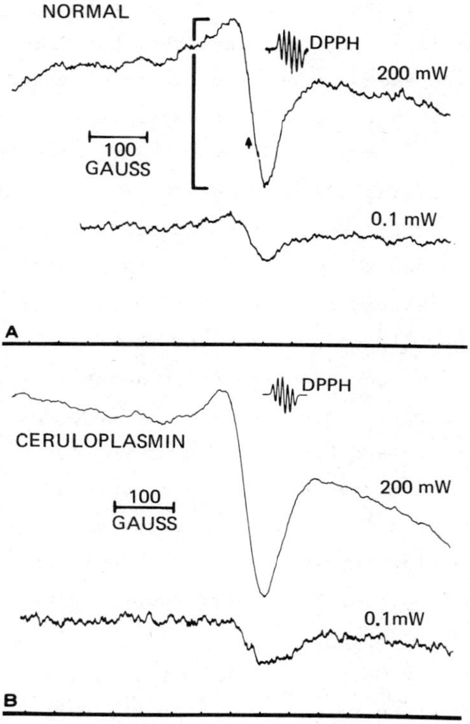

Figure 9 (A) ESR spectrum of normal human control plasma is frozen solution at 77°K. Microwave power levels are as indicated. Other instrumental conditions were as follows: magnetic field scan rate 1000 G in 8 min; time constant 3 sec; magnetic field modulation amplitude, 6 G peak-peak at a frequency of 100 kHz. Vertical bar, peak-peak height used to measure "ESR signal heights"; arrow, position of g = 2.05. (B) ESR spectrum of purified ceruloplasmin solution at a concentration of 75 mg/100 ml. Experimental conditions are the same as for part A. The differences in noise between the two spectra are due to the effect of the bubbling nitrogen in the sample Dewar. (From Ref. 67.)

to the presence of the tumor. The finding of an increased g = 2.05 signal in cancer patients was initially of great interest, but further studies revealed that this increased signal was due to elevated ceruloplasmin levels [67] (Fig. 9). A review of the literature indicated that elevated ceruloplasmin levels had been previously noted in a number of pathological conditions, including cancer [68,69]. Ceruloplasmin is an "acute phase nonspecific reaction substance". Like the erythrocyte sedimentation rate and the C-reactive protein level [69], the ceruloplasmin level is elevated in a number of different pathological conditions. Both the oxidase activity and the copper content of the ceruloplasmin provide convenient means to measure ceruloplasmin without ESR spectroscopy. In a few laboratories, however, such ESR studies have continued and have yielded useful correlations of ceruloplasmin levels with activity of certain types of malignancies such as Hodgkin's disease [70-73]. ESR is one of the best techniques for measuring transferrin—so if both ceruloplasmin and transferrin measurements are required, ESR spectroscopy may be the method of choice [70,73,74].

Quite recently there have been reports suggesting that ceruloplasmin might not exist in the same environment or be in the same state in cancer patients as compared to normal controls [75]. Horn et al. report an apparent change in ESR signals attributable to ceruloplasmin, when blood is held at room temperature for 1-12 hours. Under these circumstances the intensity of the ESR signal due to ceruloplasmin in the blood of cancer patients appears to increase more than that of ceruloplasmin from controls. Probably this is due to conversion of nonparamagnetic copper to paramagnetic copper in the ceruloplasmin molecule. The reproducibility of this work as well as the cause of these changes and their occurrence in other disease states remain to be established.

5. *Significance of ESR studies of blood*

As indicated at the onset of this section, if significantly different and reproducible changes do occur in the blood of cancer patients and can be detected by ESR spectroscopy, this information can be of practical importance and also may have theoretical value. To be useful, the differences must be not only reproducible but also relalively specific for cancer vs. other pathological conditions. Unfortunately most published studies have not demonstrated reproducible changes or have not used suitable controls. The use of healthy volunteers is generally not a suitable control because nonspecific changes due to illness may occur. Also, there often has been significant difference in the age of the controls vs. that of the cancer patients. Changes in ceruloplasmin levels may be useful, but their routine determination does not require ESR spectroscopy. Therefore, at this time, ESR studies of blood remain an interesting topic but of unproven value.

III. INTERPRETATION AND SIGNFICANCE OF EXPERIMENTAL FINDINGS IN ESR INVESTIGATIONS OF CANCER

A. Theories Closely Linking Free Radicals and Cancer

These are the most general but least defined of the theories. In general, they have led to the prediction that concentrations of free radicals increase with the development of most or all types of cancer [76-82]. These theories imply that free radical scavengers should be effective antitumor agents [52,77,83,84]. There are actually two interrelated but distinct concepts involved in theories closely linking free radicals with cancer: the first is that free radical reactions cause the molecular changes characteristic of cancer; the second is that free radicals have a special role in the physiology or metabolism of malignant cells. The latter implies that interference with free radical reactions should affect the growth of tumors, whereas the former suggests that interfering with free radical reactions ought to prevent the development of tumors.

The question of a close association of free radicals with the origin of cancer has really not been adequately investigated. The involvement of free radicals, even as the proximate cause of cancer, may be such that the radical events are quite transient; therefore, studies to investigate these theories must specifically focus on the cells that become malignant at the crucial time that they become malignant. As a minimum, therefore, successful studies would require the use of synchronized cell culture in which a large percentage of the cells become malignant at one time. Such studies have not yet been carried out.

The role of free radical reactions in the growth and division of cancer cells is considered separately in Section III.C.

B. Theories That Carcinogens Act by a Free Radical Mechanism

A more specific hypothesis is that some carcinogens, especially chemical carcinogens, act via free radical mechanisms. Although there are other possibilities, it is most likely that such a mechanism would involve oxidation of the carcinogen by removal of an electron forming a free radical. The carcinogen free radical will attack a critical component of the cell initiating changes that lead to malignant transformation. Such reactions are considered in detail in other chapters in this book. In general, it seems that some carcinogens may indeed act by such a mechanism but others clearly do not. In any event, the results of the published ESR studies described in this chapter do not bear directly on this type of hypothesis. Even if studies do demonstrate free radical intermediates of carcinogens, it is difficult to go from this to prove that the free radical initiates the reaction which causes malignant transformation.

C. Theories Relating Free Radicals to Regulation of Cell Growth and Division

At least two specific theories have been put forth attempting to relate the level of free radicals to cell division. The first is a direct consequence of Szent-Györgyi's theory of cancer in which he postulates formation of charged transfer complexes with proteins as the key element in regulation of cell multiplication [85-87]. Formation of such a charge transfer complex could generate free radicals [88]. He postulates that cancer results from a lack of such formation; therefore tumors should have decreased levels of free radicals. The data cited here are consistent with such a theory but certainly does not prove it. Reports from Szent-Györgyi's laboratory suggest that mixtures of substances capable of forming charge transfer complexes with structural proteins such as collagens result in free radical formation [85].

Burlakova has postulated that low levels of intracellular free radicals are associated with increased cell division and vice versa [56]. She further links the concentration of intracellular free radicals with what she terms antioxidant activity of the cells. That is, an inverse relationship is proposed between the concentration of antioxidants and the concentration of free radicals. Her laboratory has published a large body of literature supportive of this hypothesis [58-60,89]. The low concentration of free radicals found in most malignant cells is consistent with this hypothesis.

D. Theories of Nonspecific Free Radical Pathology

It has been postulated that a number of pathological processes, including cancer, result from unregulated free radical reactions within a cell [90-92]. Such reactions should convert or destroy normal constituents of the cell, leading to a variety of pathological states. The origin of the free radicals, it is presumed, can either be free radicals which escape from normal metabolic reactions (e.g., due to decreases in antioxidants or disruption of intracellular membranes) or free radicals formed by other chemical reactions or physical processes such as ionizing radiation. These theories do not seem to predict a particular type of change in the levels of detectable free radicals in cells.

E. Empirical Correlations of Experimental Findings with Cancer

Most of the ESR studies that have been performed either do not bear upon the various theories that link free radicals with cancer or provide only limited support for them. It is quite possible, however, to use the data empirically to follow patients, for screening or diagnosis, or for understanding basic aspects of carcinogenesis. In this section we consider such empirical correlations.

1. Free radicals in vivo

By in vivo free radicals we refer to free radicals intrinsic to cells but not induced by the sample preparation process itself. That is, we are concerned with signals seen in quickly frozen (nonlyophilized) or unfrozen preparations. In general, free radical levels are low in tumors, although there are a few exceptions to this rule. There is no evidence that the decreased concentration of free radicals is due to a depression of a specific type of free radical in the tumor. Free radical levels in benign tumors show a decrease similar to that of malignant tumors. The measurement of free radicals in tumors requires tissue biopsy, and it seems definite—at least at this time—that the microscopic findings on such samples are far more specific for diagnosis and differential diagnosis than is the level of free radicals [62].

2. Paramagnetic trace elements in tumors

No clear-cut correlation exists between any ESR signal associated with paramagnetic trace metals and the presence of malignant cells. Correlations do exist between the NO-iron-protein signals and the occurrence of necrosis of tissues. There are some data that suggest some of the observed changes in paramagnetic metals might also be associated with redox changes occurring in malignant cells, but at this time this relationship is too uncertain to be useful. The recently reported changes in Mn^{2+} levels may prove to be a reliable indication of redox changes, but so far they have been studied in only two types of experimental tumors [22,23].

3. Paramagnetic trace elements in blood

No unique signals have been found in blood of animals or humans having malignant tumors. Ceruloplasmin increases in blood from animals with cancer, but the increase is modest and can be measured by means other than ESR. Preliminary reports suggest that the stability of the oxidation state of the copper in ceruloplasmin correlates with the presence of tumors [75]. These data, however, are still quite preliminary. Studies correlating transferrin levels and, in many cases, ceruloplasmin levels also, with cancer development have been conducted. But as yet there are no general trends which appear useful for screening or diagnosis of cancer.

4. Lyophilized samples

Emanuel and co-workers [52] have reported the most interesting empirical correlations between tumor development and the free radical signal of lyophilized preparations. There seems to be no doubt that most or all of the signals seen in the lyophilized samples are artifactual, but this does not invalidate consistent empirical correlations. If such correlations do exist, the most important information they would yield would probably be on the nature of the precursors that give rise of the lyophilization-induced signals. Recent data, especially from our laboratory, suggest that under rigidly controlled conditions the empirical correlations that have been observed may be less general than originally reported. Some progress has been made in understanding the nature of the lyophilization-induced ESR signal, and this information should eventually enable adequate judgement to be made as to the potential value of these changes for the detection or understanding of cancer.

The findings on lyophilized blood are even less straightforward. Reports of apparent correlations between free radicals in lyophilized blood and the presence of cancer have been made several times but not confirmed. It is possible that preparation techniques will be developed that will give sufficient reproducibility and also demonstrate the existence of correlations between the magnitude of the ESR signal and the presence of cancer. If such correlations are found, they will presumably be based on changes in oxidation-reduction equilibria in blood.

IV. CONCLUSIONS

A large number of studies have been performed on complex biological systems determining paramagnetic changes in relationship to the development of cancer. The molecular origin of many of the signals (especially those associated with metals) that have been observed has been determined, and these findings have increased our basic knowledge of changes that occur in cells and tissues under a variety of circumstances. To date they have not offered significant new insights into the nature or treatment of cancer. The existing data indicate that we are unlikely to find broad general relationships between cancer and either the nature or the concentration of free radicals in tissues. It is still possible that useful empirical correlations may be found between certain ESR parameters and the presence of cancer, but to date no such correlations have been unequivocably demonstrated. The most promising areas of research appear to be the use of ESR studies to detect and understand the nature of redox changes that occur in cancer cells. The general trend of experimentation in this field seems to be away from studies involving whole-tissue preparations and toward more simplified systems, including model systems. Such studies have more promise of obtaining results that can be interpreted on a fundamental basis.

ACKNOWLEDGMENTS

I gratefully acknowledge the support of N.I.H. (RR-01008) and the National Foundation for Cancer Research.

REFERENCES

1. N. J. D. Dodd, *Metal Ions in Biological Systems: Carcinogenicity* (H. Siegel, Ed.), Vol. 10. Marcel Dekker, New York (1980).
2. H. M. Swartz, *Adv. Cancer Res. 15*:227-252 (1972).
3. H. M. Swartz, J. Bolton, and D. Borg (Eds.), *Biological Applications of Electron Spin Resonance*. Wiley, New York, 1972, Chaps. 1-3.
4. D. C. Borg, in *Free Radicals in Biology* (W. Pryor, Ed.), Vol I. Academic Press, New York, 1976, Chap. 3.
5. H. M. Swartz, in *Biological Applications of Electron Spin Resonance* (H. M. Swartz, J. R. Bolton, and D. C. Borg, Eds.) Wiley, New York, 1972, Chap. 4.
6. J. R. Mallard and M. Kent, *Nature 210*:588-591 (1966).
6a. B. Commoner, J. J. Heise, B. B. Lippincott, R. E. Norbert, J. V. Passonneau, and J. Townsend, *Science 126*:57-63 (1972).
7. V. M. Chumakov, L. P. Kalinichenko, and A. E. Kalmanson, *Biofizika 11*:910-913 (1966).
8. A. Kalmanson, L. Lipchina, and A. Chetverikov, *Biofizika 6*:410-422 (1961).
9. I. G. Kharitonenkov, *Biofizika 12*:224-232 (1967).

10. I. G. Kharitonenkov, A. E. Kalmanson, A. G. Chetverikov, and L. A. Blyumenfel'd, *Biofizika 9*:172-179 (1964).
11. B. Commoner and T. C. Hollocher, *Proc. Nat. Acad. Sci. U.S. 46*:405-416 (1960).
12. T. C. Hollocher and B. Commoner, *Proc. Nat. Acad. Sci. U.S. 46*: 416-425 (1960).
13. O. N. Brzhevskaya, V. S. Marinov, O. S. Nedelina, and E. M. Sheksheyev, *Biofizika 12*:354-356 (1967).
14. H. M. Swartz, J. C. Darin, and J. D. Lewis, *Proc. 1st Eur. Biophys. Congr. 2*:557-561 (1971).
15. A. G. Chetverikov and A. Vanin, *Biofizika 13*:255-259 (1968).
16. B. Commoner and J. L. Ternberg, *Proc. Nat. Acad. Sci. U.S. 47*: 1374-1384 (1961).
17. M. Schara, M. Sentjurc, and T. Kozelj, *J. Mag. Res. 6*:628-632 (1972).
18. H. Abe and Y. Jurata, *Gann 73*:65-75 (1974).
19. J. R. Mallard and M. Kent, *Nature 204*:1192-1194 (1964).
20. H. M. Swartz, C. Mailer, S. Ambegaonkar, and W. E. Antholine, *Cancer Res. 33*:2588-2595 (1973).
21. J. Duchesne and A. Van de Vorst, *Proc. Belg. Acad. Sci.* pp. 433-448 (1970) .
22. H. M. Swartz, W. E. Antholine, and B. A. Reichling, *Phys. Med. Biol. 23*:235-244 (1978).
23. P. L. Gutierrez, H. M. Swartz, and E. J. Wilkinson, *Br. J. Cancer 39*:330-336 (1979).
24. H. B. Demopoulos, W. Landgraf, P. S. Duke, and H. Tai, *Lab. Invest. 15*:1652-1658 (1966).
25. P. Duke, W. Landgraf, A. Mitamura, and H. J. Demopoulous, *J. Nat. Cancer Inst. 37*:191-198 (1966).
26. R. Sarna and H. M. Swartz, *Folia Histochem. Cytochem. 16(14)*: 275-286 (1978).
27. R. C. Sealy, C. C. Felix, J. S. Hyde, and H. M. Swartz, in *Free Radicals in Biology* (W. Pryor, Ed.), Vol. IV. Academic Press, New York, 1980, p. 210-251.
28. T. Cole and C. P. Hodgkinson, *Technical Report*, Ford Motor Company, Detroit, 1965.
29. T. F. Slater and J. W. R. Cook in *Cytology Automation* (D. M. D. Evans, Ed.). Livingstone, Edinburgh and London, 1970, pp. 108-120.
30. M. J. Brennan, T. Cole, and J. A. Singley, *Proc. Soc. Exp. Biol. Med. 123*:715-718 (1966).
31. N. M. Emanuel, A. N. Saprin, V. A. Shabalkin, L. E. Kozlova, and K. E. Kruglykova, *Nature 222*:165-167 (1969).
32. A. J. Vithayathil, J. L. Ternberg, and B. Commoner, *Nature 207*. 1246-1249 (1965).
33. J. Duchesne, R. Goutier, R. Marechal, and A. Van de Vorst, *Phys. Med. Biol. 20*:305-309 (1975).
34. A. N. Saprin, Ye. A. Minenkova, L. G. Nagler, Yu. S. Kaznackeyev, S. A. Kruglyak, A. Kruglyakova, and N. Emanuel, *Biofizika 12*:1099-1102 (1967).
35. H. M. Swartz and P. L. Gutierrez, *Science 198*:936-938 (1977).
36. P. L. Gutierrez and H. M. Swartz, *Br. J. Cancer 39*:24-34 (1979).

37. A. L. Segre, A. Benedetto, T. Eremanko, P. Volpe, A. DiNola, and F. Conti, *Biochim. Biophys. Acta 497*:615-621 (1976).
38. C. C. Felix, J. S. Hyde, T. Sarna, and R. C. Sealy, *J. Am. Chem. Soc. 100*:3922-3926 (1978).
39. R. A. Floyd, A. Brondsdon, and B. Commoner, *Ann. N.Y. Acad. Sci. 222*:1077-1086 (1973).
40. W. Vaughan, J. Henry, and B. Commoner, *Biochim. Biophys. Acta 32*:159-162 (1973).
41. N. Dodd, *Br. J. Cancer 28*:257-262 (1973).
42. N. Dodd and J. M. Giron-Conland, *Br. J. Cancer 32*:451 (1975).
43. J. M. Silcock and N. Dodd, *Br. J. Cancer 34*:550-555 (1976).
44. N. Dodd and J. M. Silcock, *Br. J. Cancer 38*:612 (1978).
45. B. H. Bielski and H. W. Richter, *Ann. N.Y. Acad. Sci. 258*: 231-238 (1975).
46. J. C. Woolum and B. Commoner, *Biochim. Biophys. Acta 201*: 131-140 (1970).
47. R. W. Chiang, J. C. Woolum, and B. Commoner, *Biochim. Biophys. Acta 257*:452-460 (1972).
48. A. F. Vanin, A. N. Osipov, L. N. Kubrina, D. S. Burbaev, and R. Nalbandyan, *Studia Biophys. 49*:13 (1975).
49. B. Commoner, J. C. Woolum, B. H. Senturia, and J. L. Ternberg, *Cancer Res. 30*:2091-2097 (1970).
50. T. Maruyama, N. Kataoka, S. Nagase, W. Nakada, H. Sato, and H. Sasaki, *Cancer Res. 31*:179-184 (1971).
51. A. F. Vanin, L. V. Vakhnina, and A. G. Chetverikov, *Biofizika 15*:1044-1051 (1970).
52. N. M. Emanuel, *Q. Rev. 9*:283-308 (1976).
52a. R. J. Heckley, in *Biological Applications of Electron Spin Resonance* (H. M. Swartz, J. R. Bolton, and D. C. Borg, Eds.). Wiley, New York, 1972, Chap. 5.
53. A. G. Chetverikov, L. A. Blyumenfel'd, and G. V. Fomin, *Biofizika 10*:476-486 (1965).
54. F. K. Truby and J. W. Goldzeiher, *Nature 182*:1371-1372 (1958).
55. A. F. Vanin, A. Chetverikov, and L. A. Blyumenfel'd, *Biofizika 13*:66-69 (1968).
56. Ye. B. Burlakova and Ye. M. Molochkina, *Biofizika 13*:443-448 (1968).
57. *Ye. B. Burlakova, Proc. 4th Int. Biophys. Cong. Symposial Papers* pp. 14-62 (1973).
58. Ye. B. Burlakova, N. I. Dubinskaya, Ye. V. Koperina, and N. P. Pal'mina, *Biofizika 11*:1008-1012 (1966).
59. Ye. B. Burlakova, *Biofizika 16*:39-43 (1971).
60. Ye. B. Burlakova, *Biofizika 16*:201-205 (1971).
61. H. M. Swartz, *CIBA Foundation Symposium No. 67: Submolecular Biology and Cancer*. Elsevier/Excerpta Medica, Amsterdam, 1979, pp. 102-124.
62. H. M. Swartz, *J. Mag. Res. 29*:393-396 (1978).
63. H. M. Swartz and J. Weisner, *Radiology 104*:209-210 (1972).
64. L. Pavlova and A. R. Livenson, *Biofizika 10*:169-171 (1965).
64. S. C. Lillicrap and B. C. Wilson, *Br. J. Radiol. 47*:601-602 (1974).

65. E. V. Klochko, O. A. Kovalenko, V. I. Pinzur, K. E. Kruglykova, V. M. Chibrykyn, T. A. Kassinsky, and N. M. Emanuel, *Trans. Moscow Soc. Naturalists 32*:120-124 (1970).
66. D. W. Nebert and H. S. Mason, *J. Cancer Res. 31*:179-184 (1963).
67. C. Mailer, H. M. Swartz, M. Konieczny, S. Ambegaonkar, and V. L. Moore, *Cancer Res. 34*:637-642 (1974).
68. N. R. Hughes, *Aust. J. Exp. Biol. Med. Sci. 50*:97-107 (1972).
69. I. Sternlieb and I. H. Scheinberg, *Proc. N.Y. Acad. Sci. 94*: 71-76 (1963).
70. M. Foster, A. Dawson, T. Pocklington, and L. Fell, *Clin. Radiol. 28*:23-28 (1977).
71. M. Foster, L. Fell, T. Pocklington, F. Akinsete, A. K. Dawson, J. M. Hutchison, and J. R. Mallard, *Clin. Radiol. 28*:15-22 (1977).
72. M. Foster, T. Pocklington, J. Miller, and J. Mallard, *Br. J. Cancer 28*:340-348 (1973).
73. T. Pocklington and M. A. Foster, *Br. J. Cancer 36*:369-373 (1977)
74. M. Foster, T. Pocklington, and J. R. Mallard, *Phys. Med. Biol. 22*:531-542 (1977).
75. R. A. Horn, E. J. Friesen, R. L. Stephens, W. R. Hedrick, and J. D. Zimbrick, *Cancer Res.* in press (Mar. 1979).
76. A. M. Brues and E. S. G. Barron, *Ann. Rev. Biochem. 20*:343-366 (1951).
77. N. M. Emanuel, *Izv. Akad. Nauk SSSR (Biol.) 6*:773-778 (1974).
78. N. M. Emanuel and A. N. Saprin, *Proc. 1st Eur. Biophys. Congr. 6*:269-273 (1971).
79. A. Fitzhugh, *Nature 118*:783 (1953).
80. M. J. Lyons, J. F. Gibson, and D. J. Ingram, *Nature 181*:1003-1004 (1958).
81. B. S. Oppenheimer, E. T. Oppenheimer, A. P. Stout, T. Danishefsky, and F. R. Eirich, *Nature 118*:783 (1953).
82. B. S. Oppenheimer, E. T. Oppenheimer, T. Danishefsky, A. P. Stout, and F. R. Eirich, *Cancer Res. 15*:333-340 (1955).
83. N. M. Emanuel and T. E. Lepatova, *Proc. Acad. Sci. USSR 130*: 221-222 (1960) .
84. K. K. Georgieff, *Science 173*:537-539 (1971).
85. H. A. Pohl, P. R. Gascoyne, and A. Szent-Györgyi, *Proc. Nat. Acad. Sci. U.S. 74*:1558-1560 (1977).
86. A. Szent-Györgyi, *Acta Biochim. Biophys. Acad. Sci. Hung. 8*: 117-127 (1973).
87. A. Szent-Györgyi, I. Isenberg, and S. Baird, *Proc. Nat. Acad. Sci. U.S. 46*:1444-1449 (1960).
88. P. B. Ayscough, *Electron Spin Resonance in Chemistry*. London, 1967, p. 413.
89. Ye. B. Burlakova and N. P. Pal'mina, *Biofizika 11*:258-262 (1966).
90. H. V. Boenig, *J. Am. Geriat. Soc. 14*:1211-1220 (1966).
91. D. Harman, *Rad. Res. 16*:753-763 (1962).
92. W. A. Pryor (Chairman), Symposium on Free Radicals in Biology. *Fed. Proc. 32*:1859-1908 (1973).

4
MULTIPLE MECHANISMS OF ACTIVATION IN AROMATIC HYDROCARBON CARCINOGENESIS

Ercole Cavalieri, Eleanor Rogan, and Robert Roth*

Eppley Institute for Research in Cancer
University of Nebraska Medical Center
Omaha, Nebraska

I. INTRODUCTION

It is hypothesized that chemical carcinogens interact covalently with macromolecular cellular targets to initiate the series of events leading to cancer. Polycyclic aromatic hydrocarbons (PAH) are not active per se but require some type of metabolic activation by monooxygenase enzymes for the manifestation of their carcinogenic activity. In recent years it has been proposed that the ultimate carcinogenic forms of PAH are vicinal diol epoxide derivatives [1]. The "bay region" theory of carcinogenicity has been developed from theoretical calculations which predict an unusually high ease of formation of benzylic carbonium ions derived from bay region epoxides [2-4]. A considerable amount of data supporting this concept is available for unsubstituted as well as alkyl-substituted PAH. It must be recognized, however, that the complexity of PAH carcinogenesis cannot be understood by postulating only diol epoxides as critical carcinogenic intermediates. Comprehension of this process must include recognition of multiple mechanisms of activation and the balance between detoxifying pathways and those leading to carcinogenic metabolites. While it is logical to assume that a predominant mechanism of activation for each hydrocarbon triggers tumor formation, it is equally plausible that the mode of activation may change, depending on the nature of the activating enzymes present in the target tissues of various animals.

*Present affiliation: Midwest Research Institute, Kansas City, Missouri

To determine the molecular mechanism(s) of activation for each hydrocarbon, studies of carcinogenicity, mutagenicity, enzymology, metabolism, and binding in the target organs should be combined. Carcinogenicity studies are a powerful tool and represent the ultimate test in establishing whether or not a certain hypothetical mechanism of tumor initiation might be correct. These activities are quantitatively compared in two main kinds of experiments. In one study, the carcinogenicity of a parent compound is compared to its hypothetical proximate and ultimate intermediates. In such experiments, it is expected that proximate metabolites should be more active than the parent compound and, in turn, the ultimate more active than the proximate. It is possible, however, that when these intermediates are too reactive, the expected results will not be obtained. The other study compares the relative carcinogenic activity of closely related compounds in which steric and electronic properties differ slightly. The introduction of substituents designed to block a specific activation or deactivation pathway should produce an alteration in carcinogenicity. In this case, interpretation of results is commonly clouded by ambiguities arising from simultaneous alterations in physical and electronic properties, molecular geometry, generation of new metabolic pathways specific to the substituent functionality, and modification of the balance between detoxifying and activating pathways.

Mutagenesis studies can be very informative if DNA is the critical cellular target of PAH carcinogenesis. Mutagenicity tests are especially important for an ultimate carcinogenic form of a certain hydrocarbon, because in this case no metabolic activation is required. When instead the mutagenicity of parent compounds or proximate carcinogens which require metabolic activation is tested, questions can arise as to whether or not the activating system is a pertinent model for producing biologically important carcinogenic metabolites.

The enzymology involved in the metabolism of PAH includes the aryl hydrocarbon monooxygenase system, epoxide hydrase, and glutathione S-epoxide transferase [5]. Since monooxygenase enzymes have been found to be one-electron acceptors [6], it is possible that they activate PAH by one-electron oxidation forming radical cations [7]. The particular reactive metabolites formed depend upon the specificity of the monooxygenases present in a tissue at a given time [8] and on the relative activities of the other enzymes involved in the metabolism of PAH.

Metabolic studies represent another powerful tool in revealing the presence of a presumed critical metabolite and in determining the efficiency of conversion of a certain hydrocarbon to such a metabolite. Most of the published studies of PAH have been conducted in liver, which is not a target organ. The problem remains as to whether the metabolic pathways in liver and a target organ are qualitatively and quantitatively the same. Recent advances in the analysis of metabolic products are enabling investigators to perform studies in the appropriate target organs, and the significance of presumed critical metabolites will have to await publication of such results.

Binding studies of PAH to DNA and RNA provide an excellent tool for substantiating or eliminating hypothetical mechanisms of PAH activation. If binding occurs via a substitution reaction, loss of tritium from a position in specifically labeled [^{3}H,^{14}C]PAH can be a measure of binding at that position. However, this interpretation can be incorrect if tritium is lost through formation of a metabolite which subsequently binds to DNA at another position. Determination of the structure of the covalent DNA-PAH adduct provides a probe of the mechanism(s) of PAH activation. In this case, however, several shortcomings can be envisioned: One is the need for complete recovery of the DNA from the target tissue. Second is the need for complete hydrolysis of the DNA to monomeric units. Third is that instability of certain adducts can hamper their isolation and even prevent recognition of their presence. The profile of adducts present in cells appears to change with time after treatment, raising new questions about the relevance of certain adducts to the carcinogenic process.

In this chapter we present evidence that multiple mechanisms of activation are responsible for the carcinogenic activity of the various PAH. The reactive intermediates that will be considered are arene oxides, radical cations, and hydroxy derivatives at the benzylic carbon atom of alkyl-substituted PAH. We have evaluated the possible contribution of these intermediates to the carcinogenic activity of the different PAH. From all the evidence, the complexity of the various factors contributing to PAH carcinogenicity is apparent.

II. EXPERIMENTAL EVIDENCE

A. Arene Oxides

1. *Benzo[a]pyrene*

Activation, metabolism and tumorigenicity have been studied most thoroughly for benzo[a]pyrene (BP) (Fig. 1). Most of these studies suggest that the hydrocarbon undergoes metabolic activation via formation of BP 7,8-oxide, hydration of the arene oxide to BP 7,8-dihydrodiol*, and epoxidation of the 7,8-dihydrodiol to form the 9,10-oxide. This line of thought was initiated after Borgen et al. [9] observed that BP 7,8-diol was bound to DNA more extensively than BP itself when these compounds were incubated with DNA and microsomes. Subsequently Sims et al. [1] proposed that the BP diol epoxide was the reactive intermediate responsible for binding of BP to DNA in rodent cells cultured in the presence of BP.

When binding of BP to DNA was catalyzed by rat liver microsomes, the only adduct formed was with deoxyguanosine [10]. This adduct was postulated to correspond to the trans addition of only one enantiomer, since enzymatic formation of BP diol epoxide is stereoselective [11-13]. Somewhat lower stereospecificity was observed in binding of BP to DNA of BHK21 cells [14,15], human and bovine bronchial explants [16] and hamster embryo cell culture

*For all dihydrodiols and diol epoxides, unless otherwise stated, all compounds are racemic mixtures where optical enantiomers are possible. (See the abbreviations list at the end of this chapter before the references.)

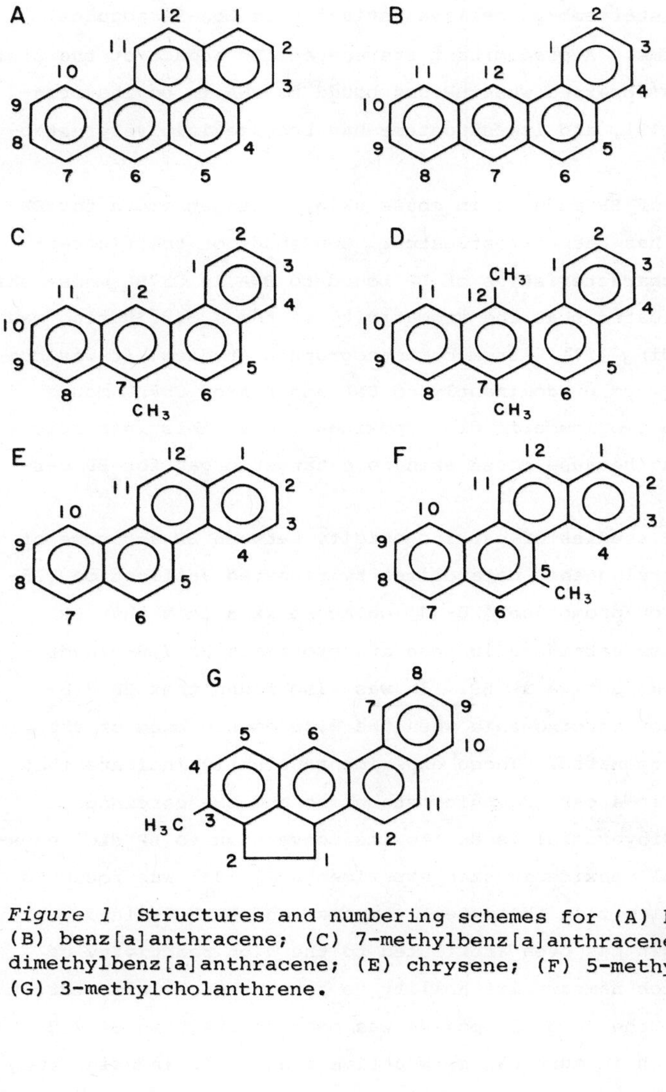

Figure 1 Structures and numbering schemes for (A) benzo[a]pyrene; (B) benz[a]anthracene; (C) 7-methylbenz[a]anthracene; (D) 7,12-dimethylbenz[a]anthracene; (E) chrysene; (F) 5-methylchrysene; and (G) 3-methylcholanthrene.

[17,18]. In hamster embryo cells a variation in DNA-BP adducts occurred over time. A predominant stereospecific adduct of the diol epoxide has been observed when BP was bound to RNA in bovine bronchial explants [19], and its structure has been definitively established [20-22].

The nature of BP adducts in mouse skin, a target organ for BP carcinogenesis, has been investigated. One study of the fluorescence spectral characteristics of BP bound to DNA in C57BL mouse skin after 24 hr indicates that the aromaticity of the 7,8,9,10-ring was lost during binding [23]. From chromatographic studies, it was concluded that BP binds predominantly to DNA and RNA of C57BL mouse skin through the two isomeric diol epoxides [24]. This is particularly significant because mouse skin is a target organ for BP carcinogenesis.

Comparative studies of carcinogenicity between BP and some of its proximate carcinogenic metabolites by repeated application [25-27] and initiation-promotion [28-31] on mouse skin show that BP 7,8-oxide was less carcinogenic than its proximate BP 7,8-dihydrodiol, which was as active as BP. It was also found that BP 7,8-dihydrodiol is not carcinogenic when the 9,10 double bond of the molecule is hydrogenated. These experiments clearly indicate that BP diol epoxide is a carcinogenic metabolite and the carcinogenicity of BP 7,8-dihydrodiol is due to its conversion to BP diol epoxide. The BP diol epoxide in skin experiments [29-31] was found to be much less active than the parent compound and BP 7,8-dihydrodiol. This result has been attributed to the high reactivity of the epoxide, which hampers its ability to reach the cellular target sites. However, the BP diol epoxide was more active than BP 7,8-dihydrodiol, which in turn was more active than BP in inducing lung adenomas and malignant lymphomas in newborn mice [32,33]. In this case, the BP diol epoxide can be considered the principal ultimate carcinogenic form.

Both diastereomers of BP diol epoxides are among the most potent mutagens ever tested in *Salmonella typhimurium* and cultured mammalian cells [34-48]. More detailed studies of the enantiomers of the diastereomers indicated that these compounds have high but different levels of mutagenicity [39].

2. Benz[a]anthracene series: benz[a]anthracene, 7-methylbenz[a]anthracene, 7,12-dimethylbenz[a]anthracene, and 3-methylcholanthrene

BA (Fig. 1) is very weakly carcinogenic [40] but shows consistent tumor-initiating activity [41]. In the case of BA, extensive studies of mutagenicity in bacteria and mammalian cells [42-44] and carcinogenicity in initiation-promotion experiments [44-46] clearly indicates that BA 3,4-diol-1,2-oxide, which has the epoxide on the saturated angular benzo ring forming part of the bay region of the hydrocarbon, is the ultimate carcinogenic metabolite of BA. This conclusion has been reached from the fact that BA 3,4-dihydrodiol was found to be more mutagenic and carcinogenic than the parent compound and other dihydrodiol derivatives. In turn, the corresponding diol epoxides of BA 3,4-dihydrodiol were more mutagenic and carcinogenic than BA 3,4-dihydrodiol.

For BA-7-CH_3 (Fig. 1), studies of metabolism, binding, carcinogenicity in mouse skin, in vitro transformation and mutagenesis in mammalian cells, and mutagenicity in bacteria indicate that BA-7-CH_3 3,4-diol-1,2-oxide is an important ultimate carcinogenic metabolite. Mouse skin maintained in short-term organ culture metabolized BA-7-CH_3 to all of its possible trans-dihydrodiols [47]. Fluorescence spectra of DNA from BA-7-CH_3-treated skin of C57BL mice were consistent with binding of BA-7-CH_3 3,4-diol-1,2-oxide but did not rule out a bound form which retained the BA aromaticity [48]. Analysis of [^{3}H]BA-7-CH_3-DNA adducts from C57BL mouse skin in vivo [47] indicates that activation may occur by formation of the BA-7-CH_3 diol epoxide, although definitive identification of adducts has not been made. In all these studies a significant and sometimes vastly predominant portion of the tritium in degraded [^{3}H]BA-7-CH_3-DNA samples chromatographed on Sephadex LH20 is in the early, unanalyzed fractions. Thus it is possible that adducts other than those arising from the diol epoxide were present but undetected.

The relatively high tumor-initiating activity of BA-7-CH_3 3,4-dihydrodiol, when compared to the parent compound and the other possible trans-dihydrodiols, provides strong support for the important role of the 3,4-diol-1,2-oxide derivative in determining the carcinogenic activity of BA-7-CH_3 [49]. Similar results demonstrating the relatively high biological activity of the 3,4-dihydrodiol were obtained in tests of microsome-mediated mutagenesis with *S. typhimurium* TA98 [50], as well as in mutagenicity tests with Chinese hamster V79 cells [51]. Furthermore, this derivative was more active in malignant transformation of M2 mouse fibroblasts than the parent compound and other dihydrodiol compounds [51].

Fluorescence spectral analysis of DMBA*-DNA adducts in mouse skin suggests that the saturated angular benzo ring is involved in the binding, although these data do not exclude the presence of a fully aromatic benzanthracene ring system bound to DNA [52]. The same spectral fluorescence analysis has been used to investigate the possible structure of DMBA-DNA nucleoside adducts in cultured mouse [53] and hamster embryo cells [54]. The results are consistent with a mechanism involving the formation of diol epoxide in the 1,2,3,4-ring. Chromatographic analysis of DMBA-nucleoside adducts in mouse [55,56] and hamster [57] embryo cells and mouse skin [55] points to binding of the same DMBA diol epoxide. However, a significant amount of DMBA is contained in the unanalyzed, early fractions from Sephadex LH20 chromatography and the possibility exists that other kinds of DMBA adducts are present.

Additionally, metabolism of DMBA in 3-methylcholanthrene (MC)-treated mouse skin has been studied using high-pressure liquid chromatography, and the major metabolite formed was BA-7-CH_2OH-12-CH_3 [58]. No clear indication from these studies, however, could be gained in regard to a possible carcinogenic pathway of activation for DMBA.

---

*DMBA is 7,12-dimethylbenz[a]anthracene.

Some preliminary evidence for activation of MC (Fig. 1) by formation of a diol epoxide has been presented by King et al. [59] in studies of MC-DNA adducts in mouse embryo cells, as well as of DNA adducts formed by rat liver microsomes. In metabolic studies of MC with rat liver microsomes, the formation of the trans MC-9,10-dihydrodiol could not be detected [60]. However, when the predominant metabolite of MC, namely, MC-1-OH, was further metabolized, the corresponding 9,10-dihydrodiol was found. This metabolism in vitro would imply that the formation of an ultimate carcinogen from MC requires three oxidative steps, namely, hydroxylation at C-1, dihydrodiol formation at C-9 and C-10, and final epoxidation at C-7 and C-8. Binding studies to DNA in mouse embryo cells suggest that this diol epoxide could indeed play a role in the carcinogenesis of MC in vivo without the preliminary formation of MC-1-OH [59].

3. *Chrysene and 5-methylchrysene*

Chrysene (Fig. 1) is a weak carcinogen [40] which manifests tumor-initiating activity [61,62]. The stronger mutagenic and tumor-initiating activity of chrysene 1,2-dihydrodiol relative to the parent compound and its other dihydrodiols suggests that the 1,2-dihydrodiol is a proximate carcinogenic metabolite of chrysene and that the chrysene 1,2-diol-3,4-oxide, in which the epoxide group forms part of the bay region in the molecule, is likely to be an ultimate carcinogenic metabolite of chrysene [63,64].

Among the methylchrysenes, 5-methylchrysene (Fig. 1) manifests a potent carcinogenic activity [61,62]. To study the structural features required for the carcinogenicity of this compound, a series of 5-methylchrysenes fluorinated at various positions was tested for mutagenicity in Salmonella typhimurium TA 100 [65] and for tumor-initiating activity in mouse skin [66]. The compounds fluorinated at the 1, 3, or 12 position showed decreased mutagenic and carcinogenic activities, indicating that these positions are involved in the metabolic activation of 5-methylchrysene. These data suggest further that the 1,2-diol-3,4-oxide may be the ultimate carcinogen of 5-methylchrysene.

B. Hydroxymethyl Derivatives

Numerous studies have shown that the major products in the metabolism of alkyl-substituted PAH are the hydroxylated derivatives at the alkyl side chain. It is also well-documented that the carcinogenicity of these oxygenated derivatives is generally lower than that of the parent compounds. This has been observed for the two monohydroxymethyl derivatives of DMBA [67-70], for MC-1-OH [71-73] and for BP-6-CH_2OH [74,75], suggesting that such metabolites are not important proximate carcinogens of the parent compounds, as was earlier assumed, although it is not excluded that this oxidative pathway might contribute to some extent to the carcinogenic activity of these alkyl-substituted PAH.

In the case of BA-7-CH_3, recent data ([76] and unpublished results from our laboratory) provide evidence that hydroxylation at the methyl group might be an important pathway. The relative carcinogenicity of BA-7-CH_3 and some of its derivatives at the methyl group was quantitatively compared in a dose-response experiment by repeated application on mouse skin. As previously observed [77], BA-7-CH_2CH_3 was much less active than BA-7-CH_3, suggesting that the methyl group might be involved in a metabolic pathway leading to carcinogenesis. The stronger activity of BA-7-CH_2OH relative to that of BA-7-CH_3 and the formation of BA-7-CH_2OH as a predominant metabolite of BA-7-CH_3 in mouse skin suggest that BA-7-CH_2OH is likely to be an important proximate carcinogen of BA-7-CH_3. Furthermore, the stronger carcinogenic potency of BA-7-CH_2OCOCH_3 relative to that of BA-7-CH_2OH indicates that the activity of the latter compound may be related to the formation of an ester which is able to act as an ultimate reactive alkylating compound. This point is substantiated by the much stronger carcinogenicity of BP-6-$CH_2OSO_3^-$ in relation to its parent alcohol in a similar skin experiment at three dose levels ([75] and unpublished results from our laboratory).

Sulfate esters of BA-7-CH_2OH and BP-6-CH_2OH have been found to be mutagenic toward *Salmonella typhimurium* TA 98 without microsomal activation. The compounds showed an activity comparable to the very strongly mutagenic BP diol epoxide. In contrast, the parent alcohols and nonbenzylic esters were virtually inactive (our unpublished results). Therefore sulfate esters of alkyl-substituted PAH might be ultimate carcinogenic compounds if sulfate conjugation can occur in a target organ.

There is strong evidence that the sulfate ester of N-OH-2-AAF, which is formed in rat liver, is the ultimate carcinogenic metabolite of acetylaminofluorene in liver [78,79]. In the case of PAH the only data available concerning sulfate ester formation come from the presence of BP-3-yl sulfate in human, hamster, and rat lung cells cultured with BP [80], suggesting the presence of sulfotransferase activity in these biological systems. To our knowledge the study of sulfotransferase activity in the skin has been limited to steroids, which have been found to form sulfates in human skin [81,82].

The biological formation of reactive benzylic esters is not limited to sulfates but also includes phosphates. In this context it has been found that BP-6-CH_2OH and BA-7-CH_2OH-12-CH_3 bind efficiently to DNA in vitro in the presence of ATP alone [83,84], suggesting that binding occurs through the intermediacy of a phosphate ester. It should be mentioned here that pyrrolic esters of pyrrolizidine alkaloids [85,86] and benzylic esters of safrole [87-89] have been implicated as the ultimate carcinogenic metabolites for these compounds.

C. Radical Cations

One-electron oxidation of PAH to radical cations has been postulated as an important pathway of activation leading to carcinogenesis [7,90,91]. A general scheme illustrating one-electron oxidation of a hydrocarbon and subsequent binding to a nucleophile is outlined in Figure 2. Removal of one electron from the π system generates a radical cation in which the charge distribution varies with the structure of the hydrocarbon. The position(s) of highest charge density are the most susceptible to nucleophilic attack. In Path 1 (Fig. 2) the nucleophile attacks the carbon atom on the hydrocarbon

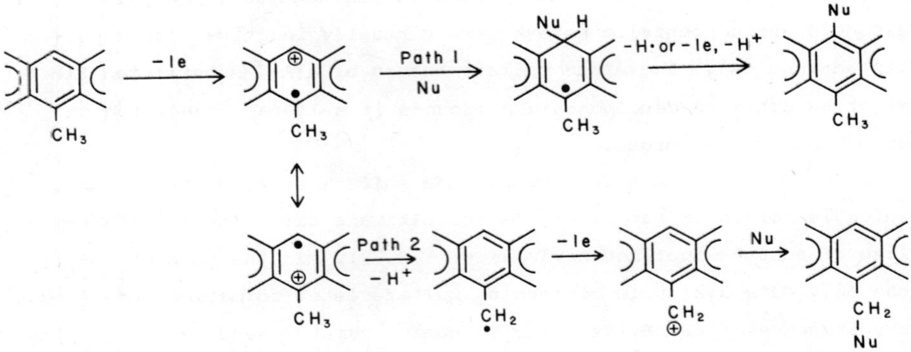

Figure 2 Mechanism of trapping of a PAH radical cation by a nucleophile (Nu).

nucleus. The resulting radical may then lose a hydrogen atom or be oxidized to an arenonium ion with loss of a proton to complete the substitution reaction. When charge density at the carbon atom adjacent to the methyl group is appreciable, loss of a methyl proton (Fig. 2, Path 2) yields a benzylic radical which is rapidly oxidized to a benzylic carbonium ion with subsequent trapping by a nucleophile.

The hypothesis that radical cations are potentially important intermediates is based on several lines of evidence which include the strong electron-donating properties of PAH and the presence in vivo of several potential oxidants in the form of metal ion-containing enzymes. The PAH-activating monooxygenases contain hexacoordinated iron (Fe^{3+}) in the form of cytochrome P_{450}, which has been recently demonstrated to be a one-electron acceptor [6]. It has been demonstrated that Fe^{3+} will oxidize BP, DMBA and MC to their respective radical cations, which are then trapped by various nucleophiles [91-93]. Furthermore, the ability of iodine to induce one-electron oxidation of PAH has been clearly established [90,94-97]. In addition, covalent binding of PAH to DNA [98,99] occurs in the presence of iodine, which is known to generate PAH radical cations by one-electron oxidation [92,96].

Table 1 Binding of BP and DMBA to DNA Without Activated Oxygen

	Binding of [^{14}C]hydrocarbon (pmole bound/μmole of DNA)		AHH (pmole product/mg of protein)
	BP	DMBA	
Complete	11.0 ± 0.6	10.2 ± 2.3	400-665
-NADPH	2.6 ± 1.1	3.4 ± 1.2	0(<0.5)
+CO	1.8 ± 0.4	1.8 ± 0.2	0-4.9
+CO -NADPH	2.1 ± 0.8	—	—
Background[a]	0.6 ± 0.2	0.3 ± 0.3	

[a]Background binding was measured as 0-min and 20-min incubations with substrate added just before stopping the reaction and with incubation at 0°.

Binding of PAH radical cations to DNA is expected to require close proximity among the activating enzymes, hydrocarbon, and cellular target. The presence of monooxygenase enzymes in the nuclear envelope [100,101] satisfies this prerequisite. Another indirect line of evidence stems from the fact that limited binding of PAH to DNA is observed under conditions which prevent oxygenation. In these experiments binding of BP and DMBA to DNA in isolated rat liver nuclei was studied and compared to nuclear AHH under several reaction conditions [102]. As shown in Table 1, binding occurred when NADPH was deleted or when CO was bubbled through the reaction mixture, although no aryl hydrocarbon hydroxylase (AHH) could be detected in an assay which was more sensitive than the binding assay. Since CO binds to cytochrome P_{450} and inhibits monooxygenase activities including AHH [103,104], the partial inhibition of binding by CO offers strong evidence that cytochrome P_{450} is responsible for such activation. The mechanism of activation appears not to include the usual oxygenation of PAH. A possible alternative mechanism is one-electron oxidation catalyzed by cytochrome P_{450}. In this mechanism the $Fe^{3+}$of the cytochrome is reduced to Fe^{2+} by an electron donated by the PAH. Since in the absence of NADPH or in the presence of CO the reduced cytochrome P_{450} cannot be recycled to the oxidized form, binding reaction stops at the observed reduced level.

In the case of BP, nucleophilic trapping of its radical cation in chemical experiments occurs almost exclusively at C-6 [93-96,105, 106], which is the site of highest charge density. In a comparative study of BP vs. some derivatives substituted at C-6, it was found that the carcinogenic activity of these BP derivatives on mouse skin was less than that of BP or nonexistent [75]. These data are consistent with activation via the BP diol epoxide or one-electron oxidation, although other possible modes of activation cannot be excluded from this experiment. The carcinogenicity of mono-, di-, and trimethylBPs [107] might also be interpreted in terms of these two mechanisms of activation. In this series, the 2-, 3-, 4-, 11-, and 12-methylBPs, as well as the 1,2-, 1,3-, 1,4-, 2,3-, 4,5-, and 3,12-dimethylBPs showed tumorigenic activity similar to that of BP, whereas the 1,6- and 3,6-dimethylBPs and 1,3,6-trimethylBP were relatively less potent.

One study of fluorescence data on BP bound to DNA in the skin of ddN mice 24 hr after treatment concludes that the entire BP-conjugated ring structure remained intact [108]. However, the best evidence for formation and binding of PAH radical cations comes from studies using specifically labeled [^{3}H, ^{14}C]PAH, although it has not been definitively proven that loss of tritium from a position indicates binding at that position. Using this technique in studies of the binding of BP to DNA of cultured mouse kidney cells, Blackburn et al. [106] found evidence for binding of BP at C-6 or C-1, the positions predicted for activation by one-electron oxidation. Similarly, DMBA appeared to bind to DNA in Vero (monkey kidney) cells at the 7-methyl position [109], a position postulated to be reactive with the DMBA radical cation [97]. The picture was somewhat confused by the finding that similar experiments did not show binding of BP at C-6 in microsome-catalyzed binding to denatured DNA [110] or poly(G) [111] or to DNA in cultured mouse embryo cells [110].

A comprehensive study of binding with [^{3}H, ^{14}C]BPs was conducted and included rat liver microsome-catalyzed binding to native and denatured calf thymus and rat liver DNA, binding to DNA in purified

Table 2 Loss of Tritium During Binding of BP to DNA and RNA[a]

System	Loss of tritium from indicated position in [^3H,^{14}C]BP (%)						
	6	1,3,6		1,3	6,7		7
In vitro:							
Microsome + poly(G)	0	21	(9)	16			
Microsome + denatured DNA	0	19	(2)	3			
Microsome + native DNA	42	44	(42)	42	20	(8)	0
Nuclei	69	45	(52)	37	30	(22)	10
In vivo:							
Skin DNA	77	67	(55)	37	29	(29)	17
Skin RNA	79	67	(55)	35	26	(29)	17

[a]Numbers in parentheses are losses calculated from the [6- and 1,3-^3H]BPs or the [6- and 7-^3H]BPs as described in Ref. 112.

rat liver nuclei, and binding to DNA and RNA in mouse skin epidermis in vivo [112]. The results are summarized in Table 2, and several conclusions can be reached. The nature of the microsome-catalyzed binding depended upon the state of the DNA. With denatured DNA and poly(G), no binding occurred at C-6 and little, if any, at the 1 and 3 positions. With native DNA, almost half of the microsome-catalyzed binding occurred at the 1, 3, and 6 positions. In isolated rat liver nuclei, binding at C-6 predominated and the total loss of tritium at C-1, C-3, and C-6 was approximately 100%, indicating that such a loss might correspond to binding through a substitution reaction. Binding of BP by formation of the radical cation with reaction at the 1, 3 or 6 position appears to account for all of the binding to DNA in isolated nuclei. Similar results were found with binding to DNA and RNA in CD-1 mouse skin. Total binding at C-1, C-3, and C-6 was equivalent to substitution of one position, indicating that all of the binding occurred among these three positions. Binding at C-6, the most favored position for reaction of the radical cation, was predominant. These results offer strong, though not conclusive, evidence that in a target organ in vivo BP binds predominately at C-6 and activation of the hydrocarbon might occur by one-electron oxidation.

These data are in contrast with those obtained from C57BL mouse skin (see above), in which BP bound primarily through the diol epoxide [24]. One should observe, however, that in the two experiments the potentially important factors of mouse strain and of contact time of the hydrocarbon were different. More explicit experiments are needed to determine whether or not these results are reconcilable.

Nucleophilic substitution of 3-methylcholanthrene (MC) via its radical cation generated by iodine [97] and by Mn^{3+} (our unpublished results) occurs specifically at the 1-carbon atom. Comparative carcinogenicity studies of MC, MC-1-OH, MC-2-OH, MC-1-one, and MC-2-one by repeated application showed that MC and MC-2-OH were the strongest carcinogens, MC-2-one was somewhat weaker, MC-1-OH was the weakest carcinogen, and MC-1-one was noncarcinogenic [73]. These results are consistent with activation of MC by one-electron oxidation and subsequent binding of MC through the 1-carbon atom.

One-electron oxidation of N-OH-2-AAF, a proximate carcinogen of 2-acetylaminofluorene, generates a nitroxyl free radical which leads to the formation of the active carcinogens nitrosofluorene and N-acetoxy-2-acetylaminofluorene. The sequence of these activating steps has been demonstrated by Bartsch et al. [113,114] and Floyd et al. [115-117] for model system peroxidases including horseradish (HRP), lactoperoxidase, hematin, and methemoglobin. Recently the activation of N-OH-2-AAF via the one-electron oxidation mechanism has been observed with peroxidase from a target organ—rat mammary gland [118]. Since mammary cells are also a target organ for PAH carcinogenesis, we have begun studies of activation of PAH by peroxidase with two objectives in mind: (a) activation by purified HRP may constitute a model system for studying one-electron oxidation of PAH, and (b) mammary cell peroxidase might activate the PAH (DMBA, MC, and BP) carcinogenic in that organ. Cytochrome P_{450} might catalyze the activation of PAH, since cytochrome P_{450} has been shown to have peroxidase activity [119-121]. HRP catalyzes the binding of BP and MC to calf thymus DNA [122]. Optimal reac-

Table 3 Horseradish Peroxidase/H_2O_2 Catalyzed Binding of BP to DNA[a]

Position of tritium in [^3H, ^{14}C]BP	[^{14}C]BP bound (pmole/µmole DNA)	[^3H/^{14}C](% original ratio)
6	47.9 ± 7.2[b] (6)	5.7 ± 1.3
1,3,6	79.5 ± 9.1 (7)	81.4 ± 1.7
7	53.0 ± 11.9 (12)	88.4 ± 1.7
Control incubations	0.2 ± 0.1 (6)	—

[a]Reaction mixtures containing 0.2 mg HRP, Type II, 4 mM DNA, 30 µM [^3H, ^{14}C]BP, 5 x 10^{-4}M H_2O_2 and Sörenson buffer, pH 7.0, to 1 ml were incubated at 37° for 30 min in the dark. The reaction was stopped and DNA purified as previously described. Control incubations did not contain H_2O_2.

[b]Standard error of measurement. The number in parentheses is the number of determinations.

tion conditions are shown in Table 3, which contains the results of experiments studying loss of tritium from [^3H, ^{14}C]BPs during HRP-catalyzed binding to DNA. Loss of tritium was almost complete (94%) from [6-^3H, ^{14}C]BP, suggesting that activation of BP may occur by one-electron oxidation.

The ability of a PAH radical cation to form covalent bonds with nucleophilic groups of cellular macromolecules should depend on several factors which include: (a) ease of formation of these reactive intermediates; (b) geometry of the hydrocarbons; (c) close spatial relationship between the activating enzyme, PAH, and cellular target; (d) proper orientation of the PAH radical cations in their initial physical interaction with the cellular target; and (e) high charge localization of the PAH radical cations at the most positive carbon atom(s).

The ease of formation of radical cations is dependent on the ionization potential of the hydrocarbons, since these intermediates are obtained by removal of an electron from PAH. Ionization potentials (IPs) of unsubstituted and alkyl-substituted alternant PAH and their carcinogenic activities are presented in Table 4 [123,124]. A low IP for PAH and localization of the positive charge in their radical cations are the essential prerequisites for the possible

Table 4 Ionization Potentials of Aromatic Hydrocarbons (Energies in eV)

Name	Structure	Charge-transfer complex data with chloranil[a]	Electronic absorption data[b]	Polarographic oxidation[c]	Carcinogenicity[d]
Benzene				9.21	-
Phenanthrene	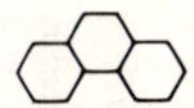	8.19	8.08	8.03	-
Triphenylene	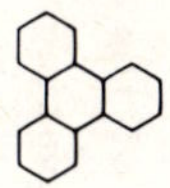	8.15	8.17	8.10	-
Naphthalene	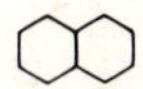	8.13	8.16	8.09	-
Benzo[c]phenanthrene	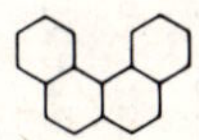	7.93	7.86		+
4-Methylbenzo[c]phenanthrene	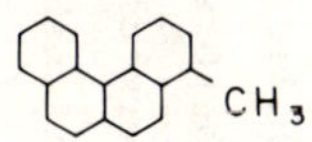	7.91			++
1-Methylbenzo[c]phenanthrene	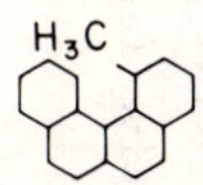	7.91			+

3-Methylbenzo[c]phenanthrene	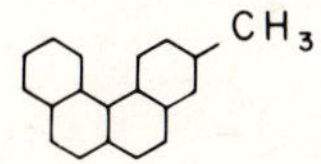	7.91			++
2-Methylbenzo[c]phenanthrene	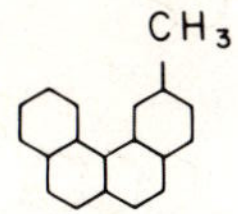	7.88			+
6-Methylbenzo[c]phenanthrene	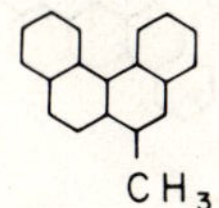	7.86			++
5-Methylbenzo[c]phenanthrene	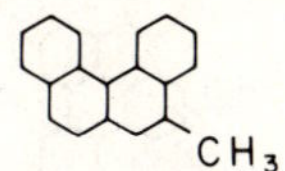	7.84			+++
Chrysene	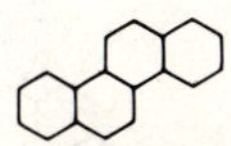	CA. 7.8	7.83	7.81	±
5-Methylchrysene	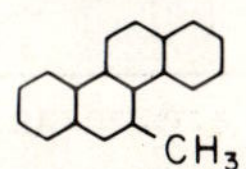	CA. 7.7			+++

Table 4 (Continued)

Name	Structure	Charge-transfer complex data with chloranil[a]	Electronic absorption data[b]	Polarographic oxidation[c]	Carcinogenicity[d]
Picene	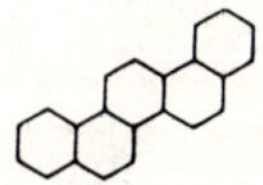		7.75	7.78	-
Dibenzo[e,l]pyrene	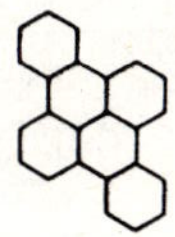		7.75		-
Dibenz[a,c]anthracene	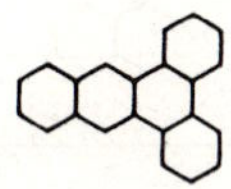	CA. 7.7	7.60	7.66	+
Dibenz[a,j]anthracene	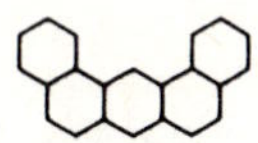		7.58	7.68	+
Dibenz[a,h]anthracene	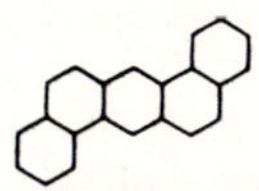		7.58	7.57	+++

Benzo[e]pyrene		7.62	7.73	7.69	-
6-Methylchrysene	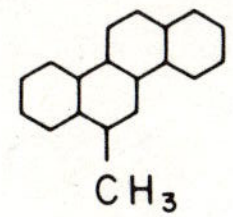	7.59			±
Coronene	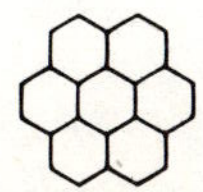	7.57	7.65	7.63	-
Benz[a]anthracene	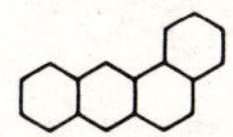	7.54	7.53	7.56	±
1-Methylbenz[a]anthracene	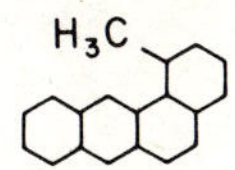	7.50			-
6-Methylbenz[a]anthracene	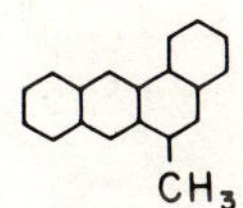	7.50		7.51	++
Pyrene	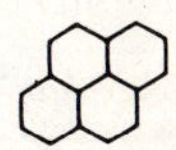	7.50	7.72	7.53	-

Table 4 (Continued)

Name	Structure	Charge-transfer complex data with chloranil[a]	Electronic absorption data[b]	Polarographic oxidation[c]	Carcinogenicity[d]
11-Methylbenz[a]anthracene	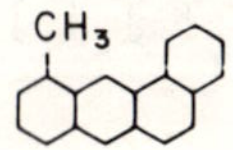	7.48		7.50	-
2-Methylbenz[a]anthracene	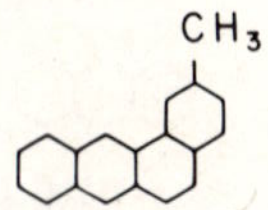	7.46		7.50	-
8-Methylbenz[a]anthracene	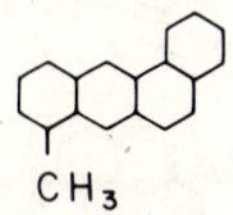	7.46		7.49	++
5-Methylbenz[a]anthracene	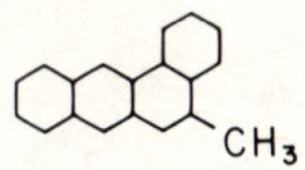	7.46		7.51	-
6,8-Dimethylbenz[a]anthracene	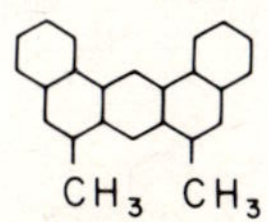	7.45			+++

2-Methylpyrene	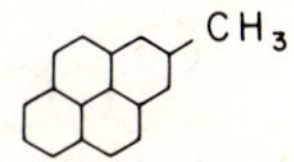	7.45			-
Anthracene	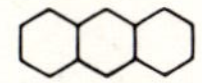	7.43	7.43	7.43	-
12-Ethylbenz[a]anthracene	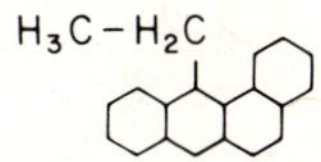	7.43			
4-Methylpyrene	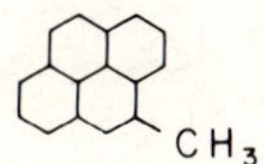	7.41			-
7-Ethylbenz[a]anthracene	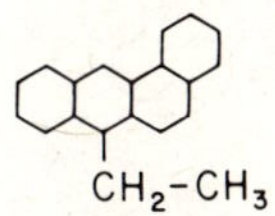	7.39			+
12-Methylbenz[a]anthracene	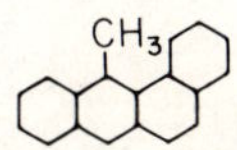	7.38		7.40	++
7-Methylbenz[a]anthracene	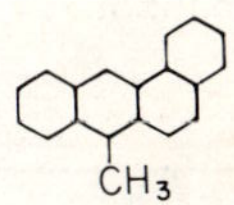	7.37		7.41	+++

Table 4 (Continued)

Name	Structure	Charge-transfer complex data with chloranil[a]	Electronic absorption data[b]	Polarographic oxidation[c]	Carcinogenicity[d]
1-Methylpyrene	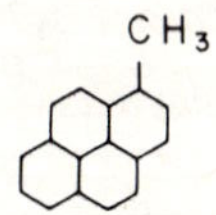	7.36			-
Dibenzo[a,e]pyrene	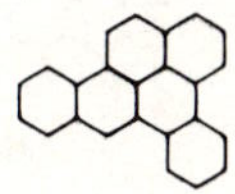	7.35	7.41	7.31	+++
6,12-Dimethylbenz[a]anthracene	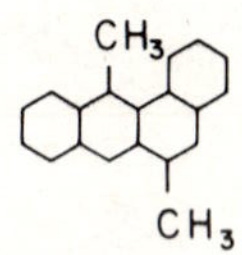	7.33			++++
Benzo[ghi]perylene		7.32	7.35	7.31	-
Tribenzo[a,e,i]pyrene	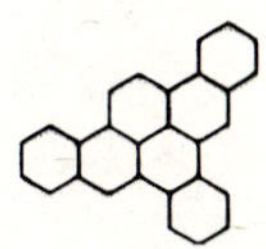	7.31	7.38		++

6,8,12-Trimethylbenz[a]anthracene	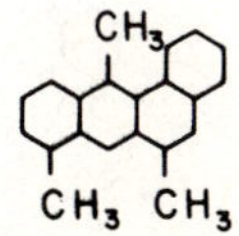	7.30			++++
Dibenzo[a,i]pyrene	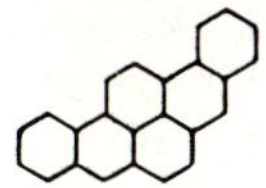	7.25	7.30		++++
Dibenzo[a,l]pyrene	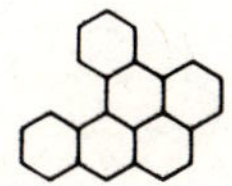		7.27		+++
Benzo[a]pyrene	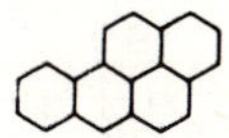	7.23	7.37	7.21	++++
7,12-Dimethylbenz[a]anthracene	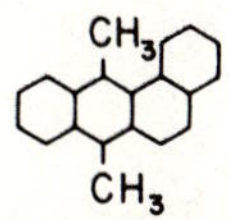	7.22		7.23	+++++
2-Methylbenzo[a]pyrene	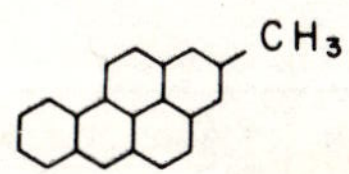	7.19			++++

Table 4 (Continued)

Name	Structure	Charge-transfer complex data with chloranil[a]	Electronic absorption data[b]	Polarographic oxidation[c]	Carcinogenicity[d]
4-Methylbenzo[a]pyrene	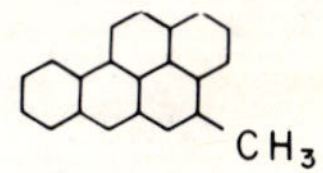	7.16			++++
1,3-Dimethylcholanthrene	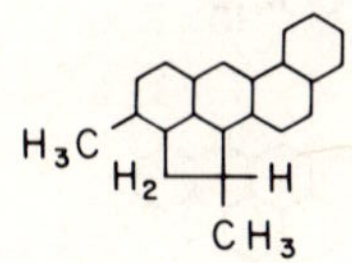	7.15			++
11-Methylbenzo[a]pyrene	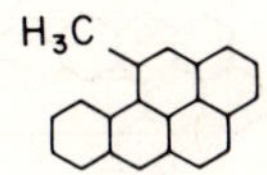	7.14			++++
9,10-Dimethylanthracene	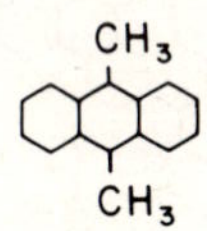	7.14		7.10	+
12-Methylbenzo[a]pyrene	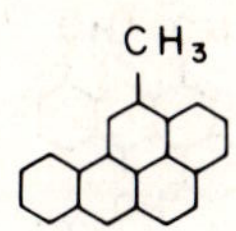	7.14			++++

2,3-Dimethylcholanthrene	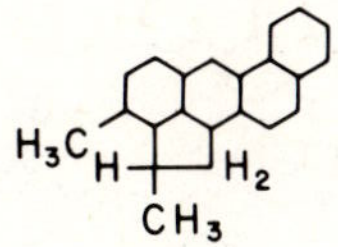	7.13		
3-Methylcholanthrene	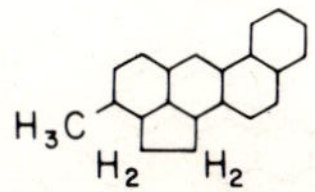	7.12	7.10	++++
4,5-Dimethylbenzo[a]pyrene	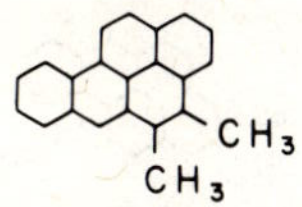	7.10		++++
6-Ethylbenzo[a]pyrene	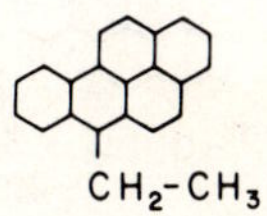	7.09		-
3-Methylbenzo[a]pyrene	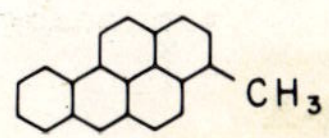	7.09		++++

Table 4 (Continued)

Name	Structure	Charge-transfer complex data with chloranil[a]	Electronic absorption data[b]	Polarographic oxidation[c]	Carcinogenicity[d]
1,2-Dimethylbenzo[a]pyrene	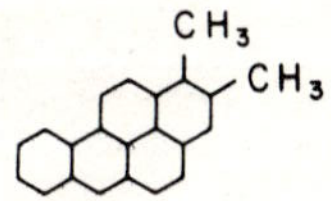	7.08			++++
6-Methylbenzo[a]pyrene	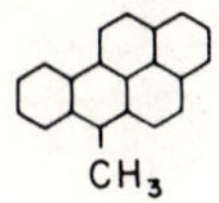	7.08			+++
3,12-Dimethylbenzo[a]pyrene	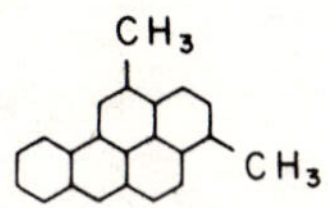	7.06			++++
Perylene	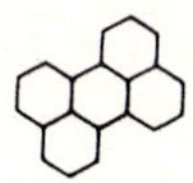	7.06	7.11	7.07	-
1,3-Dimethylbenzo[a]pyrene	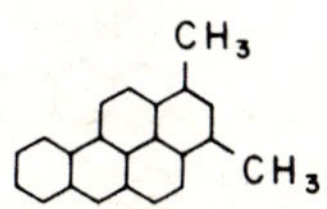	7.06			++++

1,4-Dimethylbenzo[a]pyrene	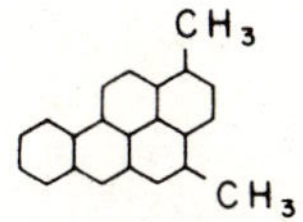	7.05			++++
1,6-Dimethylbenzo[a]pyrene	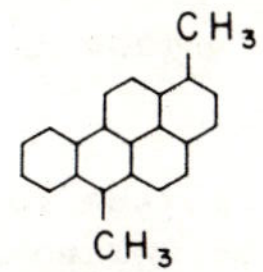	7.00			+++
3,6-Dimethylbenzo[a]pyrene	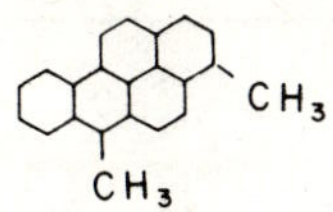	6.98			+++
Dibenzo[a,h]pyrene	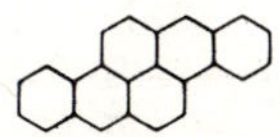	6.97	7.04		++++
Anthanthrene	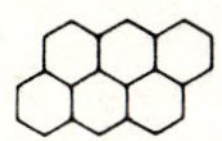	6.96	7.11		±
Naphthacene	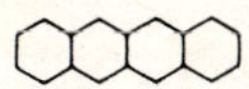	6.96	6.95	6.96	-

Table 4 (Continued)

Name	Structure	Charge-transfer complex data with chloranil[a]	Electronic absorption data[b]	Polarographic oxidation[c]	Carcinogenicity[d]
6-Methylanthanthrene	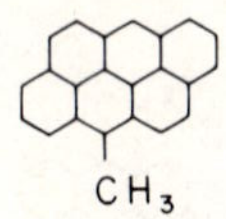	6.85			++

[a]The IP were calculated from the equation IP = 1.228 E_{CT} + 5.038 (eV), where the E_{CT} energy corresponds to the maximum absorption of the charge transfer band of the complex with chloranil in chloroform observed for each compound. The equation was obtained from a least squares plot of the IP of several PAH from polarographic oxidation (c) against the experimentally determined E_{CT} ($r > 0.99$).

[b]From Ref. 123.

[c]From Ref. 124.

[d]Key: extremely active +++++; very active, ++++; active, +++; moderately active, ++; weakly active, +; very weakly active, ±; inactive, -.

activation through the one-electron oxidation mechanism. Since there is evidence that the carcinogenicity of BA-7-CH_3 (IP: 7.37 eV) is related to the formation of a diol epoxide and possibly to the conversion to BA-7-CH_2OH with subsequent formation of a reactive alkylating ester, it is suggested that radical cations might be important ultimate carcinogenic forms for unsubstituted and methyl-substituted alternate PAH with IPs below ~7.37 eV.

Some PAH, such as benzo[ghi]perylene, perylene, anthanthrene, and naphthacene have low IPs and are noncarcinogenic. The common feature of the radical cations of these four PAH is a minimal or nonexistent difference between the most positively charged carbon atom and the second most positively charged on the same peripheral edge of the molecule (Table 5) [125]. This difference, which measures the localization of positive charge at a given carbon atom, is appreciable for some of the most potent carcinogens and might be critical in the reactivity toward cellular nucleophiles.

Table 5 Difference Between Charge at the Most Positive Carbon Atom and Second Most Positive Carbon Atom of Radical Cations on the Same Peripheral Edge[a]

Radical cation	Charge		Carcinogenicity[d]
Benzo[a]pyrene	0.112[b]	0.0564[c]	++++
Dibenzo[a,l]pyrene	0.112		+++
Dibenzo[a,e]pyrene	0.099		+++
Tribenzo[a,e,i]pyrene	0.093		++
Dibenzo[a,i]pyrene	0.090	0.0585	++++
Dibenzo[a,h]pyrene	0.084	0.0508	++++
Anthanthrene	0.034	0.0031	±
Perylene	0.025	0.0000	-
Benzo[g,h,i]perylene	0.024		-
Naphthacene	0.000	0.0000	-

[a]"Peripheral edge" has been defined by Pullman et al. [125] as the sequence of peripheral carbon atoms in which no atom is separated from the others by three or more consecutive secondary carbon atoms. This corresponds intuitively to not "turning the corner" of the molecule.

[b]Data in this column from Ref. 125.

[c]Data from A. Liakus, G. A. Gallup, and E. Cavalieri, unpublished results.

[d]See key to symbols in footnote d of Table 4.

III. DISCUSSION

The most investigated hydrocarbon thus far is BP. Several lines of evidence, as pointed out earlier, implicate the BP diol epoxide an important metabolically activated form for BP carcinogenicity. In the induction of lung adenomas and malignant lymphomas in newborn mice, the increase in carcinogenicity in the series BP, BP 7,8-dihydrodiol, and BP diol epoxide strongly suggests that the diol epoxide is the principal ultimate reactive species [32,33]. On mouse skin the similar carcinogenic activity of BP and BP 7,8-dihydrodiol in dose-response studies by repeated application [25-27] and initiation-promotion [28-31] leaves a serious question that this is the sole proximate carcinogen of BP. The optically active (-)BP 7,8-dihydrodiol has proven slightly more carcinogenic than BP as a tumor initiator [30], but this finding does not provide an adequate solution to the problem. The fact that BP 7,8-dihydrodiol is not several times more active than BP on skin has been attributed to the higher polarity of the former compound, which might impede its reaching the target sites critical for carcinogenesis and/or could more easily facilitate its excretion. The more potent activity of the proximate carcinogens BA 3,4-dihydrodiol [44,45] and chrysene 1,2-dihydrodiol [64] when compared to the parent compounds BA and chrysene, respectively, refutes this speculation.

Double-label experiments in which binding of BP to DNA in mouse skin was accompanied by loss of tritium primarily at C-6 [112] suggest that activation of BP may occur through formation of its radical cation.

Recent carcinogenicity experiments, however, cast serious doubt on the ability of the diol epoxide and/or the radical cation to explain completely the carcinogenicity of BP. In fact, it has been reported that when a methyl group was substituted at the 7 position of BP, the induction of sarcomas in rats decreased slightly compared to the parent compound [126]. With a methyl group at the 10 position, the carcinogenicity was found to be almost nil in induction of sarcomas [126] and moderately reduced in an initiation-promotion experiment on mouse skin [127]. In contrast 7,10-dimethylBP failed to elicit a carcinogenic response in both experiments [126,127]. Since it is rather unlikely that enzymatic epoxidation can efficiently occur at a double bond involving a substituted carbon atom [2], the carcinogenicity of the 7- and 10-methylBPs does not seem likely to result from formation of the bay region diol epoxide. The weak carcinogenicity of 10-methylBP is not compatible with activation by one-electron oxidation either. All the results presented suggest that additional mechanisms of metabolic activation are required to obtain a complete picture of BP carcinogenicity.

For chrysene and 5-methylchrysene there exists strong evidence that the 1,2-diol-3,4-oxide is an important ultimate reactive form [64,66]. Evidence points to the 3,4-diol-1,2-oxide as the ultimate form of the borderline carcinogen BA [44-46]. Nonetheless, the strong or moderate sarcomagenic activity of 4-fluoroBA in rats or mice, respectively, and its moderate carcinogenic activity in repeated application experiments on mouse skin [128] suggest multiple mechanisms of activation for BA.

Similar to activation of BA, activation through the bay region diol epoxide has been suggested to be important for BA-7-CH_3 carcinogenesis [47]. The lack of carcinogenic activity of 4-methylBA-7-CH_3 [129] and 5-fluoroBA-7-CH_3 is consistent with the above-proposed mechanism, since the former compound is substituted at the angular benzo ring and the latter contains a fluoro group at the peri position adjacent to the angular ring, which has been suggested to impair the metabolic activation [66]. Similarly, the carcinogenic activity of 6-fluoroBA-7-CH_3, 9-fluoroBA-7-CH_3, and 10-fluoroBA-7-CH_3 [126, 128] is compatible with activation by formation of the diol epoxide. However, the carcinogenic activity of 2-fluoroBA-7-CH_3 [126] and 3-fluoroBA-7-CH_3 [128] cannot be explained in the context of this mechanism of activation. Hydroxylation of the methyl group and subsequent esterification cannot be excluded from the possible mechanisms of activation of BA-7-CH_3. This hypothesis is based on the stronger carcinogenic activity of BA-7-CH_2OH compared to BA-7-CH_3, and BA-7-CH_2OCOCH_3 compared to BA-7-CH_2OH ([76] and our unpublished results). Furthermore, BA-7-$CH_2OSO_3^-$ was found to be a very strongly mutagenic compound in bacteria without metabolic activation (our unpublished results).

The activity of the very potent DMBA has also been attributed to the formation of an ultimate diol epoxide, as in BA and BA-7-CH_3 [53-57]. Methylation of DMBA at the 2 or 3 position completely suppressed its carcinogenic activity [129,130]. The same negative effect is also reported for 1-fluoroDMBA and 2-fluoroDMBA [126]. The lack of carcinogenicity of 2-fluoroDMBA does not parallel the moderate carcinogenicity of 2-fluoroBA-7-CH_3 (see above). However, methyl and fluoro substitution at the 4 position of DMBA [126] produced no detectable loss and a moderate loss of carcinogenicity, respectively, when compared to the parent compound. These results are in contrast to those obtained with 4-methylBA-7-CH_3 (see above), which is devoid of carcinogenic activity.

Other mechanisms of activation, such as one-electron oxidation [7,109], might participate in the carcinogenic activity of DMBA, but thus far not enough evidence has been gathered to reach any conclusions.

Some preliminary evidence for MC also indicates that the diol epoxide is an ultimate carcinogenic form [59]. Results of comparative studies of the carcinogenicity of MC, MC-1-OH, MC-2-OH, MC-1-one, and MC-2-one by repeated application on mouse skin are consistent with activation of MC by one-electron oxidation [73], although other activating mechanisms cannot be excluded.

Additional data indicating multiple mechanisms of activation of PAH stem from carcinogenicity studies of the anthanthrene series. In this case, anthanthrene was found to be inactive by subcutaneous injection in rats whereas the mesoanthracenic 6-methyl- and 6,12-dimethylanthanthrene exhibited sarcomagenic activity [131]. In a recent dose-response study at two dose levels by repeated application in CD-1 mice, only 6-methylanthanthrene was carcinogenic whereas anthanthrene and 6,12-dimethylanthanthrene were inactive (our unpublished results). Similarly, 9,10-dimethylanthracene was shown to be a weak carcinogen [132], while the parent compound anthracene is not. Since hydroxylation at a benzylic carbon atom represents a major metabolic route of alkyl-substituted PAH, it is logical to speculate that further conjugation to reactive benzylic esters might be an important pathway for the carcinogenicity of these compounds. For the aforementioned hydrocarbons, which have a low IP, an alternative pathway of activation might be one-electron oxidation and subsequent binding of the hydrocarbon to a cellular nucleophile through the activated methyl group (Fig. 2). The involvement of the methyl group in the metabolic pathway leading to carcinogenesis is also suggested by the weak or nonexistent carcinogenicity of the ethyl derivatives. This is the case with 7,12-diethylbenz[a]anthracene vs. DMBA [133], with BA-7-CH_2CH_3 vs. BA-7-CH_3 [76,77], with MC-1-CH_3 vs. MC (our unpublished results), and with BP-6-CH_2CH_3 vs. BP-6-CH_3 (our unpublished results).

IV. CONCLUSIONS

The complexity of PAH carcinogenesis can be best understood in terms of multiple mechanisms of activation. The data presently available indicate the important role of diol epoxides as ultimate carcinogenic metabolites. The relevancy of other intermediates has received thus far less attention, and their precise roles cannot be assessed on the basis of the presently limited experimental evidence. It is apparent that for most alternate PAH, however, multiple mechanisms of activation exist and the predominance of one mechanism over another may change depending on the nature of the activating enzymes present in the various target organs.

ACKNOWLEDGMENTS

We wish to thank Mr. K. Saugier for valuable technical assistance, Dr. M. Newman for providing samples of some methylbenz[a]anthracenes, Drs. D. Nagel and L. Wallcave for samples of methylchrysenes and methylbenzo[c]phenanthrenes, and Dr. G. Daub for samples of methylbenzo[a]pyrenes. This work was supported by Public Health Service contract N01 CP33278 from the National Cancer Institute, NIH.

ABBREVIATIONS

AHH, aryl hydrocarbon hydroxylase
BA, benz[a]anthracene
BA-7-CH_2OCOCH_3, 7-acetoxymethylbenz[a]anthracene
BA 3,4-dihydrodiol, trans-3,4-dihydroxy-3,4-dihydrobenz[a]anthracene
BA-7-CH_3 3,4-dihydrodiol, trans-3,4-dihydroxy-3,4-dihydro-7-methylbenz[a]anthracene
BA 3,4-diol-1,2-oxide, trans-3,4-dihydroxy-1,2-epoxy-1,2,3,4-tetrahydrobenz[a]anthracene
BA-7-CH_3 3,4-diol-1,2-oxide, trans-3,4-dihydroxy-1,2-epoxy-1,2,3,4-tetrahydro-7-methylbenz[a]anthracene
BA-7-CH_2CH_3, 7-ethylbenz[a]anthracene
BA-7-CH_2OH, 7-hydroxymethylbenz[a]anthracene
BA-7-CH_2OH-12-CH_3, 7-hydroxymethyl-12-methylbenz[a]anthracene
BA-7-CH_3, 7-methylbenz[a]anthracene
BP, benzo[a]pyrene
BP 7,8-dihydrodiol, trans-7,8-dihydroxy-7,8-dihydrobenzo[a]pyrene
BP diol epoxide, trans-7,8-dihydroxy-9,10-epoxy-7,8,9,10-tetrahydrobenzo[a]-pyrene
BP-6-CH_2OH, 6-hydroxymethylbenzo[a]pyrene
BP-6-$CH_2OSO_3^-$, 6-hydroxymethylbenzo[a]pyrene sulfate ester
chrysene 1,2-dihydrodiol, trans-1,2-dihydroxy-1,2-dihydrochrysene
chrysene 1,2-diol-3,4-oxide, trans-1,2-dihydroxy-3,4-epoxy-1,2,3,4-tetrahydrochrysene
chrysene-5-CH_2OH, 5-hydroxymethylchrysene
DMBA, 7,12-dimethylbenz[a]anthracene
DMBA diol epoxide, trans-3,4-dihydroxy-1,2-epoxy-1,2,3,4-tetrahydro-7,12-dimethylbenz[a]anthracene
HRP, horseradish peroxidase
IP, ionization potential
MC-9,10-dihydrodiol, trans-9,10-dihydroxy-9,10-dihydro-3-methylcholanthrene
MC-1-OH, 1-hydroxy-3-methylcholanthrene
MC-2-OH, 2-hydroxy-3-methylcholanthrene
MC, 3-methylcholanthrene
MC-1-one, 3-methylcholanthrene-1-one
MC-2-one, 3-methylcholanthrene-2-one
N-OH-AAF, N-hydroxy-2-acetylaminofluorene
PAH, polycyclic aromatic hydrocarbon(s)

REFERENCES

1. P. Sims, P. L. Grover, A. Swaisland, K. Pal, and A. Hewer, *Nature 252*:326-328 (1974).
2. D. M. Jerina and J. W. Daly, in *Drug Metabolism* (D. V. Parke and R. L. Smith, Eds.). Taylor & Francis, London, 1976, pp. 15-33.
3. D. M. Jerina, R. Lehr, M. Schaefer-Ridder, H. Yagi, J. M. Karle, D. R. Thakker, A. W. Wood, A. Y. H. Lu, D. Ryan, S. West, W. Levin, and A. H. Conney, in *Origins of Human Cancer*, Book B (H. H. Hiatt, J. D. Watson, and J. A. Winsten, Eds.). Cold Spring Harbor Laboratory, N.Y., 1977, pp. 639-658.
4. D. M. Jerina, R. E. Lehr, H. Yagi, O. Hernandez, P. Dansette, P. G. Wislocki, A. W. Wood, R. L. Chang, W. Levin, and A. H. Conney, in *In Vitro Metabolic Activation and Mutagenesis Testing* (F. J. DeSerres, J. R. Fouts, J. R. Bend, and R. M. Philpot, Eds.). North Holland Publ., Amsterdam, 1976, pp. 159-177.
5. J. W. DePierre and L. Ernster, *Biochim. Biophys. Acta 473*:149-186 (1978).
6. J. A. Peterson, R. E. White, Y. Yasukochi, M. L. Coomes, D. H. O'Keeffe, R. E. Ebel, B. S. S. Masters, D. P. Ballou, and M. J. Coon, *J. Biol. Chem. 252*:4431-4434 (1977).
7. E. Cavalieri, R. Roth, and E. G. Rogan, in Carcinogenesis, Vol. I. *Polynuclear Aromatic Hydrocarbons. Chemistry, Metabolism and Carcinogenesis* (R. I. Freudenthal and P. W. Jones, Eds.). Raven Press, N.Y., 1976, pp. 181-190.
8. W. E. Fahl, C. R. Jefcoate, and C. G. Kasper, *J. Biol. Chem. 253*:3106-3113 (1978).
9. A. Borgen, H. Darvey, N. Castagnoli, T. T. Crocker, R. E. Rasmussen, and I. Y. Wang, *J. Med. Chem. 16*:502-506 (1973).
10. T. Meehan, K. Straub, and M. Calvin, *Nature 269*:725-727 (1977).
11. S. K. Yang, D. W. McCourt, P. P. Roller, and H. V. Gelboin, *Proc. Nat. Acad. Sci. U.S. 73*:2594-2598 (1976).
12. D. R. Thakker, H. Yagi, A. Y. H. Lu, W. Levin, A. H. Conney, and D. M. Jerina, *Proc. Nat. Acad. Sci. U.S. 73*:3381-3385 (1976).
13. D. R. Thakker, H. Yagi, H. Akagi, M. Koreeda, A. H. Y. Lu, W. Levin, A. W. Wood, A. H. Conney, and D. M. Jerina, *Chem.-Biol. Interactions 16*:281-300 (1977).
14. H. W. S. King, M. R. Osborne, F. A. Beland, R. C. Harvey, and P. Brookes, *Proc. Nat. Acad. Sci. U.S. 73*:2679-2681 (1976).
15. M. R. Osborne, M. H. Thompson, E. M. Tarmy, F. A. Beland, R. G. Harvey, and P. Brookes, *Chem.-Biol. Interactions 13*:343-348 (1976).
16. A. M. Jeffrey, I. B. Weinstein, K. W. Jennette, K. Grzeskowiak, K. Nakanishi, R. G. Harvey, H. Autrup, and C. Harris, *Nature 269*:348-350 (1977).

17. W. M. Baird and L. Diamond, *Biochem. Biophys. Res. Commun. 77*: 162-167 (1977).
18. V. Ivanovic, N. E. Geacintov, H. Yamasaki, and I. B. Weinstein, *Biochemistry 17*:1597-1603 (1978).
19. I. B. Weinstein, A. M. Jeffrey, K. W. Jennette, S. H. Blobstein, R. G. Harvey, C. Harris, H. Autrup, H. Kasai, and K. Nakanishi, *Science 193*:592-595 (1976).
20. A. M. Jeffrey, K. W. Jennette, S. H. Blobstein, I. B. Weinstein, F. A. Beland, R. G. Harvey, H. Kasai, I. Miura, and K. Nakanishi, *J. Am. Chem. Soc. 98*:5714-5715 (1976).
21. K. Nakanishi, H. Kasai, H. Cho, R. G. Harvey, A. M. Jeffrey, K. W. Jennette, and I. B. Weinstein, *J. Am. Chem. Soc. 99*: 258-260 (1977).
22. H. Yagi, H. Akagi, D. R. Thakker, H. D. Mah, M. Koreeda, and D. M. Jerina, *J. Am. Chem. Soc. 99*:2358-2359 (1977).
23. P. Daudel, M. Duquesne, P. Vigny, P. L. Grover, and P. Sims, *FEBS Lett. 57*:250-253 (1975).
24. M. Koreeda, P. D. Moore, P. G. Wislocki, W. Levin, A. H. Conney, H. Yagi, and D. M. Jerina, *Science 199*:778-781 (1978).
25. W. Levin, A. W. Wood, H. Yagi, P. M. Dansette, D. M. Jerina, and A. H. Conney, *Proc. Nat. Acad. Sci. U.S. 73*:243-247 (1976).
26. W. Levin, A. W. Wood, H. Yagi, D. M. Jerina, and A. H. Conney, *Proc. Nat. Acad. Sci. U.S. 73*:3867-3871 (1976).
27. W. Levin, A. W. Wood, P. G. Wislocki, J. Kapitulnik, H. Yagi, D. M. Jerina, and A. H. Conney, *Cancer Res. 37*:3356-3361 (1977).
28. I. Chouroulinkov, A. Gentil, P. L. Grover, and P. Sims, *Br. J. Cancer 34*:523-532 (1976).
29. T. J. Slaga, A. Viaje, D. L. Berry, and W. Bracken, *Cancer Lett. 2*:115-121 (1976).
30. W. Levin, A. W. Wood, R. L. Chang, T. J. Slaga, H. Yagi, D. M. Jerina, and A. H. Conney, *Cancer Res. 37*:2721-2725 (1977).
31. T. J. Slaga, W. M. Bracken, A. Viaje, W. Levin, M. Yagi, D. M. Jerina, and A. H. Conney, *Cancer Res. 37*:4130-4133 (1977).
32. J. Kapitulnik, W. Levin, A. H. Conney, H. Yagi, and D. M. Jerina, *Nature 266*:378-380 (1977).
33. J. Kapitulnik, P. G. Wislocki, W. Levin, H. Yagi, D. M. Jerina, and A. H. Conney, *Cancer Res. 38*:354-358 (1978).
34. C. Malaveille, H. Bartsch, P. L. Grover, and P. Sims, *Biochem. Biophys. Res. Commun. 66*:693-700 (1975).
35. F. F. Newbold and P. Brookes, *Nature 261*:52-54 (1976).
36. E. Huberman, L. Sachs, S. K. Yang, and H. V. Gelboin, *Proc. Nat. Acad. Sci. U.S. 73*:607-611 (1976).
37. A. W. Wood, P. G. Wislocki, R. L. Chang, W. Levin, A. Y. H. Lu, H. Yagi, O. Hernandez, D. M. Jerina, and A. H. Conney, *Cancer Res. 36*:3358-3366 (1976).
38. P. G. Wislocki, A. W. Wood, R. L. Chang, W. Levin, H. Yagi, O. Hernandez, D. M. Jerina, and A. H. Conney, *Biochem. Biophys. Res. Commun. 68*:1006-1012 (1977).

39. A. W. Wood, R. L. Chang, W. Levin, H. Yagi, D. R. Thakker, D. M. Jerina, and A. H. Conney, *Biochem. Biophys. Res. Commun. 77*:1389-1396 (1977).
40. J. C. Arcos and M. F. Argus, *Chemical Induction of Cancer*, Vol. II, Pt. A. New York, Academic Press, 1974, p. 18.
41. T. J. Slaga, G. T. Bowden, J. D. Scribner, and R. K. Boutwell, *J. Nat. Cancer Inst. 53*:1337-1340 (1974).
42. A. W. Wood, W. Levin, A. Y. H. Lu, D. Ryan, S. B. West, R. E. Lehr, M. Schaefer-Ridder, D. M. Jerina, and A. H. Conney, *Biochem. Biophys. Res. Commun. 72*:680-686 (1976).
43. A. W. Wood, R. L. Chang, W. Levin, R. E. Lehr, M. Schaefer-Ridder, J. M. Karle, D. M. Jerina, and A. H. Conney, *Proc. Nat. Acad. Sci. U.S. 74*:2746-2750 (1977).
44. T. J. Slaga, E. Huberman, J. K. Selkirk, R. G. Harvey, and W. M. Bracken, *Cancer Res. 38*:1699-1704 (1978).
45. A. W. Wood, W. Levin, R. L. Chang, R. E. Lehr, M. Schaefer-Ridder, J. M. Karle, D. M. Jerina, and A. H. Conney, *Proc. Nat. Acad. Sci. U.S. 74*:3176-3179 (1977).
46. W. Levin, D. R. Thakker, A. W. Wood, R. L. Chang, R. E. Lehr, D. M. Jerina, and A. H. Conney, *Cancer Res. 38*:1705-1710 (1978).
47. B. Tierney, A. Hewer, C. Walsh, P. L. Grover, and P. Sims, *Chem.-Biol. Interactions 18*:179-193 (1977).
48. P. Vigny, M. Duquesne, H. Coulomb, C. Lacombe, B. Tierney, P. L. Grover, and P. Sims, *FEBS Lett. 75*:9-12 (1977).
49. I. Chouroulinkov, A. Gentil, B. Tierney, P. Grover, and P. Sims, *Cancer Lett. 3*:247-253 (1977).
50. C. Malaveille, B. Tierney, P. L. Grover, P. Sims, and H. Bartsch, *Biochem. Biophys. Res. Commun. 75*:427-433 (1977).
51. H. Marquardt, S. Baker, B. Tierney, P. L. Grover, and P. Sims, *Int. J. Cancer 19*:828-833 (1977).
52. P. Vigny, M. Duquesne, H. Coulomb, B. Tierney, P. L. Grover, and P. Sims, *FEBS Lett. 82*:278-282 (1977).
53. R. C. Moschel, W. M. Baird, and A. Dipple, *Biochem. Biophys. Res. Commun. 76*:1092-1098 (1977).
54. V. Ivanovic, N. E. Geacintov, A. M. Jeffrey, P. P. Fu, R. G. Harvey, and I. B. Weinstein, *Cancer Lett. 4*:131-140 (1978).
55. C. A. H. Bigger, J. E. Tomaszewski, and A. Dipple, *Biochem. Biophys. Res. Commun. 80*:229-235 (1978).
56. A. Dipple and J. A. Nebzydoski, *Chem.-Biol. Interactions 20*: 17-26 (1978).
57. W. M. Baird and L. Diamond, *Chem.-Biol. Interactions 20*:181-190 (1978).
58. J. DiGiovanni, T. J. Slaga, D. L. Berry, and M. R. Juchau, *Drug Metab. Dispos. 5*:295-301 (1977).
59. H. W. S. King, M. R. Osborne, and P. Brookes, *Int. J. Cancer 20*: 564-571 (1977).
60. D. R. Thakker, W. Levin, A. W. Wood, A. H. Conney, T. A. Stoming, and D. M. Jerina, *J. Am. Chem. Soc. 100*:645-647 (1978).
61. D. Hoffman, W. E. Bondinell, and E. L. Wynder, *Science 183*:215-216 (1974).

62. S. S. Hecht, W. E. Bondinell, and D. Hoffman, *J. Nat. Cancer Inst. 53*.1121-1133 (1974).
63. A. W. Wood, W. Levin, D. Ryan, P. E. Thomas, H. Yagi, H. D. Mah, D. R. Thakker, D. M. Jerina, and A. H. Conney, *Biochem. Biophys. Res. Commun. 78*.847-854 (1977).
64. W. Levin, A. W. Wood, R. L. Chang, H. Yagi, H. D. Mah, D. M. Jerina, and A. H. Conney, *Cancer Res. 38*:1831-1834 (1978).
65. S. S. Hecht, M. Loy, R. Mazzarese, and D. Hoffman, *J. Med. Chem. 21*:38-44 (1978).
66. S. S. Hecht, N. Hirota, M. Loy, and D. Hoffman, *Cancer Res. 38*: 1694-1698 (1978).
67. E. Boyland, P. Sims, and C. Huggins, *Nature 207*:816-817 (1965).
68. E. Boyland and P. Sims, *Int. J. Cancer 2*:500-504 (1967).
69. D. N. Wheatley and M. S. Inglis, *Br. J. Cancer 22*:122-127 (1968).
70. J. W. Flesher and K. L. Sydnor, *Cancer Res. 31*:1951-1954 (1971).
71. M. J. Shear and J. Leiter, *J. Nat. Cancer Inst. 2*:99-113 (1941).
72. P. Sims, *Int. J. Cancer 2*:505-508 (1967).
73. E. Cavalieri, R. Roth, C. Grandjean, J. Althoff, K. Patil, S. Marsh, and D. McLaughlin, *Chem.-Biol. Interactions 22*: 69-81 (1978).
74. F. Dewhurst, D. A. Kitchen, and G. Calcutt, *Br. J. Cancer 26*: 506-508 (1972).
75. E. Cavalieri, R. Roth, C. Grandjean, J. Althoff, K. Patil, S. Liakus, and S. Marsh, *Chem.-Biol. Interactions 22*:53-67 (1978).
76. R. Roth, C. Grandjean, and E. Cavalieri, *Proc. Am. Assoc. Cancer Res. 19*:203 (1978).
77. J. Pataki and C. Huggins, *Cancer Res. 29*:506-509 (1969).
78. J. R. DeBaun, J. Y. Rowley, E. C. Miller, and J. A. Miller, *Proc. Soc. Exp. Biol. Med. 129*:268-273 (1968).
79. J. R. DeBaun, E. C. Miller, and J. A. Miller, *Cancer Res. 30*: 577-595 (1970).
80. G. M. Cohen, S. M. Haws, B. P. Moore, and J. W. Bridges, *Biochem. Pharmacol. 25*:2561-2570 (1976).
81. I. Faredin, I. Toth, A. G. Fazekus, K. Kokai, and M. Julesz, *J. Endocrinol. 41*:295-296 (1968).
82. D. L. Berliner, J. R. Pasqualini, and A. J. Gallegos, *J. Invest. Dermatol. 50*:220-224 (1968).
83. L. K. Tay, K. L. Sydnor, and J. W. Flesher, *Proc. Am. Assoc. Cancer Res. 19*:181 (1978).
84. J. W. Flesher and L. K. Tay, *Proc. Fed. Am. Soc. Exp. Biol. 31*: 749 (1978).
85. C. C. J. Culvenor, D. T. Downing, J. A. Edgar, and M. V. Jago, *Ann. N.Y. Acad. Sci. 163*:837-847 (1969).
86. A. R. Mattocks and I. N. H. White, *Chem.-Biol. Interactions 3*: 383-396 (1971).
87. P. Borchert, J. A. Miller, E. C. Miller, and T. K. Shires, *Cancer Res. 33*:590-600 (1973).
88. P. G. Wislocki, P. Borchert, J. A. Miller, and E. C. Miller, *Cancer Res. 36*:1686-1695 (1976).
89. P. G. Wislocki, E. C. Miller, J. A. Miller, E. C. McCoy, and H. S. Rosenkranz, *Cancer Res. 37*:1883-1891 (1977).

90. M. Wilk, W. Bez, and J. Rochlitz, *Tetrahedron 22*:2599-2608 (1966).
91. J. Fried in *Chemical Carcinogenesis* (P. O. P. Ts'o and J. Di-Paolo, Eds.) Pt. A., Dekker, New York, 1974, pp. 197-215.
92. E. M. Menger, R. B. Spokane, and P. D. Sullivan, *Biochem. Biophys. Res. Commun. 71*:610-616 (1976).
93. E. Cavalieri and R. Auerbach, *J. Nat. Cancer Inst. 53*:393-397 (1974).
94. J. Rochlitz, *Tetrahedron 23*:3043-3048 (1967).
95. M. Wilk and W. Girke, *J. Nat. Cancer Inst. 49*:1585-1597 (1972).
96. W. Caspary, B. Cohen, S. Lesko, and P. O. P. Ts'o, *Biochemistry 12*:2649-2656 (1973).
97. E. Cavalieri and R. Roth, *J. Org. Chem. 41*:2679-2684 (1976).
98. S. A. Lesko, P. O. P. Ts'o, and R. S. Umans, *Biochemistry 8*: 2291-2298 (1969).
99. H. D. Hoffman, S. A. Lesko, Jr., and P. O. P. Ts'o, *Biochemistry 9*:2594-2604 (1970).
100. C. B. Kasper, *J. Biol. Chem. 246*:577-581 (1971).
101. E. G. Rogan and E. Cavalieri, *Biochem. Biophys. Res. Commun. 58*: 1119-1126 (1974).
102. E. G. Rogan, P. Mailander, and E. Cavalieri, *Proc. Nat. Acad. Sci. U.S. 73*:457-461 (1976).
103. D. Y. Cooper, S. Levin, S. Narasimhulu, O. Rosenthal, and R. W. Estabrook, *Science 147*:400-402 (1965).
104. D. W. Nebert and H. V. Gelboin, *J. Biol. Chem. 243*:6242-6249 (1968).
105. L. Jeftic and R. N. Adams, *J. Am. Chem. Soc. 92*:1332-1337 (1970).
106. G. M. Blackburn, P. E. Taussig, and J. P. Will, *J. Chem. Soc. Chem. Commun.* pp. 907-908 (1974).
107. A. Lacassagne, F. Zajdela, N. P. Buu-Hoi, and O. Chalvet, *Int. J. Cancer 3*:238-243 (1968).
108. M. Kodama and C. Nagata, *Biochemistry 14*:4645-4650 (1975).
109. G. M. Blackburn, A. J. Flavell, P. E. Taussig, and J. P. Will, *J. Chem. Soc. Chem. Commun.* pp. 358-359 (1975).
110. H. W. S. King, M. H. Thompson, M. R. Osborne, R. G. Harvey, and P. Brookes, *Chem.-Biol. Interactions 12*:425-428 (1976).
111. T. Meehan, D. Warshawsky, and M. Calvin, *Proc. Nat. Acad. Sci. U.S. 73*:1117-1120 (1976).
112. E. Rogan, R. Roth, E. Cavalieri, P. Katomski, and J. Benderson, *Chem.-Biol. Interactions 22*:35-51 (1978).
113. H. Bartsch and E. Hecker, Biochim. *Biophys. Acta 237*:567-578 (1971).
114. H. Bartsch, J. A. Miller, and E. C. Miller, *Biochim. Biophys. Acta 273*:40-51 (1972).
115. R. A. Floyd, L. M. Soong, and P. L. Culver, *Cancer Res. 36*: 1510-1519 (1976).
116. R. A. Floyd, L. M. Soong, R. N. Walker, and M. Stuart, *Cancer Res. 36*:2761-2767 (1976).
117. R. A. Floyd and L. M. Soong, *Biochim. Biophys. Acta 498*:244-249 (1977).
118. D. L. Reigh, M. Stuart, and R. A. Floyd, *Experientia 34*:107-108 (1978).

119. E. G. Hrcay and P. J. O'Brien, *Arch. Biochem. Biophys. 147*:14-27 (1971).
120. E. G. Hrcay and P. J. O'Brien, *Arch. Biochem. Biophys. 147*:28-35 (1971).
121. E. G. Hrcay, H. G. Jonen, A. Y. H. Lu, and W. Levin, *Arch. Biochem. Biophys. 166*:145-151 (1975).
122. E. Rogan, P. Katomski, R. Roth, and E. Cavalieri, *Proc. Fed. Am. Soc. Exp. Biol. 37*:750 (1978).
123. G. Brieglieb, *Angew. Chem. (Int. Ed.) 3*:617-632 (1964).
124. E. S. Pysh and N. C. Yang, *J. Am. Chem. Soc. 85*:2124-2130 (1963).
125. B. Pullman, A. Pullman, R. Umans, and B. Maigret, *Physicochemical Mechanisms of Carcinogenesis* (E. D. Bergmann and B. Pullman, Eds.), Vol. 1. Israel Academy of Science and Humanities, Jerusalem, 1969, pp. 325-338.
126. R. G. Harvey and F. B. Dunne, *Nature 273*:566-568 (1978).
127. S. S. Hecht, N. Hirota, and D. Hoffman, *Cancer Lett. 5*:179-183 (1978).
128. J. A. Miller and E. C. Miller, *Cancer Res. 23*:229-239 (1963).
129. J. Pataki, C. Duguid, P. W. Rabideau, H. Husiman, and R. G. Harvey, *J. Med. Chem. 14*:940-945 (1971).
130. M. S. Newman and W. M. Hung, *J. Med. Chem. 20*:179-181 (1977).
131. A. Lacassagne, N. P. Buu Hoi, and F. Zajdela, *C. R. Acad. Sci. (Paris) 246*:1477-1480 (1958).
132. W. Lijinsky and U. Saffiotti, *Ann. Ital. Dermatol. Clin. Sper. 19*:34-44 (1965).
133. C. R. Huggins, J. Pataki, and R. G. Harvey, *Proc. Nat. Acad. Sci. U.S. 58*:2253-2260 (1967).

5

SPIN BIOASSAYS WITH NUCLEIC ACIDS

Albert M. Bobst

University of Cincinnati
Cincinnati, Ohio

I. INTRODUCTION

It is the purpose of this chapter to review and discuss the application the nitroxide-labeling technique to the field of nucleic acid chemistry. So far the labeling can be roughly divided into two groups, as is shown in Figure 1. The first group includes nitroxide radicals covalently bound to small ligands (Fig. 1A). The ligands chosen for these spin-labeling purposes are known to strongly interact in the unlabeled state with single- or double-stranded nucleic acids. Usually, the ligands have aromatic character and intercalate with nucleic acids. The second group includes nitroxide radicals covalently attached to nucleic acids without site specificity (first-generation spin-labeled nucleic acids), or site specificity with respect to the nucleic acid building block (second-generation spin-labeled nucleic acids), or finally site specificity with respect to nucleotide and nucleic acid sequence (third-generation spin-labeled nucleic acids) (Fig. 1B). The nitroxide radicals in the first group will be referred to as *nucleic acid spin probes* and in the second group as *spin-labeled nucleic acids*, $1(N)_n$. This distinction has already been systematically applied earlier [1]. Spin-labeled nucleic acids have been shown to be useful for monitoring nucleic acid-nucleic acid interactions as well as binding of proteins to nucleic acids. The latter interaction process has led to the development of a spin assay to measure relative affinities of proteins to

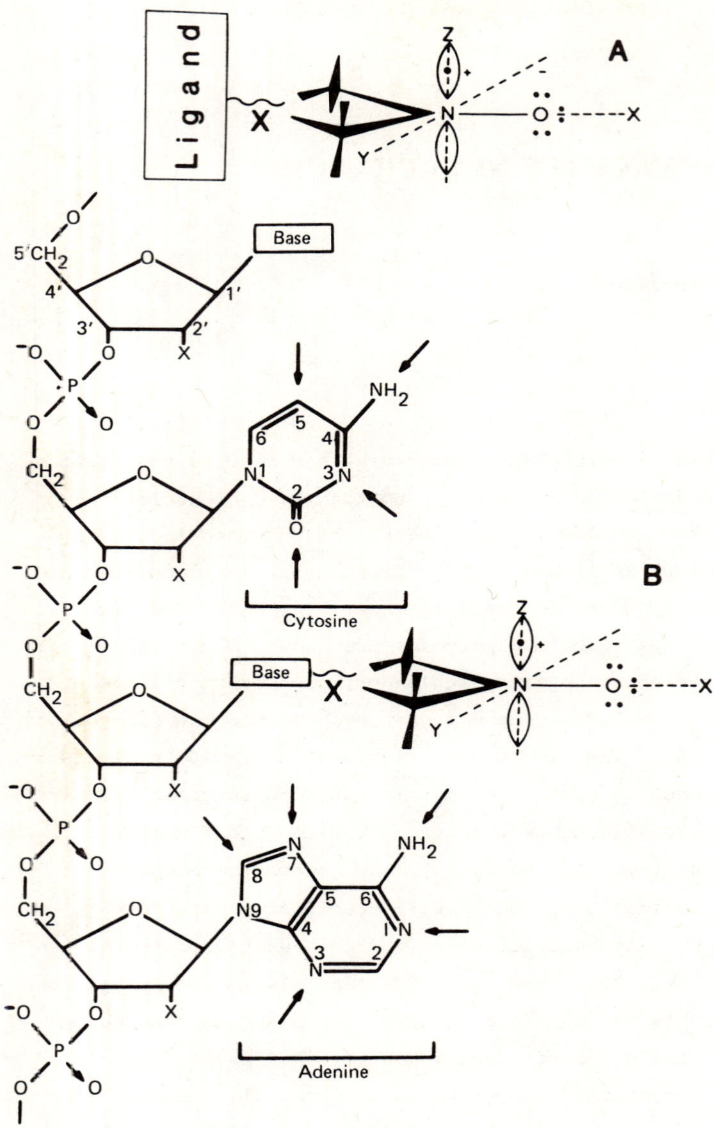

Figure 1 (A) Nitroxide moiety covalently linked via chain x to a nucleic acid binding ligand. (B) Segment of polynucleotide (RNA, x = OH; DNA, x = H, DNA) spin-labeled by covalently attaching nitroxide moiety via chain x to the nucleic acid base.

nucleic acids. Such assays can be applied to all proteins known to interact with nucleic acids, such as DNA unwinding proteins and polymerases. Among the polymerases, reverse transcriptase, believed to be present in leukemic blood cells, is a suitable candidate because of its probably strong affinity for certain polynucleotide analogs as determined from inhibition studies.

II. BINDING OF SPIN PROBES TO NUCLEIC ACIDS

One of the first applications of nitroxide radicals to nucleic acids was the surface probing of polynucleotides with the spin probe tempol [2]. Tempol was incorporated noncovalently into the nucleic acid matrix by combining an aqueous solution of nitroxide radicals with native or denatured calf thymus DNA as well as with yeast tRNA. Various lyophilized tempol-nucleic acid complexes gave essentially the same ESR spectra characteristic of strongly immobilized spin probes. Changing the relative humidity of the tempol-containing nucleic acid samples had a pronounced sample specific effect on the ESR spectra. Subsequently, tempol was used to study DNA preparations of varying water content as a function of temperature [3]. Arrhenius plots of the temperature-dependent data displayed breakpoints which showed a small dependence on the water content of the DNA preparations and occurred at about 20° and 65°. The high-temperature breakpoint is believed to reflect base pair melting, whereas the low temperature transition, so far only observed by ESR, may indicate the presence of structural rearrangement of the nucleic acid matrix. It is interesting to note that the low-temperature transition was subsequently also observed with nitroxide radicals covalently bound to nucleic acids. (For a further discussion of the low-temperature transition in DNA, see Section III.B.)

A series of spin-labeled molecules was later synthesized which, based on the structure of the unlabeled analogs, should show great affinity for nucleic acids. Since the binding of these spin-labeled molecules is of a noncovalent nature, they are referred to as spin probes following the classification mentioned earlier (see

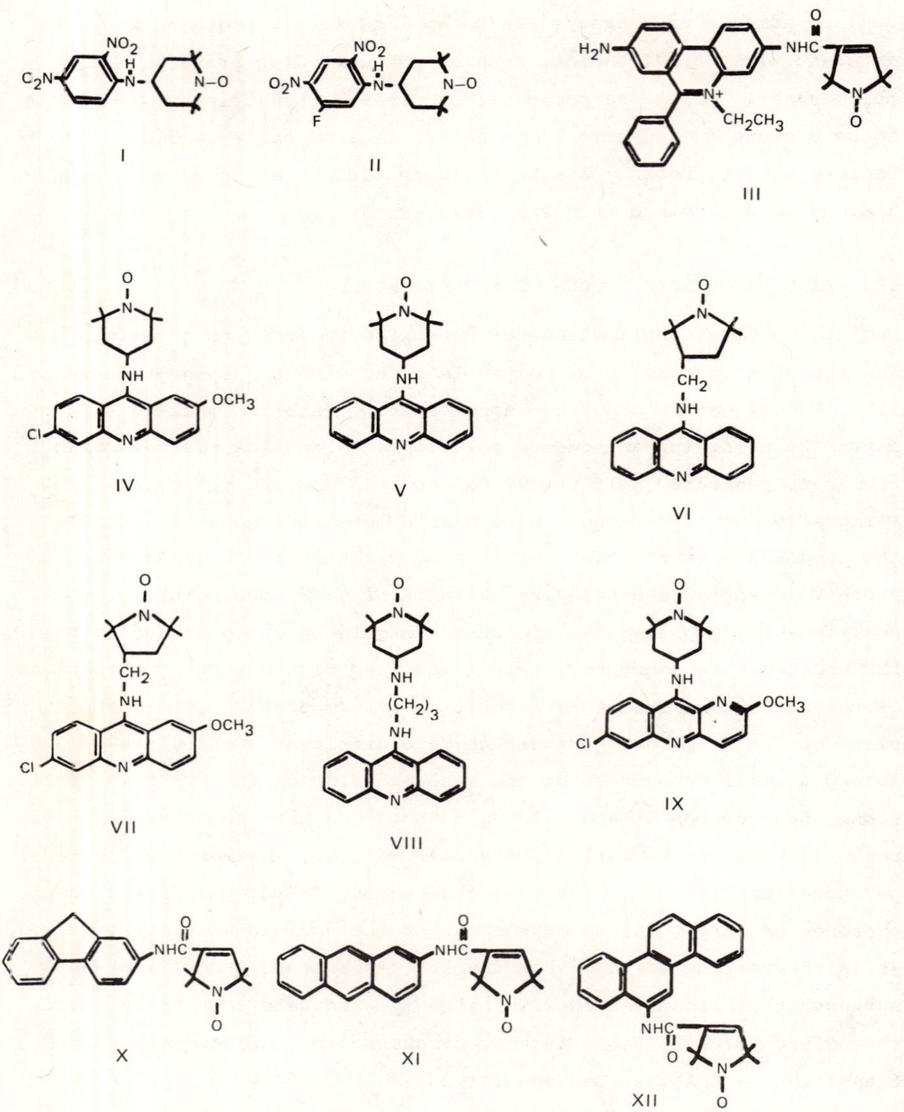

Figure 2 Nucleic acid spin probes. The probes are identified as: I, DNBSL; II, FDNBSL; III, ethidium bromide, IV-VIII, aminoacridines; IX, aminoazaacridine; X, aminofluorene; XI, aminoanthracene; XII, aminochrysene. (Modified from Ref. 1).

Fig. 1). The chemical structure of such probes used so far is given in Figure 2. It is apparent from Figure 2 that all nucleic acid spin probes synthesized so far contain either a five- or six-membered stable nitroxide radical, which is, however, not rigidly attached to the moiety known to display nucleic acid affinity. Incidently, the lack of rigidity can complicate a rigorous theoretical interpretation of the ESR data as is discussed below.

Some of the intercalation characteristics of spin-labeled ethidium bromide III were analyzed [4], and subsequently the structure of III was also given [5]. The synthesis and biological properties of the spin-labeled 9-aminoacridines IV, V, VI, VII, and VIII, as well as of the spin-labeled aminoazaacridine IX, were reported [6]. The acridine spin probes caused unwinding of calf thymus DNA, and those probes containing both 6-chloro- and 2-methy-oxy-substituents were less toxic to leukemia L1210 in static culture than were the corresponding unsubstituted analogs.

Several aromatic amine carcinogens have also been used as spin probes. Hong and Piette, in 1976, studied the intercalation properties of the spin-labeled carcinogens X, XI, and XII by ESR [5]. More recently the same investigators synthesized spin-labeled nitrobenzenes in order to analyze the nitrobenzene-DNA intercalation [7]. Of several derivatives prepared and examined, the 2,4-dinitrobenzene analogs I and II showed the strongest binding with DNA by intercalation.

An analysis of the ESR data indicates that upon binding of the spin probes I to XII to double-stranded DNA the motion of the nitroxide moiety is highly restricted. The resulting rigid limit spectra will provide valuable information about the binding mechanism, especially if one succeeds in getting a spin probe-DNA complex where the orientiation of the DNA fibers with respect to the magnetic field is well defined. However, a thorough analysis of a rigid limit spectrum is only possible if the spin probe is rigidly bound to the nucleic acid matrix and the geometrical relationship between the nitroxide moiety and the ligand is known. If these requirements are fulfilled, the rigid limit spectrum is directly dependent on the angle of orientation between the T and g tensors of the nitroxide moiety and the DNA fiber geometry. Analysis of the data is done according to Eq. (1):

$$\mathcal{H} = \beta \cdot H \cdot g \cdot S + S \cdot T \cdot I \tag{1}$$

where $\mathcal{H}$ is the spin Hamiltonian, g and T are the tensors of the nitroxide radical, S and I are the electron and nuclear spin operators, respectively, and β is the Bohr magneton. For instance, the ESR spectrum of the DNA-spin probe III complex showed an angular dependence of the orientation relative to the applied magnetic field [5]. In that study, it was assumed that the nitroxide radical is coplanar with the ethidium bromide ring. With this assumption, it was possible to conclude, based on the orientation dependence of the ESR spectrum, that spin probe III intercalates between the base pairs in the DNA duplex. The ESR data also seemed to indicate that the z axis of the nitroxide moiety is parallel to the DNA fiber axis. In the case of the spin probes I and II, the structural relationship between reporter group and ligand is believed to be better defined, since space-filling atomic models show that the nitroxide group can no longer freely rotate around the amine linkage due to steric hindrance of the 2-nitro group and the 6-hydrogen atom [7]. Actually, the correlation time for the nitroxide moiety seems to indicate that the reporter group does not move relative to the ligand, suggesting that the nitroxide radical wedges itself tightly into the helix. The anisotropic ESR data suggest that the nitroxide is positioned at a 45° angle with respect to the aromatic ring of the ligand.

Occasionally, Scatchard plots have been calculated to determine the mode of binding between spin probes and nucleic acids. The concentration of the unbound spin probes can be easily measured from the amplitude of the low-field ESR signal. In the case of the aminoacridine probe IV, the results suggested the presence of two independent binding sites on DNA [8]. Hong and Piette [7] used the Scatchard analysis to follow the binding of spin probes I and II to calf thymus DNA and observed that the fluoro-analog II binds covalently to DNA.

The stability of spin probe-DNA complexes was determined using ESR by monitoring the intensity of the unbound label as a function of temperature. Hong and Piette [5] found for the spin probe III-DNA complex that the probe dissociates in a cooperative fashion. The results also indicated that upon renaturation the probe is only partially reassociated into the complex. Furthermore, denaturation studies with spin-labeled and unlabeled ethidium bromide revealed that the value of the binding constant is considerably lower in the presence of the nitroxide radical. With the acridine probes IV and V it was observed that the melting curves were reversible provided the samples were not held at temperatures above 80° for any significant period of time [8]. Using the acridine probes it was also possible to show that the T_m values determined by ESR agree reasonably well with those measured spectrophotometrically.

Finally, although the structural relationship between reporter group and ligand moiety is better defined in the spin probes I and II than in the remaining ones (III to XII), all currently used nucleic acid spin probes carry a relatively bulky nitroxide group which certainly modifies some, if not all, binding characteristics of the parent ligand. Ideally, the geometry of the p_z orbital of the nitroxide relative to the ligand should be fixed and the size of the structure carrying the N-O moiety should be minimal. Unfortunately, these requirements necessitate the design of novel spin probes which cannot be simply obtained through some alkylation of acylation reactions of existing ligands. Nevertheless, it has been possible so far to clearly establish that nitroxides per se are not reacting with the major building blocks of nucleic acids and that the nitroxide can report binding events between nucleic acids and ligands. As was pointed out earlier [5], working with probes such as X, XI, and XII could be useful for investigating the sequence of molecular events in a complex process such as carcinogenesis.

III. SPIN-LABELED NUCLEIC ACIDS

A. Practical and Theoretical Considerations

Smith and Yamane were the first to show that some spin-label reagents can covalently bind to highly polymerized yeast RNA, to unfractionated tRNA from yeast, and to calf thymus DNA [9,10]. At that time, exhaustive dialysis was used to remove noncovalently bound nitroxide radicals from the non-site-specific spin-labeled nucleic acids (first generation spin-labeled nucleic acids). Non-site-specific labeling of nucleic acids with nitroxide radicals containing a reactive group was subsequently attempted by several other groups [11-22]. Because it is difficult to chemically introduce nitroxide radicals site specifically into nucleic acids containing only major bases, site-specific labeling studies have been attempted on some well-selected tRNAs. However, because of the danger of some background labeling with chemical modifications, careful fractionating of the tRNA is usually required [12]. More recently, some spin-label reagents and procedures for specific spin-labeling of several tRNAs by chemical means have been described [23]. But, in general, most of the spin-labeled nucleic acids obtained by chemical modification are considered as first-generation spin-labeled nucleic acids because of the uncertainty of the location of the reporter group on the nucleic acid matrix [1].

A new era in the spin-labeled nucleic acid field started when second-generation spin-labeled nucleic acids were synthesized by enzymatic means. For that purpose, spin-labeled nucleic acid building blocks were designed which were acceptable as an enzyme substrate for incorporation into a nucleic acid chain. It was found that the spin-labeled nucleoside diphosphate ppRUGT, ppRU-Glycolamido-TEMPO, can be copolymerized with the uridine-5'-diphosphate, ppU, to form the synthetic polynucleotide $(RUGT,U)_n$, the weight average of which depended on the reaction conditions [24]. Polynucleotide phosphorylase (PNPase) from *Micrococcus luteus* or *Escherichia coli* was used for the copolymerization step. It should be emphasized that since PNPase cannot transcribe a template, with this

Figure 3 A CPK atomic model showing the molecular structure of $(RUGT,U)_n$ base paired with the trinucleotide segment AAG; the nitroxide radical is in the lower left of the chain.

enzyme one can only achieve nitroxide labeling specificity with respect to the nucleic acid building block. Spin-labeled nucleic acids with this degree of specificity are referred to as second-generation spin-labeled nucleic acids. In Figure 3 an Ealing CPK atomic model of $(RUGT,U)_n$ base paired with the trinucleotide AAG is shown. The nitroxide moiety can be seen on the left of the polymeric strand, and it should be apparent that incorporation of a nitroxide radical at every fiftieth to hundreth residue will have a minimal effect on most physicochemical characteristics of the nucleic acid system. Synthesis of third-generation spin-labeled nucleic acids, i.e., specifically labeled with respect to building block and nucleic acid sequence, has not yet been achieved, but work is in progress in this laboratory to reach this goal.

For practical reasons it is also of primary importance to be certain that the spin-labeled polynucleotide contains no free spin contamination, and for reasons of sensitivity it is desirable that the reporter group tumbles in the motional narrowing region. The former goal was achieved by designing a sensitive free spin test [1,25]; and the latter goal by selecting a nitroxide radical attached to the nucleic acid matrix via an appropriate number of bonds (designed in Fig. 1 by x).

The theory has been well worked out for calculating correlation times (τ values) for nitroxide radicals. The nitroxide tumbles fairly rapidly in the motional narrowing region ($\tau = 10^{-10}$ sec), and the ESR spectra of such weakly immobilized radicals can be readily analyzed. Correlation times can be calculated using the approach developed for rapidly isotropically tumbling nitroxide radicals [26]. For that purpose Eq. (2) is solved thus:

$$T_2(m)^{-1} = \tau\left\{\frac{[3I(I+1) + 5m^2]b^2}{40} + \frac{4}{45(\Delta\gamma H)^2} - \frac{4}{15b\Delta\gamma Hm}\right\} + x \qquad (2)$$

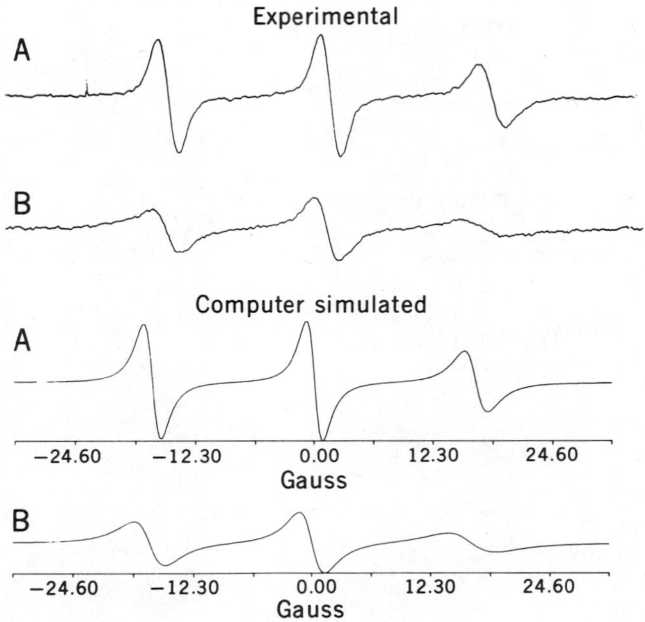

Figure 4 Comparison of experimental and computer-simulated ESR spectra of spin-labeled poly(dUfl) (Spectrum A) and spin-labeled poly(dUfl)·poly(A) (Spectrum B). (Reproduced with permission from A. M. Bobst, T. K. Sinha, and Y. C. E. Pan, *Science 188*:153-155. Copyright 1975 by the American Association for the Advancement of Science.)

where H is the applied field strength in gauss; I = 1 is the nuclear spin quantum number for ^{14}N;

$$b = (4\pi/3)(A_{zz} - A_{yy}),\ \Delta\gamma = (-|\beta|/h)[g_{zz} - 1/2(g_{xx} + g_{yy})];$$

and x stands for contributions from other broadening mechanisms that are m independent. Here, τ values are usually calculated from the equation quadratic in m only, and this is done with Eq. (3):

$$\tau = 5.47 \times 10^{-10}\ \Delta H_o[\sqrt{h_o/h^{-1}} + \sqrt{h_o/h^{+1}} - 2] \text{ sec} \qquad (3)$$

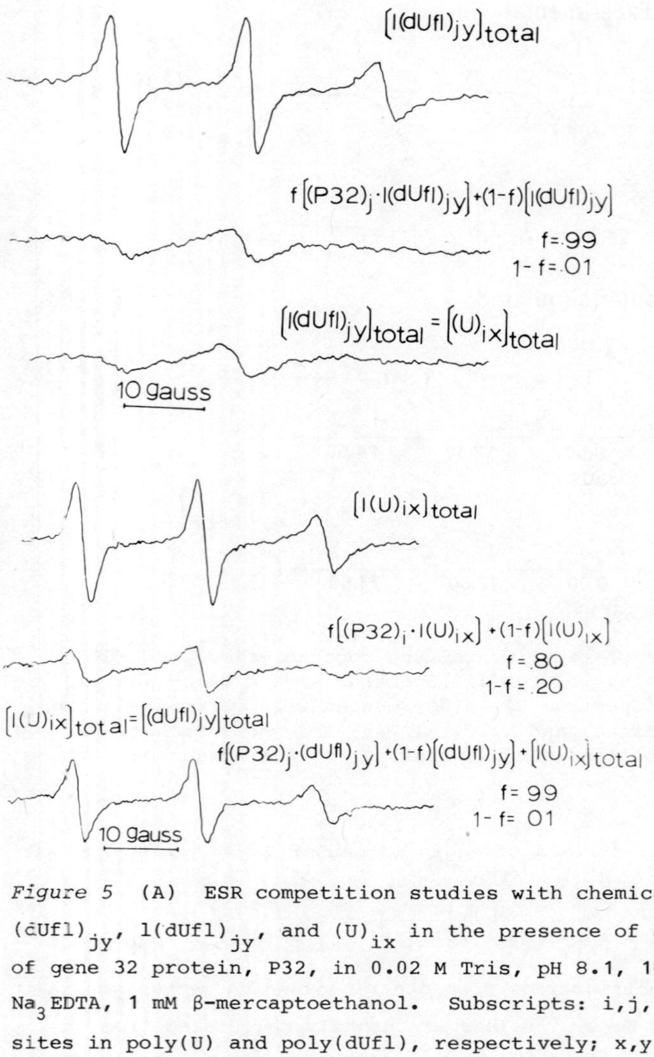

Figure 5 (A) ESR competition studies with chemically spin-labeled $(dUfl)_{jy}$, $l(dUfl)_{jy}$, and $(U)_{ix}$ in the presence of a limiting amount of gene 32 protein, P32, in 0.02 M Tris, pH 8.1, 10% glycerol, 1 mM Na_3EDTA, 1 mM β-mercaptoethanol. Subscripts: i,j, number of binding sites in poly(U) and poly(dUfl), respectively; x,y, number of residues per binding site in poly(U) and poly(dUfl), respectively. (B) ESR competition studies with chemically spin-labeled $(U)_{ix}$, $l(U)_{ix}$, and $(dUfl)_{jy}$ in the presence of a limiting amount of gene 32 protein, P32. Buffer and definitions of i, j, x, y are the same as in part A.

Equation (3) is valid for correlation times shorter than approximately 5×10^{-9} sec; h_{+1}, h_o, h_{-1} are the peak heights in arbitrary units, and ΔH_o is the peak separation of the center peak in gauss. With this equation correlation times can be conveniently calculated using a small programmable pocket calculator. This is particularly useful for determining approximate τ values at various temperatures to get Arrhenius plots.

With the aid of computer simulations a more rigorous approach for calculating tumbling times of spin-labeled nucleic acids has been adopted. For that purpose the formalism of Freed et al. [27] and Goldman et al. [28] has been used. Experimental and computer-simulated ESR spectra of spin-labeled nucleic acid systems are shown in Figure 4 [29]. The details for the simulation have been discussed elsewhere [1]. Spectrum A was simulated by choosing an isotropic reorientation value of 3.5×10^{-10} sec for τ while anisotropic reorientation had to be assumed for the base-paired complex. The $\tau_{R\parallel}$ and $\tau_{R\perp}$ chosen for simulating spectrum B were 6.3×10^{-10} and 22.1×10^{-10} sec, respectively. The change in ESR line shape upon base pairing is quite significant, and such spectral changes can be expected when poly(A) tails are titrated with spin-labeled $(U)_n$ or its analog $(dUfl)_n$. The change in line shape is even more pronounced for complex formation between spin-labeled nucleic acids and proteins (Figs. 5 and 6).

B. Binding and Competition Studies with Nucleic Acid Binding Proteins

Kamzolova et al. [18] and Sukhorukov et al. [30] attempted to clarify the question of whether complex formation between *E. coli* RNA polymerase and T2 DNA results in conformational changes of the DNA. ESR results seemed to indicate that upon addition of RNA polymerase to spin-labeled T2 DNA at 37° the nitroxide label tumbles faster; in contrast, formation of the RNA polymerase-T2 DNA complex at 2° left the mobility of the reporter group unchanged. The results are

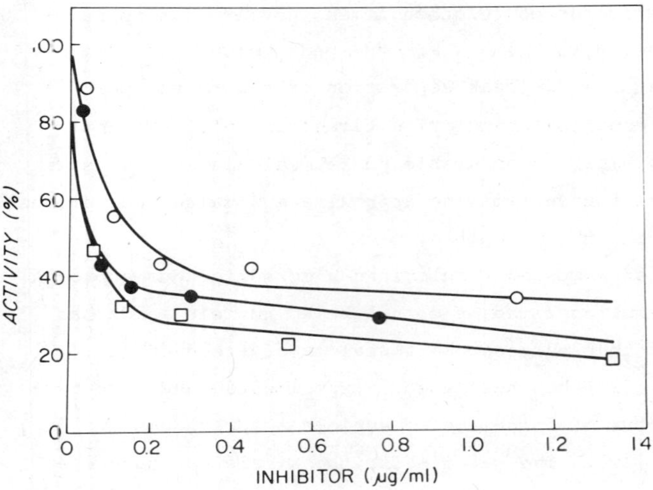

Figure 6 Inhibition of AMV reverse transcriptase activity by $(RUGT,U)_n$ (□), $1(U)_n$ (o), and (U)n (●). Concentration of enzyme 480 ng/ml; concentration of $(C)_n \cdot (dG)_{12-18}$ 2.7 μg/ml (0.4-0.8K_m). (From Ref. 37.)

believed to indicate the existence of a conformational change of the DNA in the RNA polymerase-T2 DNA complex above 20°. Although this is a provocative observation, it should be interpreted with caution since an increase in spin mobility could also be explained by assuming strong nuclease activity at 37° [1]. In a different study it was observed that RNA polymerase activity decreased when spin-labeled *E. coli* DNA was used as template [31]. Two possible explanations were given for the loss of activity: either the nitroxide mechanically hinders the movement of the polymerase along the template or the nitroxide covalently attached to the DNA prevents the formation of the form of the DNA believed to be required for RNA polymerase binding. Although it is apparent that a considerable amount of more work is needed for a better understanding of such complex interactions, the potential of spin-labeled nucleic acids to gain a better insight into a polymerase template binding process is evident.

A different approach using spin-labeled nucleic acids to analyze nucleic acid protein complexes has been applied in this laboratory. It is based on competition experiments with nucleic acid binding proteins in the presence of spin-labeled and unlabeled nucleic acids. So far, such competition experiments have been used to determine the affinity of T4 gene 32 protein (P32) for various nucleic acids [24,29]. Typical experimental ESR data observed during the course of competition studies with P32 are shown in Figure 5A,B. The top spectrum in Figure 5A illustrates free spin-labeled $(dUfl)_{jy}$. Upon addition of a limiting amount of protein to $1(dUfl)_{jy}$, the second spectrum is recorded. Addition of unlabeled $(U)_{ix}$ at a concentration equal to that of $1(dUfl)_{jy}$ gives the bottom spectrum, which does not differ in terms of line shape from the second one. Since the ESR results indicate no displacement with $(U)_{ix}$ it can be concluded that P32 has a considerably greater affinity for $(dUfl)_{jy}$ than for $(U)_{ix}$. In Figure 5B the control experiment is shown which consists of forming the complex first with spin-labeled $(U)_{ix}$, $1(U)_{ix}$, and then competing with unlabeled $(dUfl)_{jy}$. It is obvious from the line shape of the bottom spectrum that $1(U)_{ix}$ was displaced by $(dUfl)_{jy}$, thereby confirming the affinity relationship determined with the experiment shown in Figure 5A. With this strategy it was possible to show unequivocally that a significant difference exists in the binding constants of various P32-nucleic acid complexes. The following P32 affinity for single-stranded polynucleotides was determined with first-generation spin-labeled nucleic acids [29]:

$$(dT)_n > (dUfl)_n > (A)_n > (U)_n \simeq (dA)_n$$

Some of the aforementioned competition studies were repeated with the second-generation spin-labeled nucleic acid $(RUGT,U)_n$ [24], and the same affinity relationship established earlier was determined for the polynucleotides studied. In the meantime, temperature-dependent CD studies have confirmed that P32 binds more tightly to $(dT)_n$

than to $(dA)_n$ [32]. Since it was generally assumed that unwinding proteins such as P32 belong to the class of "non-specific-site binding" proteins, the conclusion, originally based on ESR experiments with spin-labeled nucleic acids, concerning the nucleic acid specificity of P32 was unexpected.

C. Inhibition of Reverse Transcriptase

Although a limited number of nitroxide radicals (on the average about 1 nitroxide/100 nucleotides) are covalently bound to the nucleic acid matrix, it seemed desirable to use a sensitive biological assay to determine whether the reporter group has a detectable effect on the interaction of nucleic acids with proteins. Reverse transcriptase was chosen as nucleic acid binding protein because earlier studies had shown that the biological assay for the DNA polymerase activity is sensitive to differences in nucleic acid structure. It was reported that $(U)_n$ exerts an inhibitory effect by competing for the template binding site of avian myeloblastosis virus (AMV) reverse transcriptase [33], and it was also observed that substitution in the 2'-position of $(U)_n$ by a fluoro- [34] or an azido-group [35] enhances the inhibition of virus production. Therefore, AMV reverse transcriptase activity and the kinetics of the enzyme inhibition were determined in the presence of a series of spin-labeled and unlabeled polynucleotides and polynucleotide analogs [36,37]. This was accomplished by measuring the effective doses for 50% inhibition of activity (ED_{50}) and monitoring the kinetics of inhibition with varying substrate concentrations. Figure 6 displays the ED_{50} for $(U)_n$, spin-labeled $(U)_n$ (first generation), and $(RUGT,U)_n$ (second generation). It is apparent that the ED_{50} is essentially the same for $(U)_n$ and $(RUGT,U)_n$ and only slightly different for the chemically modified $(U)_n$. From a series of such studies it was concluded that the ED_{50} and the kinetic patterns of inhibition are similar for the spin-labeled and unlabeled polynucleotide inhibitors studied [37]. Thus, the presence of a limited number of spin-labels covalently bound to the nucleic acid matrix has little effect, if any, on reverse transcriptase-nucleic acid complexes. It has also been noticed recently that even at a spin-label-to-nucleotide ratio of 3:100, the spin-labeled nucleic acid shows the same characteristics as the unlabeled one with respect to the reverse transcriptase assay; however, increasing the ratio of labeling to 6% or higher gave enhanced inhibition [38]. The enhancement could not be reversed by DTT, and therefore it is suggested that the increased inhibition with highly spin-labeled $(U)_n$ is not caused by simple oxidation of thiol group(s) in reverse transcriptase, although it has been reported earlier that this enzyme contains sensitive thiol group(s) [39].

D. Determination of (A)n Tracts

Spin-labeled $(U)_n$, $1(U)_n$, as well as spin-labeled $(dUfl)_n$, $1(dUfl)_n$, have been used in a novel bioassay for detecting and titrating $(A)_n$ tracts in RNAs by ESR [40,41]. $(A)_n$ tracts have been found in eukaryotic messenger RNAs and in 70S RNA of tumor viruses at the 3'-OH end of the RNA.

The assay is based on the pronounced change in the ESR line shape upon base pairing of the spin-labeled single-stranded polynucleotides with the complementary $(A)_n$ tract as shown earlier in Figure 4. The line shape change allows one to directly follow the titration, and for convenience the ratio h_{-1}/h_o is plotted as a function of nanomoles of $(A)_n$ added. Examples of such titration results are shown in Figure 7A,B. In Figure 7A, we see that $1(dUfl)_n$ was titrated with a mixture of R17 RNA and $(A)_n$. R17 RNA containing no long $(A)_n$ sequences was added as a control. Indeed, the same ESR ratio and breakpoint were obtained without the addition of R17 RNA (Fig. 7B). Furthermore, it was established for $1(dUfl)_n$, as well as $1(U)_n$, that the ratio h_{-1}/h_o remains constant after hybridization is completed. The ESR titration assay was successfully applied for the detection and quantitation of $(A)_n$

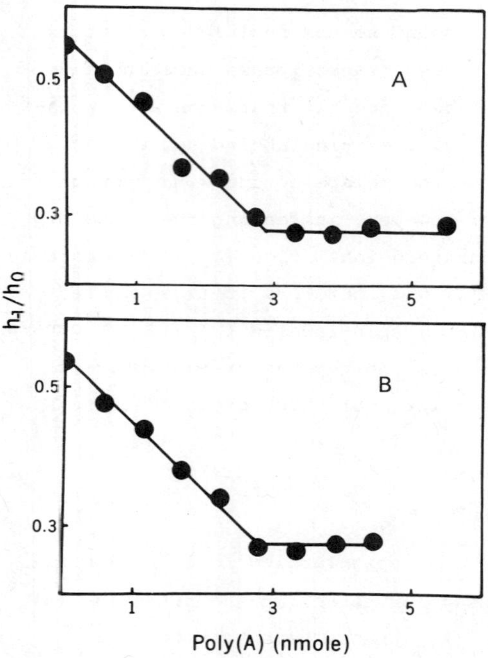

Figure 7 ESR tritration study of $(A)_n$ with spin-labeled $(dUfl)_n$ in 0.01 M Tris·HCl, 0.01 M NaCl, 0.001 M EDTA, pH 7.5, at 26°C. (A) Variation of the ratio of high field (h_{-1}) to center field (h_o) hyperfine components of 2.8 nmol of spin-labeled $(dUfl)_n$ upon titration with a mixture consisting of R17 RNA and $(A)_n$ at a molar ratio of 9 to 1, respectively. (B) Variation of the ratio of h_{-1} to h_o of 1.8 nmol of spin-labeled $(dUfl)_n$ upon titration with $(A)_n$. (Reproduced with permission from A. M. Bobst, T. K. Sinha, and Y. C. E. Pan, *Science 188*:153-155. Copyright 1975 by the American Association for the Advancement of Science.)

tracts in rabbit globin mRNA [41]. No more than 36 µg of $(A)_n$-containing RNA was required for the titration, and with second-generation spin-labeled nucleic acids, as well as the new TM ESR cavity, the amount of RNA needed will be even less. Finally, as was pointed out earlier, the assay also has potential for direct quantitation of RNA-containing tumor viruses [41].

E. Interaction with Cell Monolayers

In light of the interferon-inducing activity of some nucleic acids, interaction studies between polynucleotides and mammalian cells have been of particular interest. Usually such studies have been carried out with radio-labeled single- or double-stranded polynucleotides and native RNA. However, obtaining experimental data with radio-labeled nucleic acids usually requires extensive cell manipulation, leading to cell damage. Recently, an ESR approach was reported for following the interaction of Vero cell monolayers with the nucleic acids $(A)_n$, $(dUfl)_n$, $(U)_n$, and $(A)_n \cdot (U)_n$, all of which were spin-labeled chemically [42]. In contrast to interaction studies using radio-labeled nucleic acids, information obtained with spin-labeled nucleic acids did not require extensive cell manipulation. In fact, with the ESR approach it was possible to monitor the interaction under optimum cell growth conditions.

In Figure 8 the procedure is outlined for obtaining cell monolayers which can be directly monitored in the ESR cavity. Cell monolayers were grown on quartz plates, and the well of a quartz tissue cell was filled with fresh aliquots of culture medium containing spin-labeled nucleic acids. The temperature was controlled for the temperature-dependent studies by placing the E-4 ESR in a small, well-insulated room which served with its heating and cooling facilities essentially the purpose of a walk-in incubator. With these arrangements, an interesting temperature barrier for the interaction between nucleic acids and Vero cells was discovered. Namely, at 26° or above, spin-labeled nucleic acids interact significantly with a Vero cell surface whereas at 20° no such interaction is observed by ESR. The temperature barrier was attributed to a temperature-dependent phase transition of membrane components or cell surface products, and it was concluded that a major nucleic acid-Vero cell surface interaction occurs only at high temperature.

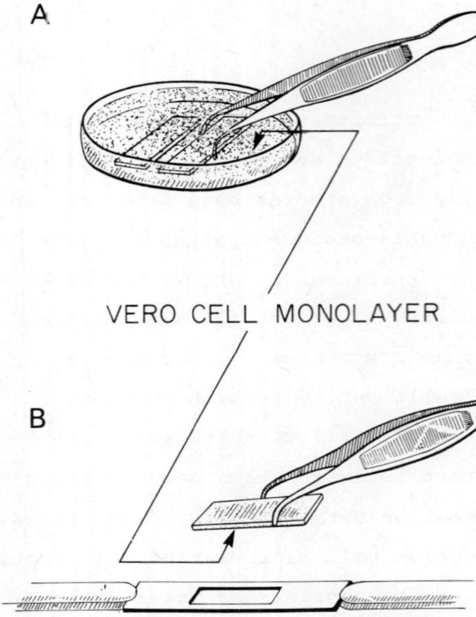

Figure 8 Procedure illustrating the approach taken to prepare Vero cell monolayers in a suitable form for ESR measurements in a multipurpose cavity. (From Ref. 1.)

IV. CONCLUSIONS

The spin-labeling approach as applied to nucleic acids has great potential in view of recent advancements in this field. In view of the complexity of nucleic acid chemistry, the application of spin-labeling to nucleic acid related problems has been slower than in other areas of biochemistry. However, valuable information has already been obtained with nucleic acid spin probes as well as with first- and second-generation spin-labeled nucleic acids. Table 1 outlines some applications for spin-labeled nucleic acids, and Table 2 summarizes the advantages of spin assays vs. isotopic assays. Clearly, the creative potential of the spin-labeling approach has not yet been fully exploited, and the further development of new nucleic acid spin probes and spin-labeled nucleic acids will continue to provide a powerful means for investigating complex processes such as carcinogenesis.

Table 1 Biological Applications with Spin-Labeled Nucleic Acids

Type of interaction	Examples	Further applications	Competing methods
Nucleic acid-nucleic acid	Detection and determination of average $(A)_n$ tail size in HbmRNA [41]	Titration of oncogenic viral RNA	Radiolabeled nucleic acids in combination with filtration
Nucleic acid-protein	Competition studies with gene protein 32 [24,29]	Inhibition studies of reverse transcriptase by nucleic acids; characterization of nucleic acid binding proteins	Radiolabeled nucleic acids in combination with filtration

Table 2 Advantages of Spin Assay vs. Isotopic Assay

1. Rapid and simple
2. Potential of automation
3. No large space requirement
4. No safety hazard
5. Long shelf life of spin-labeled nucleic acids
6. High sensitivity (μg range with nucleic acid binding proteins)
7. Straightforward interpretation of data (no problems with background)

ACKNOWLEDGMENTS

I wish to thank Dr. Patricia G. Warwick for carefully reading the manuscript and Mrs. E. V. Bobst for helping to develop the spin-labeled nucleic acid field in this laboratory. This work was supported in part by CA 15717, awarded by the National Cancer Institute, NIH.

REFERENCES

1. A. M. Bobst, in *Spin Labeling* (L. J. Berliner, Ed.), Vol. II. Academic Press, New York, 1979, pp. 291-345.
2. B. I. Sukhorukov, A. M. Vasserman, L. A. Kozlova, and A. L. Buchachenko, *Dokl. Akad. Nauk SSSR 177*:454-457 (1967).
3. B. I. Sukhorukov and L. A. Kozlova, *Biofizika 15*:539-541 (1970).
4. L. H. Piette, *Fed. Proc. 33*:Abstr. No. 832 (1974).
5. S. J. Hong and L. H. Piette, *Cancer Res. 36*:1159-1171 (1976).
6. B. K. Sinha, R. L. Cysyk, D. B. Millar, and C. F. Chignell, *J. Med. Chem. 19*:994-998 (1976).
7. S. J. Hong and L. H. Piette, *Arch. Biochem. Biophys. 185*:307-315 (1978).
8. B. K. Sinha and C. F. Chignell, *Life Sci. 17*:1829-1836 (1975).
9. I. C. P. Smith and T. Yamane, *Proc. Nat. Acad. Sci. U.S. 58*:884-887 (1967).
10. I. C. P. Smith and T. Yamane, in *Recent Developments of Magnetic Resonance in Biological Systems* (S. Fujiwara and L. H. Piette, Eds.). Hirokawa, Tokyo, 1968, pp. 45-100.
11. B. M. Hoffman, P. Schofield, and A. Rich, *Proc. Nat. Acad. Sci. U.S. 62*:1195-1202 (1969).
12. P. Schofield, B. M. Hoffman, and A. Rich, *Biochemistry 9*:2525-2533 (1970).
13. A. S. Girshovich, M. A. Grachev, D. G. Knorre, V. P. Kumarev, and V. I. Levintahl, *FEBS Lett. 14*:199-202 (1971).
14. A. M. Bobst, *Proc. 23rd Int. Congr. Pure and Appl. Chem., Macromolecular Preprint* Vol. II. Boston, 1971, pp. 1228-1235.
15. A. M. Bobst, *Biopolymers 11*:1421-1433 (1972).
16. R. I. Artyukh, G. B. Postnikova, B. I. Sukhorukov, and S. G. Kamzolova, *Biokhimiya 37*:902-906 (1972).
17. A. I. Petrov, G. B. Postnikova, and B. I. Sukhorukov, *Izv. Akad. Nauk SSSR, Ser. Khim. 6*:1453-1454 (1972).
18. S. G. Kamzolova, A. I. Kolontarov, L. I. Elfimova, and B. I. Sukhorukov, *Dokl. Akad. Nauk SSSR 208*:245-247 (1973).
19. E. M. Mil', S. K. Zavriev, G. L. Grigoryan, and K. E. Kruglyakova, *Dokl. Akad. Nauk SSSR 209*:217-220 (1973).
20. S. Zavriev, G. Grigoryan, I. Krilova, and P. Karmanov, *Biochem. Biophys. Res. Commun. 54*:123-126 (1973).

21. S. K. Zavriev, G. L. Grigoryan, P. A. Karmanov, G. A. Trveliashvili, and E. G. Rozantsev, *Molek. Biol. 8*:302-309 (1974).
22. W. J. Caspary, J. J. Greene, L. M. Stempel, and P. O. P. Ts'o, *Nucleic Acid Res. 3*:847-861 (1976).
23. M. Caron and H. Dugas, *Nucleic Acid Res. 3*:19-34 (1976).
24. A. M. Bobst and P. F. Torrence, *Polymer 19*:115-117 (1978).
25. Y. C. E. Pan, *A spin-label approach for studying nucleic acid-nucleic acid and nucleic acid-protein interactions.* Ph.D. Thesis, University of Cincinnati, 1975.
26. T. J. Stone, T. Buckman, P. L. Nordio, and H. M. McConnell, *Proc. Nat. Acad. Sci. U.S. 54*:1010-1017 (1965).
27. J. H. Freed, G. V. Bruno, and C. F. Polnaszek, *J. Phys. Chem. 75*:3385-3399 (1971).
28. Ś. A. Goldman, G. V. Bruno, C. F. Polnaszek, and J. H. Freed, *J. Chem. Phys. 56*:716-735 (1972).
29. A. M. Bobst and Y. C. E. Pan, *Biochem. Biophys. Res. Commun. 67*:562-570 (1975).
30. B. I. Sukhorukov, S. G. Kamzolova, A. I. Kolontarov, and A. I. Petrov, *Biofizika 18*:377-378 (1973).
31. S. K. Zavriev, G. L. Grigoryan, L. E. Minchenkova, and V. I. Ivanov, *Molek. Biol. 8*:775-783 (1974).
32. T. Greve, M. F. Maestre, H. Moise, and Hosoda, *Biochemistry 17*:887-893 (1978).
33. R. J. Erickson, B. Janik, and R. G. Sommer, *Biochem. Biophys. Res. Commun. 52*:1475-1482 (1973).
34. R. J. Erickson and J. C. Grosch, *Biochemistry 13*:1987-1993 (1974).
35. E. DeClercq, A. Billian, T. Hobbs, P. F. Torrence, and B. Witkop, *Proc. Nat. Acad. Sci. U.S. 72*:284-288 (1975).
36. P. E. Warwick, A. Hakam, and A. M. Bobst, *Biophys. J. 21*:209a (1978).
37. P. E. Warwick, A. Hakam, and A. M. Bobst, *Nucleic Acid Res. 5*:2525-2534 (1978).
38. P. E. Warwick, A. Hakam, E. V. Bobst, and A. M. Bobst, *Proc. Nat. Acad. Sci. U.S. 77*:4574-4577 (1980).
39. M. Gorecki and A. Panet, *Biochemistry 17*:2438-2442 (1978).
40. A. M. Bobst, *Fed. Proc. 33*:Abstr. No. 1481 (1974).
41. A. M. Bobst, T. K. Sinha, and Y. C. E. Pan, *Science 188*:153-155 (1975).
42. S. Kouidou, T. K. Sinha, B. Janik, and A. M. Bobst, *Biochim. Biophys. Acta 520*:210-218 (1978).

6

RADIATION-INDUCED BINDING OF RADIOSENSITIZING DRUGS TO DNA

J. Cadet and R. Téoule

Département de Recherche Fondamentale
Laboratoire de Radiobiochimie
Grenoble, France

I. INTRODUCTION

Chemical radiosensitizers are substances which increase the response of living cells to ionizing radiation. The forms of life which are susceptible to chemical sensitization are very varied, ranging from viruses to mammals. The notion of sensitization is thus widely linked to the parameters used for the various biological systems investigated. Survival ratio for mammals, plaque-forming ability for viruses, and colony-forming ability for *Escherichia coli* are useful indices frequently chosen. Furthermore, a large number of vital biomolecules can be involved in these processes. Consequently, sensitization is a very complex phenomena which may occur at different steps of the progression of irradiation damage. Although the predominant mechanism of action in vivo can only be speculated upon, many experiments support the idea that DNA is the main target of the radiation in cells [1]. Many types of radiosensitizers seem to influence the effects of radiation either by sensitizing DNA in the cells at the time of irradiation or by changing the capability of the cell to repair DNA damage. This latter aspect will not be discussed further herein.

II. RADIOSENSITIZERS

A. Oxygen Effects

Recent studies including rapid mixing experiments and gas explosion techniques suggest that events taking place on DNA in the millisecond time scale are of considerable importance in the former type of sensitization process [2,3]. Radiation energy is absorbed in about 10^{-16} sec, giving rise to excited and ionized molecules. Free DNA radicals produced in several ways, e.g., OH, H, or e^-_{aq} addition, chain breaks, electron transfer, or bond rupture, all of which can be detected in vitro by electron spin resonance (ESR) spectroscopy, may be considered as transient lesions. Oxygen which gives peroxides with these radical species prevents the restitution reaction of the initial biopolymer. The peroxides are further decomposed into rather stable substances, and the DNA damage may be considered as "fixed" [3-5]. This is the reason why lesions produced in hypoxic conditions are less lethal than those produced in oxygenated conditions. Solid tumors are poorly vascularized, and to overcome this limitation irradiation in hyperbaric conditions has been tried. Unfortunately, the results have been rather disappointing.

B. Similarity of Radiosensitizers and Oxygen

The use of radiosensitizers which mimic the action of oxygen and fix the DNA radicals seems to be a more promising approach. Emmerson and Howard-Flanders [6] selected stable radical nitroxyls as potential radiosensitizers because these substances possess, as does oxygen, an unpaired electron. Stable free radicals such as triacetonamine-N-oxyl (TAN), norpseudopelletierine-N-oxyl (NPPN), and 2,2,6,6-tetramethyl-4-piperidinol-N-oxyl (TMPN) (Fig. 1), which contain an N-oxyl group, are electron affinic [7,8]. In vitro experiments have revealed that TAN, NPPN, and TMPN are able to react with OH-induced DNA transients in two ways: either (a) by an oxidation-reduction giving rise to reduced radiosensitizers (TAN-H) or (b) by binding [9,10].

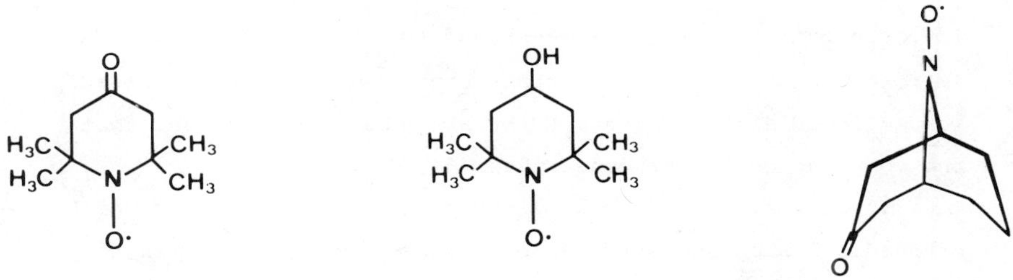

2, 2, 6, 6 - tetramethyl-4-piperidone-N-oxyl (TAN)

2,2,6,6-tetramethyl-4-piperidinol-N-oxyl (TMPN)

Norpseudopelletierine-N-oxyl (NPPN)

Figure 1 Structures of three N-oxyl radicals.

If such substances reached the DNA within the poorly oxygenated tumor cells, the enhancement of the radiation damage could be similar to that of oxygen. Unfortunately, the possible chemical use of the nitroxyl stable radicals has not yet appeared feasible because at nontoxic concentrations low enhancement ratios have been reported for mammalian cells both in vitro as well as in vivo. The free radical nitroxyls are rapidly decomposed by the tissue and blood [11]. The experiments reported on tissue cultures give a rather complex picture of the action mechanism of these substances. TMPN exhibits a net protective effect on hypoxic human cells line of NH1K 3025 cultivated in vitro at doses below 400 rads and a net sensitizing effect above this value [12]. On the basis of "target theory" it has been suggested that double or multiple events should be of importance in cell lethality. It is well recognized in cancer therapy that alkylating agents having two or more functional sites are more efficient than the monofunctional compounds in their antitumor properties. Taking advantage of these facts, a nitroxyl biradical RO-03-6061 bis (2,2,6,6-tetramethyl-1-oxyl-4-piperidinyl) succinate having two unpaired electrons has been tested [13]. Comparison between mono- and biradical at equimolecular concentration mixtures indicate that the biradical removed the shoulder region of the hypoxic cell survival curves [14] and had a much greater effect on the sensitization of V79 Chinese hamster cells irradiated in a nitrogen atmosphere [13] than did the monoradical. The D_o value (dose required to reduce survival by a factor of 0.37

in the exponential region) was found to be 330 rad vs. 535 rad for TMPN at 0.2 mM. However, further study showed that the biradical is destroyed after mixing with human red blood cells and that the compound no longer sensitizes [14].

It is worth noting that electron-affinic substances which are not radical substances are able to sensitize the cells in the absence of oxygen [15]. Two promising drugs have been recently proposed: 5-nitroimidazole and 2-nitroimidazole. The latter is the most efficient for experimental carcinoma, giving a therapeutic gain of 2.1 when 1 g/kg is administered intraperitoneally [16]. However, in the search for more clinically useful radiosensitizers, it is highly desirable to investigate in depth the relation between chemical structure and sensitization.

C. General Aspects of Radiosensitization

The biological effects of the aforementioned classes of radiosensitizers are mainly explained by their capability to induce new types of DNA chemical modification under γ- or x-irradiation. Three types of chemical radiosensitization mechanism have been suggested. One hypothesis ascribes the sensitization effects to the redox properties of drugs which would lower the probability of recombination of ion radicals produced by the direct interaction of ionizing radiations with a vital target molecule [17]. Other hypotheses deal with the oxidation and/or binding reactions of radiosensitizers with radiation-induced DNA transients resulting from the addition or abstraction reactions with water radiolysis species, namely, OH radicals, aqueous electrons (e_{aq}^-) and H atoms [18,19]. The knowledge of the sensitizer effects on the radiation chemistry of DNA components is an approach to a better understanding of the mechanism of cellular radiosensitization.

Electron transfer from electron adduct of pyrimidine and purine DNA components to most of the electron-affinic radiosensitizers was shown to occur quantitatively [18-21]. However, oxidation of neutral pyrimidine radicals which have a higher value of kinetic redox potential compared to those of the aforementioned radiation-induced anionic transients [22] seems to be a minor process [20,23, 24] except for H-adduct pyrimidine radicals [23,25] and α-hydroxy sugar radicals [26]. The occurrence of competitive binding reactions to the neutral transient has been demonstrated as well by pulse radiolysis experiments as by final product determinations. The latter aspect will be discussed in detail in the following section.

III. PULSE RADIOLYSIS

Rapid mixing experiments have permitted determination of the rate constant for the binding reaction of most radiosensitizing drugs to neutral DNA component radicals. The radiolysis of water produces the following reactive species [27]:

$$H_2O \rightarrow e^-_{aq}\ (2.6),\ OH\ (2.7),\ \text{and}\ H\ (0.6)$$

where the numbers in parenthesis are the G values (yield for an absorbed dose of 100 eV) of the radicals. In most of the investigations the aqueous deaerated solutions were saturated with N_2O so as to convert aqueous electrons to OH radicals

$$N_2O + e^-_{aq} \rightarrow N_2 + OH^- + OH$$

The net result is a rapid removal of e^-_{aq} ($t_{1/2} = 2.5 \times 10^{-9}$ sec) and a doubling of the OH yield. The concentration of DNA component was generally chosen to be as low as 5×10^{-4} M to minimize competing reactions of substrate with e^-_{aq}, hence preventing the formation of significant amounts of radiation-induced electron pyrimidine and purine addition transients.

A. N-oxyl Radicals

The most popular free nitroxide radicals used as radiosensitizers are the 2,2,6,6-tetramethyl-4-piperidone-N-oxyl (TAN), 2,2,6,6-tetramethyl-4-piperidinol-N-oxyl (TMPN), and norpseudopelletierine-N-oxyl (NPPN).

1. Bases and nucleosides

The first pulse radiolysis study of the interaction of TAN with OH-thymine adduct was reported by Emmerson and Willson [28]. Aqueous solutions of thymine (6 x 10^{-4} M) were irradiated with 1.8 MeV electron pulses in the presence of saturated nitrous oxide (2 x 10^{-2} M). The hydroxyl radicals are rapidly scavenged by thymine to give a mixture of 5-hydroxy-5,6-dihydrothymin-6-yl and 6-hydroxy-5,6-dihydrothymin-5-yl radicals [29-31]. In the absence of a second solute, these transients which absorb in the range 300-400 nm decayed by a second-order process with a rate constant = 2.2 x 10^9 M^{-1} sec^{-1} [32]:

thymine + $\dot{O}H$ → thymine-OH (TOH) k = 5.2 x 10^9 M^{-1} sec^{-1} [33].

TOH + TOH → products such as 5,6-dihydroxy-5,6-dihydrothymine [34,35].

When thymine is irradiated in the presence of TAN (3.75 x 10^{-4} M), the transient decays more rapidly by a first-order process with a half-life inversely proportional to the TAN concentration [36,37]. From these results the occurrence of a pseudomonomolecular binding reaction was suggested between TAN and OH-thymine radical adduct with a bimolecular rate constant = 3.5 x 10^8 M^{-1} sec^{-1} [28].

The second-order rate constant for the addition reaction between N-oxyl radicals and most of the OH-induced transients in cytosine, adenine, and guanine fall in the range 1-7 x 10^8 M^{-1} sec^{-1} (Table 1). The low reactivity of the OH-adenine radical ($k < 10^7$ M^{-1} sec^{-1}) may be due to the inaccessibility of the site of the purine radical to TMPN and TAN [37].

The deoxyribo- and ribonucleosides show a nearly similar behavior [38]. The sterically unhindered NPPN molecule reacts between two to four times more rapidly with radiation-induced neutral nucleoside radicals than do piperidine nitroxides. The lower reactivity of TMPN compared to that of TAN probably implies a conformational change within the piperidine ring. The six-membered ring of TMPN is in a chair form, whereas that of TAN adopts a twistboat conformation in solution [39]. The decrease in the reactivity of N-oxyl radicals with transients of thymidine and

Table 1 Second-Order Rate Constants[a] of the Reaction Between N-Oxyl Radicals and Hydroxyl Adduct of the Four DNA Bases

DNA bases	TMPN	TAN	NPPN
Thymine	2.3	3.2	6.1
Cytosine	3.1	3.7	7.0
Adenine	—[b]	—[b]	5.7
Guanine		1.7	5.3

[a]Rate constants are expressed in moles per second, multiplied by 10^{-8}.
[b]The value is less than 10^7 M^{-1} sec^{-1}.
Source: From Ref. 37.

cytidine with respect to the corresponding bases may be explained in terms of a protective effect owing to the OH scavenging by the osidic moiety. These data also suggest that the binding reaction of nitroxides to hydrogen abstraction sugar transient seems to be only a minor process.

2. *DNA: role of conformation*

Bimolecular reactions were shown to occur between the three N-oxyl radicals and radiation-induced OH radical of denaturated DNA in aqueous solution saturated with N_2O [39a,40]. Table 2 demonstrates that the binding ability of free nitroxide radical decreases in the order NPPN > TAN > TMPN, as expected if the accessibility to the radical site is the limiting step of the binding reaction.

Steric factors play a major role in native DNA. The kinetics of the reaction of NPPN with irradiated native DNA in N_2O-bubbled aqueous solution suggest a simple bimolecular mechanism [40,41]. The sevenfold decrease in the related rate constant compared to that of denaturated DNA (Table 2) may be accounted for by the low binding ability of the nitroxide to the sugar radical and/or the decrease of the yield of base radicals. The reactions of the hindered piperidine-N-oxyl radicals were postulated to be more complex. It was recently reported that the decay of native DNA hydroxyl adduct in the presence either of TMPN or NPPN contains at least two components, the lower one being independent of the concentration of N-oxyl radicals [41a].

Table 2 Bimolecular Rate Constants[a] for the Reaction Between N-Oxyl Radicals and Hydroxyl DNA Component Adducts

OH-pyrimidine radical	TMPN	TAN	NPPN	References
Thymine	2.6	3.5	6.4	36
Thymidine	0.9	1.1	4.2	38
Thymidine-5'-monophosphate	0.7	1.0	3.7	40
Denaturated DNA	0.6	0.8	2.0	40
Native DNA	0.44	—[b]	0.41	41[a]

[a]The rate constants k' are given in moles per second, multiplied by 10^{-8}.
[b]Not determined. The value must lie in the range 0.40 to 0.44 (E. M. Fielden, personal communication).
Source: Data from Refs. 36, 38, 40, and 41a.

B. Electron-Affinic Radiosensitizers

Heterocyclic electron-affinic radiosensitizers including quinones, nitrofurans, nitroimidazoles, and nitrobenzenes possess a conjugated system of π bonds. These various drugs have in common the ability to transfer an electron from electron-pyrimidine and purine adducts at a diffusion control rate [20,23,25]. More recent investigations have shown that competitive binding reaction between neutral pyrimidine radicals and radiosensitizers occurs to a significant extent.

The rate constant for the addition reaction between nifuroxime and OH-induced radical of thymine, adenine, uracil, and dihydrouracil were assumed to be in the range 0.8 - 2 x 10^9 M^{-1} sec^{-1}. For deoxyribose, one electron oxidation is two times more important than adduct radical formation [23].

The interaction of para-nitroacetophenone (PNAP) has been studied both by optical and conductimetric pulse radiolysis (Fig. 2). This latter technique is very suitable for the analysis of ionic species and confirmed the lack of any significant oxidation of OH-uracil adduct in the presence of PNAP. The transient absorption spectrum produced in N_2O saturated solution of uracil (9.7 x 10^{-4} M) and PNAP (5 x 10^{-5} M) is very complex. Theoretical calculations

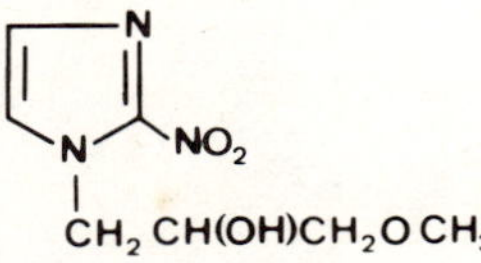

1-(2-hydroxy-3-methoxy-propy P)-2-nitroimidazole (Ro-07-0582)

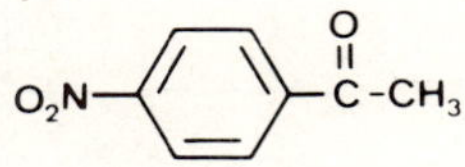

p-nitroacetophenone (PNAP)

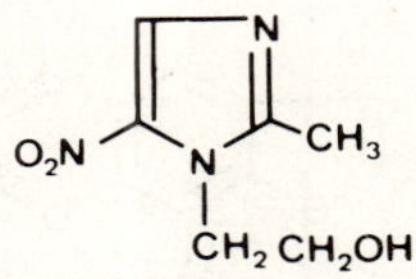

metronidazole ("flagyl")

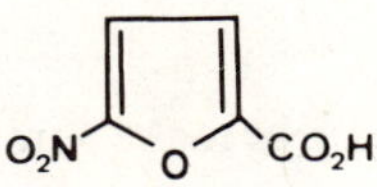

5-nitro-2-furoic acid

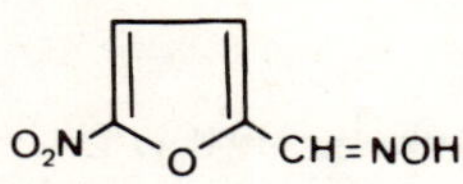

anti-5-nitro-2-furaldoxime (nifuroxime)

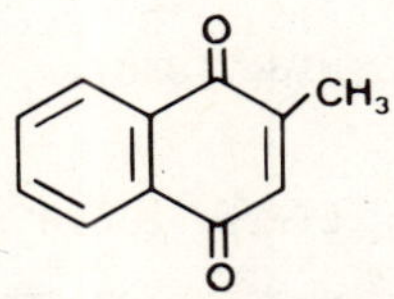

2-methyl-1,4-naphtoquinone (menadione)

Figure 2 Structures of various electron-affinic radiosensitizers.

showed that below 370 nm the most important contribution to this spectrum is given by an addition product between 5(6)-hydroxy-5,6-dihydrouracil-6(5)-yl radical and PNAP. In a similar way binding reaction was also postulated to occur between PNAP and 5,6-dihydrouracil-5(6)-yl radical [41].

C. Other Radiosensitizers

1. *N-Ethylmaleimide*

The mechanisms by which N-ethylmaleimide (NEM) acts as a radiosensitizer may be related to its sulfhydryl-binding ability, which prevents the chemical repair process, and/or to interactions with radiation-induced DNA radicals (Fig. 3). This latter aspect as studied by pulse radiolysis showed that an efficient electron-transfer reaction occurs from radical anions of various pyrimidines to NEM [26,42]. However the presence of NEM has no detectable effect on the decay rate of 5(6)-hydroxy-5,6-dihydrothymin-6(5)yl radical, indicating a rate constant lower than 10^7 M^{-1} sec^{-1} [26, 42]. Since covalent binding reactions were shown to occur between NEM and thymine-OH adduct (vide infra), this apparent lack of reac-

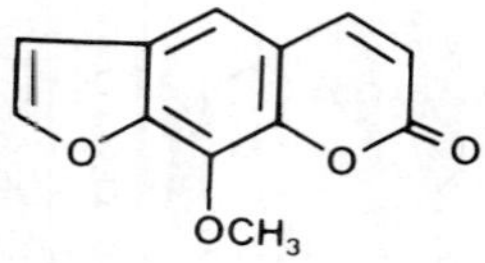

8-methoxypsoralen (8-MOP)

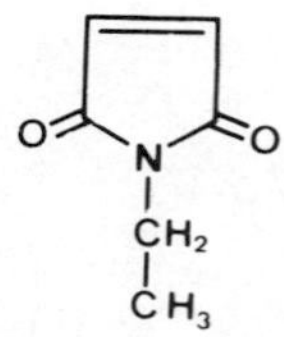

N-ethylmaleimide (NEM)

Figure 3 Structures of 8-methoxypsoralen (8-MOP) and N-ethylmaleimide (NEM).

tivity could be explained in terms of the similar ultraviolet (UV) absorption spectra for both the radical NEM adduct and the initial pyrimidine transient. Hydrogen abstraction produces deoxyribose radicals which react with NEM by both an electron transfer process and a binding reaction across the ethylenic bond of NEM, giving rise to an α-amino radical [26].

2. *Psoralens*

The bifunctional furocoumarin 8-methoxypsoralen (8-MOP), which is used as a skin-photosensitizing agent in the treatment of psoriasis, was shown to selectively sensitize anoxic T7 bacteriophage to the lethal action of γ-rays [43,44,]. It was concluded from recent pulse radiolysis studies that OH-addition radicals of thymine, thymidine, and cytosine react with 8-MOP (Fig. 3) with a rate constant of $\simeq 4 \times 10^9\ M^{-1}\ sec^{-1}$ [44a]. The binding reaction of 8-MOP to the neutral pyrimidine radical was postulated to be three times more important than the oxidation process leading to the formation of the electron adduct of 8-MOP. The former reaction could be considered as the first step in the formation of cross-links between adjacent DNA strands. However, it is unlikely that these radiation-induced lesions are identical with the photo-addition reaction between the pyrimidine ring and both furyl and pyrone functional groups of 8-MOP.

IV. STEADY-STATE RADIOLYSIS

The bimolecular reactions between radiation-induced DNA radicals and some radiosensitizers give rise directly or through secondary processes to final stable covalent addition products.

A. Covalent Binding to DNA and Related Compounds

The extent of the radiation-induced binding of [^{3}H]TAN (2×10^{-4} M) to DNA (1 mg/ml) was determined by measuring the radioactivity of the DNA fraction after purification on a Sephadex G-25 column [45]. Covalent complex formation which was only observed to occur in oxygen-free solutions showed a linear dependence upon the absorbed dose of radiation. The degree of TAN binding to purine deoxyribonucleotides was found to be negligible, suggesting that the deoxyribose moiety is not involved in the formation of significant amount of [^{3}H]TAN-addition products [46]. Several lines of available evidence indicate that the binding reaction is located on the pyrimidine bases within the DNA. This reaction is initiated by OH radical since the yield of binding was increased by a factor of 2 when the irradiations were carried out in N_2O-saturated solutions compared to pure deaerated solutions.

NEM has been observed to attach itself to DNA on irradiation in oxygen-free aqueous solutions [47]. The yield of NEM bound to DNA increases linearly with increase of the ratio DNA/NEM. Since these two solutes compete for the water radiolysis species, it may be concluded that the covalent attachment of NEM results mainly from the interaction of the maleimide with a DNA radical [47]. As the binding yield increases by a factor of 2 when N_2O saturated solutions are used it may be assumed that only OH-induced DNA radicals show a binding ability [42]. Steady-state radiolysis of deaerated solutions of [^{3}H]thymine (10^{-2} M) in the presence of [^{14}C]NEM (5×10^{-4} M) gave rise to a major product which contained both types of radioactive label [42]. The occurrence of a binding reaction within the deoxyribose moiety has also been suggested.

5-Nitro-2-furaldehyde semicarbazone (nitrofurazone), an electron-affinic radiosensitizer used clinically, was shown to form covalent complexes with various biological compounds on irradiation in the absence of oxygen [48]. However, it was demonstrated that DNA is 14- to 17-fold more reactive than proteins in the adduct formation. Radiation-induced attachment of $[^{14}C]$nitrofurazone to the TCA fraction of hamster cells was observed to occur in anoxic conditions.

The main radiation-induced degradation products of thymine (10^{-3} M) in oxygen-free aqueous solutions contaning various radiosensitizers (5×10^{-4} M) have been separated by two-dimensional thin-layer chromatography [49]. One class of these compounds corresponds to the 5- and/or 6-hydroxylated dihydrothymine derivatives, previously characterized in the irradiated deaerated aqueous solutions of thymine [50,51]. Significant amounts of new specific pyrimidine derivatives, identified as radiosensitizer addition products, were also formed in the presence of N-oxyl radicals and NEM (Table 3). The binding yield decreases in the order TAN > TMPN > NEM as expected from the reported rate constants for their bimolecular rate constants with OH addition thymine radical (vide supra). PNAP was shown to be unable to form detectable amounts of stable covalently linked adduct with uracil on steady-state radiolysis [52].

B. Chemical Structure of the Stable Adducts

1. Triacetoneamine-N-oxide addition products

The structural assignment of the different TAN addition products with radiation-induced thymine radicals provides further information on these reactions. The characterization of the two major adducts as the trans isomers of structures [1] and [2] (Fig. 4) on the basis of their $[^{13}C]$NMR (nuclear magnetic resonance) and mass spectrometry data [53] indicates that both 6-hydroxy-5,6-dihydrothymin-5-yl and 5-hydroxy-5,6-dihydrothymin-6-yl radicals are involved in the binding reactions. The unexpected derivative [1] is unstable in aqueous solution and gives rise to 5-hydroxymethyluracil subsequent to a characteristic Cope rearrangement reaction

Table 3 Radiation[a]-Induced Degradation of Thymine in Oxygen-Free Aqueous Solution Containing Various Radiosensitizers[b]

Radiosensitizers	Thymine decomposition	Common products[c]	Addition products
TAN	0.79	0.26	0.53
TMPN	0.81	0.39	0.42
NEM	1.02	0.86	0.16
Menadione bisulphite	0.91	0.84	0.06
Metronidazole	1.02	0.95	0.07
Naphtaquinone	0.88	0.84	0.04
5-Nitro-2-furoic acid	1.32	1.29	0.03

[a]G values for an absorption dose of 450 krad from ^{60}Co.
[b]Concentrations 1 mM for thymine and 0.5 mM for the radiosensitizers.
[c]Mostly dihydrothymine derivatives [51].
Source: Data from Ref. 55.

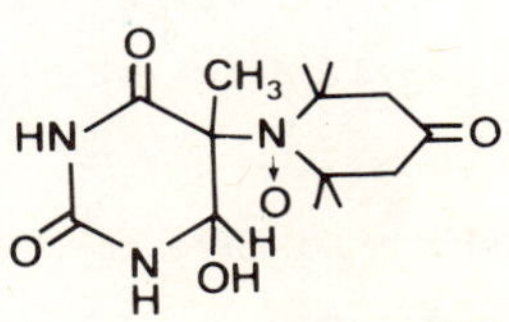

[1]

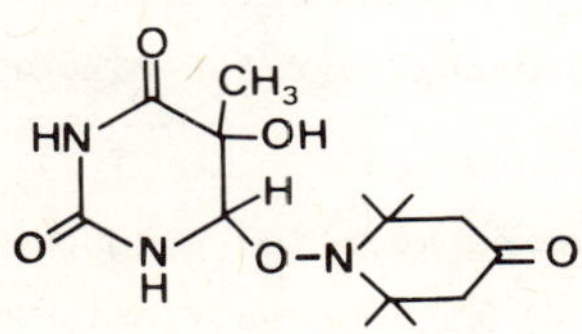

[2]

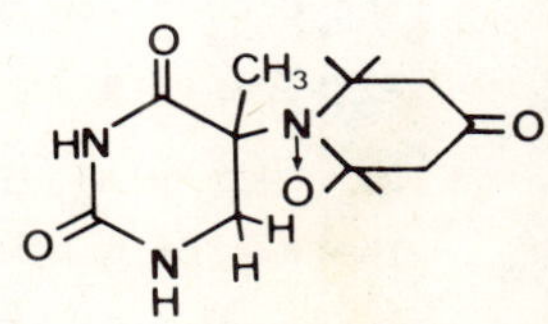

[3]

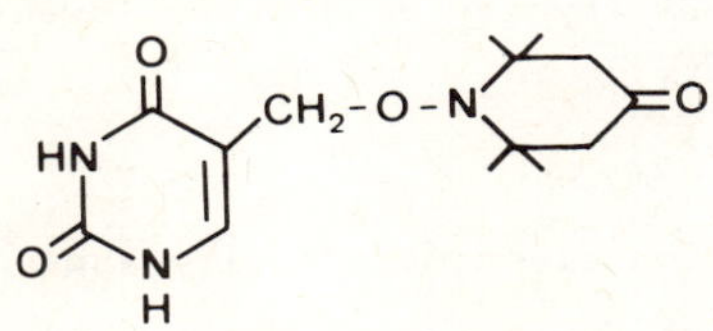

[4]

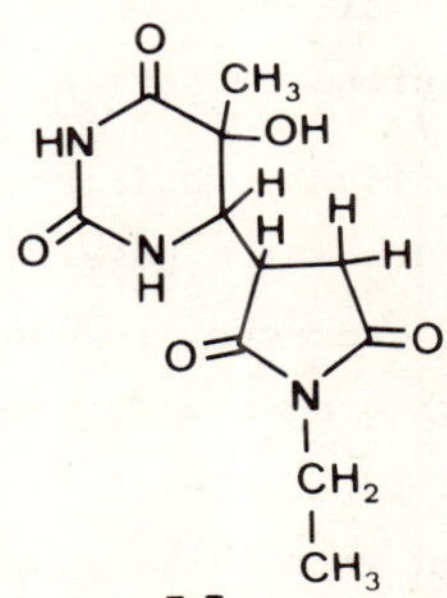

[5]

Figure 4 Main covalently addition products of TAN and NEM with radiation-induced thymine radicals. (From Refs. 53 and 55.)

Table 4 G Values of the Radiation-Induced[a] 6(5)-Hydroperoxy-5(6)-hydroxy-5,6-dihydrothymine in Aqueous Oxygenated Solutions of Thymine

Hydroperoxides	G values at two thymine concentrations 10 mM[b]	2 mM[c]
cis-6-Hydroperoxy-5-hydroxy-5,6-dihydrothymine	0.22	0.32
cis-5-Hydroperoxy-6-hydroxy-5,6-dihydrothymine	0.04	0.08
trans-6-Hydroperoxy-5-hydroxy-5,6-dihydrothymine	0.42	0.67
trans-5-Hydroperoxy-6-hydroxy-5,6-dihydrothymine	0.08	0.17

[a] γ-rays from ^{60}Co.
[b] Absorption dose 2520 krad [60].
[c] Absorption dose 300 krad [5].
Source: Data from Refs. 5 and 60.

[54]. The presence of the two minor adducts [3] and [4] suggests the occurrence of a bimolecular reaction between TAN and, respectively, (uracilyl) 5-methyl and 5,6-dihydrothymin-5-yl radicals.

2. *N-Ethylmaleimide adducts*

The two main NEM-thymine addition products were shown to have the structure [5] (Fig. 4) as deduced from [^{1}H]-NMR and mass spectrometry results [55]. The formation of these two isomers, which interconvert rapidly in solution, may be explained by the initial reaction of NEM with 5-OH-thymine adduct, followed by a disproportionation reaction.

3. *Thymine hydroperoxides*

Oxygen which possesses both a high electron affinity and a biradical character is considered to be the most efficient radiosensitizing agent. Oxygen shows a behavior similar to that of TAN or NEM in its reaction with radiation-induced radicals. While electron pyrimidine adducts are readily oxidized in aerated solutions, oxygen binds to neutral radicals at a diffusion controlled rate [56]. The subsequent peroxy radicals give rise to the final hydroperoxides through a secondary reaction involving a reduction step [57]. The cis and trans isomers of 6-hydroperoxy-5-hydroxy and 5-hydroperoxy-6-hydroxy-5,6-dihydrothymine were shown to be the main peroxides (Table 4) in γ-irradiated aerated solution of thymine [58,59], the trans-6-hydroperoxide being preponderant [60]. The minor peroxides, namely, 6- and 5-hydroperoxy-5,6-dihydrothymine and 5-hydroperoxymethyluracil have also been characterized [61,62]. The mutagenic action of cis-6-hydroperoxy-5-hydroxy-5,6-dihydrothymine was recently demonstrated on transforming DNA of *Haemophilus influenzae* [63].

V. CONCLUSION

The mechanism of chemical radiosensitization of DNA and related compounds by various chemicals which have been reviewed in this article are well substantiated both by pulse radiolysis experiments from the kinetic aspect and by the final product determination. However, the biological modes of action of these substances remain largely unknown. Attempts should be made to characterize the radiation products of DNA with special emphasis on the radiosensitizer DNA adducts, since related compounds were shown to be involved in aging, mutagenesis and carcinogenesis [64].

ACKNOWLEDGMENT

The authors would like to thank Mrs. Arlette Mannent for her secretarial help in preparing this manuscript.

REFERENCES

1. J. Hüttermann, W. Köhnlein, R. Téoule, and A. J. Bertinchamps, (Eds.), *Molecular Biology, Biochemistry and Biophyics*, Vol. 27. Springer-Verlag, New York, 1978.
2. I. Johansen and E. Boye, *Nature 255*:740-742 (1975).
3. O. Sapora, E. M. Fielden, and P. S. Loverock, *Radiat. Res. 64*: 431-442 (1975).
4. J. D. Chapman, A. P. Reuvers, J. Borsa, and C. L. Greenstock, *Radiat. Res. 56*:291-306 (1973).
5. R. Téoule and J. Cadet, in *Molecular Biology, Biochemistry and Biophysics* (J. Hüttermann, W. Köhnlein, R. Téoule, and A. J. Bertinchamps, Eds.), Vol. 27. Springer-Verlag, New York, 1978, pp. 171-203.

6. P. T. Emmerson and P. Howard-Flanders, *Radiat. Res. 26*:54-62 (1965).
7. G. E. Adams and D. L. Dewey, *Biochem. Biophys. Res. Commun. 12*: 473-477 (1963).
8. M. Simic and E. L. Powers, *Int. J. Radiat. Biol. 26*:87-90 (1974).
9. F. F. Nakken, T. Sikkeland, and T. Brustad, *FEBS Lett. 8*:33-36 (1970).
10. E. Wold and T. Brustad, *Acta Chem. Scand. B29*:781-786 (1975).
11. W. Schimmack, U. Deffner, and M. D. Michailov, *Int. J. Radiat. Biol. 30*:393-397 (1976).
12. E. O. Pettersen, R. Oftebro, and T. Brustad, *Int. J. Radiat. Biol. 26*:523-534 (1974).
13. B. C. Cooke, E. M. Fielden, M. Johnson, and C. E. Smithen, *Radiat. Res. 65*:152-162 (1976).
14. B. C. Millar, E. M. Fielden, and C. E. Smithen, *Radiat. Res. 69*: 489-499 (1977).
15. R. L. Willson, W. A. Cramp, and R. M. J. Ings, *Int. J. Radiat. Biol. 26*:557-569 (1974).
16. J. Denekamp and S. R. Harris, *Radiat. Res. 61*:191-203 (1975).
17. G. E. Adams and M. S. Cooke, *Int. J. Radiat. Biol. 15*:457-471 (1969).
18. C. L. Greenstock, G. E. Adams, and R. L. Willson, in *Radiation Protection and Sensibilisation* (H. L. Moroson and M. Quintiliani, Eds.). Taylor & Francis, London, 1970, pp. 65-71.
19. P. T. Emmerson in *Advances in Radiation Chemistry* (M. Burton and J. L. Magee, Eds.). Wiley (Interscience), New York, 1972.
20. G. E. Adams, C. L. Greenstock, J. J. van Hemmen, and R. L. Willson, *Radiat. Res. 49*:85-95 (1972).
21. D. W. Whillans, G. E. Adams, and P. Neta, *Radiat. Res. 62*:407-421 (1975).
22. P. S. Rao and E. Hayon, *J. Am. Chem. Soc. 97*:2986-2989 (1975).
23. C. L. Greenstock and I. Dunlop, *Radiat. Res. 56*:428-440 (1973).
24. H. B. Michaels, E. J. Rasburn, and J. W. Hunt, *Radiat. Res. 65*: 250-267 (1976).
25. M. Simic and E. Hayon, *Int. J. Radiat. Biol. 22*:507-511 (1972).
26. E. Hayon and M. Simic, *Radiat. Res. 50*:464-478 (1972).
27. E. J. Hart, *Radiat. Res. Rev. 3*:285-304 (1972).
28. P. T. Emmerson and R. L. Willson, *J. Phys. Chem. 72*:3669-3671 (1968).
29. M. G. Ormerod and B. B. Singh, *Int. J. Radiat. Biol. 10*:533-538 (1966).
30. C. Nicolau, M. McMillan, and R. O. C. Norman, *Biochim. Biophys. Acta 174*:413-422 (1969).
31. C. Lagercrantz, *J. Am. Chem. Soc. 95*:220-225 (1973).
32. L. S. Myers Jr., M. L. Hollis, L. M. Theard, F. C. Peterson, and A. Warnick, *J. Am. Chem. Soc. 92*:2875-2882 (1970).
33. G. Scholes, P. Shaw, R. L. Willson, and M. Ebert, in *Pulse Radiolysis* (M. Ebert, J. P. Keene, A. J. Swallow, and J. H. Baxendale, Eds.). Academic Press, New York, 1965, pp. 151-164.

34. J. F. Ward, *Adv. Radiat. Biol.* *5*:181-239 (1975).
35. J. Cadet, J. Ulrich, and R. Téoule, *Tetrahedron* pp. 2057-2061 (1975).
36. P. T. Emmerson, E. M. Fielden, and I. Johansen, *Int. J. Radiat. Biol.* *19*:229-236 (1971).
37. T. Brustad, H. Bugge, W. B. G. Jones, and E. Wold, *Int. J. Radiat. Biol.* *22*:115-129 (1972).
38. E. Wold and T. Brustad, *Int. J. Radiat. Biol.* *24*:153-160 (1973).
39. R. Brière, H. Lemaire, and A. Rassat, *Bull. Soc. Chim. Fr.* pp. 3273-3283 (1965).
39a. T. Brustad, W. B. G. Jones, and E. Wold, *Int. J. Radiat. Biol.* *24*:33-43 (1973).
40. P. B. Roberts and E. M. Fielden, *Int. J. Radiat. Biol.* *20*:363-371 (1971).
41. H. B. Michaels and J. W. Hunt, *Radiat. Res.* *72*:32-47 (1977).
41a. P. O'Neill, T. C. Jenkins, and E. M. Fielden, *Radiat. Res.* *82*:55-64 (1980).
42. J. F. Ward, I. Johansen, and J. Aasen, *Int. J. Radiat. Biol.* *15*:163-170 (1969).
43. D. Becker, L. I. Grossweiner, J. Oviada, and J. L. Redpath, *Int. J. Radiat. Biol.* *28*:195-198 (1975).
44. J. L. Redpath and M. L. Tortorello, *Int. Radiat. Biol.* *32*:465-469 (1977).
44a. J. L. Redpath, J. Ihara, and L. K. Patterson, *Int. J. Radiat. Biol.* *33*:309 (1978).
45. T. Brustad, W. B. G. Jones, and K. F. Nakken. *Int. J. Radiat. Phys. Chem.* *3*:55-59 (1971).
46. K. F. Nakken, T. Brustad, and E. Tamas, *Advances in Chemical Radiosensitization*. International Atomic Energy, Vienna, 1974, pp. 37-53.
47. I. Johansen, J. F. Ward, K. Siegel, and A. Sletten, *Biochem. Biophys. Res. Commun.* *33*:949-954 (1968).
48. J. D. Chapman, A. P. Reuvers, J. Borsa, A. Petkau, and D. R. McCalla, *Cancer Res.* *32*:2630-2632 (1972).
49. J. Cadet, M. Guttin-Lombard, and R. Téoule, *International Symposium: Protein and Other Adducts to DNA*. Williamsburg, Va., 1975, Abstr. NAM-8.
50. C. Nofre and A. Cier, *Bull. Soc. Chim. Fr.* pp. 1326-1333 (1966).
51. J. Cadet and R. Téoule, *Int. J. Appl. Radiat. Isot.* *22*:273-280 (1971).
52. A. J. Varghese, *Int. J. Radiat. Biol.* *28*:477-484 (1975).
53. J. Cadet, M. Berger, M. Guttin-Lombard, and M. J. Bobenrieth, *Tetrahedron* pp. 2743-2748 (1979).
54. J. Cadet and R. Téoule, *J. Radiat. Res.* *18*:93-101 (1977).
55. J. Cadet, M. Guttin-Lombard, and R. Téoule, *Int. J. Radiat. Biol.* *30*:1-11 (1976).
56. G. Scholes, *Photochemistry and Photobiology of Nucleic Acids*, (S. Y. Wang, Ed.), Vol 1. Academic Press, New York, 1976, pp. 521-577.
57. G. Scholes, in *Radiation Chemistry of Aqueous Systems* (G. Stein, Ed.). Wiley (Interscience), New York, 1968, 259-285.

58. B. Ekert, *Radiobiologie appliquée* (J. Loiseleur, Ed.), Vol 1. Gauthier-Villars, Paris, 1966, pp. 301-352.
59. R. Latarjet, B. Ekert, and P. Demerseman, *Radiat. Res. Suppl.* *4*:192-212 (1963).
60. J. Cadet and R. Téoule, *Biochem. Biophys. Res. Commun.* *59*:1047-1052 (1974).
61. J. Cadet and R. Téoule, *Biochim. Biophys. Acta* *238*:8-26 (1971).
62. B. S. Hahn and S. Y. Wang, *J. Org. Chem.* *41*:567-568 (1976).
63. H. F. Thomas, R. M. Herriot, B. S. Hahn, and S. Y. Wang, *Nature* *259*:341-342 (1976).
64. K. C. Smith, *Aging, Carcinogenesis and Radiation Biology: The Role of Nucleic Acid Addition Reactions*. Plenum Press, New York, 1976.

7

ACTIVATION OF GUANYLATE CYCLASE BY N-NITROSO CARCINOGENS AND RELATED AGONISTS: EVIDENCE FOR A FREE RADICAL MECHANISM

Frederick R. DeRubertis and Patricia A. Craven

Veterans Administration Hospital and
University of Pittsburgh
Pittsburgh, Pennsylvania

I. INTRODUCTION

The two major naturally occurring cyclic nucleotides of mammalian tissues, adenosine 3':5'-cyclic monophosphate (cAMP) and guanosine 3':5'-cyclic monophosphate (cGMP), are thought to act as intracellular mediators in the expression of the biological actions of a variety of external stimuli. Numerous studies have suggested possible roles for cAMP and cGMP in the control of cell growth and transformation. Results of these studies have been well summarized in several recent reviews [1-4]. Despite extensive investigations in this area over the past decade, delineation of the biological roles of cAMP and cGMP in the regulation of cell growth and transformation remains incomplete. Based on current data [1-4], several inferences seem warranted: (a) alterations in cyclic nucleotide metabolism represent only one of the many factors capable of influencing cell growth characteristics; (b) the effects of cyclic nucleotides are likely integrated with those of other growth regulatory factors, such as calcium; and (c) the impact of an alteration in cyclic nucleotide metabolism on proliferation is not identical in different phases of the growth cycle of the same cell. Within these limitations, there are observations to suggest that cGMP could serve as a positive signal for cell growth and transformation. Several laboratories have observed that addition of cGMP to cultured lymphocytes stimulates DNA synthesis and transformation to blast form [5-11], while cAMP inhibits these processes [6,11-15]. Moreover, cGMP is elevated in a number of tumors [16-19] and in several rapidly growing cell lines, some of which demonstrate a concurrent reduction in cAMP [16-22]. However, present data [1-4,23-25] are not consistent with a simple stereotyped stimulatory (cGMP) or inhibitory (cAMP) or inhibitory (cAMP) effect of each nucleotide on cell growth, as proposed earlier [26].

Any change in cyclic nucleotide metabolism that plays a primary role in the initiation of the process of malignant transformation would be expected to occur early after exposure of target cells to transforming stimuli. Recent studies have demonstrated very marked and rapid activation of the guanylate cyclase-cGMP system of mammalian tissues by a number of chemical carcinogens including N-nitroso compounds, hydrazine, and butadiene diepoxide [27-35]. This response is especially striking with N-nitroso compounds [27,32]. These agents produce rapid increases in cGMP in several tissues without concomitant alterations in cAMP [27,31]. The role of this acute enhancement of cellular accumulation of cGMP in the expression of the oncogenic action of N-nitroso and other carcinogens is a question of obvious importance, but one which is still unanswered. The present review will focus on the properties and regulation of the guanylate cyclase-cGMP system, with emphasis on recent observations suggesting that N-nitroso carcinogens and other agonists of guanylate cyclase alter enzyme activity by a free radical or oxidative mechanism. This unique pathway for enzyme regulation may permit guanylate cyclase stimulation by the active metabolites of a wide variety of structurally distinct chemical carcinogens [36].

II. REGULATION OF cGMP

cGMP was first detected as a constituent of rat urine in 1963 [37], and subsequently was identified in mammalian tissues [38,39]. The cGMP content of most tissues is much lower (10% or less) than that of cAMP (0.1 to 1.0 pmol/mg wet weight), a factor which until recently limited the study of mechanisms regulating cGMP economy in some intact cell systems [40]. This problem has now been largely resolved by modification of the radioimmunoassay procedure for cGMP (succinylation or acetylation of the nucleotide), which greatly enhances assay sensitivity [41-44]. These procedures permit detection of femtomole quantities of cGMP [42-44].

A number of naturally occurring humoral and neurohumoral agents have been shown to enhance the accumulation of cGMP in target cells. These agents include vasoactive compounds and secretagogues (cholinergic and α-adrenergic stimuli, bradykinin, histamine, serotonin, insulin, and secretin) [45,46]. Increases in cGMP and responses to these agents occur rapidly (seconds to minutes) and generally follow a peak-decline pattern. However, causal relationships between increases in cGMP and the well-recognized physiological actions of these agents to alter cellular secretory activity, ionic transport, and/or the contractile state have not been established.

A. Ca^{2+} Dependence

Both the basal accumulation of cGMP and expression of the actions of many humoral agents to increase cellular cGMP is highly dependent upon calcium, a relationship first recognized by Schultz and co-workers in studies of the rat ductus deferens [47] and subsequently observed in many other tissues [48-52]. Exclusion of extracellular Ca^{2+} reversibly reduces basal cGMP accumulation and abolishes cGMP responses to agents such as cholinergic stimuli [48,52]. The changes induced by extracellular Ca^{2+} exclusion are not mimicked by exclusion of Mg^{2+} [52], and in many tissues the blunted cGMP responses occur without concomitant alterations in basal or hormone-induced changes in cAMP [48,52]. Several observations suggest, but do not prove, that Ca^{2+}-dependent alterations in cGMP accumulation involve transmembrane fluxes of this cation. The divalent cation ionophore A23187 increases cGMP in several tissues in the presence but not in the absence of extracellular Ca^{2+} [50-56].

Moreover, in rat renal cortical slices incubated in the presence of extracellular Ca^{2+}, tetracaine and verapamil lower basal cGMP and abolish nucleotide responses to carbamylcholine and other Ca^{2+}-dependent agonists [52,57]. The effect of tetracaine and verapamil on cGMP may be linked to the inhibitory actions of these agents on transmembrane Ca^{2+} flux [58,59]. Alterations in the extracellular concentrations of K^+ and Na^+ have also been shown to alter cellular cGMP metabolism, but the influence of these cations appears to be Ca^{2+}-dependent [2,48,60,61]. Thus, increases in cellular cGMP accumulation induced by high extracellular concentrations of K^+ occur only in the presence of Ca^{2+} and are thought to be linked to changes in Ca^{2+} flux during membrane depolarization [48].

As already noted, N-nitroso carcinogens significantly increase accumulation of cGMP in mammalian tissues. In striking contrast to the Ca^{2+}-dependent action of physiological agonists of cGMP, enhancement of cGMP by N-nitroso compounds does not require the presence of extracellular Ca^{2+} [27,28,31,57]. Table 1 shows the effects of extracellular Ca^{2+} exclusion on basal and stimulated cGMP and cAMP accumulation in rat renal cortex. Carbamylcholine, histamine, and N-methyl-N'-nitro-N-nitrosoguanidine (MNNG) all significantly increase cGMP accumulation in tissue incubated in buffer which contains a physiological concentration of Ca^{2+}. Of these test agents, only histamine produces a concomitant increase in tissue cAMP. Incubation of tissue in Ca^{2+}-free buffer reduces basal cGMP and abolishes cGMP responses to carbamylcholine and histamine. By contrast, under these same conditions, MNNG still increases cGMP 35-fold over basal. As also shown in Table 1, basal- and histamine-induced increases in cAMP are not altered by Ca^{2+} exclusion. Several other agents share with N-nitroso compounds the capacity to increase cellular cGMP in the absence of extracellular Ca^{2+}. These include sodium azide, hydroxylamine, nitroprusside, nitrite and the carcinogen 4-nitroquinoline-1-oxide [27-29,56,62,63]. The differential requirement for Ca^{2+} of N-nitroso carcinogens and related agents compared to those of physiological agonists of cGMP may reflect fundamental differences in the mechanisms of nucleotide stimulation.

Table 1 Extracellular Calcium Requirements for Stimulation of cGMP and cAMP Accumulation in Rat Renal Cortex[a]

	cGMP pmoles/g wet weight		cAMP pmoles/mg wet weight	
	$+Ca^{2+}$	$-Ca^{2+}$	$+Ca^{2+}$	$-Ca^{2+}$
None	155 ± 17	65 ± 6	2.25 ± 0.32	2.56 ± 0.37
Carbamyl-choline (50 μM)	883 ± 91*	71 ± 7	2.61 ± 0.34	2.73 ± 0.35
Histamine (0.1 mM)	486 ± 50*	68 ± 7	8.49 ± 0.90*	8.73 ± 0.94*
MNNG (1 mM)	4780 ± 472*	2254 ± 243*	2.77 ± 0.38	2.86 ± 0.34

[a]Slices of rat renal cortex were incubated initially for 20 min at 37° in 2 ml of complete (2.0 mM Ca^{2+}) or Ca^{2+} free Krebs Ringer bicarbonate buffer. All buffer solutions contained 5 mM glucose, 0.5 mM EGTA, and 2 mM 3-isobutyl-1-methylxanthine (IBMX); 95% O_2 and 5% CO_2 served as the gas phase. Test agents were then added at the final concentrations shown and incubations continued for an additional 5 min. cGMP and cAMP were assayed in the same slice extract. Each value represents the mean ± SE of values pooled from three to five separate experiments.

*$p < 0.001$ vs. no additions.

Source: Data from Ref. 57.

B. O_2 Dependence

Further evidence for the existence of important differences between the actions of physiological agonists of cGMP and those of N-nitroso carcinogens is provided by an examination of the O_2 requirements of the cGMP responses to these agents [28,57]. As shown in Table 2, O_2 deprivation reversibly lowers basal cGMP accumulation in renal cortical slices and abolishes their responsiveness to carbamylcholine and histamine. By contrast, MNNG produces marked increases in cGMP in O_2-deprived tissue. With many agonists of cGMP, O_2 and Ca^{2+} dependence are correlated. Thus nitroprusside, nitrite, and hydroxylamine enhance cGMP accumulation in O_2 and Ca^{2+}-deprived renal cortical slices, whereas bradykinin and the ionophore A23187 require both Ca^{2+} and O_2 [57]. A similar interdependent influence of O_2 and

Table 2 Oxygen Requirements for Stimulation of cGMP Accumulation in Rat Renal Cortex[a]

Test agent (conc.)	Gas phase O_2 content [cGMP (pmoles/g wet weight)] 95% (20 min)	0% (20 min)	0% (first 20 min) 95% (final 20 min)
None	173 ± 19	76 ± 8	152 ± 17
Carbamylcholine (50 μM)	794 ± 86*	83 ± 9	671 ± 78*
Histamine (0.1 mM)	512 ± 58*	71 ± 9	433 ± 46*
MNNG (1 mM)	4318 ± 468*	2543 ± 274*	3812 ± 419*

[a]Renal cortical slices were incubated in complete Krebs Ringer bicarbonate buffer for 20 min with a gas phase consisting of 95% O_2 + 5% CO_2 or 95% N_2 + 5% CO_2. Where indicated, some slices were deprived of O_2 for an initial 20 min and then incubated in the presence of 95% O_2 for a final 20 min. Test agents were present for the final 5 min of the incubation. All buffers contained 5 mM glucose and 2 mM IBMX. Each value represents the mean ± SE of values pooled from three to five separate experiments.

*$p < 0.001$ vs. corresponding value with no additions.

Source: Data from Ref. 57.

Ca^{2+} on basal cGMP and responses to several agonists was described earlier by Clyman and co-workers in studies conducted in human umbilical artery [54]. This correspondence between O_2 and Ca^{2+} requirements does not pertain for NaN_3. The latter agent requires O_2 for expression of its stimulatory action on tissue cGMP accumulation [28,57] but is effective in the absence of Ca^{2+} [62, 63]. Studies of azide action on guanylate cyclase in broken cell systems (see below) imply that molecular O_2 may be required for oxidation of azide to nitric oxide, a potential active intermediate in enzyme activation [64,65]. By contrast, O_2 is not required for the release of NO from MNNG or for the expression of MNNG actions on guanylate cyclase [28].

The basis for the interdependent requirements of Ca^{2+} and O_2 for the expression of the action of some physiological agonists of cGMP is at present uncertain. Agents such as carbamylcholine and histamine whose actions on cGMP accumulation in intact cells require the presence of Ca^{2+} and O_2 fail to stimulate guanylate cyclase activity in broken cell preparations [45]. Nevertheless, it is thought that these agents increase cellular accumulation of cGMP by enhancing nucleotide generation rather than inhibiting degradation, since they are effective in the presence of potent inhibitors of cGMP phosphodiesterase activity [27,28]. These agents may enhance cGMP generation in intact cells indirectly through Ca^{2+}- and O_2-dependent processes, which are not subsequently expressed under conditions of enzyme assay in broken cell preparation. In this regard, it is of interest that recent observations by Mittal and Murad have suggested that free radicals derived from O_2 may activate guanylate cyclase [66]. Soluble platelet guanylate cyclase is also activated by the endoperoxides PGG_2 and PGH_2 as well as several fatty acid hydroperoxides [67,68]. Superoxide, hydroxyl radicals, and singlet oxygen are formed or utilized in the generation of these peroxides [69-71]. Generation of fatty acid peroxides or endoperoxides by the action of phospholipase and then lipoxygenase or cyclo-oxygenase on cellular lipids [72,73] requires both Ca^{2+} and O_2 [74]. Many agents which increase cellular cGMP levels (carbamylcholine, serotonin, norepinephrine) have been found to increase membrane phospholipase A_2 activity [75]. Thus, endoperoxides, fatty acid peroxides, and/or radicals derived from O_2 whose generation is dependent upon Ca^{2+} and O_2 could serve as endogenous mediators of guanylate cyclase activation in cells exposed to agents such as carbamylcholine. This sequence of events might explain the dual Ca^{2+} and O_2 requirements for expression of the action of certain agents on cGMP. While such considerations remain speculative, it is clear that the conditions necessary for expression of the actions of N-nitroso carcinogens on cGMP differ markedly from those required for most currently recognized humoral agonists of the nucleotide. Accordingly, the interaction of N-nitroso compounds with the guanylate cyclase-cGMP system and the local regulation of this interaction by ionic and other factors may differ in several fundamental aspects from those of endogenous agonists.

III. GUANYLATE CYCLASE: PROPERTIES AND ACTIVATION

Guanylate cyclase, the enzyme catalyzing the conversion of GTP to cGMP and pyrophosphate, was first identified in 1969 [39,76-78]. The basic properties of this enzyme system have been summarized in several recent reviews [45,56,79,80], and only selected aspects will be discussed here. Unlike adenylate cyclase activity which is found predominantly, if not exclusively, in the particulate cell fraction, guanylate cyclase is present in both the soluble and particulate cell fractions [45,56,79,80]. The subcellular distribution of enzyme activity varies widely from tissue to tissue. It is predominantly soluble in liver [81], platelets [82], and rat renal cortex [83], but predominantly particulate in intestinal mucosa [30,39], fibroblasts [84], and rat renal inner medulla [83]. The observation that detergents increase the activity of the particulate form of the enzyme to a greater extent than the soluble form implies that much of the membrane-bound activity is latent, at least under in vitro assay conditions [79,80,85]. The properties of soluble and particulate forms of the enzyme differ in several respects [45,56,79,80], including apparent size (soluble, 150,000; particulate, 300,000), sensitivity to inhibition by ATP (greater with soluble enzyme), free Mn^{2+} requirements (less for particulate enzyme from some sources), and responses to Ca^{2+} (usually stimulates soluble and inhibits particulate activity). Recently, particulate guanylate cyclase has been purified from sea urchin sperm [86]. Antibodies to this preparation inhibited mammalian particulate but not soluble activity [87]. Some sulfhydryl blocking agents also differentially inhibit particulate and soluble activity in rat liver [88]. For example, concentrations of $CdCl_2$ which markedly suppress crude and partially purified soluble enzyme activity fail to alter hepatic particulate activity. Similarly, nitrite, a potent agonist of soluble guanylate cyclase from rat kidney, does not alter particulate activity from this organ [83]. Such observations suggest, but do not establish, the existence of structural and functional differences in soluble and particulate activities. The relative contribution of each form to cGMP production in intact cells is not known. An important contribution of particulate activity has been suggested by (a) relative resistance of this form of the enzyme to inhibition by physiological concentrations of ATP [70]; (b) a correlation between steady-state cGMP levels and the amount of enzyme activity in the particulate fraction from different regions of the kidney [83]; and (c) prominent membrane localization of cGMP by immunocytochemical staining in several tissues [89,90].

As noted earlier, humoral agents which increase cGMP accumulation in intact cells fail to activate either soluble or particulate guanylate cyclase when added directly to subcellular fractions, possibly reflecting an indirect mechanism of enzyme activation. This dissocation between the responses of intact and broken cell preparations is in striking contrast to the adenylate cyclase-cAMP system. Hormonal enhancement of cAMP accumulation in intact target cells correlates well with hormonal activation of membrane-bound adenylate cyclase in subcellular preparations [91]. Until very recently, agents which could both increase cGMP in an intact cell preparation and activate guanylate cyclase in subcellular preparations were lacking. This limited the delineation of the general mechanisms for enzyme activation in subcellular preparations that might have potential relevance in the intact cell. The observation of Kimura and co-workers [62,92] that azide activated guanylate cyclase in subcellular systems and increased cGMP in several intact cell systems provided an important experimental probe for studies in this area. Use of this and related compounds has led to new insights into the mechanism of guanylate cyclase activation.

Questions concerning the biological relevance of studies of guanylate cyclase activity conducted in broken cell preparations, have, in part, also arisen because of the conditions required for optimal expression of enzyme activity. Unlike adenylate cyclase acitivity, which is well expressed with either Mg^{2+} or Mn^{2+} serving as divalent metal cofactor [91], both the soluble and particulate forms of guanylate cyclase show a marked preference for Mn^{2+} [76, 77]. Activity expressed with either Mg^{2+} or Ca^{2+} serving as sole divalent cation ranges from less than 10% to 40% of that with Mn^{2+}. Moreover, the concentration of Mn^{2+} required for optimal expression of guanylate cyclase activity greatly exceeds cellular concentrations of this cation [93]. Some insight into this problem has been provided recently by examination of the divalent cation requirements of activated forms of guanylate cyclase. These forms demonstrate a markedly reduced dependence on free Mn^{2+} and effective utilization of Mg^{2+} as divalent metal cofactor [28,63,83,92,94,95]. Indeed, with Mg^{2+} as sole cation, proportional enzyme activation over basal greatly exceeds that observed with Mn^{2+} [28,63,83,92]. Further, use of Mn^{2+} in some instances has prevented detection of enzyme responses apparent when Mg^{2+} was employed as metal cofactor [95]. These considerations emphasize the importance of examining enzyme responses in the presence of Mg^{2+}. The development of sensitive and relatively simple techniques for assay of guanylate cyclase activity [42,95,96] have permitted reliable detection of the low basal rates of cGMP formation found with Mg^{2+}. Effective utilization of Mg^{2+} as metal cofactor has now been demonstrated after activation of guanylate cyclase with a number of chemically diverse agonists [28, 63,83,92,94] and may represent a general property of activated forms of the enzyme.

A. Spontaneous Activation of Soluble Guanylate Cyclase: Early Indication for a Redox Mechanism

The observation of Bohme et al. [82] of spontaneous activation of platelet soluble guanylate cyclase in air provided an early clue that redox mechanisms were involved in enzyme activation. Similar spontaneous activation of soluble activity was also observed in rat lung [97] and uterus [98] and mice and guinea pig spleen cells [99]. The spontaneous increase in the activity of these preparations is prevented by dithiothreitol. In studies of soluble activity from rat lung, White and co-workers [100] demonstrated that spontaneous activation was enhanced by increasing the O_2 tension of incubates and blocked by O_2 deprivation. The thiol alkylator N-ethylmaleimide (NEM) also blocked spontaneous activation [98]. Addition of exogenous H_2O_2 plus the catalase inhibitor KCN or a peroxide-generating system activated the enzyme [100]. These and other observations led to the suggestion that spontaneous activation may be related to H_2O_2 generation [100]. Similar characteristics of spontaneous activation of rat uterus guanylate cyclase were reported by Haddox and co-workers [99]. Moreover, studies of the process of enzyme inactivation which occurs later and can also be prevented by dithiothreitol led to the proposal that more than one oxidizable thiol site regulates enzyme activity.

B. Activation by Oxidants

A number of agents with oxidant activity have been shown to stimulate guanylate cyclase in broken cell preparations. The agents include periodate [101], dehydroascorbate [99], peroxides [67], and epoxides [36]. In general, enzyme activation by these agents is suppressed by reducing agents and radical scavengers. In recent studies, Mittal and Murad [66] have observed activation of partially purified soluble rat hepatic guanylate cyclase by the formation of H_2O_2 from endogenous superoxide anion present in guanylate cyclase reaction mixtures. Nevertheless, while the action of dismutase was inhibited by catalase and several radical scavengers, it was not mimicked by H_2O_2. Activation by the dismutase was enhanced by addition of nitrate reductase, which may catalyze the formation of hydroxyl radical from superoxide and H_2O_2. It was suggested that the activation of guanylate cyclase by addition of the dismutase [102] was related to formation of hydroxyl free radicals by the Haber-Weiss or Fenton reactions [103]. These observations are consistent with a redox or free radical mechanism of enzyme activation and could have a bearing on O_2- and Ca^{2+}-dependent regulation of cGMP in intact cells. However, at present the findings implicating the hydroxyl radical as the specific activating moiety are quite circumstantial and inconclusive.

C. Activation by N-Nitroso Carcinogens and Related Nitro Compounds

Activation of guanylate cyclase by a free radical mechanism was first proposed as a result of studies of the effects of N-nitroso carcinogens [27], agents with a known capacity for radical formation [104]. A number of carcinogens and noncarcinogenic N-nitroso compounds activate guanylate cyclase from tissues of humans and experimental animals [27,28,30-33]. The most potent agonists of enzyme activity in subcellular preparations are compounds such as MNNG and N-methylnitrosourea which are capable of nonenzymatic decomposition to reactive intermediates [105-107]. Such agents activate soluble guanylate cyclase from several tissues at micromolar concentrations [28,31]. Unlike spontaneous or azide activation of guanylate cyclase [65,100], expression of MNNG action does not require molecular O_2 [28]. Similarly, azide but not MNNG action on partially purified guanylate cyclase system requires catalase [65, 108] (see Table 5 in Section II.F). Studies with MNNG clearly indicated that guanylate cyclase activation is mediated by a labile decomposition product of this compound [28]. Figure 1 shows the temporal relationship between spontaneous and cysteine induced decomposition of MNNG and the capacity to activate soluble guanylate cyclase of rat liver. Solutions originally containing 100 μM MNNG retain full capacity for maximal activation of guanylate cyclase after complete decomposition of the MNNG, assessed spectrophotometrically by loss of the nitroso group. Capacity for guanylate cyclase activation was subsequently lost in a time-dependent fashion (Fig. 1). Cysteine, which is known to accelerate MNNG decomposition [105,107], also accelerated loss of capacity for guanylate cyclase activation (Fig. 1).

D. Effects of Antioxidants and Radical Scavengers

A number of antioxidants and radical scavengers inhibit guanylate cyclase activation by N-nitroso compounds [28,109]. As shown in Figure 2, a high concentration of dithiothreitol reversibly inhibits MNNG activation of hepatic soluble guanylate cyclase. Similar effects are observed with high concentrations of cysteine, glutathione, and ascorbate. The inhibitory effects of these agents may reflect direct chemical interactions with reactive decomposition products of MNNG, protection of oxidizible tissue sites involved in enzyme activation or both [28,30,31]. It is of interest that butylated hydroxyanisole (BHA), butylated hydroxytoluene (BHT), and retinoids inhibit activation of guanylate cyclase by MNNG and other carcinogenic agonists [109]. These agents have putative anticarcinogenic activity [110-113]. As shown in Figure 3, BHA and retinol reverse activation of guanylate cyclase mediated by 100 μM MNNG and prevent reactivation by 3 mM MNNG. A relationship between BHA and retinoid inhibition of guanylate cyclase activation and suppression of the oncogenic effects of MNNG [114; also DeRubertis and Craven, unpublished data] and other N-nitroso compounds [113] is possible but is unproven. The ability of these and other antioxidants and radical scavengers to suppress guanylate activation by azide, nitrite, nitroprusside, hydroxylamine, and superoxide dismutase [28, 66,109,116] support an oxidative or free radical mechanism of enzyme activation. However, the complex influence of antioxidants on enzyme activation deserves emphasis. Under appropriate condi-

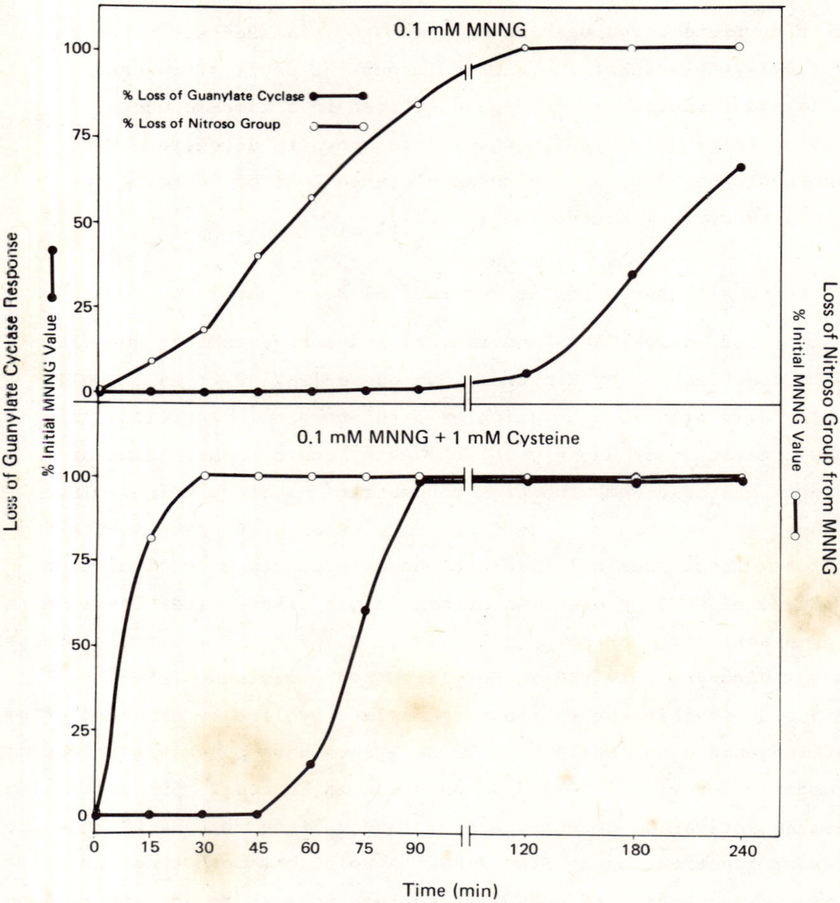

Figure 1 Temporal relationship between spontaneous and cysteine-induced decomposition of N-methyl-N'-nitro-N-nitrosoguanidine (MNNG) and loss of capacity to activate soluble hepatic guanylate cyclase. MNNG (0.1 mM) was initially incubated at 37° in Tris/sucrose buffer (pH 7.4) with or without 1 mM cysteine. At the incubation times shown, loss of the nitroso group from MNNG was determined spectrophotometrically at 400 nm, while the capacity of other aliquots of these same mixtures to activate guanylate cyclase in the 100,000 x *g* supernatant fraction of liver was assayed concomitantly. Basal enzyme activity in the experiment shown was 42 ± 4 pmol cGMP min^{-1} mg^{-1} protein and activity in response to a stock solution of 0.1 mM MNNG, maintained at 0-4° throughout to prevent spontaneous decomposition before use, was 3017 ± 325. (From DeRubertis and Craven [28].)

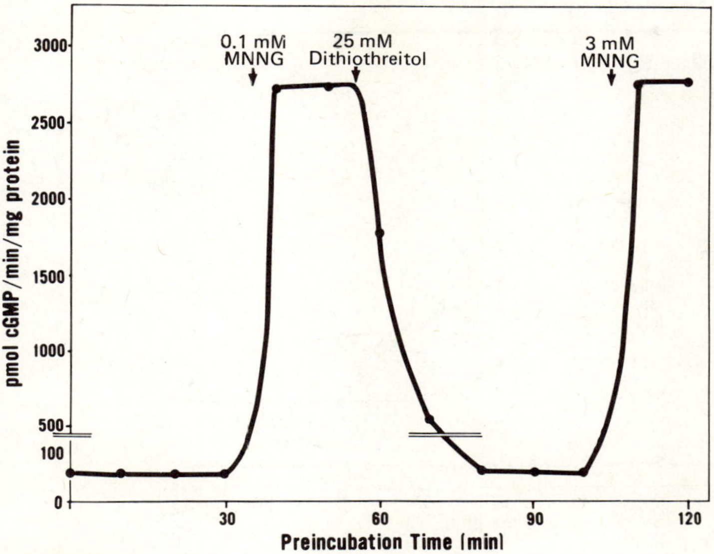

Figure 2 The effects on guanylate cyclase activity of sequential additions of MNNG and dithiothreitol to the 100,000 x *g* soluble fraction of hepatic homogenates. Additions of MNNG or dithiothreitol were made to the same hepatic supernatant fraction maintained at 0-4° at the times indicated by the arrows. A 25 μl aliquot of the preincubation mixture was removed for assay at 10-min intervals starting at zero time and continuing to the end of the 120-min preincubation period. (From DeRubertis and Cravén [28].)

tions, effects of thiols and ascorbate on the stimulation of partially purified soluble hepatic guanylate cyclase of rat liver by MNNG and related agonists are clearly biphasic [28,64; and see below]. These observations may reflect interactions of reducing agents at more than one step in the pathway of enzyme activation.

E. Effects of Sulfhydryl Blockers

Inhibition of basal guanylate cyclase activity from several tissues by sulfhydryl blockers is well recognized [76,117-119]. More recent studies have demonstrated that these agents also suppress

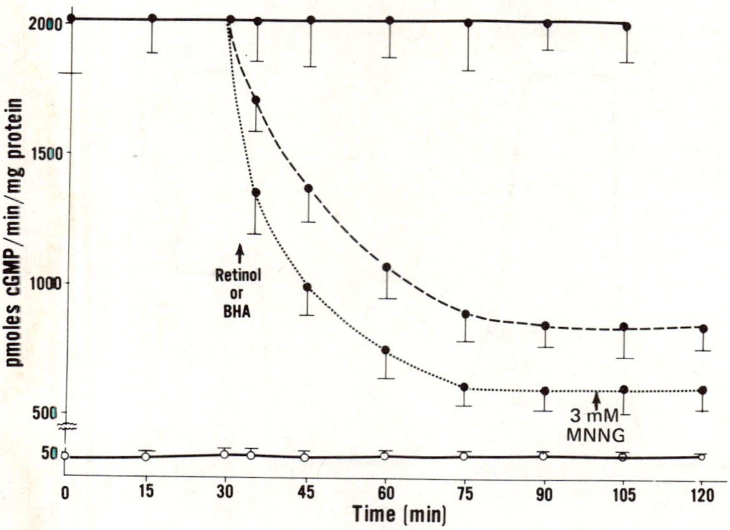

Figure 3 Hepatic supernatants were preincubated from 0 to 120 min at 0-4° with no addition (0) or 0.1 mM MNNG (●—●). At 30 min, retinol (●---●) or butylated hydroxyanisole (BHA) (●····●) at a final concentration of 1 mM was added to an aliquot of some of the control and MNNG-treated hepatic supernatants. The incubation was then continued for an additional 70 min. At 100 min, those incubation mixtures that contained retinol or butylated hydroxyanisole received an additional 3 mM MNNG, and the incubation was continued to 120 min. Guanylate cyclase activity was determined on aliquots of the preincubation mixtures at the times indicated. Addition of retinol or butylated hydroxyanisole had no effect on guanylate cyclase activity in control preparations. (From Ref. 109.)

enzyme activation and cGMP accumulation in intact cells mediated by N-nitroso compounds, nitrite, azide, nitroprusside, and other agonists [28,67,88,99]. Table 3 shows the effects of a number of structurally distinct thiol inhibitors on basal, MNNG, nitroprusside, nitrite, and azide responsive soluble hepatic guanylate cyclase. As shown, $CdCl_2$, p-hydroxymercuribenzoate, N-ethylmaleimide, arsenite, and iodoacetamide suppress both basal activity and responses to the agonists. Such observations suggest that free thiol groups may be the oxidizable sites critical for enzyme activation. The ability of arsenite and $CdCl_2$ to suppress guanylate cyclase activation is of particular interest (Table 3), since these

Table 3 The Effects of Thiol Inhibitors on Basal Soluble Hepatic Guanylate Cyclase Activity and Its Responsiveness to Several Agonists[a]

Initial additions (10 min at 0°C)	Final additions (10 min at 37°C; pmoles min^{-1} mg^{-1} protein) None	0.1 mM MNNG	10 mM nitroprusside	10 mM $NaNO_2$	0.1 mM NaN_3
None	47 ± 6	2914 ± 312	1235 ± 158	1321 ± 185	924 ± 105
0.2 mM $CdCl_2$	5 ± 0.7*	11 ± 2	6 ± 0.8*	5 ± 0.7*	5 ± 0.6*
0.5 mM p-hydroxymercuribenzoate	4 ± 0.5*	26 ± 4*	17 ± 2*	12 ± 2*	9 ± 1*
1 mM N-ethylmaleimide	8 ± 1*	58 ± 7*	28 ± 4*	33 ± 4*	19 ± 3*
50 mM arsenite	11 ± 2*	762 ± 93*	429 ± 57*	388 ± 48*	286 ± 37*
50 mM iodoacetamide	16 ± 2*	1346 ± 161*	781 ± 92*	643 ± 91*	532 ± 63*

[a]The 100,000 x *g* soluble fraction of hepatic homogenates was initially incubated for 10 min at 0° with the concentration of each of the thiol inhibitors indicated. A final addition of test agent was then made to the preincubation mixture where indicated and the preincubation continued for an additional 10 min at 37° to abolish the 2- to 8-min lag periods of enzyme activation observed with azide, nitrite, and nitroprusside.

*$p < 0.01$ vs. corresponding value with no initial addition.

Source: From Ref. 88.

agents are thought to interact specifically with closely juxtaposed thiol groups [120-125]. The evidence for such a mechanism is more compelling in the case of arsenite. Potentiation of arsenite inhibition by an equimolar concentration of 2,3-dimercaprol [or British anti-Lewisite (BAL)] is characteristic but not proof of a vicinal dithiol mechanism of interaction [120-124]. The latter action of BAL is thought to occur by formation of a cyclic dithiol-arsenite complex with enhanced accessibility to protein hydrophobic regions [121]. Moreover, a molar excess of BAL reverses arsenite and $CdCl_2$ inhibition, presumably by competing with the protein dithiol sites for the inhibitors [123,124]. Monothiols substitute poorly if at all for BAL in either potentiating the effects of equimolar arsenite or in reversing arsenite inhibition when added in molar excess [120-122,124]. Figure 4 compares the effects of arsenite alone and arsenite plus equimolar BAL on basal- and MNNG-responsive soluble hepatic guanylate cyclase activities. Over the concentration range shown, BAL alone had no effect on basal enzyme activity or the response to 0.1 mM MNNG. However, equimolar BAL markedly potentiated the inhibitory effects of arsenite (Fig. 4), whereas equimolar mercaptoethanol was without effect [88]. Addition of excess BAL also reversed inhibition of both crude and partially purified hepatic soluble activity by 0.5 mM arsenite plus BAL or 0.2 mM $CdCl_2$. By contrast, excess mercaptoethanol was relatively ineffective in reversing inhibition mediated by both arsenite-BAL and $CdCl_2$ [88]. These observations are consistent with a role for closely juxtaposed thiol groups in the expression of basal guanylate cyclase activity and response to MNNG and other agonists. Accordingly, SH to S—S transformation represents a potential mechanism of enzyme activation. Several observations have suggested that more than one critical dithiol site is involved in the regulation of guanylate cyclase activity [88]. However, since studies have not yet been conducted with a homogeneous preparation of the enzyme, these sites could reside on either the enzyme per se or on other macromolecular tissue components required for expression of enzyme activity.

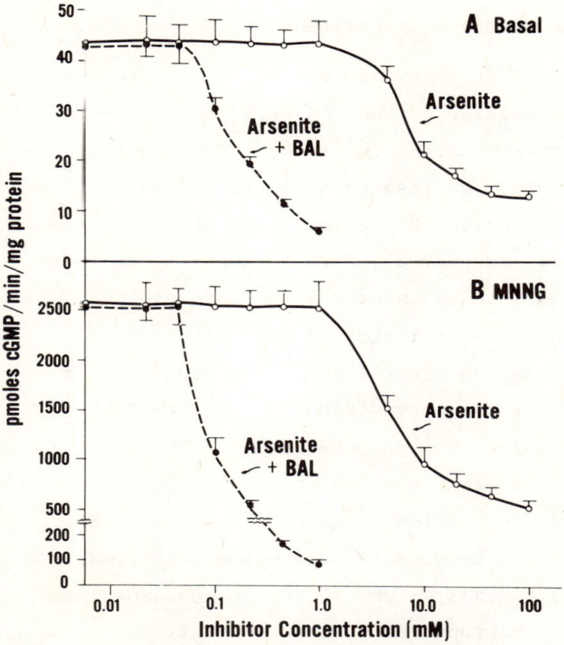

Figure 4 The concentration response relationship between arsenite vs. arsenite plus 2,3-dimercaprol and basal soluble hepatic guanylate cyclase activity (panel A) or MNNG stimulated guanylate cyclase activity (panel B). The 100,000 x *g* hepatic supernatant was preincubated for 10 min at 0-4° with indicated concentrations of arsenite alone (o—o) or with equimolar arsenite plus 2,3-dimercaprol (●-----●). At the end of the 10-min preincubation, a 25 μl aliquot of supernatant fraction was added to a 50 μl aliquot of guanylate cyclase reaction mixture with (panel B) or without (panel A) MNNG at a final concentration of 0.1 mM. Preincubation of hepatic supernatants with 0.01 to 1 mM 2,3-dimercaprol alone had no effect on basal or 0.1 mM MNNG stimulated guanylate cyclase activities. (From Ref. 88.)

F. Role of NO and NO-Heme in Enzyme Activation

Studies with MNNG indicated that guanylate cyclase activation was mediated by a labile decomposition product of this compound (Fig. 1). Concurrent with these observations, Murad and co-workers demonstrated the activation of guanylate cyclase by NO [64]. Since N-nitroso compounds, azide, nitroprusside, NH_2OH, and other nitro compounds which activate guanylate cyclase may all form NO under appropriate conditions [105-107,126-129], these investigators proposed that NO represented the common proximate activator of guanylate cyclase derived from all the precursor agonists [130]. The reduction of catalase-H_2O_2 (compound I) by NaN_3, resulting in the formation of NO is well known [126,128,131]. The proposed role of NO as proximate reactive species would explain the requirement of O_2 and catalase for the expression of NaN_3 action on guanylate cyclase [65, 108]. By contrast, MNNG and nitroprusside decompose nonenzymatically to yield NO [105-107,132,133]. The latter process does not require O_2. Since MNNG and nitroprusside do not require O_2 or catalase to increase guanylate cyclase activity [28], these observations are consistent with a role for NO as mediator of MNNG and nitroprusside action. Nevertheless, specific physicochemical identification of the proximate reactive moiety responsible for enzyme activation by MNNG and related agonists has been lacking. Based on the capacity of nitroprusside and NO to produce small increases in the activity of a partially purified preparation of soluble guanylate cyclase, it has been suggested that NO interacts directly with the enzyme without a requirement for other tissue components in the expression of its action [130,134,135]. Since NO oxidizes SH groups under anaerobic conditions [105], direct interaction of this moiety with thiol sites regulating enzyme activity remains a plausible mechanism of enzyme activation. However, when soluble hepatic guanylate cyclase activity is partially purified (140-fold) by DEAE* cellulose and agarose chromatography [136], the responsive-

*DEAE, diethylaminoethyl.

ness of the enzyme to NO, MNNG nitroprusside, nitrite, and azide is markedly reduced or abolished, compared to that observed in the hepatic supernatant fraction [137] (Table 4). Responses to NO, MNNG, nitroprusside, and nitrite, but not to azide, are restored by addition of heat-treated hepatic supernatant fraction to the purified enzyme. The heated supernatant contained no assayable guanylate cyclase or catalase activity. These findings suggested that heat-stable soluble tissue components other than the enzyme were required for expression of the action of NO, MNNG, nitrite, and nitroprusside. As shown in Table 5 [136a], addition of 5 μM free hematin or an equimolar concentration of protein-bound heme in the

Table 4 Effect of Partial Purification on Responsiveness of Hepatic Soluble Guanylate Cyclase to MNNG, NO, and Nitroprusside[a]

	Enzyme preparation (nMoles cGMP min^{-1} mg^{-1} protein)		
Test agent	Supernatant fraction	Partially purified enzyme	Partially purified enzyme + heated supernatant
None	0.038 ± 0.004	5.8 ± 0.5	5.5 ± 0.4
MNNG (1 mM)	2.1 ± 0.2*	5.7 ± 0.5	97 ± 8*
Nitroprusside (1 mM)	2.1 ± 0.3*	5.2 ± 0.6	89 ± 10*
NO	2.2 ± 0.3*	5.8 ± 0.5	104 ± 11*
$NaNO_2$ (10 mM)	0.51 ± 0.07*	5.6 ± 0.5	43 ± 5*
NaN_3 (1 mM)	0.76 ± 0.09*	5.0 ± 0.6	5.2 ± 0.6

[a]The 100,000 x *g* hepatic supernatant fraction was assayed without treatment or after heating at 95° for 15 min. The heated preparation was devoid of guanylate cyclase and catalase activity. Soluble enzyme was partially purified by DEAE cellulose chromatography, ammonium sulfate precipitation and agarose chromatography [136], in the presence of dithiothreitol and ethylenediaminetetraacetate(EDTA). Where indicated, the partially purified enzyme (5 μg protein) was preincubated for 1 min at 0° with the heated soluble fraction (2 mg protein). A 25 μl aliquot of each enzyme preparation was then added to 50 μl standard guanylate cyclase reaction mixtures (1 mM GTP, 4 mM $MnCl_2$) containing no further additions, MNNG, nitro prusside, $NaNO_2$, or NaN_3 at the final concentration shown. Where indicated, NO was delivered to the final assay tube at a rate of 1 ml/min for 2 sec under anaerobic conditions. Results are mean ± SE of determinations pooled from six separate experiments.

*$p < 0.01$ compared to corresponding value with no test agent.

Table 5 Effects of Hematin and Heme Proteins on Responsiveness of Partially Purified Guanylate Cyclase to MNNG, NO, and NaN_3[a]

Addition to partially purified soluble enzyme	Guanylate cyclase activity (nmoles cGMP min^{-1} mg^{-1} protein			
	Basal	MNNG	NO	NaN_3
None	5.2 ± 0.5	5.3 ± 0.6	5.4 ± 0.4	5.0 ± 0.6
Hematin	5.1 ± 0.6	102 ± 11*	118 ± 15*	4.8 ± 0.5
Hemoglobin	5.2 ± 0.6	97 ± 11*	111 ± 13*	5.3 ± 0.6
Methemoglobin	4.9 ± 0.5	105 ± 10*	124 ± 14*	5.2 ± 0.5
Catalase	5.2 ± 0.5	98 ± 10*	91 ± 9*	21 ± 3*
Inactivated catalase	5.3 ± 0.6	107 ± 10*	105 ± 12*	4.9 ± 0.6

[a]Soluble hepatic guanylate cyclase activity was purified 140-fold by ammonium sulfate precipitation, DEAE cellulose, and agarose chromatography [136]. Dithiothreitol was present at a concentration of 1 mM. Final concentrations of hematin, hemoglobin, methemoglobin, and catalase were 5 μM with respect to hematin or heme. Where indicated, catalase was inactivated by heating for 15 min at 95°. MNNG and NaN_3 were present at a final assay concentrate of 1 mM. NO was delivered as described in Table 4.

*$p < 0.01$ compared to corresponding value with no addition to the partially purified enzyme.

Source: Adapted from Craven and DeRubertis [137].

form of hemoglobin, methemoglobin, or intact or heat-inactivated catalase also restored the responsiveness of partially purified hepatic soluble guanylate cyclase to MNNG or NO. Similar results were obtained with nitroprusside or nitrite (not shown). By contrast, of the heme proteins tested in Table 5, only intact catalase effectively restored azide responsiveness. Protoprophyrin IX, dimethyl ester, $FeCl_2$, $FeCl_3$, the stable ADP-$FeCl_3$ complex, free and complexed $CuCl_2$, and $ZnCl_2$ all failed to enhance partially purified guanylate cyclase responsiveness to MNNG or NO.

These results implied that heme or hemeprotein may be specifically required for expression of the actions of NO and MNNG on guanylate cyclase. The studies shown in Table 5 were conducted in the presence of 1 mM dithiothreitol. Figures 5 and 6 demonstrate the potentiating effect of an appropriate concentration of a reducing agent on the responsiveness of partially purified enzyme to MNNG. In the presence of either 1 mM dithiothreitol (Fig. 5) or ascorbate (Fig. 6) plus hematin, responses of partially purified guanylate cyclase to MNNG, as well as to NO, nitroprusside, and nitrite [137], were greatly enhanced. In the presence of 1 mM dithiothreitol, the minimal concentration of hematin required for restoration of MNNG responsiveness fell from 1 μM to 20 nM, while the concentration of hematin required for optimal expression of MNNG action was reduced from 5.00 to 0.25 μM (Fig. 5). Similarly, addition of ascorbate with hematin (Fig. 6) potentiated MNNG-responsive activity compared to that observed with hematin alone, a change noted over a wide range of MNNG concentrations (1 μM to 3 mM). At 1 mM, ascorbate alone (no hematin) did not alter MNNG-responsive activity of partially purified guanylate cyclase while the combined addition of concentrations of 1 mM ascorbate and 5 μM hematin had no effect on basal activity. However, analogous to observations in crude hepatic supernatant fraction (Fig. 2), sufficiently high concentrations of dithiothreitol or ascorbate suppressed the responses of the partially purified enzyme activity to MNNG. These complex effects suggest that reducing agents may influence the process of

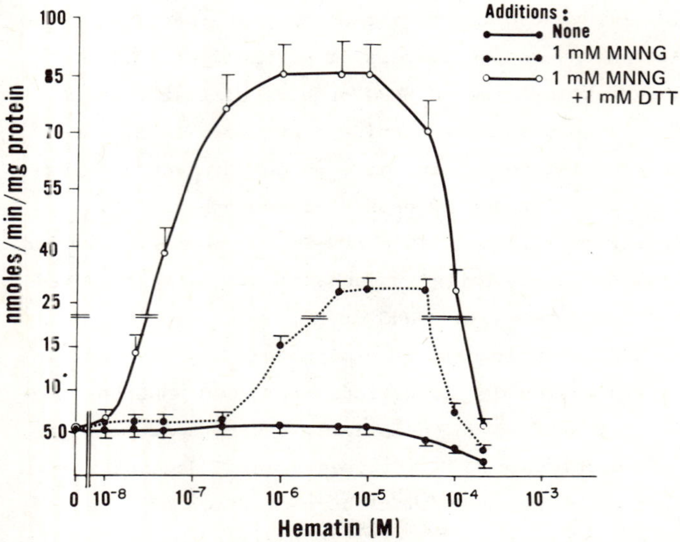

Figure 5 The effect of dithiothreitol on the concentration vs. response relationship between hematin- and MNNG-stimulated soluble hepatic guanylate cyclase activity. Partially purified guanylate cyclase was preincubated for 1 min at 0° with the indicated concentration of hematin (0 to 2.5 x 10^{-4} M) or hematin plus 1 mM dithiothreitol. At the end of the preincubation period, a 25 µl aliquot of the mixture was added to a 50 µl aliquot of guanylate cyclase assay mixture which contained no further additions of test agents or 1 mM MNNG. Dithiothreitol (1 mM) had no effect on enzyme activity measured in the absence of MNNG over the range of hematin concentrations tested. Results shown are means ± SE of duplicate determinations pooled from three separate experiments. (From Craven and DeRubertis [137].)

guanylate cyclase activation at more than one step. As is also shown in Figure 5, 50 µM or greater hematin suppressed basal and MNNG responsive guanylate cyclase activity compared to values at the lower hematin concentrations. Similar biphasic concentration dependent effects are observed with hemoglobin [136]. Several reports have emphasized only the inhibitory actions of hemoglobin on guanylate cyclase activities [134,135,138]. However, Hidaka and Asano [67] have demonstrated that hemoglobin has biphasic effects on the activation of platelet guanylate cyclase by fatty acid peroxides

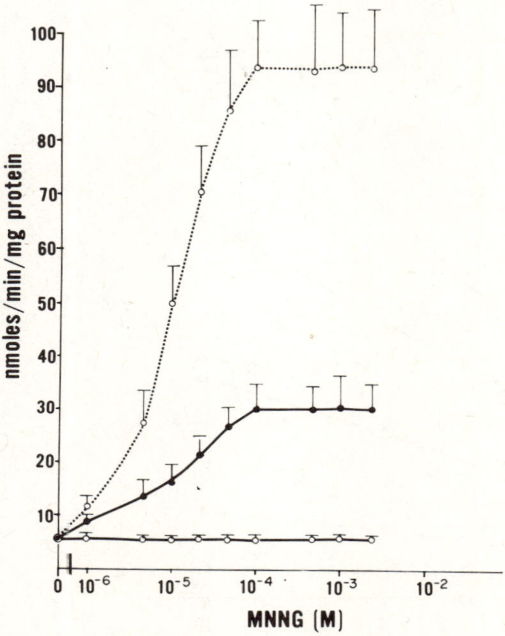

Figure 6 Concentration response relationship between MNNG and partially purified hepatic soluble guanylate cyclase activity. The enzyme was preincubated for 1 min at 0° with no additions (o—o), 5 μM hematin (●—●), or 5 μM hematin plus 1 mM ascorbate (o••••o). A 25-μl aliquot of the preincubation mixture was then added to a 50-μl aliquot of guanylate cyclase reaction mixture which contained MNNG at the indicated final concentration. Results shown are means ± SE of duplicate determinations from three separate experiments. (From Craven and DeRubertis [137].)

and enhances maximal responses to these agents. Thus, heme or heme proteins may be required for optimal expression of the action of many agonists of guanylate cyclase.

The conditions (heme + reducing agent) which restore responses of partially purified guanylate cyclase to MNNG, NO, nitrite, and nitroprusside may lead to the generation of paramagnetic nitrosyl heme complexes [104,132,133,139]. Accordingly, this complex was evaluated as a potential active intermediate responsible for enzyme stimulation. Figure 7 shows the effects on partially purified hepatic guanylate cyclase of nitrosyl-hemoglobin, which had been pre-

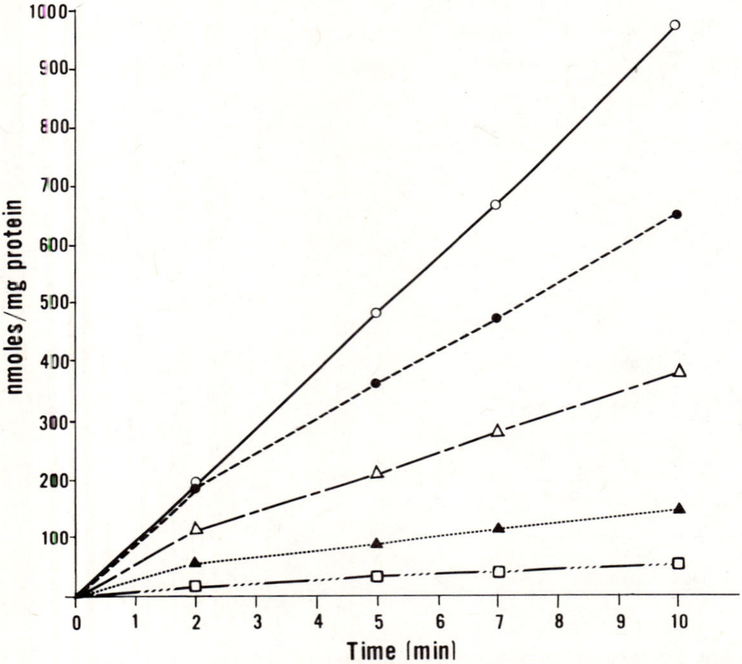

Figure 7 Effect of various concentrations of NO-hemoglobin on the time course of generation of cGMP by partially purified guanylate cyclase activity. Buffer (□) or NO-hemoglobin was added to the partially purified enzyme preparation to give a final concentration of 100 μM (o), 10 μM (●), 5 μM (Δ), or 0.1 μM (▲). (From Craven and DeRubertis [137].)

formed quantitively by the reaction of deoxyhemoglobin and NO under anaerobic conditions [139]. NO-hemoglobin markedly stimulates enzyme activity when tested over a wide concentration range (0.1 to 500 μM). The action of preformed NO-hemoglobin, unlike that of MNNG, NO, and nitroprusside, is optimally expressed without addition of a reducing agent. Storage of NO-hemoglobin solutions at -70° for 24 hr or in an O_2 atmosphere at 19° for 3 hr results in approximately 25% and 50% loss of capacity to activate guanylate cyclase, respectively. Incubation in N_2 for 3 hr at 19° was without effect. Loss of biological action on the enzyme correlated well with reductions in NO-hemoglobin content of the test solutions, as assessed concomitantly from ESR spectroscopy and as reported previously from other laboratories [140]. When incubated in the presence of low concentrations of NO-hemoglobin, the rate of cGMP generation from guanylate cyclase was nonlinear (Fig. 7). By contrast, under anaerobic conditions cGMP formation was linear at all stimulatory concentrations of NO-hemoglobin. These results implied an inhibitory effect of O_2 on NO-hemoglobin action, possibly related to oxidation of NO.

As shown in Figure 8, effects of NO-hemoglobin on partially purified guanylate cyclase differed qualitatively from those of O_2 or CO-hemoglobin. Over the same concentration range (0.1 to 500 μM) which leads to NO-hemoglobin activation of the enzyme, O_2 and CO-hemoglobin were either without effect or were inhibitory. Changes in the divalent cation requirements of partially purified guanylate cyclase were analogous after enzyme activation by either preformed NO-hemoglobin or the combination of MNNG, hematin, and a reducing agent. Thus, the proportional increases over basal in response to NO-hemoglobin or MNNG were fivefold greater with Mg^{2+} as sole divalent cation (100 x basal) than with Mn^{2+} (20 x basal). Activation by either preformed NO-hemoglobin or MNNG was also blocked by NEM and arsenite, indicating that the requirement for free thiol groups involves a step distal to generation of nitrosyl-heme [137].

The conditions required for restoration of the responsiveness of purified guanylate cyclase to MNNG or related agonists and the stimulatory action of preformed NO-hemoglobin are both consistent with a role for the latter complex as an obligate intermediate in enzyme activation [136]. Figure 9 shows the electron spin resonance (ESR) spectra of test solutions whose capacity to activate guanylate cyclase was examined concurrently. The ESR spectra of preformed NO-hemoglobin (Fig. 9A) and di-t-butylnitroxide (DTNB; Fig. 9F) are identical to those reported previously [139,141] and are shown for comparison. The characteristic ESR spectra of NO-hemoglobin was identified in incubation mixtures containing deoxyhemoglobin and cysteine, plus MNNG (Fig. 9B), nitroprusside (Fig.

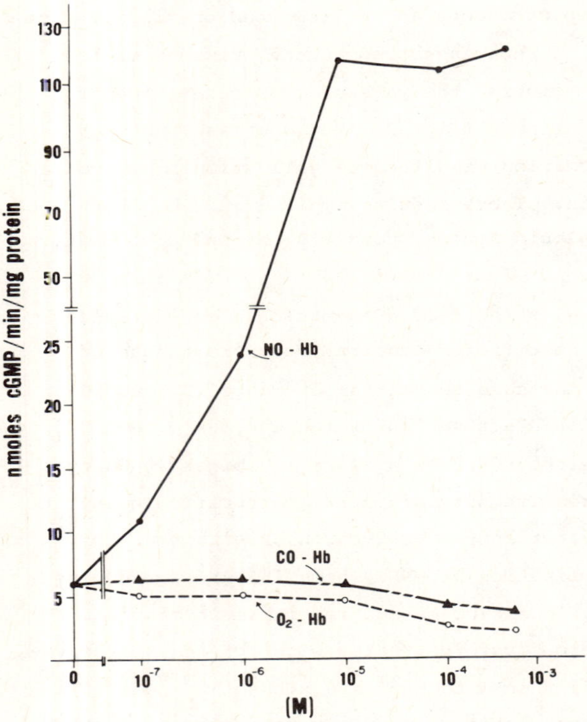

Figure 8 Comparison of the effects of NO (●—●), O_2 (o----o), and CO (▲-···-▲) hemoglobin on the activity of partially purified guanylate cyclase. Reactions were conducted for 2 min.

9D), or nitrite (Fig. 9E). In incubation mixtures containing nitroprusside and cysteine without added hemoproteins (Fig. 9C) the ESR spectrum of the pentacyanonitrosoferrate anion radical $[Fe(CN)_5NO]^{-3}$ was identified [142]. Accordingly, in incubations containing nitroprusside, cysteine, and deoxyhemoglobin, the amount of NO-hemoglobin formed was determined by subtraction of the spectra obtained in the absence of deoxyhemoglobin. As shown in Table 6, the test solutions containing NO-hemoglobin detectable by ESR spectroscopy all activated guanylate cyclase. NaN_3, which failed to form NO-hemoglobin under these conditions was ineffective. In the presence of cysteine, conversion of deoxyhemoglobin to NO-hemoglobin by reaction with MNNG

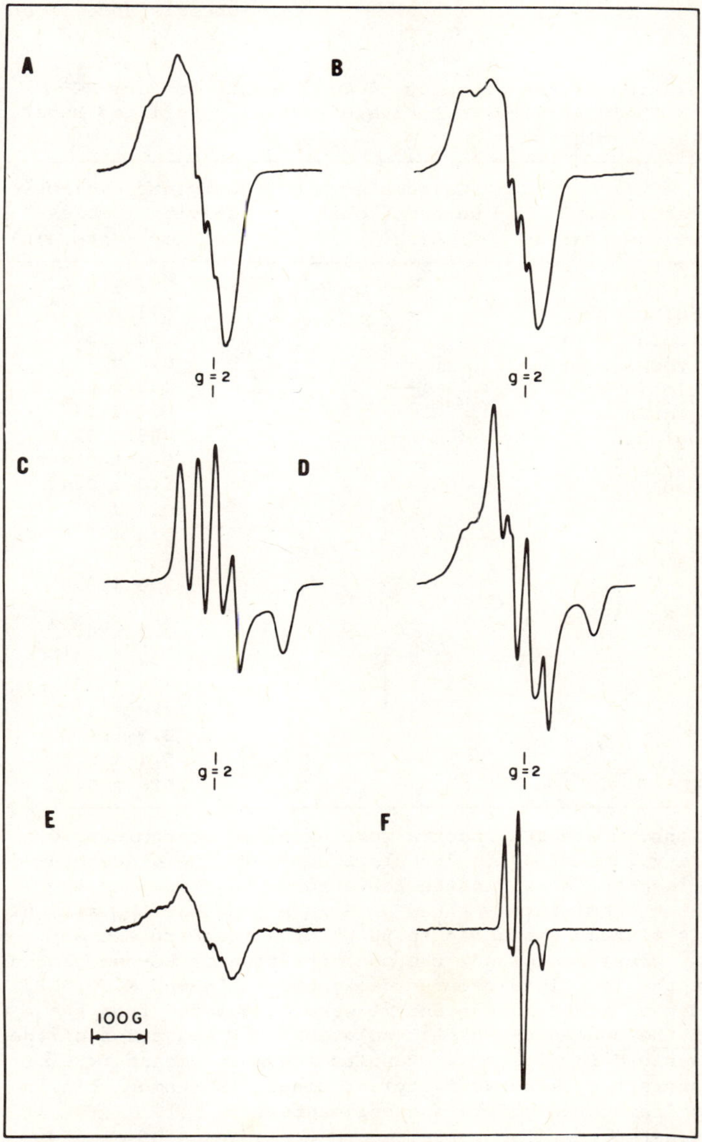

Figure 9 ESR spectra of guanylate cyclase test agents examined at −180°: (A) 0.5 mM NO-hemoglobin preformed by the reaction of NO with deoxyhemoglobin (instrument gain = 3.2×10^3); (B) 0.5 mM deoxyhemoglobin, 10 mM cysteine, 10 mM MNNG (gain = 2×10^3); (C) 10 mM cysteine, 10 mM nitroprusside (gain = 1.25×10^3); (D) nitroprusside; (E) 0.5 mM deoxyhemoglobin, 10 mM cysteine, 10 mM $NaNO_2$ (gain = 8×10^3); (F) 10 mM di-t-butylnitroxide in ethanol. All mixtures were prepared anaerobically. ESR spectra of mixtures B, C, D, and E were examined after a 2-min. incubation at 37°. Microwave power was 10 mW. (From DeRubertis et al. [136].)

Table 6 Correlation of the Capacity of Test Agents to Form NO-Hemoglobin with Their Ability to Activate Partially Purified Hepatic Soluble Guanylate Cyclase[a]

Additions to partially purified enzyme preparation	Calculated NO-hemoglobin (μM)	Guanylate cyclase activity (nmoles min^{-1} mg^{-1} protein)
None	—	5.2 ± 0.6
0.1 μM NO-hemoglobin	—	11 ± 1.2*
5 μM NO-hemoglobin	—	38 ± 5*
+ 45 μM deoxyhemoglobin	—	8.7 ± 1*
10 μM NO-hemoglobin	—	103 ± 10*
50 μM NO-hemoglobin	—	97 ± 11*
500 μM NO-hemoglobin	—	105 ± 12*
5 μM deoxyhemoglobin	—	5.3 ± 0.5
50 μM deoxyhemoglobin	—	3.8 ± 0.4*
+ 1 mM NP	5 ± 1	8.3 ± 1†
+ 1 mM MNNG	6 ± 1	9.2 ± 1†
+ 0.2 mM $K_3Fe(CN)_6$ + 1 mM MNNG	0	3.6 ± 0.4
50 μM deoxyhemoglobin + 1 mM cysteine	—	3.8 ± 0.4*
+ 1 mM MNNG	51 ± 10	95 ± 10†
+ 1 mM NP	49 ± 12	104 ± 11†
+ 1 mM $NaNO_2$	5 ± 1	7.9 ± 1†
+ 1 mM NaN_3	0	3.9 ± 0.4
1 mM DTBN	—	5.3 ± 0.5
1 mM $[Fe(CN)_5NO]^{3-}$	—	5.6 ± 0.5

[a]Test solutions, whose ESR spectra were examined concomitantly, were added in a volume of 20 to 180 μl of enzyme. The concentrations of each reactant shown in the table refer to those in the enzyme preincubation mixture. After a 1-min preincubation, a 25-μl aliquot of this mixture was added to 50 μl of a standard enzyme assay mixture. Where indicated, the concentration of NO-hemoglobin generated from the reactions between deoxyhemoglobin and MNNG, NP, or nitrite and then added to the enzyme were calculated from the ESR spectra of the concentrated test solutions. Preparation of the latter are described in Figure 3. Results shown are means ± SE of duplicate (guanylate cyclase activity) or single (NO-hemoglobin) determinations from three separate experiments.

*$p < 0.01$ compared to values obtained in the absence of an addition to the purified enzyme.

†$p < 0.01$ compared to values obtained with 50 μM deoxyhemoglobin in the presence or absence of 1 mM cysteine.

Source: From DeRubertis et al. [136].

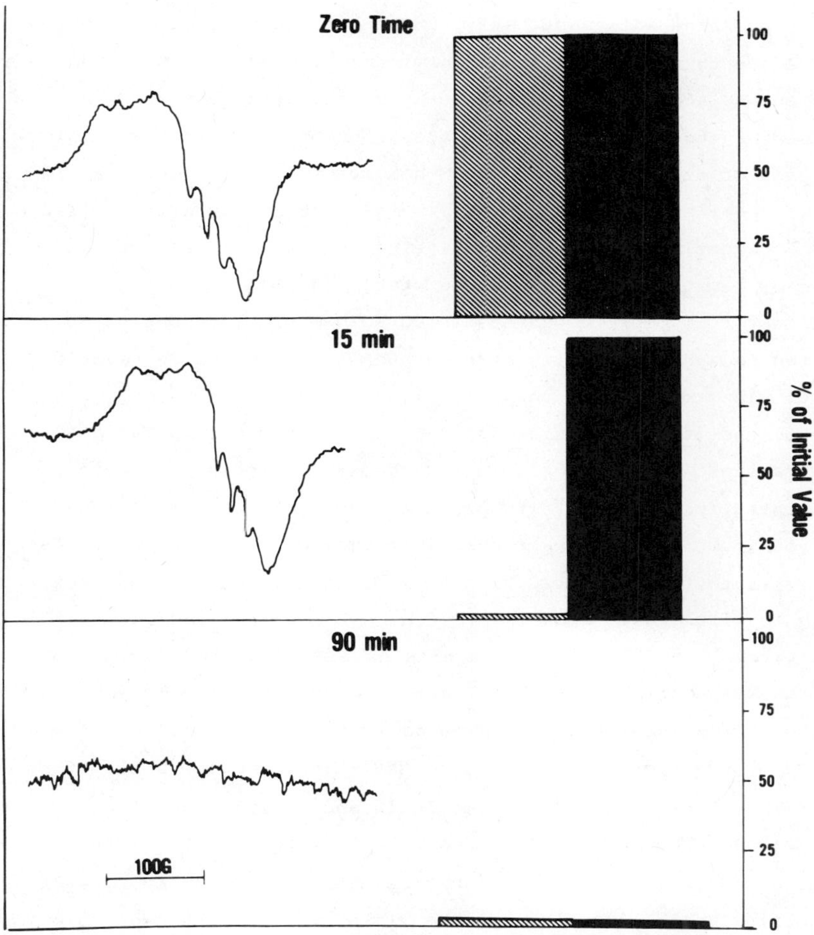

Figure 10 Relationship between capacities of MNNG solutions to generate NO-hemoglobin and activate guanylate cyclase. Nitroso group loss (cross-hatched bars) was examined at 400 nm at 0, 15, and 90 min after mixing 2 mM MNNG and 20 mM cysteine in 10 mM Tris (pH 7.6) at 37° under N_2. At 90 min the mixture was flushed with N_2 for 5 min before sampling. At each time point, an aliquot of the mixture was mixed with an equal volume of deoxyhemoglobin to give final concentrations of 0.5 mM deoxyhemoglobin, 10 mM cysteine, and 1 mM MNNG. NO-hemoglobin content of the mixture was determined by ESR spectroscopy. A second 20 μl aliquot was then added to 180 μl of a partially purified preparation of soluble hepatic guanylate cyclase to assess the capacity of the MNNG-cysteine solutions to stimulate enzyme activity (solid bars).

and nitroprusside was quantitative and the concentration of NO-hemoglobin added to the enzyme reaction mixture (50 μM) was more than sufficient to account for maximal enxyme activation based on the responses of the enzyme to a homogeneous preparation of the complex. Conversion of deoxyhemoglobin to NO-hemoglobin is less efficient in the absence of a reducing agent or in the presence of nitrite (Table 6). The potentiating effect of cysteine is likely related to maintenance of heme iron in the reduced state [132,133,139]. Under conditions of incomplete conversion to NO-hemoglobin, considerable unreacted deoxyhemoglobin is carried over into the enzyme reaction mixture. This suppresses enzyme activation by both NO-hemoglobin (Table 6) and other agonists [134-136,138]. As shown in Table 6, approximately 5 μM NO-hemoglobin (10% deoxyhemoglobin conversion) is generated from MNNG or nitroprusside (no cysteine) or nitrite plus cysteine. Activation of guanylate cyclase by these solutions was similar to that observed with 5 μM NO-hemoglobin + 45 μM deoxyhemoglobin (Table 6). DTBN and $[Fe(CN)_5NO^{-3}]$ failed to activate guanylate cyclase. Since these agents have ESR spectra in the same region as NO-hemoglobin (Fig. 1), some specificity in the action of the nitrosyl-heme complex on enzyme activity is implied. As shown in Table 6, addition of $K_3Fe(CN)_6$ to MNNG-deoxyhemoglobin solutions blocked both formation of NO-hemoglobin and stimulation of guanylate cyclase. The former action of $K_3Fe(CN)_6$ is due to oxidation of heme iron [143]. Suppression of nitrosylheme formation by $K_3Fe(CN)_6$ may explain its ability to block guanylate cyclase responses to MNNG and other nitro agonists [134].

As noted earlier (Fig. 1), activation of guanylate cyclase is mediated by a labile intermediate derived from MNNG. ESR studies indicated that the capacity of solutions of decomposed MNNG to activate guanylate cyclase correlate temporally with their ability to generate NO-hemoglobin [136] (Fig. 10). Thus, after 15 min of incubation with cysteine, MNNG is completely decomposed but the solutions retain full capacity both to generate NO-hemoglobin and to activate guanylate cyclase. At 90 min, both actions are lost (Fig. 10). These observations provide physicochemical evidence for the postulated role of NO [134] as the mediator of MNNG actions on the enzyme. Moreover, since heme or hemeprotein is required for enzyme activation, they also indicate that activation of guanylate cyclase by MNNG requires formation of the NO-heme complex rather than NO per se.

Similarly, expression of the action of NaN_3 and NH_2OH on purified preparations of guanylate cyclase requires the presence of catalase [64,65,134]. The reaction of azide or NH_2OH with catalase is known to lead to the formation of NO-catalase [128,144]. NO-catalase was preformed by reacting NO gas and catalase under anaerobic conditions. The extent of conversion of catalase to NO-catalase was assessed spectrophotmetrically (E_{580} = 16.7 μM^{-1}, cm^{-1}. A typical ESR spectrum of NO-catalase was observed and is shown (insert of Fig. 11). Addition of preformed NO-catalase activated partially purified guanylate cyclase in a concentration dependent fashion. Accordingly, this parmagnetic nitrosyl-heme complex may be the critical moiety mediating the actions of azide and NH_2OH on guanylate cyclase.

The mechanism by which the NO-heme complex activates guanylate cyclase is uncertain. Inhibition of its action by reducing agents and thiol group blockers implicate SH oxidation. NO is known to convert cysteine to cystine quantitatively under anaerobic conditions [105]. However, this agent is quite labile and rapidly converted to higher oxides of nitrogen under aerobic conditions. The NO-heme complex is much more stable, a property which may permit more effective function as an intermediate in an electron transport system leading to guanylate cyclase activation in biological systems.

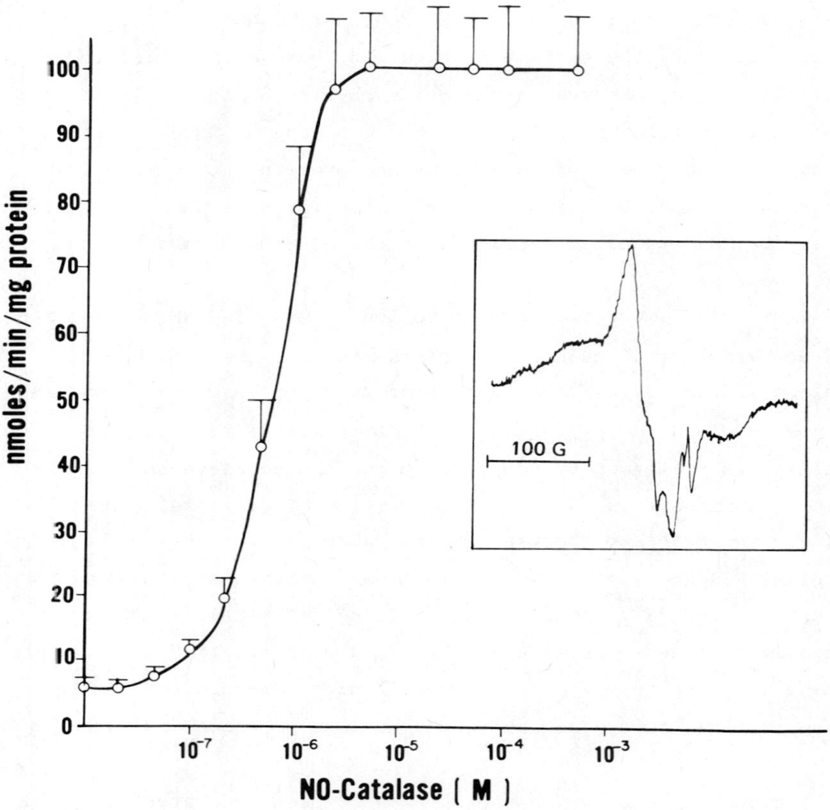

Figure 11 Concentration vs. response relationship between partially purified guanylate cyclase and NO-catalase, preformed from the reaction between NO and bovine liver catalase. ESR spectrum of the stock test solution was examined concomitantly and is shown in the insert. (From Craven and DeRubertis [136a].)

The possibility that heme groups are components of guanylate cyclase which are removed during purification cannot be excluded. This situation is recognized with L-tryptophan-2,3-dioxygenase [147-149]. Loss of heme during purification of this enzyme from liver results in loss of enzyme activity. Activity is restored by addition of hematin or hemeprotein to the purified liver preparation [147,148]. Moreover, L-tryptophan-2,3-dioxygenase has been isolated from a pseudomonad with two hemes bound per mole of enzyme, and the ESR spectral properties of the nitrosyl-heme complex of the latter enzyme have been extensively examined [149]. The heme of hemoglobin will exchange with other proteins under physiological conditions [150,151]. This exchange is enhanced by partial saturation with NO [152,153]. Thus, the ability to substitute hemoglobin for hematin in the restoration of guanylate cyclase responsiveness does not preclude binding of exchangeable heme groups to the enzyme under enzyme assay conditions. Binding of free hematin to partially purified guanylate cyclase can be readily demonstrated. After chromatographic removal of unbound hematin, the enzyme can be activated by MNNG or NO in the presence of a reducing agent [137]. However, assessment of the specificity of heme binding awaits study of a homogeneous preparation of guanylate cyclase.

G. Activation of Guanylate Cyclase by Other Carcinogens and Free Radical Species

N-Nitroso compounds were the first class of chemical carcinogens recognized to activate guanylate cyclase [27]. Subsequently, enzyme activation was demonstrated in response to hydrazine [34], butadiene diepoxide [35], and 4-nitroquinoline-1-oxide [29]. While less extensively studied, it seems likely that these compounds also enhance enzyme activity by free radical or oxidative mechanisms. Their actions are inhibited by antioxidants or radical scavengers [109]. Further, the effects of 4-nitroquinoline-1-oxide are enhanced by incubation conditions which favor its conversion to N-nitroso and/or radical intermediates [29]. Several classes of carcinogens tested at high concentrations in tissue supernatants have been reported to have no effect, or to inhibit, enzyme activity [33,154]. These agents include aromatic amines, polycyclic hydrocarbons, azo dyes, and aflatoxins. However, most of the agents reported to be ineffective or inhibitory require enzymatic conversion to an active form [36]. Since effects on guanylate cyclase were examined in tissue supernatant fractions under conditions that may not permit carcinogen activation, the results must be viewed with some caution. Thus, while the procarcinogen 2-acetylaminofluorene is without stimulatory effect on guanylate cyclase in tissue supernatant fractions [55], it is more relevant to examine the action of the N-hydroxyl derivative of this agent under conditions known to lead to formation of nitroso and radical intermediates [156-158]. Until such studies are performed with this and other procarcinogens, the extent to which structurally diverse carcinogens have a capacity to activate guanylate cyclase cannot be clearly answered.

As discussed earlier, stimulation of partially purified guanylate cyclase by MNNG requires the addition of heme plus a reducing agent. Moreover, formation of a nitrosyl-heme complex appears to be a step in the pathway leading to enzyme activation by MNNG. Nevertheless, it is possible that other radical species derived from MNNG could mediate enzyme activation by this and other N-nitroso carcinogens. Recent reports have described the trapping of two highly reactive radicals in solutions of nitrosoamines exposed to ultraviolet (UV) light [159]. Formation of one of these radicals requires O_2, whereas formation of the second radical occurs in a N_2 atmosphere [159]. Stimulation of guanylate cyclase by MNNG occurs in the absence of O_2. Thus, a radical species such as that detected by Floyd in anaerobic solutions of nitrosoamines [159] could play a role in the activation of guanylate cyclase by these agents.

IV. POTENTIAL CELLULAR MECHANISMS FOR EXPRESSION OF ACTIONS OF cGMP ON CELLULAR PROLIFERATION

Cyclic nucleotide-dependent protein kinases are thought to be the major receptors for both cAMP and cGMP [160]. Structurally distinct cAMP- and cGMP-dependent protein kinase activities have now been purified from mammalian tissues [160-165]. In contrast to the cAMP-dependent enzyme, the cGMP kinase is not dissociated into subunits by cGMP [164,165]. Both the nucleotide specificity and structural differences suggest that cAMP and cGMP kinases may mediate distinct cellular actions, although considerable overlap has been observed in the substrate specificities of the two kinases [166]. Little is presently known about the endogenous substrates for cGMP kinase activity. However, stimulation of the phosphorylation of both cell membranes [167,168] and nuclear acidic proteins [169] by cGMP has been reported. Accordingly, it is possible that cGMP could alter cell growth characteristics through effects on cell surface properties or genome activity. Several studies have also reported actions of cGMP on membrane Ca^{2+} permeability [170-174], which are presumably mediated by a phosphorylation mechanism [165,166]. Since Ca^{2+} has been implicated in the modulation of cell growth [2], effects of cGMP may be expressed indirectly via Ca^{2+}. Delineation of these possibilities awaits further investigation.

V. SUMMARY

Activation of the guanylate cyclase system of mammalian tissues appears to involve a free radical or oxidative mechanism. Among the agents capable of stimulating enzyme activity are N-nitroso compounds and other well-recognized chemical carcinogens, all of which may either form reactive oxides of nitrogen or act as oxidants. Activation of guanylate cyclase by N-nitroso compounds is inhibited by appropriate concentrations of antioxidants and radical scavengers. N-nitroso stimulation of the guanylate cyclase-cGMP system in intact and broken cell preparations requires free thiol groups but occurs in the absence of both O_2 and Ca^{2+}. Studies of partially purified hepatic soluble guanylate cyclase indicate that activation by nitrosoguanidine and related agonists, including nitric oxide, is dependent upon the presence of heme groups and probably involves formation of the paramagnetic nitrosyl-heme complex. This unique mechanism of guanylate cyclase activation may permit its stimulation by the active forms of many chemical carcinogens. The resultant acute enhancement of cGMP generation could play a role in the induction of malignant conversion.

ACKNOWLEDGMENTS

The support of the General Medical Research Service of the Veterans Administration, U. S. Public Health Service (Grants 2 R26 CA 16908 and 1 R01 AM 20187-GMB), and the Health Research and Services Foundation, Pittsburgh, Pennsylvania (Grant S-75) is gratefully acknowledged. We are indebted to Dr. David W. Pratt of the Department of Chemistry, University of Pittsburgh, for his assistance in ESR spectroscopy.

REFERENCES:

1. I. H. Pastan, G. S. Johnson, and W. B. Anderson, *Ann. Rev. Biochem. 44*:491-522 (1975).
2. M. J. Berridge, *Adv. Cyclic Nucleotide Res. 6*:1-98 (1975).
3. F. J. Chlapowski, L. A. Kelly, and R. W. Butcher, *Adv. Cyclic Nucleotide Res. 6*:245-338 (1975).
4. D. L. Friedman, R. A. Johnson, and C. E. Zeilig, *Adv. Cyclic Nucleotide Res. 7*:69-114 (1976).
5. J. F. Whitfield, J. R. MacManus, T. Youdale, and D. J. Franks, *J. Cell Physiol. 78*:355-368 (1971).
6. R. W. Gilette, G. O. McKenzie, and M. H. Swanson, *J. Reticuloendothel. Soc. 16*:289-299 (1974).
7. Y. Weinstein, D. A. Chamber, H. R. Bourne, and K. L. Melmon, *Nature 251*:352-353 (1974).
8. M. Lambert, J. Camus, and J. Christophe, *Biochem. Pharmacol. 24*:1755-1758 (1975).
9. J. Watson, *J. Exp. Med. 141*:97-111 (1975).
10. T. Diamantstein and A. Ulmer, *Nature 256*:418-419 (1975).

11. F. R. DeRubertis and T. V. Zenser, *Biochim. Biophys. Acta 428*:91-103 (1976).
12. A. Novogrodsky and E. Katchalski, *Biochim. Biophys. Acta 215*: 291-296 (1970).
13. J. M. Smith, A. L. Steiner, N. M. Newberry Jr., and C. W. Parker, *J. Clin. Invest. 50*:432-441 (1971).
14. T. Diamantstein and A. Ulmer, *Immunology 28*:113-119 (1975).
15. F. R. DeRubertis, T. V. Zenser, W. H. Adler, and T. Hudson, *J. Immunol. 113*:151-161 (1974).
16. F. W. Thomas, F. Murad, W. B. Looney, and H. P. Morris, *Biochim. Biophys. Acta 297*:564-567 (1973).
17. W. F. Criss, F. Murad, and H. Kimura, *J. Cyclic Nucleotide Res. 2*:11-19 (1976).
18. F. R. DeRubertis, R. Chayoth, and J. B. Field, *J. Clin. Invest. 57*:641-649 (1976).
19. F. R. DeRubertis and P. A. Craven, *Cancer Res. 37*:15-21 (1977).
20. J. Voorhess and E. A. Duell, *Adv. Cyclic Nucleotide Res. 5*:735-758 (1974).
21. W. Moens, A. Vokaer, and R. Kram, *Proc. Nat. Acad. Sci. U.S. 72*: 1063-1067 (1975).
22. D. Schlondorff and H. Weber, *Proc. Nat. Acad. Sci. U.S. 73*:524-528 (1976).
23. W. L. Ryan and M. L. Heidrich, *Science 162*:1484-1485 (1968).
24. J. Sheppard, *Proc. Nat. Acad. Sci. U.S. 68*:1316-1320 (1971).
25. I. H. Pastan and C. S. Johnson, *Adv. Cancer Res. 19*:303-329 (1974).
26. N. D. Goldberg, M. K. Haddox, R. Estensten, J. G. White, C. Lopez, and J. W. Hadden, in *Cyclic AMP, Cell Growth and the Immune Response* (W. Braun, L. M. Lichenstine, and C. W. Parker, Eds.). Springer-Verlag, New York, 1974, pp. 247-261.
27. F. R. DeRubertis and P. A. Craven, *Science 193*:897-899 (1976).
28. F. R. DeRubertis and P. A. Craven, *J. Biol. Chem. 252*:5804-5814 (1977).
29. F. R. DeRubertis and P. A. Craven, *J. Nat. Cancer Inst. 59*: 1741-1745 (1977).
30. F. R. DeRubertis and P. A. Craven, *Cancer 40*:3600-3608 (1977).
31. F. R. DeRubertis and P. A. Craven, *Biochim. Biophys. Acta 499*: 337-351 (1977).
32. D. L. Vesely, L. E. Rovere, and G. S. Levey, *Cancer Res. 37*: 28-31 (1977).
33. D. L. Vesely and G. S. Levey, *Proc. Soc. Exp. Biol. Med. 155*: 301-304 (1977).
34. D. L. Vesely and G. S. Levey, *Biochem. Biophys. Res. Commun. 74*:780-784 (1977).
35. D. L. Vesely and G. S. Levey, *Enzymes 23*:140-143 (1978).
36. J. A. Miller, *Cancer Res. 30*:559-576 (1970).
37. D. F. Ashman, R. Lipton, M. M. Melicorn, T. D. Price, *Biochem. Biophys. Res. Commun. 11*:330-334 (1963).

38. N. D. Goldberg, S. B. Dietz, and A. G. O'Toole, *J. Biol. Chem.* *244*:4458-4466 (1969).
39. E. Ishikawa, S. Ishikawa, J. W. Davis, and E. W. Sutherland, *J. Biol. Chem.* *244*:6371-6376 (1969).
40. A. L. Steiner, A. S. Pagliara, L. R. Chase, and D. N. Kipnis, *J. Biol. Chem.* *47*:1114-1120 (1972).
41. A. L. Steiner, C. W. Parker, and D. M. Kipnis, *J. Biol. Chem.* *247*:1106-1113 (1972).
42. J. F. Harper and G. Brooker, *J. Cyclic Nucleotide Res.* *1*:207-218 (1975).
43. E. K. Frandsen and G. Krishna, *Life Sci.* *18*:529-542 (1976).
44. T. P. Zimmerman, M. S. Winston, and L.-C. Chu, *Anal. Biochem.* *71*:79-95 (1976).
45. N. D. Golberg, R. F. O'Dea, and M. K. Haddox, *Adv. Cyclic Nucleotide Res.* *3*:155-223 (1972).
46. N. D. Goldberg and M. K. Haddox, *Ann. Rev. Biochem.* *46*:823-896 (1977).
47. G. Schultz, J. G. Hardman, K. Schultz, C. E. Baird, and E. N. Sutherland, *Proc. Nat. Acad. Sci. U.S.* *70*:3889-3893 (1973).
48. G. Schultz and J. Hardman, *Adv. Cyclic Nucleotide Res.* *5*:339-351 (1975).
49. F. A. Ferrendelli, D. A. Kinscherf, and M. M. Chang, *Mol. Pharmacol.* *9*:445-454 (1973).
50. J. Van Sande, C. DeCoster, and J. E. Dumont, *Biochem. Biophys. Res. Commun.* *62*:168-175 (1975).
51. F. R. Butcher, *Metabolism* *24*:409-418 (1975).
52. F. R. DeRubertis and P. A. Craven, *Metabolism* *25*:1113-1127 (1976).
53. R. J. Smith and L. J. Ignarro, *Proc. Nat. Acad. Sci. U.S.* *72*:108-112 (1975).
54. R. I. Clyman, A. S. Blacksin, V. C. Manganiello, and M. Vaughan, *Proc. Nat. Acad. Sci. U.S.* *72*:3883-3887 (1975).
55. R. H. Painter, F. R. Butcher, and J. N. Fain, *J. Biol. Chem.* *251*:2987-2992 (1976).
56. S. Katsuki and F. Murad, *Mol. Pharmacol.* *13*:330-341 (1977).
57. F. R. DeRubertis and P. A. Craven, *Metabolism* *27*:855-868 (1978).
58. P. Blaustein and D. E. Goodman, *J. Gen. Physiol.* *49*:1043-1063 (1966).
59. M. Kohlhardt, B. Bauer, H. Krause, and A. Fleckenstein, *Pflugers Arch. Ges. Physiol.* *335*:309-322 (1972).
60. R. J. Gullis, J. Truber, K. Fisher, C. Buchin, and B. Hamprecht, *FEBS Lett.* *59*:74-79 (1975).
61. F. R. DeRubertis and P. A. Craven, *Clin. Res.* *24*:398A (1976).
62. H. Kimura, C. Mittal, and F. Murad, *Nature* *257*:700-702 (1975).
63. F. R. DeRubertis and P. A. Craven, *J. Biol. Chem.* *251*:4651-4658 (1976).
64. S. Katsuki, W. Arnold, C. Mittal, and F. Murad, *J. Cyclic Nucleotide Res.* *3*:23-35 (1977).

65. C. Mittal, H. Kimura, and F. Murad, *J. Biol. Chem. 252*: 4384-490 (1977).
66. C. Mittal and F. Murad, *Proc. Nat. Acad. Sci. U.S. 74*: 4360-464 (2977).
67. H. Hidaka and T. Asano, *Proc. Nat. Acad. Sci. U.S. 74*: 3657-3661 (1977).
68. N. D. Goldberg, G. Graff, M. K. Haddox, J. H. Stephenson, D. B. Glass, and M. E. Moser, in *Adv. Cyclic Nucleotide Res. 9*: 101-130 (1978).
69. R. W. Egan, J. Paxton, and F. A. Kuehl, *J. Biol. Chem. 251*: 7329-7335 (1976).
70. K.-L. Fong, P. B. McCay, and J. L. Poyer, *J. Biol. Chem. 248*: 7792-7797 (1973).
71. S. Cannistraro and A. Von de Vorst, *Biochem. Biophys. Res. Commun. 74*:1177-1185 (1977).
72. R. B. Koch, *Arch. Biochem. Biophys. 125*:303-307 (1968).
73. A. Derksen and P. Cohen, *J. Biol. Chem. 250*:9342-9347 (1975).
74. R. J. Flower, *Pharmacol. Rev. 29*:33-67 (1974).
75. R. J. Gullis and C. E. Rowe, *Biochem. J. 148*:197-208 (1975).
76. J. G. Hardman and E. W. Sutherland, *J. Biol. Chem. 244*:6363-6370 (1969).
77. A. A. White and G. D. Aurbach, *Biochem. Biophys. Acta 191*:686-697 (1969).
78. G. Schultze, E. Bohme, and K. Munske, *Life Sci. 8*:1323-1332 (1969).
79. H. Kimura and F. Murad, *Metabolism 24*:439-445 (1975).
80. A. A. White, *Adv. Cyclic Nucleotide Res. 5*:353-375 (1975).
81. H. Kimura and F. Murad, *J. Biol. Chem. 250*:4810-4817 (1975).
82. E. Bohme, R. Jung, and S. Mechler, *Methods Enzymol. 38c*:199-202 (1974).
83. P. A. Craven and F. R. DeRubertis, *Biochemistry 15*:5131-5137 (1976).
84. D. Wallach and I. Pastan, *J. Biol. Chem. 251*:5802-5809 (1976).
85. J. G. Hardman, J. A. Beavo, J. P. Gray, T. D. Chrisman, W. D. Patterson, and E. W. Sutherland, *Ann. N.Y. Acad. Sci. 185*: 27-35 (1971).
86. D. L. Garbers, *J. Biol. Chem. 251*:4071-4077 (1976).
87. D. L. Garbers, *J. Biol. Chem. 253*:1898-1901 (1978).
88. P. A. Craven and F. R. DeRubertis, *Biochem. Biophys. Acta 524*: 231-244 (1978).
89. S. H. Ong, T. H. Whitley, N. W. Stowe, and A. L. Steiner, *Proc. Nat. Acad. Sci. U.S. 72*:2022-2026 (1975).
90. A. L. Steiner, S.-L. Ong., and H. J. Wedner, *Adv. Cyclic Nucleotide Res. 7*:116-155 (1976).
91. G. A. Robison, R. W. Butcher, and E. W. Sutherland, *Cyclic AMP. Academic Press*, New York, 1971.
92. H. Kimura, C. K. Mittal, and F. Murad, *J. Biol. Chem. 250*:8016-8022 (1975).
93. R. E. Thiers and B. L. Vallee, *J. Biol. Chem. 226*:911-920 (1957).

94. W. H. Frey II, B. M. Bowman, D. Newman, and N. D. Goldberg, *Fed. Proc. 35*:1750 (1976).
95. D. Wallach and I. Pastan, *J. Biol. Chem. 251*:5802-5809 (1976).
96. P. A. Craven and F. R. DeRubertis, *Anal. Biochem. 72*:455-459 (1976).
97. T. D. Chrisman, D. L. Garbers, M. A. Parks, and J. G. Hardman, *J. Biol. Chem. 250*:374-381 (1975).
98. R. C. Kraska, J. H. Stephenson, and N. D. Goldberg, *Fed. Proc. 36*:686 (1977).
99. M. K. Haddox, L. T. Furcht, M. E. Moser, C. C. Bull, J. H. Stephenson, and N. D. Goldberg, in *Cyclic Nucleotides and Regulation of Cell Growth*. Dowden, Hutchinson and Ross, Stroudsburg, PA, 1976, pp. 207-224.
100. A. A. White, K. M. Crawford, C. S. Patt, and P. J. Lad, *J. Biol. Chem. 251*:7304-7312 (1976).
101. M. R. Haddox, L. T. Furcht, S. R. Gentry, M. F. Moser, J. H. Stephenson, and N. D. Goldberg, *Nature 262*:146-148 (1976).
102. F. Haber and J. Weiss, *Proc. Roy. Soc. Edinburgh, Sect. A 147*: 332-351 (1934).
103. J. H. Boxendale, M. G. Evans, and C. S. Parks, *Trans. Faraday Soc. 42*:155 (1946).
104. C. Nagata, Y. Ioki, M. Kodama, Y. Tagashira, and M. Nakadate, *Ann. N.Y. Acad. Sci. 222*:1031-1047 (1973).
105. V. Schulz and D. R. McCalla, *Can. J. Chem. 47*:2021-2027 (1969).
106. J. D. Morrisett and H. R. Drott, *J. Biol. Chem. 244*:5083-5084 (1969).
107. P. D. Lawley and C. J. Thatcher, *Biochem. J. 116*:693-707 (1970).
108. N. Miki, M. Nagaro, and K. Kuriyamo, *Biochem. Biophys. Res. Commun. 72*:952-959 (1976).
109. P. A. Craven and F. R. DeRubertis, *Cancer Res. 37*:4088-4097 (1977).
110. L. W. Wattenberg, *Am. J. Dig. Dis. 19*:947-953 (1947).
111. L. W. Wattenberg, W. D. Laub, L. K. Lam, and J. L. Speier, *Fed. Proc. 35*:1327-1330 (1976).
112. M. B. Sporn, H. M. Dunlap, D. L. Newton, and J. M. Smith, *Fed. Proc. 35*:1332-1338 (1976).
113. M. B. Sporn, R. A. Squire, C. C. Brown, J. M. Smith, M. L. Wenk, and S. Springer, *Science 195*:487-489 (1977).
114. D. P. Chapra and L. J. Wilkoff, *J. Nat. Cancer Inst. 56*:583-589 (1976).
115. T. Narisawa, B. S. Reddy, C.-W. Wong, and J. H. Weisburger, *Cancer Res. 36*:1379-1383 (1976).
116. P. A. Craven and F. R. DeRubertis, *Fed. Proc. 35*:1907 (1977).
117. W. J. Thompson, R. H. Williams, and S. A. Little, *Biochim. Biophys. Acta 302*:329-337 (1973).
118. D. L. Garbers and J. G. Hardman, *J. Biol. Chem. 250*:2482-2486 (1975).
119. J. P. Durham, *Eur. J. Biochem. 61*:535-544 (1976).
120. W. B. Jakoby, *J. Biol. Chem. 232*:89-97 (1958).
121. A. L. Fluharty and D. R. Sanadi, *J. Biol. Chem. 236*:2772-2778 (1961).

122. M. DeLuca, G. W. Wertz, and W. D. McElroy, *Biochemistry 3*:935-939 (1964).
123. B. P. Gaber and A. I. Fluharty, *Q. Rep. Sulfur Chem. 3*:318-337 (1968).
124. A. M. Spiegel, E. M. Brown, and G. D. Aurbach, *J. Cyclic Nucleotide Res. 2*:393-404 (1976).
125. L. A. Stocken and R. H. S. Thompson, *Biochem. J. 40*:535-548 (1948).
126. D. Keilen and E. F. Hartree, *Nature 173*:720-723 (1954).
127. J. C. Woolum and B. Commoner, *Biochim. Biophys. Acta 201*:131-140 (1970).
128. P. Nicholls, *Biochem. J. 90*:331-343 (1964).
129. E. Freese and E. B. Freese, *Biochemistry 4*:2419-2433 (1965).
130. W. P. Arnold, C. K. Mittal, S. Katsuri and F. Murad, *Proc. Nat. Acad. Sci. U.S. 74*:3203-3207 (1977).
131. H. Theorell and A. Ehrenberg, *Arch. Biochem. Biophys. 41*:462-474 (1952).
132. R. P. Smith and H. Kruszyma, *J. Pharmacol. Exp. Ther. 191*:557-563 (1974).
133. B. Benko and Vuk-Pavlovic, *Biochem. Biophys. Res. Commun. 71*:1303-1307 (1967).
134. C. Mittal and F. Murad, *J. Cyclic Nucleotide Res. 3*:381-392 (1977).
135. C. K. Mittal, W. P. Arnold, and F. Murad, *J. Biol. Chem. 253*:1266-1271 (1978).
136. F. R. DeRubertis, P. A. Craven, and D. W. Pratt, *Biochem. Biophys. Res. Commun. 83*:158-167 (1978).
136a. P. A. Craven and F. R. DeRubertis, *J. Biol. Chem. 254*:8213-8222 (1979).
137. P. A. Craven and F. R. DeRubertis, *J. Biol. Chem. 253*:8433-8443 (1978).
138. N. Miki, Y. Kawabe, and K. Kuriyama, *Biochem. Biophys. Res. Commun. 75*:851-856 (1977).
139. H. Kon, *J. Biol. Chem. 243*:4350-4357 (1968).
140. Q. H. Gibson and F. J. M. Roughton, *J. Physiol. (London) 136*:507-526 (1957).
141. O. H. Griffith, D. W. Cornell, and H. M. McConnell, *J. Chem. Phys. 43*:2909-2912 (1965).
142. J. B. Raynor, *Nature 201*:1216-1217 (1964).
143. E. Antonini and M. Brunori, *Frontiers Biol. 21*:31-33 (1971).
144. T. Yonetani, H. Yamamoto, J. E. Erman, J. S. Leigh Jr., and G. H. Reed, *J. Biol. Chem. 247*:2447-2455 (1972).
145. T. Yonetani, H. R. Drott, J. S. Leigh, G. H. Reed, M. R. Waterman, and T. Asakura, *J. Biol. Chem. 245*:2998-3003 (1970).
146. B. M. Hoffman and D. H. Petering, *Proc. Nat. Acad. Sci. U.S. 67*:637-641 (1970).
147. P. Feigelson and O. Greengard, *J. Biol. Chem. 236*:153-157 (1961).
148. K. Tokuyama and W. E. Knox, *Biochim. Biophys. Acta 81*:201-204 (1964).

149. Y. Henry, Y. Ishimura, and J. Peisack, *J. Biol. Chem.* *251*:1578-1581 (1976).
150. A. Rossi-Fanelli and E. Antonini, *J. Biol. Chem.* *243*:PC 4-5 (1960).
151. H. F. Bunn and J. H. Jandl, *J. Biol. Chem.* *243*:465-475 (1968).
152. H. Kon, *Biochim. Biophys. Acta* *379*:103-113 (1975).
153. P. Hille, G. Palmer and J. S. Olson, *J. Biol. Chem.* *252*:403-405 (1977).
154. D. L. Vesely and G. S. Levey, *Clin. Res.* *25*:502A (1977).
155. F. R. DeRubertis and P. A. Craven (unpublished observations).
156. R. A. Floyd, L. M. Soong, R. N. Walker, and M. Stuart, *Cancer Res.* *36*:2761-2765 (1976).
157. R. A. Floyd and L. M. Soong, *Biochem. Biophys. Res. Commun.* *74*:79-84 (1977).
158. R. A. Floyd, L. M. Soong, and P. L. Culver, *Cancer Res.* *36*:1510-1519 (1976).
159. R. A. Floyd, *Fed. Proc.* *37*:Abstr. 2510 (1978).
160. H. G. Nimmo and P. Cohen, *Adv. Cyclic Nucleotide Res.* *8*:145-266 (1977).
161. D. A. Walsh, J. P. Perkins, and E. G. Krebs, *J. Biol. Chem.* *243*:3763-3765 (1968).
162. J. F. Kuo and P. Greengard, *J. Biol. Chem.* *244*:3417-3419 (1969).
163. J. F. Kuo, W. N. Juo, M. Shoji, C. W. Davis, V. L. Serry, and T. E. Donnelly, *J. Biol. Chem.* *251*:6287-6293 (1976).
164. T. M. Lincoln, W. L. Dills, and J. D. Corbib, *J. Biol. Chem.* *252*:4269-4285 (1977).
165. G. N. Gill, K. E. Holdy, G. M. Walton, and C. B. Kanstein, *Proc. Nat. Acad. Sci. U.S.* *73*:3918-3922 (1976).
166. T. M. Lincoln and J. D. Corbin, *Proc. Nat. Acad. Sci. U.S.* *74*:3239-3243 (1977).
167. J. E. Casnellie and P. Greengard, *Proc. Nat. Acad. Sci. U.S.* *71*:1891-1895 (1974).
168. H. R. deJonge, *Nature* *262*:590-593 (1976).
169. E. M. Johnson and J. M. Hadden, *Science* *187*:1198-1200 (1974).
170. G. Schultz, K. Schultz, and J. G. Hardman, *Metabolism* *24*:429-437 (1975).
171. K. D. Schultz, K. Schultz, and G. Schultz, *Nature* *265*:750-751 (1977).
172. J. W. Kebaian, A. L. Steiner, and P. Greengard, *J. Pharmacol. Exp. Ther.* *193*:474-488 (1975).
173. H. Nawrath, *Nature* *262*:509-511 (1976).
174. W. Trautwein and G. Trube, *Pflugers Arch. Ges. Physiol.* *366*:293-295 (1976).

8
FREE RADICALS DURING APPEARANCE AND GROWTH OF TUMORS

N. M. Emanuel
Institute of Chemical Physics
Academy of Sciences of the USSR
Moscow, USSR

I. INTRODUCTION

Michaelis, half a century ago, published his potentiometric titration data on reduction-oxidation (redox) reactions of some organometallic complexes [1]. He demonstrated that two-electron transfer in the oxidation-reduction of complex ligands proceeded through two one-electron steps to form semiquinone radical intermediates. Ten years later Michaelis stated that "all organic redox systems invariably involve two-electron transfer, and a radical obviously is an intermediate between oxidized and reduced forms of the system" [2]. The formation of a "semiquinone" can be seen from the reduction of a simple quinone, p-benzoquinone.

All subsequent investigations on the most important biochemical reactions involving metal-containing enzymes led to the conclusion that many of these reactions were redox in nature, involving electron transfer, and that free radicals were intermediates in the reaction [3). The development of free radical biology in its present sense was first started by an observation of Commoner and colleagues [4] demonstrating free radicals in lyophilized animal and plant tissues by the electron spin resonance (ESR) technique (Fig. 1).

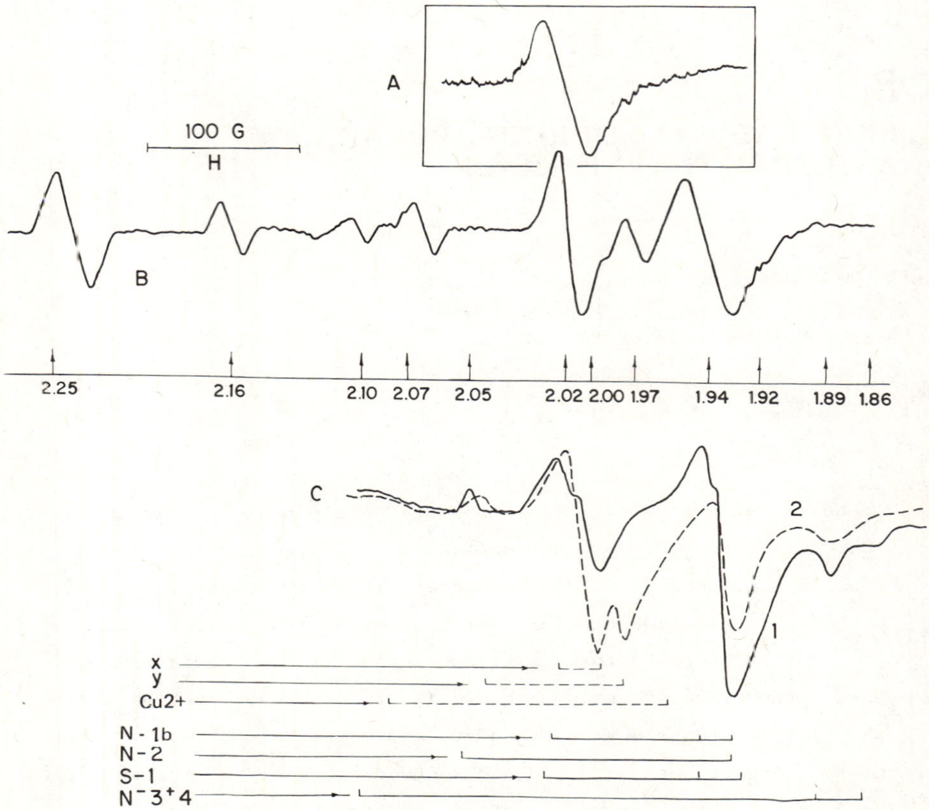

Figure 1 ESR spectra of different biological tissues: (A) Lyophilized samples of leaves from barley seedlings, field size is not indicated, at room temperature. (B) Frozen samples of mouse liver tissue recorded at T 77°K, by X-band spectrometer. (C) Frozen samples of bull heart mitochondria, recorded at 12°K. Curves: 1, initial mitochondria; 2, same sample after keeping in air at 271°K for 20 min. (Part A from Ref. 4; parts B and C from Refs. 5 and 6, respectively.)

Since then, a great number of experiments have demonstrated that ESR signals are present in various cells and tissues. Free radicals have been detected in normal and pathological tissue as well as after treatment with various physical and chemical agents.

In general, $g \simeq 2.00$ signals of organic free radicals having various intensities have been recorded. These signals have usually been ascribed to the semiquinone free radical form of various biochemically important compounds such as naphthoquinones, flavins, ubiquinones, coenzymes, and vitamins. However, with the development of the ESR technique, attention has been directed to other signals in the ESR spectra which correspond to other paramagnetic centers, such as metal ion complexes (Fe, Mn, Mo) involved in electron transfer and in energy conversion in mitochondria, microsomes, and other organelles.

A characteristic ESR spectrum of frozen liver tissue (77°K) is presented in Figure 1B [5]. It shows signals attributable to cytochrome P_{450} ($g = 2.25$), Fe-S complexes ($g = 1.94$), other metal complexes ($g = 1.97$; 2.07; 2.10; 2.16) and free radicals. Some tissues display ESR signals with g values of 2.10, 2.03, 4.0 and 6.0.

Special procedures, for instance, decreasing the temperature of the sample to below 100°K, have allowed more detailed interpretation of the ESR spectra by resolving superimposed signals having almost equivalent g factors. Figure 1C shows an ESR spectrum recorded at 12°K in the 1.80 to 2.15 g-value range. Signals of certain iron- and sulfur-containing centers functioning in the electron transfer chain of ox-heart mitochondria were identified [6]. At present a great number of paramagnetic centers in various tissues, cells, cell organelles, and individual biomacromolecules have been identified, and data on their structure and environment have been obtained [7-9].

The kinetic approach to the study of the changes in paramagnetic properties of tissues undergoing transition to different stages has become important in the use of ESR techniques. This approach has greatly extended the information yielded by such studies, and also provided more precise results and ruled out random conclusions.

Data on the kinetic changes in the content of specific free radicals in animal tissue during the course of leukemia [10] were first published in 1966. This was followed by results obtained with experimental tumors [11-14]. Normal range and extreme values in the time course of development were established. These results then allowed the interpretation of contradictory results reported by other workers on the amount of free radicals in normal tissue as well as in tissue undergoing tumor development. Moreover, a knowledge of the kinetics allowed conclusions on the biological processes involved as well as quantitative evaluation of their effects on living systems.

Work with biological materials requires a great deal of attention to sample-preparation technique in order to obtain consistent data on a particular state of the biological system. In the late 1950s and early 1960s, lyophilized tissues were used. Only singlet signals with a ΔH (peak to peak) of 8 G and a g value of approximately 2.00 were observed. Later, ESR spectra of fast-frozen samples (77°K) or of wet native tissue at room temparature yielded a singlet with g values of 2.004 to 2.005 and a ΔH of 14 to 15 G. This signal corresponds to semiquinone free radicals as well as to radicals from paramagnetic centers (see Fig. 1B).

Lyophilized samples had to be analyzed in a vacuum in order to avoid ESR signals due to humid air different from those yielded by frozen and native tissues (Fig. 2A,B). The free radical concentration was $\sim 10^{16}$ spins/g dry weight of sample. Even short contact with humid air resulted in the appearance of a signal with $\Delta H = 8G$ having an intensity an order of magnitude higher (10^{17} spins/g dry weight) than the signal with $\Delta H = 14\text{-}15$ G [8,15,16] (Fig. 2C). The narrow signal almost completely overlaps the normal free radical signal of the frozen sample. The narrow signal was attributed to ascorbic acid radicals formed in the course of lyophilization of cell protein. Thus, when analyzing the ESR spectra of free radicals in

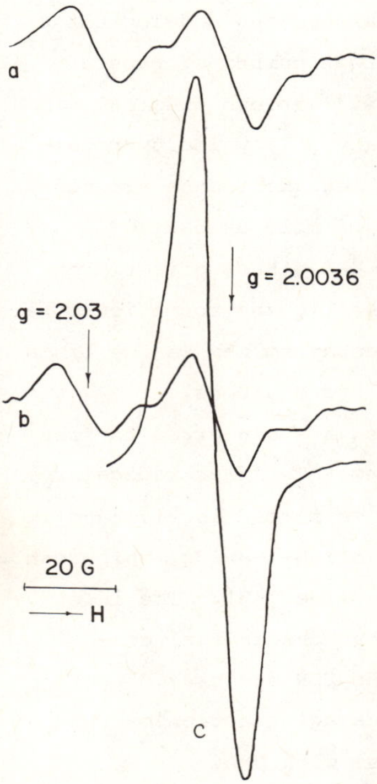

Figure 2 ESR spectra of frozen and lyophilized samples of rat liver tissue: (a) frozen sample, 77°K; (b) lyophilized sample, at room temperature; (c) lyophilized sample after 24 hr in wet air, at room temperature. (From Ref. 8.)

biological tissues, one should take into account the techniques and conditions involved in the preparation of the samples.

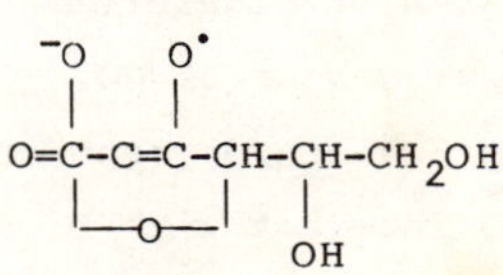

Recently the sensitivity of ESR instruments has increased to a point where it is possible to record signals from unlyophilized biological preparations. One can now bypass lyophilization with all its drawbacks, including disappearance of many signals, functional disturbances in cells, and changes in signal intensity with time and temperature.

The role of free radicals in the appearance and development of malignant tumors has been discussed in a great number of papers from many countries in the last two decades. Various chemical carcinogens can generate free radicals. Many polycyclic hydrocarbon carcinogens such as 3,4-benzpyrene and benzanthracene are converted into free radicals under conditions of mild oxidation [7,17]. These radicals will interact with DNA either directly or via active intermediate free radicals. Possibly the formation of a covalent bond with DNA by a free radical mechanism may be the basis of the carcinogenic activity of polycyclic hydrocarbons.

Carcinogens of the nitroquinoline class are converted to free radical products in the course of metabolism. The free radical species of 4-hydroxyaminoquinoline-1-oxide is formed in vivo and in vitro from the hepatocarcinogen 4-nitroquinoline-1-oxide, which can be converted later into 4-aminoquinoline-1-oxide [18]. The intermediate radical is a stronger carcinogen than the initial compounds. Matsushima et al. [19] obtained the ESR spectrum of this intermediate in solution, and studied its damaging action on DNA and its binding (or that of its metabolites) with DNA.

The carcinogenic properties of amides apparently are correlated with the appearance of N-arylhydroxamic acid, which is oxidized by a free radical mechanism to yield arylnitroso compounds and acylating agents that interact with DNA [20].

Many ESR studies have reported the presence of a large amount of paramagnetic species in tobacco smoke. The free radical content is about 10^{15} spins per gram of cigarette smoke. Tobacco smoke also contains a rather high amount of NO. Other well-known carcinogenic compounds such as 3,4-benzopyrene, benzanthracene, and benzonaphthacene may also interact with biochemical cell components by a free radical mechanism. Nonradical components of smoke and its condensate upon exposure to blood were found to interact with hemoglobin and phospholipids to form stable free radicals [17].

High concentrations of paramagnetic species have been found in chimney soot (5×10^{18} spin/g) and in the amorphous carbon black of automobile engines (2×10^{19} spin/g). These compounds may be contributing factors in the appearance of various cancers.

Various cancers and leukemias induced by radiation involve free radical events. It has been demonstrated that the interaction of radiation with living systems involves formation of free radicals and atoms [21]. Carcinogenic polycyclic hydrocarbons stimulate different biological processes that are influenced by light (photodynamic effect). The sensitizing effect of carcinogens is considered a result of their conversion to the triplet state after absorption of a quantum of light. Further conversion to primary radials can apparently lead to DNA damage and to malignant degeneration of tissues.

II. PARAMAGNETIC CENTERS IN NORMAL AND PATHOLOGICAL TISSUES

Since most oxidation-reduction reactions as well as bioenergetic processes occur in mitochondria, it would be expected that paramagnetic species would also be present mainly in mitochondria. Indeed, it has been reported that mitochondria are almost completely responsible for the g = 2.003 ESR signal observed in tissues [22-24]. The various metal complex paramagnetic centers are mainly located in the respiratory chain of the mitochondria and in microsomes. Figure 3 shows the ESR spectra of native samples of whole liver tissue and of its isolated subcellular fractions, mitochondria, microsomes, and supernatent [22]. The mitochondria have signals with g values of 4.0, 2.003 and 1.94; the microsomes, signals with g values of 2.2, 2.4, and 1.91.

The ESR signal of free radicals in native moist and frozen tissue is a singlet with a g value of 2.003 [25-29]. Some research workers report a higher g value, i.e., up to 2.005. Mainly semioxidized flavins and the semiquinone form of ubiquinones are responsible for this signal.

The aforementioned species act as electron carriers in the respiratory chain of mitochondria. The semiquione radical of the compounds in solution yield ESR signals with a complex, well-resolved hyperfine structure, evidence of the delocalizaton of the unpaired electrion via a coupled π system. However, if these com-

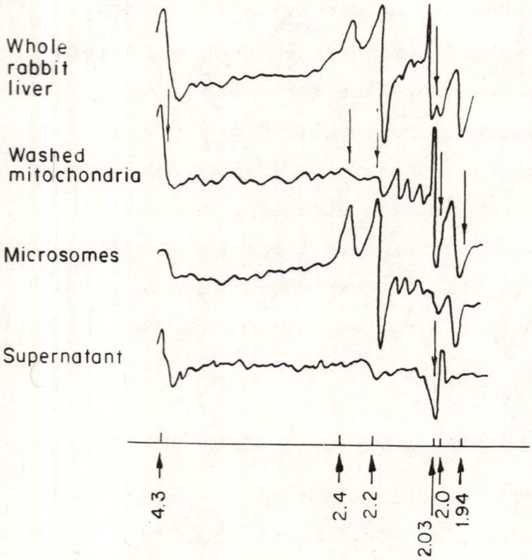

Figure 3 ESR spectra of the rabbit liver tissue and homogenates: samples represent a suspension in 0.01 M Tris-phosphate buffer, 0.2 ml, pH = 8.2 recorded at 113°K. (From Ref. 22.)

Table 1 Paramagnetic Species of Free Radical Nature in Animal Tissue

Type of species	ESR spectrum parameters	Conditions of observation	References
Semiquinone: Q_{10} (ubiquinone)	g = 2.006 ΔH = 10 G	Frozen samples	29-33
Flavins, succinate, and NADH-dehydrogenase	g = 2.003 ΔH = 14-15 G	Moist samples, room temperature	34
Flavins of the energy transport chain	g = 2.003	Moist samples, temperature	34
Radical of ascorbic acid	g = 2.003 ΔH = 8 G	Lyophilized samples in contact with moist air	16,35,36

pounds are absorbed onto protein molecules, their ESR spectra are transformed into a singlet signal characteristic of the tissue [26]. The half-width of the ESR signal from the flavin radical in the tissue is 14-15 G. For radicals of ubiquinone, the half-width is 10-11 G [29]. The half-width of the tissue ESR signal varies from 10 to 16 G depending on the ratio of flavins, ubiquinones, and other free radicals present (our unpublished data; also Refs. 7 and 24).

Table 1 lists the nature of the paramagnetic species responsible for the $g \simeq 2.00$ signal in tissue under various conditions [16,29,36].

Iron-protein complexes play an important role in the electron transport chain of mitochrondria. Some electron carriers are nonheme iron-containing proteins. These compounds are involved in respiration, coupled oxidation and phosphorylation, steroid hydroxylation, nitrogen fixation, etc. The nonheme iron content of mitochondrial membrane paramagnetic centers is higher than that of the heme compounds (cytochromes). The nonheme iron proteins contain an iron complex involving sulfhydryl groups of protein and atoms of labile sulfur [37,38].

Iron- and sulfur-containing nonheme iron complexes are, as a rule, paramagnetic in the reduced state. The ESR spectrum of these proteins usually have a component with a g value of about 1.94 and one or two other transitions depending on the symmetry of the complex. Investigations using model systems [39] have shown that the two main g values are lower than the spin value of g_e ($\Delta g = 0.06 \div 0.2$); the third value is equal to or higher than that of g_e [$\Delta g < 0.04$; and $g_{av} = (g_x + g_y + g_z)/3$ is lower than g_e ($\Delta g = 0.04 \div 0.1$)]. The signal is observed only at temperatures of less than 150°K due to enhanced broadening at higher temperatures. The ESR spectra of nonheme iron complexes in biological systems and model compounds are shown in Figure 4. Liquid helium temperature ESR observations of biological systems made possible the identification of 13 iron-sulfur centers (Fe-S centers) in the mitochondrial respiratory chain [40-44]; see Table 2 [7,39-58].

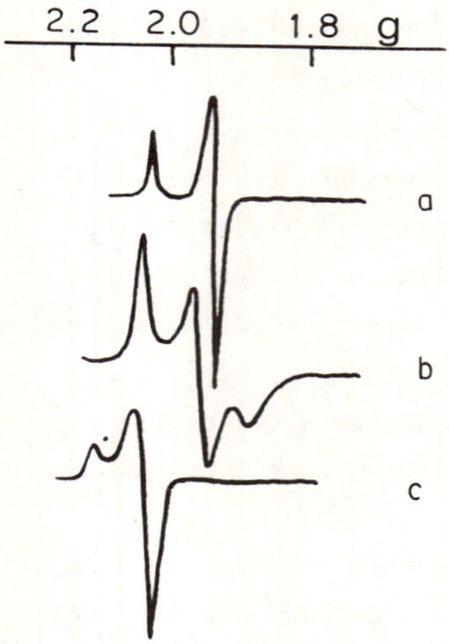

Figure 4 ESR spectra of nonheme iron complexes: (a) adrenodoxyn; (b) ferredoxyn of spinach; (c) high-potential iron-sulfur protein.

Table 2 Paramagnetic Metal Complexes in Normal Tissues

Centers	ESR spectrum parameters	Conditions of observation	Reference
Nonheme iron complexes		Observed only in the reduced state of nonhemic iron ions	39-50
I. Green complex			
N-1a	g_1 = 2.023	12°K	48,50
	$g_{2,3}$ = 1.94	12°K	
N-1b	g_1 = 2.027	12°K	48
	g_2 = 1.938		
	g_3 = 1.923		
N-2	g" = 2.054	12°K	48
	g' = 1.922		
N-3	g_1 = 2.10	12°K	48
	g_2 = 1.886		
	g_3 = 1.862		
N-4	g_1 = 2.100	12°K	48
	g_2 = unknown		
	g_3 = 1.864		
N-5 N-6	g = 2.11; 2.06; 2.03; 1.92-1.93; 1.90	12°K	48

II. Green complex			
S-1 S-2 S-3	$g_1 = 2.03$ $g_2 = 1.93$ $g_3 = 1.91$ $g_1 = 2.02$ $g'' = 1.99$	S-1 77°K S-2 20°K Oxidized state	44
III. Green Complex Rieske Protein	$g_1 = 2.026$ $g_2 = 1.887$ $g = 1.78\text{-}1.81$	150°K	
Heme iron		Observed only in the oxidized state of hemoprotein, 77°K	
P_{450} cytochrome (low-spin state)	$g_1 = 2.42$ $g_2 = 2.25$ $g_3 = 1.91$		51
P_{450} cytochrome (high-spin state)	$g_1 = 8.0$ $g_2 = 3.7$ $g_3 = 1.7$	Same as above	51
Cytochrome c peroxidase	$g_1 = 1.78$ $g_2 = 2.20$ $g_3 = 2.75$		51
Cytochrome b_5	$g_1 = 1.47$ $g_2 = 2.26$ $g_3 = 2.95$		51
Cytochrome c	$g_1 = 1.24$ $g_2 = 2.24$ $g_3 = 3.06$		51
Hemoglobin OH^-	$g_1 = 1.85$ $g_2 = 2.17$ $g_3 = 2.55$		51
Complexes of Mn^{2+}	Sextet with component intensity as a function of magnetic field $g_{AV} = 2.000$ splitting 85-90 G	77°K	52-54
Complexes of Cu^{2+}	$g_{AV} = 2.06\text{-}2.07$		7
Complexes of Mo^{5+}	$g_{AV} = 1.97$	77°K	55-58

Centers with close g values, such as N-1a, N-1b, N-2a, N-2b, N-3, N-4, N-5, N-6, S-1, and S-2 are distinguished among the Fe-S centers of first and second green complexes [39-50]. The signal from these centers were resolved by taking advantage of their differences in redox properties [48,49]. The data on paramagnetic metal complexes observed in animal tissue are presented in Table 2.

ESR signals from oxidized cytochrome P_{450} and from the heme iron complexes functioning in the electron transport chain of microsomes are also observed in frozen liver tissue. Optical absorption spectra as well as ESR spectral parameters of cytochrome P_{450} change considerably upon addition of different drugs, hormones, alcohol, and especially carcinogens and mutagens. In this regard the changes in the cytochrome P_{450} signal parameters are used as a sensitive test for changes in the energetic and detoxifying systems during hormonal and chemical carcinogenesis of tissue undergoing chemotherapy. The ESR spectra of rabbit liver microsomes treated with methylcholanthrene is shown in Figure 5. Signals corresponding to the high-spin and low-spin states of Fe^{3+} heme complex are seen [51].

The ESR signal of Mn^{2+} is a six-component hyperfine structure, the intensity of which decreases as the magnetic field strength is increased. The first component sextet signal has a g value of 2.17 (for X-band spectrometers). The ESR spectra from Mn^{2+}-containing complexes, essential for the functioning of many biological systems, have been studied [52-54].

Molybdenum complexes involved in the active center of xanthine oxidase and sulfite reductase in liver cell cytoplasm have an ESR signal with a g value of 1.97 [55-58].

The ESR spectra of living tissue represent complex combinations of individual signals from cell components in the paramagnetic state.

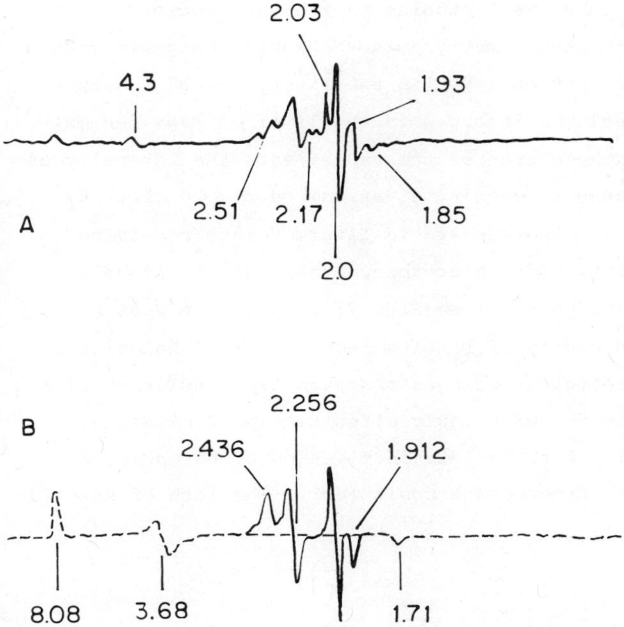

Figure 5 ESR spectra of frozen samples of rabbit liver tissue (A) and microsomes isolated from the animals after injection of methylcolanthrene (B), (recorded at 1.6°K; the amplification in the case of part B is fivefold.) The signals of high-spin state of heme complex are shown by the dotted line. (From Ref. 51.)

New ESR signals, not found in normal functioning tissues, have often been observed in tumor tissues [59]. Their appearance over a wide range of magnetic field intensity has been demonstrated [60], but the signals have not been identified. ESR signals with g values of 2.010-2.016, corresponding to metal ions, have been detected [61]. Signals with a 35-G wide superfine triplet structure were observed in some tumors and ascribed to the interaction of an unpaired electron with the ^{14}N nucleus [62). There is a consistency in the g value shifts of signals observed with tumor tissue of human patients [63].

In 1965, Commoner's group reported signals with a g value of 2.035 in tissue during early stages of malignancy [64). This signal was observed in rat liver long before the appearance of a tumor. However, in subsequent studies on the carcinogenic effect of diethylnitrosoamine and dimethylaminoazobenzene, signals with a g value of 2.035 were not observed in the liver, though interestingly the 2.035 signal was recorded in the liver of rats fed oxidized fish oil. In the course of the experiment, the intensity of the 2.03 signals changed, reaching a maximum on day 20 (Fig. 6) [65]. The 2.03 signal also appears in liver of rats receiving noncarcinogenic agents, such as acetone, ether, ethyl alcohol, decylxanthogenate, and phenol (see Fig. 7) [66-68]. A 2.03 signal appears in liver and kidney of rabbits receiving noninhalational narcotics. The histological changes observed are consistant with a degree of hypoxia but not with toxic effects. The 2.02 signal appeared very shortly after the distinct action of narcotics and persisted for a short time postnarcosis [69]. The lack of specifi-

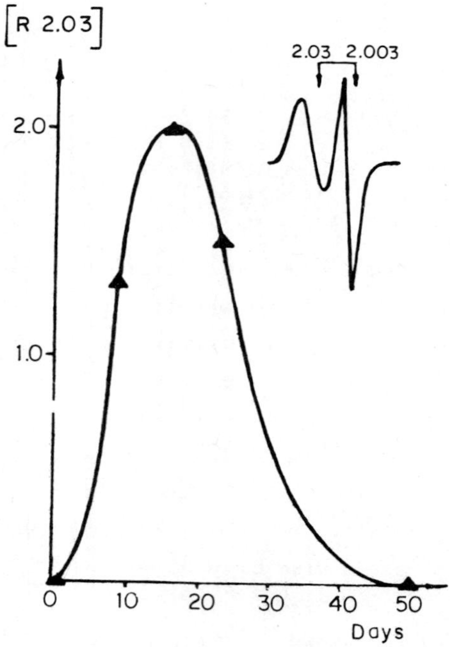

Figure 6 ESR spectra of frozen liver of rats fed oxidized fish oil, and kinetics of variations in intensity of signal with g = 2.03.

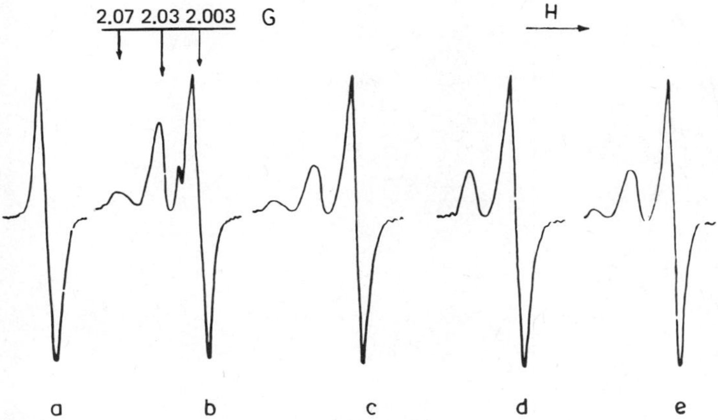

Figure 7 ESR spectra of frozen liver of rats affected by decylxanthogenate (b), ethanol (c), acetone (d), and phenol (e); control (a).

city of the 2.03 signal during the early stages of malignancy was proven by its presence in necrotic tissues [70] and in isolated tissues stored under different conditions.

Investigation of the molecular nature of the structure of the complex responsible for the 2.03 signal showed [71-75] that the signal arose from a dinitrosyl complex of nonheme iron bound to two closely located protein sulfur groups:

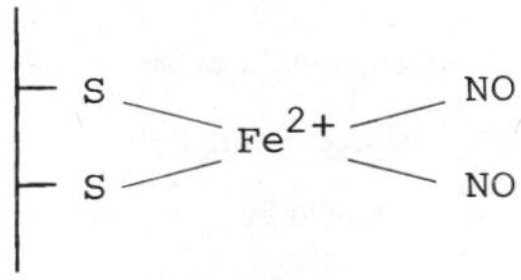

ESR signals with a triplet structure, not usually present in biological systems, were observed in many experimental tumors frozen in the state of active metabolism [76,77]. This signal is characterized by g values of 1.98, 2.03, and 2.07 and by a superfine triplet structure centered at g = 2.007 with a splitting of 15 G. Such signals have been detected in the tissue of sarcoma 45, Walker carcinosarcoma, hepatoma 22A, carcinomas of forestomach OZH_5 and PRZH, Cloudman melanoma, carcinoma 775, sarcoma 298, sarcoma

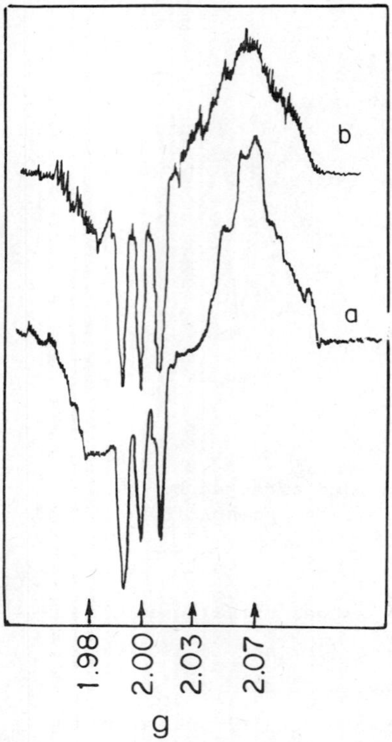

Figure 8 ESR spectra of nitrosyl iron complexes in tumors: (a) carcinosarcoma; (b) hepatoma 22a.

M-11, Gueren carcinoma, Jensen sarcoma, Pliss lymphosarcoma, sarcoma SSK, Yoshida tumor, and ascites tumor [76-79] (Fig. 8). The signals are very stable: their shape and intensity persist for many days after thawing and exposure to air at room temperature. Studies with tumor tissues and nitroxyl complexes of hemoproteins (catalases, cytochromes, hemoglobins) before and after heating at given temperatures, as well as under varying conditions of oxidation-reduction, have shown that their ESR triplet signal remained identical (Fig. 9). Thus, the triplet seems to be due to nitrosyl complexes of various hemoproteins, mainly hemoglobin.

Model nitrosyl complexes of heme-containing proteins yielded four types of ESR signals [80-85]: signal (A), with $g_1 = 2.07$, $g_2 = 2.023$, $g_3 = 1.98$ [82] with no triplet structure; signal (B), with

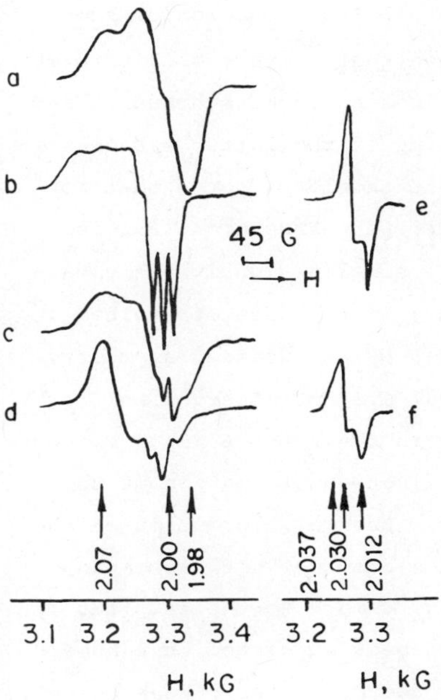

Figure 9 ESR spectra of nitrosyl iron complexes: (a-d) nitrosyl hemoglobin complexes; (e-g) nitroxyl Fe(II) complexes with cysteine and hydrosulfite. (From Ref. 80.)

g_1, = 2.1, g_2 = 2.05, g_3 = 2.008 with highly resolved triplet structure; signal (C), with g = 2.07, g = 2.01 [82] with no superfine structure; and signal (D), a broad symmetrical signal with the center at g ≃ 2.04 and half-width of 120 G [81]. The first signal is characteristic of the native state of hemoproteins; the second, of proteins with slightly disintegrated structure; the third, of hemoproteins treated with a detergent [82-85]; and the fourth is the same as that observed in lyophilized preparations of nitroxyl complexes with heme-containing proteins at room temperature. All these signals are observed in tumor tissues [73]. The signals appear in the following sequence: signal A (g = 1.98) predominates at the initial stage of tumor growth; then as it decreases the triplet structure (signal B) remains, while at a later stage, when necrosis appears, there is a signal with g = 2.04 (signal D). It has been suggested that these changes may be caused by alterations in the heme-containing protein structure. Fractionation of tumor cells into subcellular components showed that signals with triplet structure were due to metal complexes located in mitochrondria and microsomes [86]. The functional activity of the latter was distributed but not completely suppressed. It should be noted that not all tumor tissues had signals with a triplet structure. For instance, many ascites tumors had ordinary singlet signal with g values close to that of the free electron. In the case of ascites sarcoma 180, the signal is characterized by an enhanced asymmetry.

Paramagentic centers with g = 2.007 are not observed in actively proliferating tumor cells nor in their subcellular components. Also no relation has been found between formation of the complexes and the decay of tumor cells. However, formation of the complexes yielding g = 2.007 signals has been shown to accompany the development of hypoxia in the tumor, damage to the respiratory chain, and subsequent necrosis. It has been suggested that numerous hemorrhages occur in tumors in the course of their development, as witnessed by erythrocyte accumulation in tumors. Anaerobic conditions favor the formation of hemoglobin nitroxyl complexes to which the 6-coordination site of the heme iron is ligated with various substituents.

It can be concluded from the above that it is still difficult to utilize a specific ESR signal to characterize cancer in a diagnostic sense. In general, however, normal and tumor tissue can be reliably distinguished by the difference in intensity of signals in the ESR spectra. This may serve as a basis for studying the changes in paramagnetic properties of organs and tissues during carcinogenesis and tumor development.

III. RADIOBIOLOGICAL CHARACTERISTICS OF NORMAL AND TUMOR TISSUE

The action of ionizing radiation on living and model systems has been widely studied in connection with the problems of radiation damage, tumor and leukemia induction, and also with the aim of discovering efficient radiotherapeutic principles.

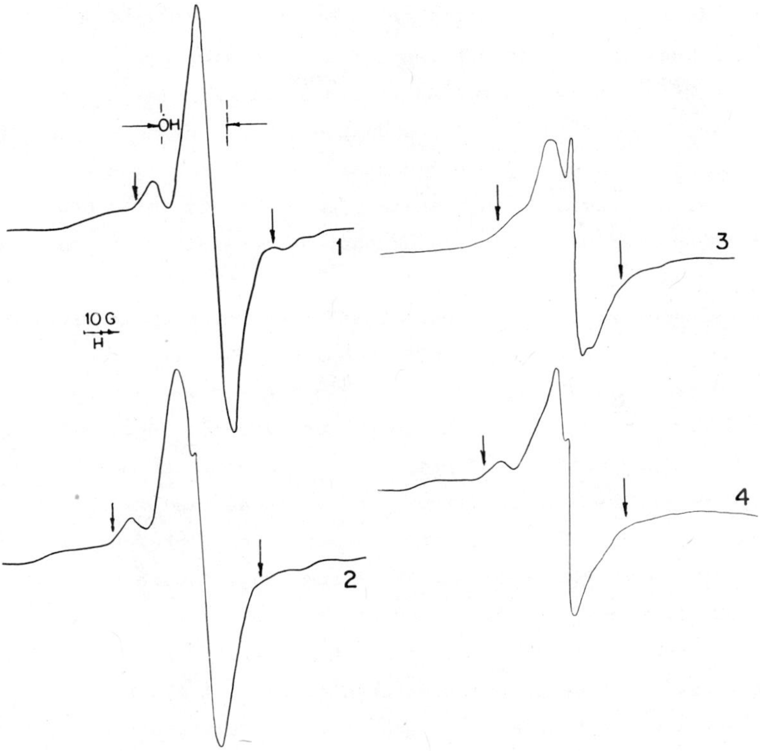

Figure 10 ESR spectra of tissues γ-radiated at −196°C: (1) normal mouse liver; (2) liver of tumor-bearing animal; (3) tumor (hepatoma 22a, on day 6 after transplantation); (4) tumor of animal fed ionol 1 hr before killing (and taking of samples). The spectra were measured under the following conditions: T = −196°C microwave power, 1 mW; amplification for 1-4 is the same and tissue weight did not differ by more than 1.5 times. Pointers indicate position of the third and fourth lines of the ESR spectra for Mn^{2+} standard in MgO.

The problems involved in using efficient sensitizers to increase tumor sensitivity over normal tissue requires a knowledge of the molecular mechanisms involved in radiation effects on individual cellular components as well as on the cell as a whole. For model systems, there are many experimental data concerning the radiation-chemical processes in proteins, polypeptides, DNA, DPN (deoxypyridinenucleotide), and polysaccharide-containing systems [87-92].

The effects of radiation on normal and tumor tissues in vitro have been compared, utilizing the ESR technique [93-95]. Frozen samples (at 77°K) were exposed to γ-radiation, after which the ESR spectra of paramagnetic species in the tissue were recorded. Figure 10 shows the ESR spectra for samples of (1) intact animal liver, (2) tumor-bearing animal liver, and (3) hepatoma 22a on the sixth day after transplantation. Each tissue type is characterized by its own individual ESR spectrum. A narrow singlet with $\Delta H = 6$ G and $g = 2.0005$ is observed only in the ESR spectra of a tumor and in the liver from tumor-bearing animals. A correlation between signal intensity and time of hepatoma development was found. Maximal signal amplitude was observed on the fifth to sixth day after transplantation.

The ESR spectra of liver, regenerating after partial hepatectomy, were obtained in order to estimate the specificity of the $\Delta H = 6$ G signal for tumor growth. A narrow singlet ESR signal, the parameters and properties of which are similar to those from hepatoma and liver of tumor-bearing animals, was observed in the case of partial hepatectomy as well. The narrow singlet with $g = 2.005$ seems to be characteristic of tissue undergoing active proliferation.

The ESR spectra of irradiated tissue analyzed on the basis of thermostability and relaxation characteristics of paramagnetic species have been recorded [94]. The ESR spectrum of a normal mouse liver obtained at various microwave power levels (77°K) is shown in Figure 11A. Signals from the radiolysis products of tissue liquid, from OH radicals, and a signal arising from radiolysis

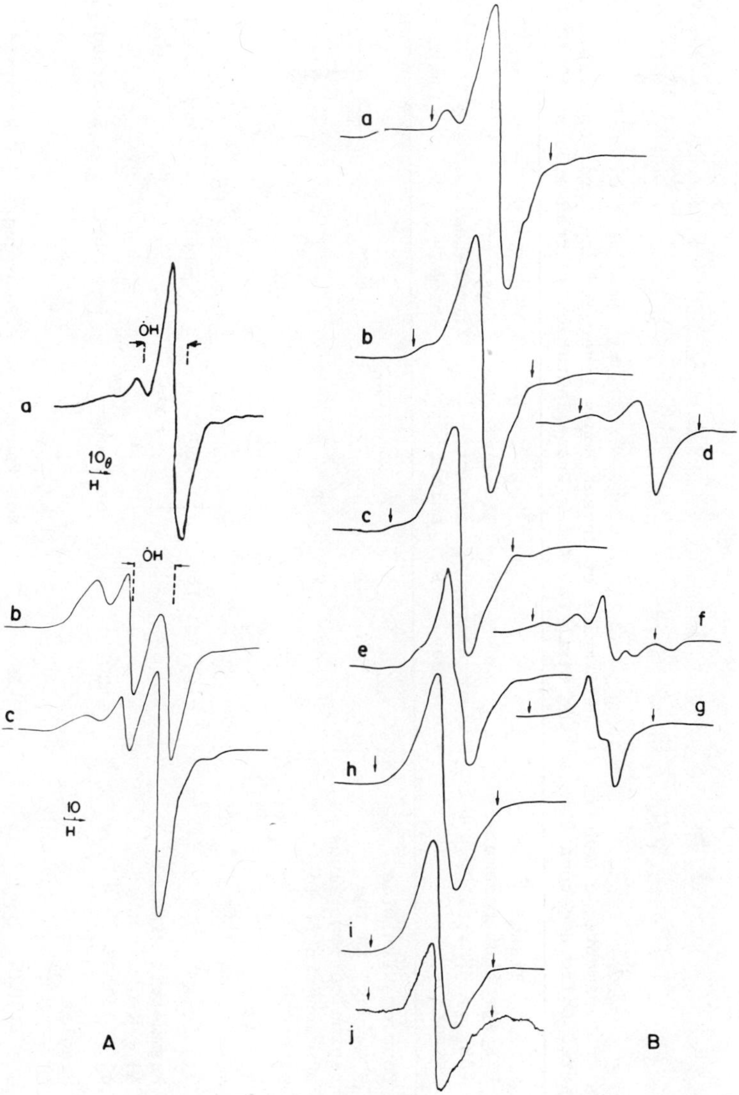

Figure 11 ESR spectra of mouse liver irradiated at -196°C with 1 Mrad. (A) measurements at various levels of microwave power: (a) 1 mVT; (b) MBT; (c) 200 mVT. Vertical pointers show the positions of g_3 and g_4 for third and fourth ESR spectrum lines from Mn^{2+} standard in MgO. (B) In the course of sample defrosting: (a) -196°C; (b) -140°C; (c) -100°C; (d) difference spectrum of b - c; (e) -70°C; (f) difference spectrum of c - e; (g) difference spectrum (-70° to 60°C); (h) -50°C (amplification is twice higher); (i) -30° (amplification is four times higher); (j) +20°C (amplification is 32 times higher). The spectra were measured at -196°C and 1 mW microwave power.

Table 3 Characteristics of Paramagnetic Centers in the Samples of Intact Mouse Liver and the Liver and Tumor of Animals with Transplanted Solid Hepatoma 22a (at the Sixth Day After Transplantation γ-Irradiated at 196°C.)

Types of paramagnetic centers	ESR spectrum parameters (ΔH_{max}, HFS, a, G, g factor)	Temperature of decay (°C)	Tissue: Liver of intact animal	Tissue: Liver of tumor-bearing animal	Tissue: Tumor
$\dot{O}H$	Doublet with wide "shoulder" in the region of low values of magnetic field strength a = 40 G $g_1 = 2.008$ $g_{11} = 2.05$	-140 to -160	Recorded to -160°C	Recorded to -140°C	Not found
Center with narrow signal (Type 1)	Asymmetric singlet $\Delta H = 6$ G $g_c = 2.0005$	-100	Not found	Recorded	Recorded
Center with narrow signal (Type 2)	Singlet $\Delta H = 21$ G $g = 2.0025$	-80	Not found	Recorded	Recorded

Peroxy (HOO•)	Asymmetric singlet $\Delta H = 46$ G $g_{11} = 2.026$ $g_{1} = 2.0014$	-90	Recorded	Not clear	Recorded
Peroxy (Type 2)	Asymmetric singlet $\Delta H = 48$ G $g = 2.03$ $g = 2.006$	-10	Not found	Not found	Recorded
R_Q	Quintet $a = 20$ G $g = 2.003$	-70	Recorded	Not found	Multiplet identity (R_Q uncertain) observed -80° + 20°
R_D	Doublet $a = 13$ G $g = 2.0025$	-50	Recorded	Recorded	Not found
	Octet	-50	Not clear	Very light intensity	Recorded
R_S	Singlet $\Delta H = 24$ G $g = 2.003$	-30	Recorded	Recorded	Not found
R_{ox}	Singlet $\Delta H = 11$ G $g = 2.0039$	-10	Recorded	Recorded	Recorded

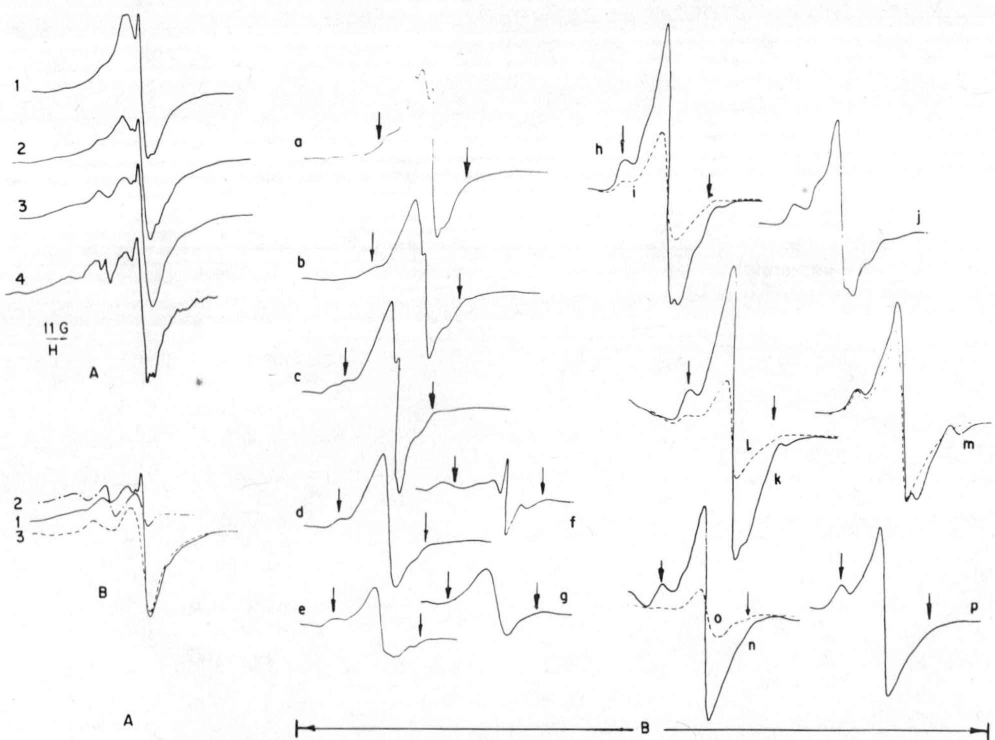

Figure 12 (A) ESR spectra of γ-irradiated (-196°C) tumors (hepatoma 22a). (a) Recording at different levels of microwave power: (1) 1 mW; (2) 5 mW; (3) 30 mW; (4) 200 mW. (b) ESR spectrum analysis recorded at 30 mW: (1) total spectrum; (2) spectrum components responsible for OH; (3) spectrum components responsible for peroxy radical ROO. (B) Changes in irradiated tumor spectra in the course of defrosting: (a) -196°C; (b) -130°C; (c) -110°C; (d) -100°C; (e) difference between c and d; (f) -80°C; (g) difference between d and f; (h) -80°C (amplification for h and i is four times higher than for a to f); (i)(dash) -50°C; (j) difference between h and i; (k) -30°C; (l) (dash) -20°C; (m) difference between k and l (amplification for k and l is 16 times higher than for a to f); (n) -20°C (amplification for n and m is two times higher than for l); (o) (dash) -10°C; (p) difference between n and o (p is shown by dashes on m).

products of other cell components having a $\Delta H = 21$ G and $g = 2.002$ have been observed [96,97]. The latter signal becomes saturated as the microwave power is increased, and only a signal from OH radicals is observed at a power higher than 100 mW (Fig. 2C). Figure 11B shows the ESR spectra recorded in the course of thawing irradiated liver. A decay of OH radicals was observed at $T \geq -160°C$. At $-140°C$ a singlet signal remains with $\Delta H = 20$ G and $g = 2.002$. From analysis of the ESR spectra ($-190°$ and $-140°C$) it can be concluded that the OH radicals disappear without interacting with the chemical components of the cell. It was found by constructing different spectra that the peroxy type signal ($\Delta H = 46$ G, $g = 2.026$, and $g = 2.0014$) disappears within the temperature range of $-140°$ to $-90°C$, and that the signal with the quintet HFS [resolution 20 G and $g = 2.003$ (R_Q)] disappears within the $-100°$ to $-70°C$ range. At $-70°C$ the spectrum represents the sum of doublets with $a = 13$ G and $g = 2.0025$ and a small amount of the quintet signal. The doublet signal (R_D) disappears within the temperature range of $-70°$ to $-50°C$. The singlet ($\Delta H = 24$ G and $g = 2.003$) remains and then disappears upon heating to $-30°C$. The paramagnetic centers observed, their properties and ESR signal parameters are given in Table 3.

A similar analysis was carried out for samples of tumor tissue, liver of tumor-bearing animals, as well as for LL tumor cell culture (see Figs. 12-14; and also Table 3). Results indicate that the radiation-induced signals and consequently, the processes occurring in the γ-irradiated tissue are different. Moreover, an octet (Figs. 13 and 14) ascribed to thymidine H-addition radicals was detected in the ESR spectra of hepatoma 22a and LL cells [98].

It was found in experiments with LL cells prepared from ascites lymphoid leukemia cells in NKLy mice [99] that the radiation-induced DNA damage occurred by trapping electrons resulting in radical anion formation of DNA bases. Protonation of the radical anion results in the formation of H adducts responsible for the octet ESR spectrum. Radiation-induced formation of thymine H adducts was studied for tissues having various radiosensitivities. DNA damage products predominated in radiosensitive tissue (spleen

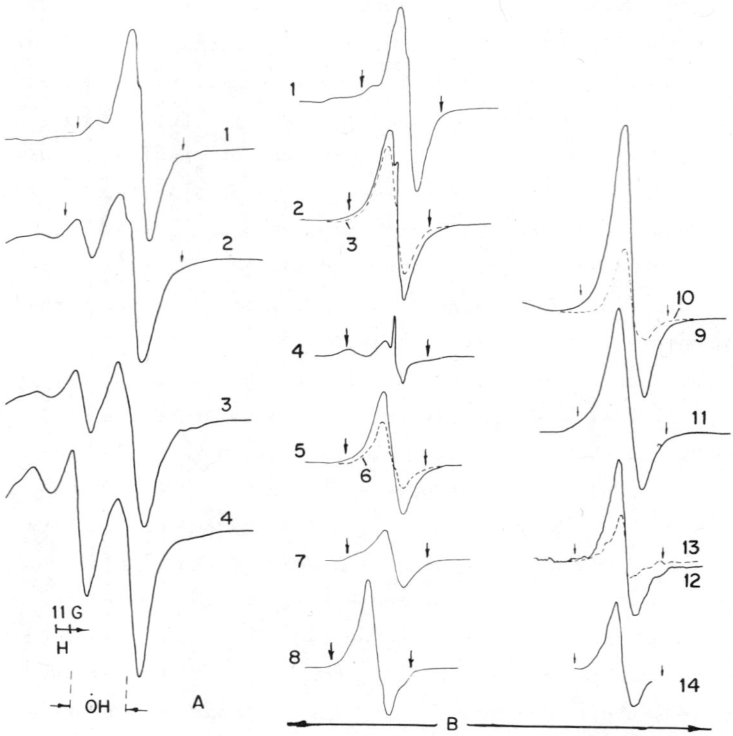

Figure 13 The ESR spectra of liver from tumor bearing animal γ-irradiated at −196°C. (A) Measurements at different levels of microwave power: (1) 1 mW; (2) 5 mW; (3) mW; (4) 100 mW. (B) Changes in the course of defrosting (microwave power = 1 mW): (1) −160°C; (2) −130°C; (3) (dash) −110°C (4) difference between 2 and 3; (5) −100°C; (6) (dash) −80°C; (7) difference between 5 and 6; (8) −70°C (x2); (9) −50°C (x4); (10) (dash) −30°C (x4); (11) difference between 9 and 10; (12) −30°C (x8); (13) (dash) −20°C; (14) difference between 12 and 13. Increase in amplification in parentheses.

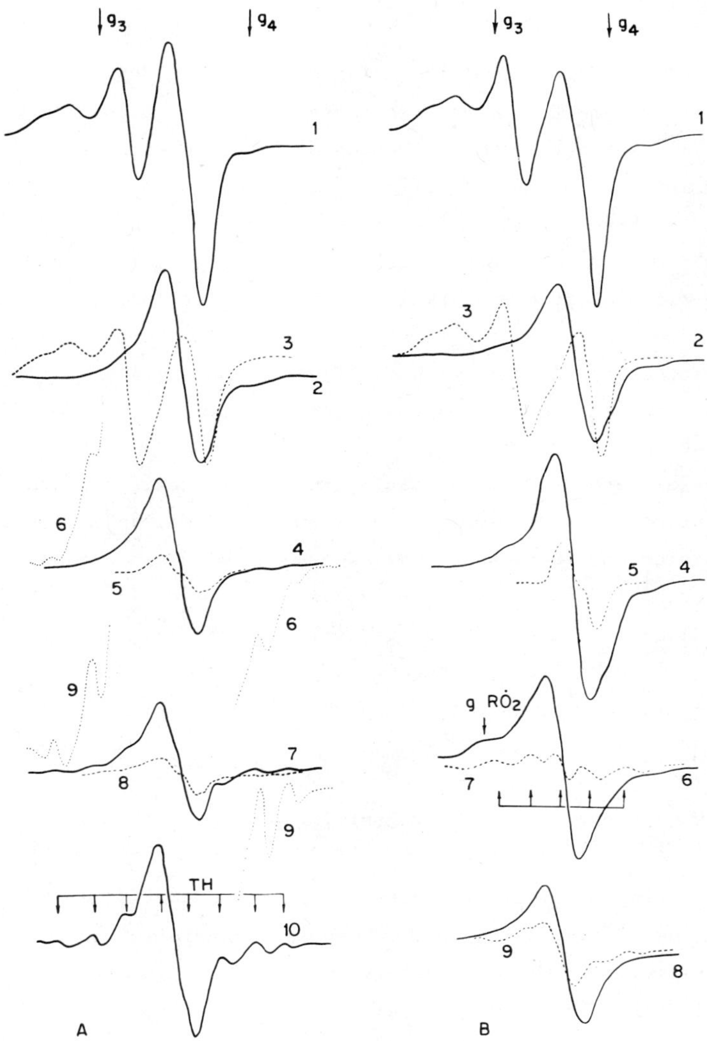

Figure 14 (A) Changes in ESR spectra of γ-irradiated LL cells during gradual defrosting of the sample: (1) -196°C; (2) -150°C; (3) difference spectrum between 1 and 2; (4) -110°C; (5) difference spectrum between 2 and 4; (6) the same as 4, but amplification is 16 times higher; (7) -70°C; (8) difference spectrum between 4 and 7; (9) the same as 7, increased 16 times by amplification; (10) -30°C, twice amplified as compared with 7. Pointers indicate components of octet signal TH. (B) Changes in the ESR spectra of γ-irradiated LL cells on exposing sample to visible light and subsequent defrosting: (1) -196°C; (2) -150°C; (3) difference spectrum between 1 and 2; (4) after exposing sample to visible light, twice amplified as compared with 2; (5) signal disappearing under the action of visible light; (6) -80°C; (7) difference spectra between 4 and 6; (8) -60°C; (9) difference spectrum between 6 and 8.

and thymus) compared to radioresistant tissue (spleen, kidney, heart). Of interest are the changes in intensity of the signal with $\Delta H = 6$ G and $g = 2.005$ after administration of ionol to animals. Antioxidant administration to animals at certain doses and at certain times before irradiation has increased the radiotherapeutic effect [100,101]. This was shown to be true for ionol administered to animals 1.0-1.5 hr before preparation of the samples [93] for ESR measurement, depending on the time when the animals were exposed to radiotherapeutic radiation [100,101]. There was a decrease in the intensity of the narrow signal observed for these samples (Fig. 10, curves c and d).

Thus, the new method of investigating the paramagnetic centers which appear during γ-irradiation of frozen tissue permit an evaluation of the differences between normal and tumor cells and establishment of distinguishing characteristics of the free radical mechanisms involved in the effects of radiation on the main chemical components of the cells. It suggests approaches to studies on radioprotective mechanisms, as well as in radiosensitization and antitumor drugs.

IV. CHANGES IN TISSUE PARAMAGNETIC PROPERTIES DURING THE COURSE OF CARCINOGENESIS

Changes in the ESR signal intensity were found in frozen tissue from various types of cancers induced by viruses, chemicals, or hormones. An increase in the free radical content was observed with hamster and rat embryonic fibroblast cells after infection with different adenoviruses. Figure 15 shows that a definite increase in intensity of the $g = 2.003$ ESR signal was observed after one day and that the maximum was reached on the fourth day. The highest free radical concentration (three times the control values) was observed in proliferating pretumor foci and primary tumor nodes (Fig. 15B) after inducing subcutaneous sarcomas in hamsters with adenovirus [12]. However, after further tumor growth, the radical content fell below normal values.

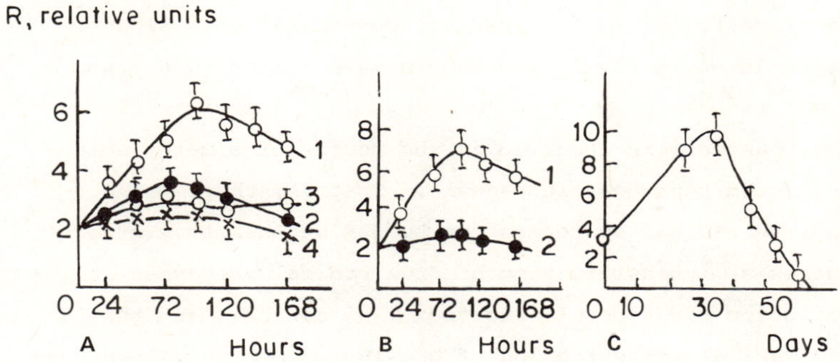

Figure 15 Free radical content (R) during viral oncogenesis: (A) In fibroblast culture of rat embryo infected with adenoviruses of the types 12(1), 3(2), and 6(3); curve 4, control. (B) In fibroblast culture of hamster embryo infected by adenovirus of type 12(1); curve 2, control. (C) In development of sarcoma induced by oncogenic virus, type 12, in hamsters.

Complex changes in tissue paramagnetic properties were found in the course of chemical carcinogenesis in animals. A decrease in the free radical concentration ($g = 2.003$ signal) by 40% as compared to normal liver was observed after the first 10 days following administration of the hepatocarcinogen diethylnitrosoamine (Fig. 16). Later the free radical content markedly increased attaining a maximum value on day 22 (50% above normal). This section of the curve is analogous to the curves of radical content obtained after toxic doses of noncarcinogenic chemical compounds such as DDT, benzene, and strongly oxidized sunflower oil [67]. The changes do not seem to be connected with any specific carcinogenic action but with a generally toxic reaction. A longer period of further increase in the $g = 2.003$ ESR signal intensity was observed after 40-50 days of diethylnitrosoamine administration, at which time the free radical concentration was close to normal. During the 90- to 110-day interval of carcinogenesis, the free radical concentration attained a maximum, exceeding normal values by 75%. This period is histologically characterized as a stage of diffusion and hyperplastic foci. During further carcinogenesis development (130-150 days), when tumors appeared in most animals, there was a decrease in the free radical concentration. Amounts were two times lower in developing tumors as compared to corresponding normal tissue.

Similar changes were observed in the course of dimethylaminoazobenzene-induced hepatocarcinogenesis. That is, there was a decrease in free radical concentration by 30% at the early stages of carcinogenesis, intensification of free radical processes at the stage of diffusion and hyperplastic foci (2-5 months), and then a reduction in radical concentration of growing tumors to values one-half that of normal (Fig. 16B). Carcinogenesis induced by 2-acetylaminofluorene resulted in a considerable increase in radical content of rat liver during the 90- to 110-day time interval [102].

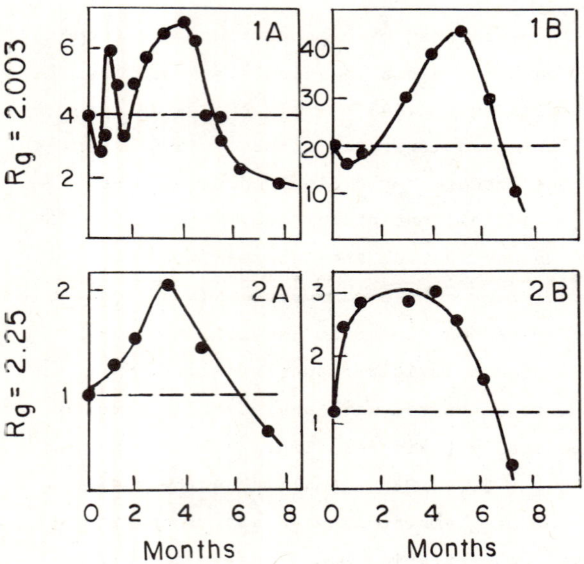

Figure 16 Free radical content (curves 1A, 1B) and P_{450} cytochrome (curves 2A, 2B) in rat liver during carcinogenesis induced by diethylnitrosoamine (A) and dimethylaminoazobenzene (B).

These observations are in good agreement with the data reported by others. A decrease in the intensity of g = 2.003 signals was observed in rat liver, muscle, and lungs during the first 15 days after administration of 3,4-benzopyrene, dimethylbenzanthracene, or 20-methylcholanthrene [103]. A decrease in the free radical content below normal was observed at early stages of carcinogenesis induced by dimethylaminoazobenzene (DAAB) and by 2-acetylaminofluorene (AAF) as well as in the resulting tumors [104]. The data are tabulated as shown here.

Tissue	Time of experiment	Relative amplitude of signal with g = 2.0045
Liver of intact rats	—	1.0
Liver of rats fed DAAB	8-30 days	0.7
Hepatoma (DAAB)	5-10 months	0.25
Liver of rats fed AAF	8-30 days	0.6-0.7
Hepatoma (AAF)	6-10 months	0.3

The phase characteristics of changes in free radical concentration were observed in connective tissue transformed by the subcutaneous implantation into the back of a rat four to six paraffin tablets containing 2 mg of benzopyrene. Sarcomas appeared within 3 months in 100% of the cases [105]. The free radical content increased, attaining maximal values within 3 months after benzo[a]-pyrene implantation, in primary tumor nodes (Fig. 17). With time the free radical concentration in growing sarcomas decreased below the normal level of free radicals in connective tissue.

Changes in the free radical content was studied in connective tissue undergoing malignant transformation after subcutaneous implantation of 20 mm Teflon plates [105]. Sarcomas appeared at various times with an average latent period of 14 months. One month after Teflon implantation, the free radical concentration in connective tissue surrounding the plate decreased approximately 45%

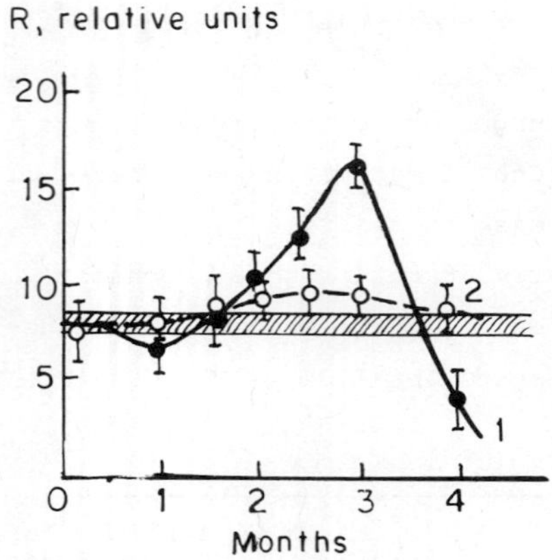

Figure 17 Free radical concentration (R) at various stages of sarcoma development, induced by benzo(a)pyrene: (1) in sarcoma; (2) in connective tissue capsule surrounding paraffin pellets (cross-hatched section)—intact subcutaneous connective tissue.

below normal connective tissue levels. After 2 to 3 months, the free radical content in Teflon implantation foci exceeded the control values by 1.6 to 1.7 times, but within 4-5 months the level was again reduced to normal values. A second rise in the free radical concentration began after the eighth to ninth month and reached a maximum by the fourteenth month. During this period, characterized by sarcoma node formation, free radical concentration increased to twice that of normal connective tissue. Further sarcoma growth resulted in a decrease of the free radical concentration to half the values of the controls.

An enhancement of free radical processes was noted during certain stages of hormone-induced tumor development [106,107]. A two-fold increase in the amount of free radicals was observed in ovarian granulocellular malignancies. These tumors were characterized by an intensive proliferation of cells developing after metastases to the spleen following double oophorectomy.

Obviously, these changes in free radical concentration of malignant tumors induced by various chemical carcinogens are evidence of the latter's specific carcinogenic action. Disturbances in the electron transport chain of mitochondria may be the cause of these changes. The administration of thyroxin may result in both an increase and a decrease of free radical concentration in the rat liver depending on the regime and dose [108,109]. The increase in radical concentration correlated with an increase in content and activity of enzymes involved in mitochondrial electron transport, i.e., of cytochrome a, cytochrome oxidase, succinate oxidase, cytochrome-c oxidase and $NADH_2$-cytochrome-c reductase. The decrease in the free radical level has been ascribed [109] to an inhibition of the energy transport chain. It has been suggested that there is a disturbance of the relation between coupling of electron transport chains and energy transport chains of liver cell mitochondria during hepatocarcinogenesis induced by 2-acetylaminofluorene and 4-dimethylaminoazobenzene [110].

Along with a change in free radical concentration during the course of chemical carcinogenesis, changes in the ESR signals related to other paramagnetic centers have also been reported. An increase in cytochrome P_{450} content (signal with g = 2.25) followed by a considerable decrease in tumors (curves 2a and 2b in Fig. 16) was observed during carcinogenesis induced by both diethylnitrosoamine and dimethylaminoazobenzene. The content of cytochrome P_{450} in developing tumors is some four to five times lower than that in normal liver. Hepatocarcinogenesis induced by 2-acetylaminofluorene [106] caused a decrease in the cytochrome P_{450} content (g = 2.25, 2.42) in the liver by the fourteenth day after beginning the experiment. Further, there was a decrease in the intensity of the signals and an appearance of the signal with g = 2.03 after 2 months. Triplet ESR signals (g = 2.00) were recorded in the developing hepatomas.

Changes in the paramagnetic properties of muscle tissue of animals receiving 3-methylcholanthrene were also studied [106]. A decrease in intensity of the ESR signal from the nonheme iron complex (g = 1.94) reached the lowest value after 1 month. A complex signal which then changed into a triplet with g = 2.007 at the terminal period of tumor development appeared within 2.5 months.

Thus malignant degeneration of tissues in the course of carcinogenesis is accompanied by changes in the content of both heme and nonheme iron paramagnetic complexes. The decrease in P_{450} cytochrome content apparently reflects a lower activity of the microsomal chain of detoxification as related to carcinogen metabolism. This may assist in the carcinogenic effect.

On the other hand, some chemicals (inducers) may induce microsomal enzymes such as the mixed function oxidases. Different chemical compounds may act as inducers. Antioxidants, for example, are inhibitors of free radical processes. These compounds, have been proposed as antitumor agents by our laboratory. In this regard, it is of interest to compare experimental data on the increase in cytochrome P_{450} concentration after administration of a free radical inhibitor (ionol) [111] with the anticarcinogenic action of this compound on tumor formation by dimethylaminoazobenzene, 3,4-benzpyrene, and dimethylbenzanthracene. The effect of ionol on hepatocarcinogenesis induced by p-dimethylaminoazobenzene was studied [112]. Experimental animals were fed 0.6% of the carcinogen plus 0.3% ionol dissolved in corn oil for 150 days. No tumors appeared in any of 30 Wistar strain rats or 14 "August" strain rats, within 12-15 months. No pretumoral changes were observed after administration of the carcinogen together with ionol (Fig. 18A).

The anticarcinogenic action of the phenolic antioxidants was later confirmed by Wattenberg [113,114], who found that ionol and butylated hydroxyanisole are inhibitors of chemical carcinogenesis induced by polycyclic hydrocarbons. These phenolic antioxidants inhibit mouse forestomach and rat mammary gland tumor development; that is, they decrease the number of animals with tumors as well as the number of tumors per animal. Similarly, butylated hydroxyanisole inhibited lung cancer induced in mice fed 7,12-dimethylbenzo[a]anthracene (DMBA), benzo[a]pyrene, urethane, and diethylnitrosoamine. The water-soluble analog of ionol (chlorhydrate 4-oxi-3,5-di-tert-butylmethylbenzamine) inhibited the induction of benzo[a]-pyrene-induced subcutaneous sarcoma in rats (Fig. 18B) [115].

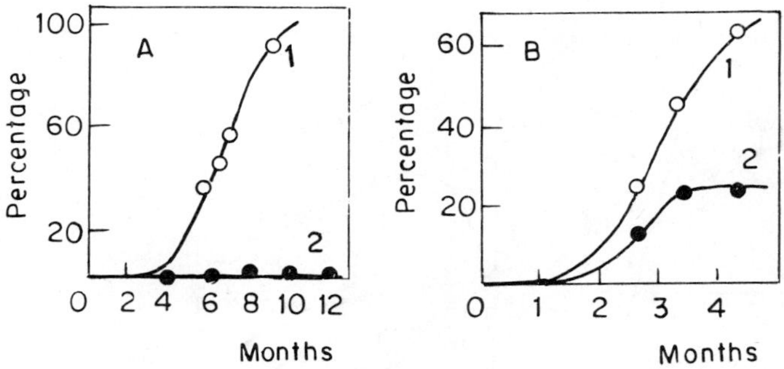

Figure 18 Kinetic curves of changes in tumor appearance frequency under the action of dimethylaminoazobenzene (A,1) and 3,4-benzpyrene (B,1). Anticarcinogenic action of ionol (A,2) and chlorhydrate of 4-oxi-3,5-di-tert-butylmethylbenzylamine (B,2).

ESR signals from liver tissue of animals on a diet including carcinogen and ionol showed at first a decrease in the free radical concentration (Fig. 19A), then an increase to normal values at the stage of diffusion and hyperplastic foci [111]. Ionol had no effect on the free radical content of liver [116].

Cytochrome P_{450} content increased by a factor of 2.5 in animals receiving DAAB, whereas in animals receiving the diet containing the carcinogen and ionol there was an almost sixfold increase in the $g = 2.25$ signal which persisted for a long time (Fig. 19B). This evidence that ionol can act as an inducer of the mixed function oxidases in liver microsomes [111] was confirmed by a study involving the administration of ionol in order to prevent the toxic effect of benzene [116].

The establishment of a pattern of changes in the paramagnetic properties of tissue as a response to the administration of chemical carcinogens allows the establishment of kinetic criteria that can be used to evaluate the action of various environmental factors on living systems. This kinetic approach has been used to study the toxic action of compounds on the change in paramagnetic properties of rat liver and brain in animals administered DDT*, benzene,

*Formal name 1,1,1-trichloro-2,2-bis(p-chlorophenyl)ethane.

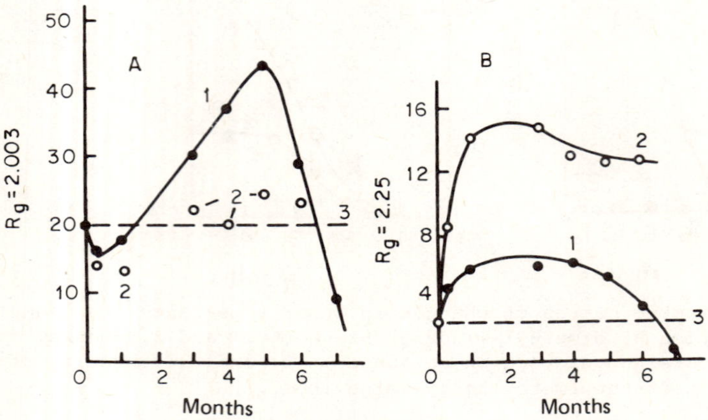

Figure 19 Concentration of free radicals (A) and P_{450} cytochrome (B) in rat liver under the action of dimethylaminoazobenzene (1) and joint action of dimethylaminoazobenzene and ionol (2); curve 3, control.

and oxidized fats [67]. The results of these experiments are shown in Figure 20A. In the case of DDT and benzene, the stationary concentration of free radicals differed significantly from normal levels even in the terminal stages of the experiment. A significant decrease in the free radical content was noted in the brain marrow.

The free radical content as well as the amount of paramagnetic metal complexes undergoes changes during carcinogenesis. The study of these changes may be of importance for the early detection of malignant diseases and for the development of a rational approach to the prevention and therapy of cancer. In addition, it may help in investigating the mechanism of action of different environmental factors upon living organisms.

V. KINETIC CHANGES IN FREE RADICAL CONTENT DURING TUMOR GROWTH

The important role of free radicals in the development of radiation disease, malignant transformation of cells, and tumor growth was suggested as early as 1957 [21,117,118]. It was also thought that

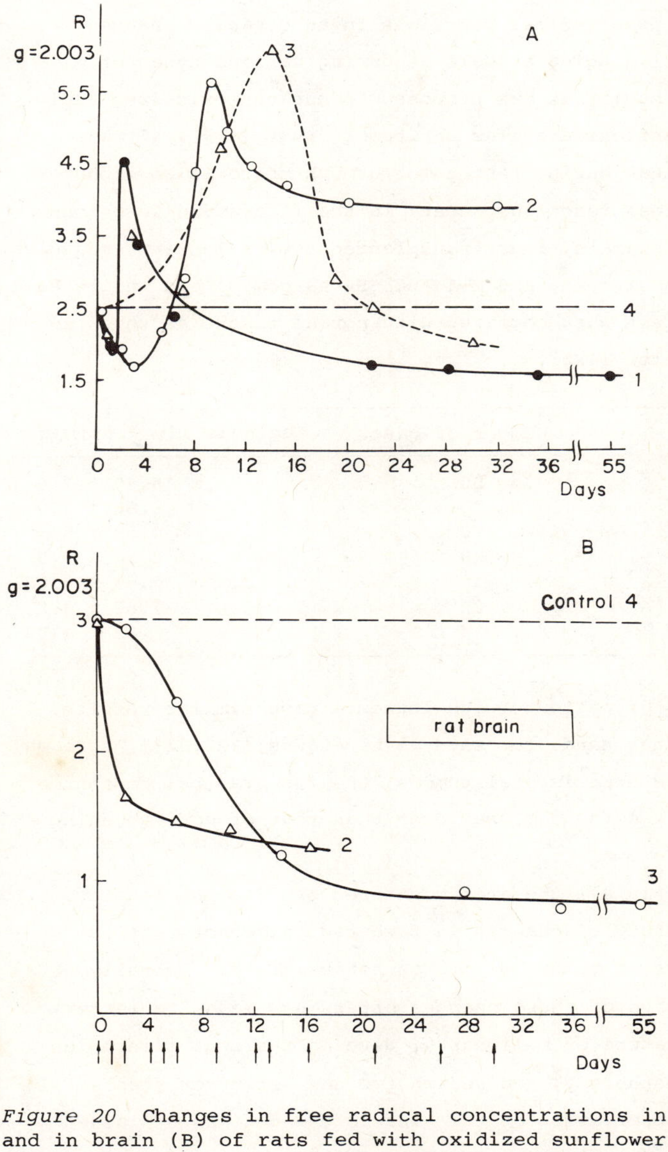

Figure 20 Changes in free radical concentrations in liver (A) and in brain (B) of rats fed with oxidized sunflower oil (1), affected with benzene (2), and DDT (3); curve 4, control.

the intensity of free radical reactions in an organism changed significantly during aging as well as during various types of stress. On this basis, it was proposed to use nontoxic free radical inhibitors (antioxidants) as antitumor drugs [119]. It was further assumed that during fast tumor growth there is an enhancement of free radical reactions. Data in the literature were contradictory. For example, four transplanted tumors in rats (rhabdomyosarcoma, lymphosarcoma, and M-1 and SSR sarcomas) had an increased free radical content when compared with normal tissue as shown in the tabulation below [120].

Tissue	Number of cases	Radicals (10^{16} Spin/g)
Sarcoma M-1	20	10.10 ± 2.0
Sarcoma SSR	20	6.80 ± 1.0
Rhabdomyosarcoma	20	4.50 ± 0.5
Muscle (normal)	20	2.54 ± 0.2
Heart (normal)	10	2.80 ± 0.02
Lymphosarcoma	20	7.60 ± 1.0
Thymus (normal)	4	1.02 ± 1.33

White blood cells of patients with leukemia gave similar results [121]. On the other hand, for several hepatomas [122,123], sarcoma 45 [124], and some experimental tumors, the free radical content of malignant tissue was found to be lower than that of corresponding normal tissue.

Nonetheless, in all the aforementioned studies the kinetic approach to the study of changes in free radical concentration during the process of tumor growth was not utilized. Actually, knowing the kinetics of these changes offers new means to better investigate the nature of malignancy, develop rational principles of antitumor chemotherapy, and search for new antitumor agents. In fact, a careful observation of the kinetic changes has explained the contradictory data on free radical content during tumor growth [125-128].

ESR spectra of tumor tissue, blood, and organs of tumor-bearing animals were obtained. The samples were prepared by lyophilization in argon and subsequent evacuation of the vials. The signal half-width was 8-10 G [125]. This ΔH value differs from those obtained with wet native and frozen samples, the signal half-width being mainly controlled by flavin radicals. Signals discussed previously [125-128] may be due to ubiquinone or to other radicals, such as ascorbic acid radicals. We observed a singlet with a ΔH of 10 G half-width in frozen samples of adenocarcinoma 775 which we attribute to ubiquinone radicals. Swartz and Guttierrez [129] also discovered signals with a different WH in frozen and lyophilized samples.

Statistical treatment of the experimental results has yielded mean values for the relative amplitudes of ESR signals, dispersion values, and confidence intervals for each series of measurement. On the basis of these calculations, kinetic curves recording the change in free radical content during tumor development were obtained. In all cases, a marked increase of free radical concentration in malignant tissue during the initial latent stage of tumor growth was found. Significantly there was a decrease in the free radical level during the terminal stage of tumor growth.

As can be seen from Figure 21 during the first stage of La leukemia development, the free radical concentration of the spleen (the most involved organ) increased, reaching maximal values by days 4-5. It is of importance that the change in free radical concentration preceded other known manifestations of leukemia development (e.g., increase of spleen weight; leukocyte and hemocytoblast counts in peripheral blood and bone marrow). At the time of animal death, the free radical concentration of the spleen was significantly below the normal level. Thus if one takes experimental samples at the final stages of malignant growth, the free radical concentration would be much lower than the normal level. This was noted by other researchers who had not performed a kinetic study. Similar trends in free radical concentration were observed in cases of ascites and solid tumors (Fig. 21).

Cells from different parts of a tumor have different degrees of viability. Cells from the central portion are poorly vascularized and die quickly, forming necrotic zones. In this connection,

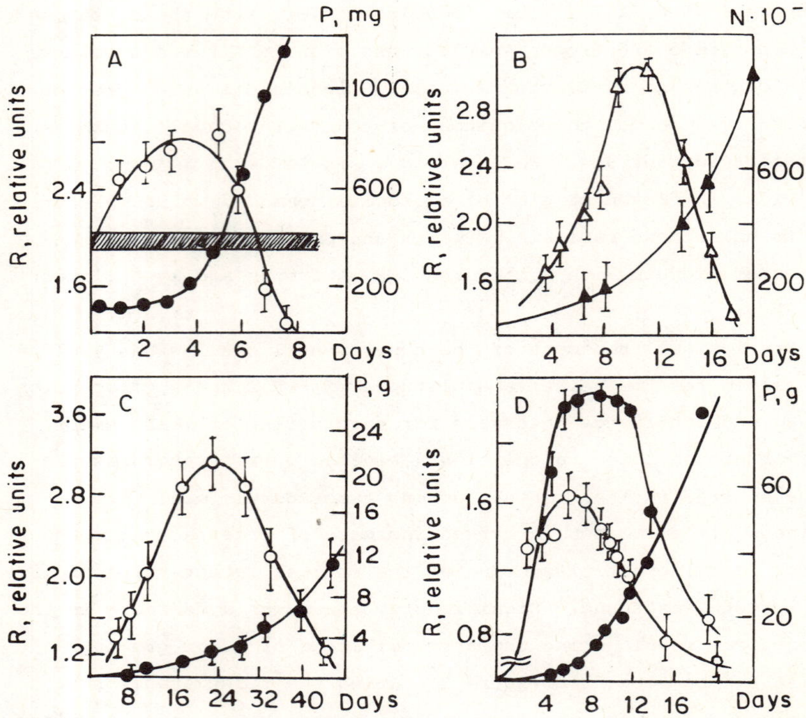

Figure 21 Kinetic curves of tumor development and changes in free radical content of lyophilized tumor samples: (A) transplanted leukemia La; (B) ascites form of sarcoma 37; (C) colid sarcoma 37; (D) Walker carcinosarcoma (peripheral and central sections, curves 1 and 2).

one would expect tissue from different tumor layers to have different levels of free radicals. Radical concentration in the peripheral zone of Walker carcinosarcoma is much higher than in the central region (see Fig. 21). Maximum radical concentration in peripheral zones of a tumor remains steady for some time (6-7 days), whereas in the central portion the concentration decreases just after reaching its maximum.

The development of a kinetic method for the investigation of free radicals in biological systems has allowed the analysis of a great number of tumors in animals and humans. It has also contributed an additional means of studying free radicals. Presently, both in Soviet and in other laboratories, about 20 tumor systems have been studied kinetically (see Table 4). The kinetics of change in free radical content during tumor growth has been studied in the United States by Wallace's group [63,130,131]. Thus, curves for changes in free radical content in the peripheral portion of actively growing tumors have been studied in three independent experiments involving rats bearing Walker carcinosarcoma. They have the same overall character and are alike on a day-to-day basis. The ESR signals have maximum amplitude points of similar value.

All authors have noted complete agreement of their results with those obtained earlier by Soviet scientists. Thus, for the fast-growing pigmented melanoma B-16, the maximum free radical content corresponds to the beginning of the highest growth rate phase.

It should be mentioned that similar changes in free-radical content of tumor were also shown by the graft copolymerization method [132]. This method is based on the ability of biogenic free radicals to initiate polymerization of the radioactive monomer (i.e. [^{14}C]acrylic acid amide) introduced into the organism. Measuring the radioactivity of a particular tissue or of separate chemical components of a cell allows one to determine the quantity of grafted monomer in a biological specimen. The degree of grafting of a labeled acrylamide in different organs and tissues of random-bred mice correlates with the quantity of free radicals as determined by the lyophilized ESR method (Fig. 22).

It has been shown by the graft copolymerization method that the change of free radical content in sarcoma 180 cells can be characterized by the extremes, the maximum occurring at the beginning of a period of highest tumor growth rate [133]. Kinetic curves of the specific radioactivity changes of grafted monomer in the lipids of sarcoma 45 and in proteins and lipids of Ehrlich ascite carcinoma cells and of Harding-Passey melanoma cells (Fig. 23) are similar [133,134] in that all have a maximum.

Table 4 Combined Data on the Investigation of Free Radicals in Experimental Tumors

No. (1)	Type of tumor (2)	Method of measurement of free radical concentration (3)	Strain of animals (4)	References (5)
1.	Leukemia	ESR (lyophilized samples)	$C_{57}B1$ Mice	125
2.	Sarcoma 37 (ascites)	Same	Random-bred mice	126
3.	Sarcoma 37 (solid)	Same	Random-bred mice	127
4.	Walker carcino-sarcoma	Same	Random-bred rats	128
5.	Walker carcino-sarcoma	ESR (frozen samples)	Random-bred rats	130
6.	Melanoma B-16	Same	$C_{57}B1$ mice	63
7.	Gardner lympho-sarcoma	Same	C_3H mice	130
8.	Sarcoma 180 (solid)	Graft copolymerization (tumor globulins)	Random-bred mice	133

9.	Sarcoma 45	Graft copolymerization (tumor lipids)	Random-bred rats	134
10.	Elrich carcinoma (ascites)	Graft copolymerization (tumor protein and lipids)	Random-bred mice	135
11.	Harding-Passey melanoma	Same	C_3HA mice	135
12.	Harding-Passey melanoma	ESR frozen samples	C_3HA mice	135
13.	Hepatoma 22A	ESR frozen samples	C_3HA mice	136
14.	Myeloid leukemia	Same	AKR/y mice	137,138
15.	Leukemia	Same	C_{57}B1 mice	139
16.	Leukemia P_{338}	Same	BDF_1 mice	139
17.	Leukemia L1210	Same	BDF_1 mice	139
18.	Mammary adenocarcinoma	Same	C_3H mice	139
19.	Cloudman melanoma	Same	DBA_2 mice	Our data
20.	Melanoma B-16	Same	BDF_1 mice	Our data

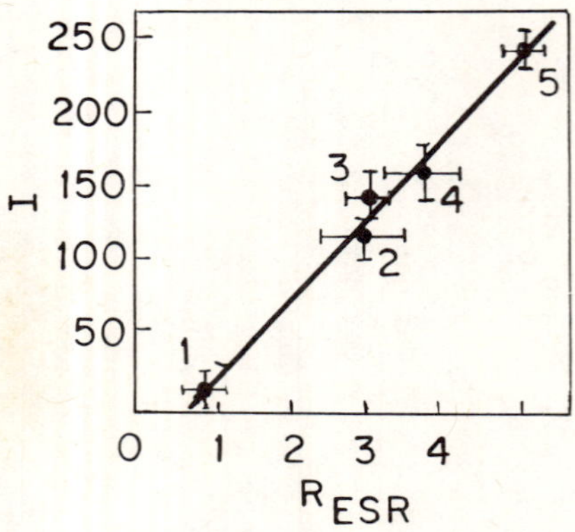

Figure 22 Correlation between free radical concentration determined by the ESR technique (I) and activity of grafted [^{14}C]acrylamide in muscle (1), kidney (2), liver (3), lung (4), and spleen (5).

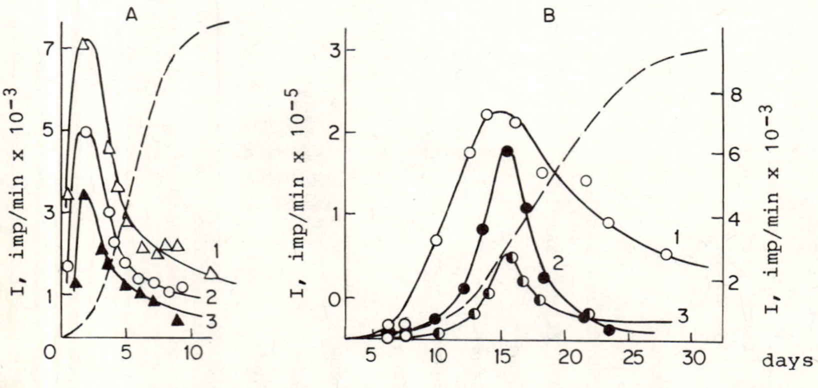

Figure 23 Changes in activity of grafted [^{14}C]acrylamide in proteins (1), free (2), and fixed (3) lipids of tumor cells during the development of Ehrlich ascites carcinoma (A) and Harding-Passey melanoma (B). Dashed line indicates kinetic curves of tumor growth.

It is essential to summarize all available data in order to establish a quantitative correlation between the kinetic parameters of tumor growth and changes in free radical content. These reports have been summarized by Emanuel [140]. For the present purpose, the exponential function $N = N_o e^t$ has been used. This allows a description of the initial phase of growth for most tumor systems. Figure 24 shows experimental points representing different tumors. Until the fifth doubling groupings are in the vicinity of the theoretical curve $\eta = 2^n$, where $\eta = N/N_o$, n = number of doublings, and N = tumor size. There is a deviation from the exponential dependence at advanced stages when the tumor shifts to the linear phase of growth, where $N = N_1 + bt$.

The intrinsic region of maximal tumor growth rate can be obtained from the tangent of an exponential, at which point the functions and their derivatives are equal:

$$N_1 + bt = Ne^{yt}$$
$$b = yNe^{yt}$$

Consequently the time of maximal growth rate is

$$t_{W_{max}} = \frac{1}{y} - \frac{N_1}{b}$$

Knowing the doubling time T for each experimental tumor, it is possible to find $n_{W_{max}}$, i.e., the number of doublings required for a tumor to reach the highest growth rate. Results given in Table 5 show that for the cases considered $n_{W_{max}}$ varies within narrow limits (5.0-6.9 doublings). The average value is 5.6 ± 0.3. The same conclusion was reached utilizing other parameters of tumor growth such as weight or volume of tumor, volume of ascites, or number of tumor cells.

Data on the change of free radical content as a function of doubling number were obtained for the relative values R/R_{max}, standardized according to the maximal concentration of free radicals in a given tumor. For this purpose, the free radical content

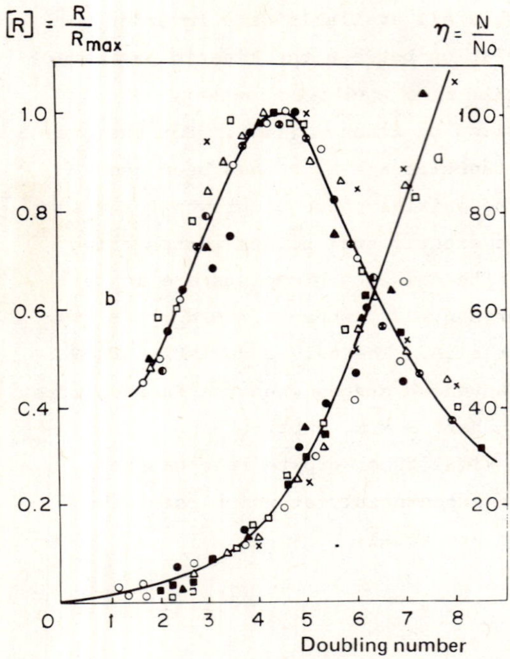

Figure 24 Computed kinetic curves of transplanted tumor growth (a) and changes in free radical content of tumor tissues (b). Various points indicate data on different tumors.

as determined by ESR of lyophilized malignant tissues was used as well as data collected by the graft copolymerization method.

Analysis of curves with regard to n vs. ln R/R_{max} coordinates shows that in all cases both branches of the exponential functions describing the change of free radical concentration can be described by using both positive and negative exponents. The doubling time was estimated at the point of anaplasia, during which the maximal content of free radicals was achieved ($n_{R_{max}}$). As can be seen from Table 5, the $n_{R_{max}}$ values vary within the limits of two doubling times (3.5-5.7 doublings); the average value is equal to 4.3 ± 0.3. The general character of the change in free radical content during tumor growth is shown in Figure 24. The average $n_{R_{max}}$ value is derived from data on various experimental tumors.

Table 5 Kinetic Parameters Characterizing the Change of Free Radical Content in Experimental Tumors

Type of tumor	ψ (days^{-1})	$t_{W_{max}}$ (days)	$n_{W_{max}}$	$n_{R_{max}}$	$\Delta = n_{W_{max}} - n_{R_{max}}$	$\frac{n_{R_{max}}}{n_{W_{max}}}$
Leukemia	0.690	5.5	5.5	4.4	1.1	0.80
Sarcoma 37 (ascites)	0.232	13.4	4.6	3.5	1.1	0.76
Sarcoma 37 (solid)	0.104	33.2	5.0	3.5	1.5	0.70
Walker carcinosarcoma	0.261	13.1	5.0	3.5	1.5	0.70
Melanoma B-16	0.263	16.1	6.2	5.0	1.2	0.81
Sarcoma 180 (solid)	0.332	14.4	6.9	5.7	1.2	0.83
Ehrlich carcinoma (ascites)	0.840	4.1	5.1	3.7	1.4	0.73
Harding-Passey melanoma	0.220	18.9	6.3	5.3	1.0	0.84
Average values			5.6 + 0.3	4.3 + 0.3	1.3 + 0.1	0.77

compare the signal amplitude of paramagnetic centers (Table 6). Using the average $n_{W_{max}}$ = 1.3, the maximal content of free radicals is reached at about one doubling earlier than the point of highest tumor growth rate.

This conclusion is obviously rather general. As discussed in previous experiments, it is confirmed only by a kinetic study of the change of free radical content in frozen samples. The relative intensity of semiquinone radicals and other paramagnetic centers was established in tumor and liver of animals bearing transplanted hepatoma 22a [136]. The kinetic curves show that the maximum occurs one doubling before the point of highest tumor growth rate (Fig. 25). Similar trends characterize the change in intensity of the g = 2.00 signal, which is due to anion radicals and indol-quinoid groups in melanin from transplanted melanomas [141]. The species responsible for the radical is:

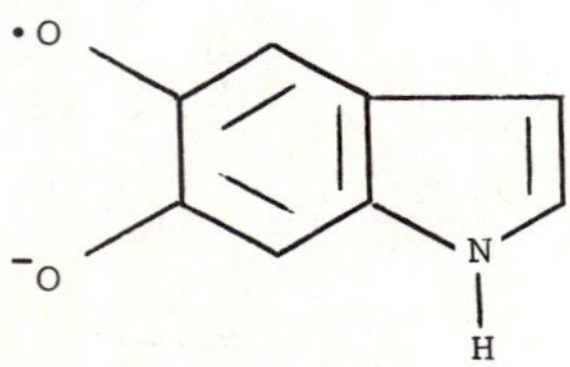

During days 4-9 of growth of adenocarcinoma 755, we have observed, using low-microwave power (0.10 mW), a free radical signal with g = 2.003 and a half-width of 10-11 G (Fig. 26). Saturation of the signal was achieved at P = 0.2 mW. Power saturation curves of samples at various stages of tumor growth demonstrated the same characteristics and therefore demonstrated the presence of free radical centers.* There was very little variation in the shape of bands as well as in the signal half-width, so it was possible to

*Since the ESR signal of free radicals is easily saturated, the recording of spectra should be performed at low-microwave power, i.e., not more than 0.2 mW (Fig. 26). At higher power levels, this signal disappears and for some tumors characteristic signals with g_{av} = 2.03 and a triplet HFS are observed.

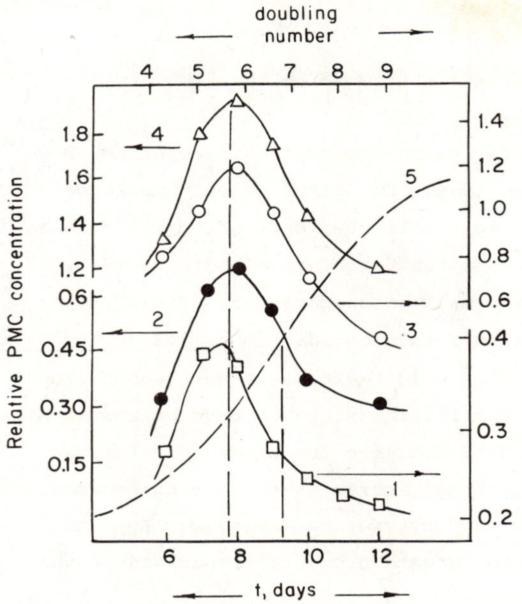

Figure 25 Changes in paramagnetic center concentration during hepatoma 22a growth: g = 2.003—tumor (1); g = 1.94—liver (2); g = 2.25—liver (3); g = 2.003—liver (4); tumor growth curve (5).

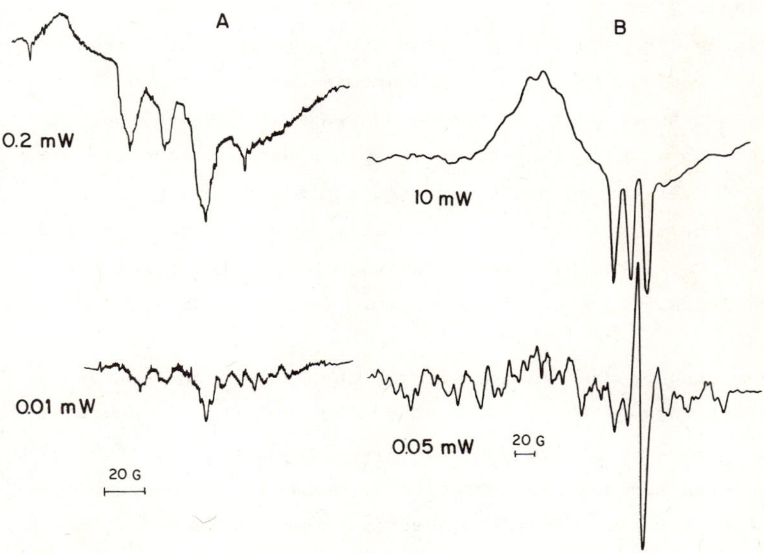

Figure 26 ESR spectra of frozen samples: (A) Walker carcinosarcoma [114] and (B) adenocarcinoma 755 (our unpublished data) at 77°K and various values of microwave power.

Figure 27 demonstrates that starting from day 5 the free radical concentration decreases.

The free radical concentration in malignant tissue at day 9 after tumor transplantation is lower, but based on our knowledge of the kinetics we know that during the first 3 days after transplantation there was an increase in the content of free radicals (Fig. 27, dotted line) up to the observed maximum on days 4-5. Similar data were obtained for myeloid leukemia and Yoshida tumor [142]. The maximum concentration of ascorbic acid radicals in wet lyophilized samples was also observed at the initial stage of growth tumor.

Swartz and Guttierrez [129], studying the change of free radical concentration in lyophilized and frozen tissue during growth of Walker carcinosarcoma, did not observe the extremes. But the absence of kinetic data on tumor growth prevents comparison of their results with our data.

The fact that the kinetic studies of free radical content shows a maximum at the initial stage of tumor growth is evidence of an increase in different types of oxidative processes in the cells at this stage.

Similarly, the study of paramagnetic properties of frozen samples of liver and spleen during leukemia development in AKR/y mice [135] demonstrated a monotonous increase of free radical concentration (signal with $\Delta H = 14$ G) in spleen and a concomitant decrease in liver throughout almost the whole leukemic process (Fig. 28). There is a monotonous decrease in the content of free radical in liver (frozen) of mice having disseminated forms and ascitic variants of experimental leukemia (see Fig. 28 and Table 7) [139].

In the case of disseminated forms, the radical concentration (R/R_o) at the time of the highest rate of growth ($t_{W_{max}}$) decreases more than in ascitic leukemia variants. Also during the terminal stage, the free radical concentration is one-half that volume during the initial stage. Further investigations are needed in order to determine the characteristic free radical kinetic curves for these tumors.

Table 6 Adenocarcinoma 755 Growth and Change of Free Radical Content in Frozen Samples of Malignant Tissue

Days after transplantation	Tumor weight P (g)	Half-width of signal ΔH (gauss)	Amplitude of ESR signal (relative units)
4	0.125 ± 0.030	9.9 ± 0.2	0.83 ± 0.13
5	0.235 ± 0.060	9.7 ± 0.4	0.94 ± 0.15
6	0.655 ± 0.150	10.9 ± 0.4	0.62 ± 0.22
7	1.160 ± 0.290	10.6 ± 0.6	0.56 ± 0.05
8	1.580 ± 0.410	10.4 ± 0.8	0.54 ± 0.12
9	3.210 ± 0.510	10.9 ± 0.5	0.39 ± 0.12

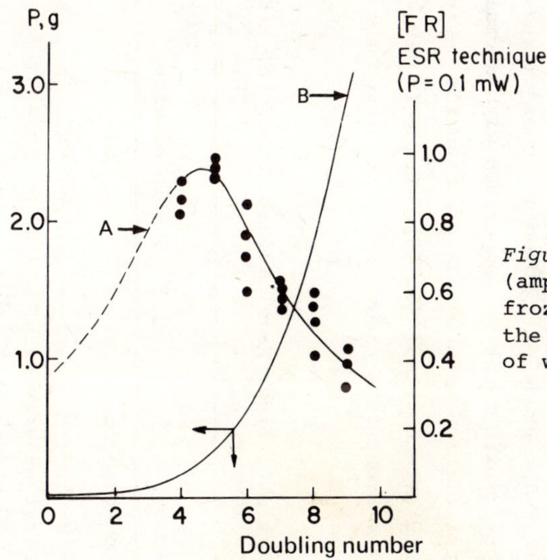

Figure 27 (A) Changes in a free radical content (amplitude of the ESR signal with g = 2.003) in frozen samples of adenocarcinoma 755 in mice during the course of tumor development. (B) Kinetic curves of weight increase for adenocarcinoma 755 in mice.

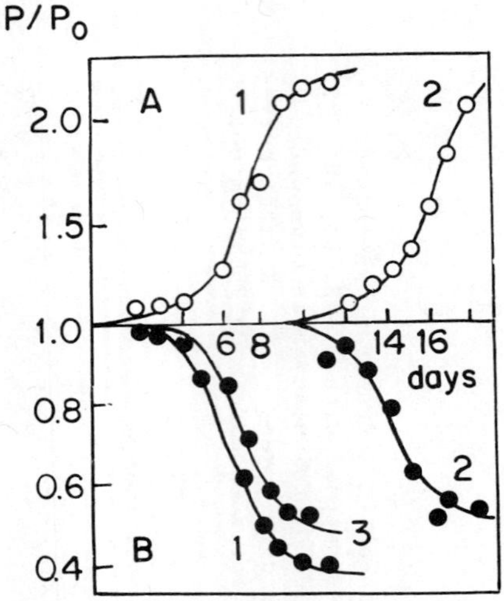

Figure 28 Changes in free radical content (B) and liver weight (A) for experimental leukemias: (1) leukemia LZ; (2) first generation spontaneous leukemia; (3) leukemia P-388 (liver weight does not change).

The study of homologous tumor tissue of hepatomas with different growth rates in comparison with normal liver, demonstrated a correlation between the paramagnetic center content of these tumors and growth rate [136,143]. Using frozen samples, it was found that the higher the growth rate, the lower the free radical (g = 2.004)

Table 7 Liver Free Radical Content of Mice with Experimental Leukemias

Tumor	Strain of mice	$t_{W_{max}}$ (days)	R/R_o
Leukemia La	$C_{57}B1$	5.0	0.7
Leukemia LZ	C_3H	7.0	0.6
First generation of spontaneous leukemia	AKR	15.0	0.6
Leukemia L1210	BDF_1	4.5	0.9
Leukemia P_{338}	BDF_1	5.5	0.9
Leukemia L5178	DBA/2	6.0	0.9

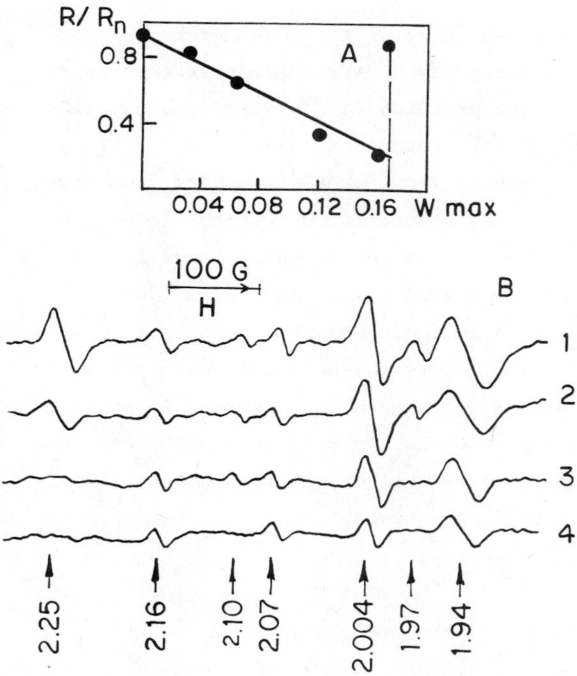

Figure 29 Paramagnetic centers in normal liver and transplanted hepatomas: (A) Relative free radical content in normal (R/R_n = 1.0), regenerating liver, and hepatomas with different growth rates; (B) ESR spectra of frozen samples of normal liver (1), minimum deviation hepatoma 46 (2), and rapidly growing hepatomas 60 and 22a (3,4).

concentration (Fig. 29A). This is in agreement with the results mentioned earlier [122,123], showing a decrease of the free radical content in hepatomas as compared with normal liver. However, the free radical content of regenerating liver, the growth rate of which is close to that of the most rapidly growing hepatoma 22a (see Fig. 29A), is not significantly different from that of normal liver. The radical content is obviously not correlated with the proliferative activity of tissue but rather depends on the degree of differentiation.

The alteration of a series of enzyme systems is accompanied by a decrease in the intensity of most ESR signals. Figure 29B demonstrates that some signals (g = 1.97, 2.10, 2.25) are absent in rapidly growing hepatomas 22a and 60.

A comparative study has been made of the paramagnetic properties of structural proteins extracted from rat liver vs. Morris hepatoma 3924A [144]. Hepatoma proteins do not yield the g = 2.002 ESR signals found in normal liver proteins.

An ESR study of mitochrondria from normal liver and from three transplanted Morris hepatomas has demonstrated that the intensity of a signal caused by a (Fe-S) complex was slightly lower in the slowly growing hepatoma 16 as compared to normal tissue [145]. There was a drastic decrease in intensity of this signal in rapidly growing hepatoma 7777. However, there was no significant difference in the ESR spectra between mitochrondria of normal liver vs. hepatoma 7800. Since hepatoma 7800 has an average growth rate, one is forced to conclude that there was no correlation between, the (Fe-S) signal and growth rate of hepatomas.

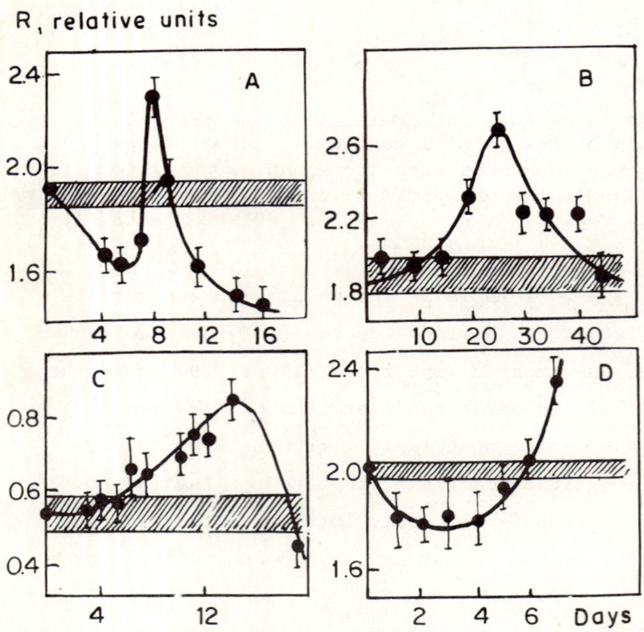

Figure 30 Kinetics of changes in free radical concentration of peripheral blood of tumor-bearing animals with solid (A) and ascites form (B) of sarcoma 37, Walker carcinosarcoma (C), and leukemia La (D).

In order to facilitate early tumor detection as well as to monitor the tumor response to therapeutic treatment, there has to be a change of radical concentration in the blood and organs of tumor-bearing animals. In solid and ascites sarcoma 37 (Fig. 30A,B) as well as in Walker carcinosarcoma (Fig. 30C), changes in free radical content were accompanied by changes in white blood cell counts [126-128]. However, these results contrasted with the change of free radical concentration in blood of animals bearing transplanted leukemia La (see Fig. 25D). During the first 4 days after a transplantation, subnormal levels of free radicals were observed, but as the leukemia developed there was an increase in the free radical content [125].

The ESR spectrum of frozen blood plasma is quite different from that of lyophilized tissue samples. The differences in the $g = 2.05$ signal are due mostly to the Cu-containing protein ceruloplasmin. Yet an increase in intensity of this signal is not a specific feature of the malignant process. Similar increases have been noticed for pathological conditions other than malignancy [146].

The free radical content of various visceral organs in tumor-bearing animals usually decreases during tumor development, as for example in the liver of animals bearing advanced ascites and solid sarcoma 37 (Fig. 31A, B). Similar results were obtained with the liver of animals bearing solid sarcoma 180 using the graft copolymerization method of free radical measurement [133]. It has also been assumed that the concentration of semiquinone radicals were subnormal in the liver of rats bearing sarcoma 45 [132] as well as in the adrenals of rats bearing Walker carcinosarcoma. But changes in the amount of free radicals in the liver of animals with Walker carcinosarcoma are complex (Fig. 31C); and, in the case of leukemia La, they become even more complex (Fig. 31D). Similar complexities have been observed in the lung of animals bearing leukemia La and in the spleen of mice bearing sarcoma 37 as well as of rats with Walker carcinosarcoma. These deviations are most likely evidence of complex tumor-host relationships as reflected by changes in free radical processes.

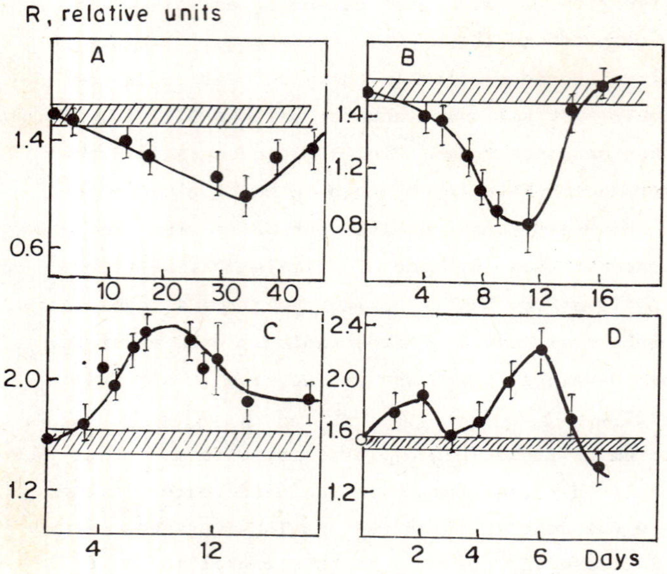

Figure 31 Free radical content in liver of tumor-bearing animals with ascites (A) and solid (B) forms of sarcoma 37, Walker carcinosarcoma (C) and leukemia La (D).

The complex character of these changes in organs of tumor bearing animals has been noticed previously [63]. It was observed that the free radical concentration in spleen and liver of rats bearing Walker carcinosarcoma did exhibit two maxima. In liver and spleen of C_3H mice bearing Gardner lymphosarcoma there was a maximum on day 6 after transplantation.

The maximum free radical content in liver, spleen, and lungs of animals bearing melanoma B-16 was observed on days 15 and 16. This essentially coincides with the highest growth rate as well as with the maximum free radical content of the tumor. Free radical concentration also increases when metastasis occurs in these organs.

At the present time, quantitative conclusions cannot be drawn concerning the radical concentration in various organs and tissue of tumor bearing animals because of the rather large amount of variable empirical data.

VI. INFLUENCE OF ANTITUMOR AGENTS ON THE FREE RADICAL CONTENT

Significant changes in the free radical reactions which accompany the development of transplanted tumors have made it possible to study the influence on these kinetic changes of various antitumor drugs belonging to the category of free radical inhibitors. The only published work of this type records the results of investigations performed by Soviet scientists using lyophilized samples. The widely used antimetabolite 5-fluorouracil (5-FU) was studied. Experiments were conducted using the transplanted leukemia La. The drug was administered beginning on day 4 after transplantation, which corresponds to the time of maximum free radical content in the spleen [148]. The 5-FU in 25 mg/kg doses had essentially no influence on the development of the leukemia process or on the level of free radicals in the spleen. Higher doses (50 mg/kg) caused a significant inhibition of leukemia development, and at the same time complex changes in the free radical concentration of spleen and liver (Fig. 32). Unlike the free radical inhibitors, 5-FU substantially changed the time course of the free radical content in intact animals causing an increase of free radicals in spleen and liver. It can be seen that 5-FU does not lead to an irreversible decrease of the free radical level. The appearance of a maximum in the free radical concentration curve of spleen from animals bearing leukemia and receiving 5-FU injection precedes the weight increase. Obviously, 5-FU exerts an indirect action on free radical processes in tumor cells.

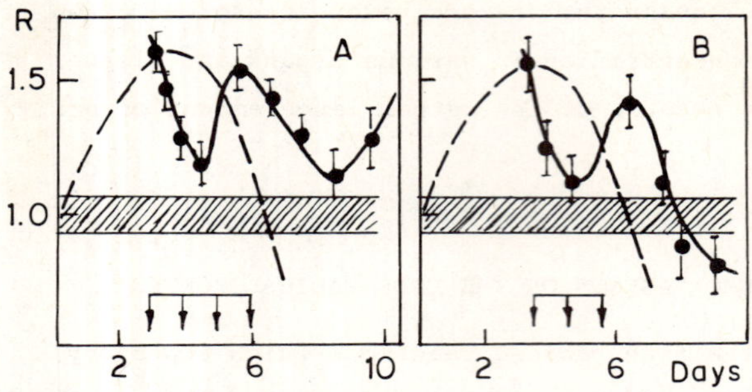

Figure 32 Changes in free radical content of lyophilized samples of spleen during leukemia La (dashed line) and under the action of 5-fluorouracyl (A) and sarcolysin (B).

The effect of the alkylating agents sarcolysin, thio-TEPA* and propylnitrosourea is in many respects similar to that of 5-FU. Thus, for example, sarcolysin at 4 mg/kg doses caused a marked inhibition of the leukemia process (χ = 1.47). As can be seen from Figure 32B, the free radical concentration at first decreased and then increased, reaching the control value by day 7. It is important to note that, as in the case of the control, this time corresponds to the beginning of the intensive development of leukemia. Similar observations during sarcolysin treatment were also noticed in liver of animals with leukemia. Thus sarcolysin, like 5-FU, does not eliminate the pathological deviation from the normal free radical level but only shifts the time.

Injection of propylnitrosourea (PNU) on the first day after transplantation of leukemia La leads to a significant inhibition of leukemia development (χ = 2.15), also causing a decrease in the level of free radicals in the spleen. At the same time, the level of free radicals in blood is essentially normal and the number of leukocytes is within normal limits. This observation allows one to see the relationship of free radical concentration and leukocyte number in peripheral blood of animals during the development of

*Thio-TEPA is triethylenethiophosporamide.

transplanted leukemia. Injection of PNU on day 5 has a more drastic antileukemic effect (χ = 2.65) and shows that the change in free radical level of spleen is similar to that brought about by sarcolysin or 5-FU. Thus with these PNU injection schedules both leukocyte number and free radical level are kept within normal limits.

Concerning the action of PNU on ascite sarcoma 37, the accumulation of ascitic fluid and of tumor cells is inhibited concurrently with a significant decrease in free radical concentration. Yet the action of PNU on ascites sarcoma 37 cells is obviously less specific than its action on leukemic cells, since the free radical concentration decreases to a lesser extent than in the case of leukemia while the time of survival does not increase significantly.

The action of thio-TEPA was studied in rats with Walker carcinosarcoma. Figure 33A demonstrates that with thio-TEPA tumor growth is two times slower than in controls (χ = 2.0). The concentration of free radicals in the peripheral parts of the tumors decreases slightly during the first period after the start of thio-TEPA injections but later increases despite the drug injections, reaching the level of radicals in controls after a lag of only 6 days. After 6 days the concentration of free radicals decreases to a level slightly below the initial value, only to increase again during renewed tumor growth. Similar changes take place in the central portion of the tumor (see Fig. 33B), the only difference being that there is a 2-day shift in the time of maximum. This may be related to the action of the weaker drug on the central portion of the tumor as compared to that on peripheral regions. Injection of thio-TEPA changes neither the total leukocyte number nor the free radical content of blood. In fact, only time shifts (about 10 days) in the maxima of leukocyte number and free radical concentration are observed, and in both cases the same trends are maintained. Attention should be drawn to the fact that the time shift values correspond to the periods of tumor inhibition induced by thio-TEPA.

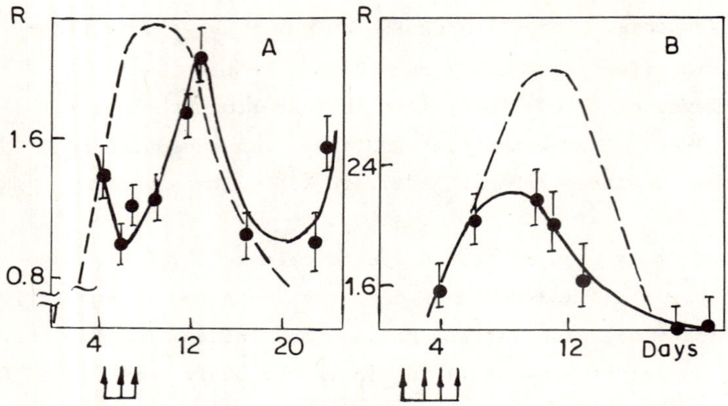

Figure 33 Kinetic curves of changes in free radical concentration in central (A) and peripheral (B) sections of Walker carcinosarcoma under the effect of thio-TEPA.

In our studies, profound changes in free radicals of spleen, liver, and adrenals were observed. In all cases, the free radical content decrease was noted along with its consequent restoration to control levels. In 10% of the animals studied, the drug brought about a cure accompanied by a normalization of the free radical content in the blood and in all the organs studied.

Emanuel and Lipatova showed in 1960 that a free radical inhibitor decreased the free radical content of Ehrlich ascites carcinoma cells [147]. Treating tumor cells with a solution of propylgallate decreased their free radical content approximately six times. Later, similar effects were achieved with propyl gallate on solid hepatoma 22 [149].

More complete information on the effect of inhibitors on the free radical level of tumors was obtained from a kinetic study of the action of one of the most widely used antioxidants, ionol (2,6-di-tert-butyl-4-methylphenol). The study began after serial injections of a 150 mg/kg dose of ionol starting on the first day after the transplantation of Leukemia La (Fig. 34A) and was pursued through subsequent injections of 10, 150, and 200 mg/kg doses at the time of maximum free radical content in the spleen of animals

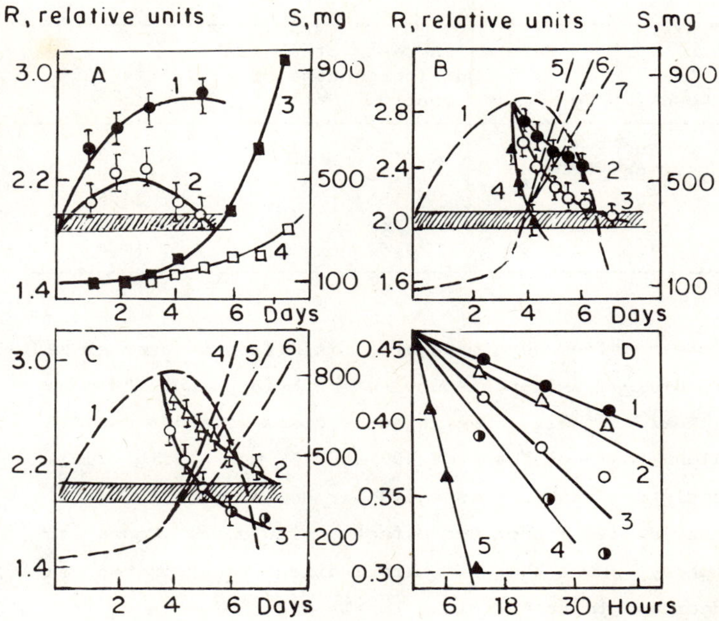

Figure 34 Variations in free radical content in spleen during leukemia under the effect of free radical inhibitors: (A) Variations in free radical content in control (1) and during administration of ionol (2): curves 3 and 4 correspond to variations in spleen weight. (B) Effect of ionol on free radical content (2-4) and leukemia development (5-7). (C) Action of "diphenyl" on variations in free radical concentration (2,3) and leukemia development (5,6). (D) Semilogarithmic changes showing consistent fall of free radical concentration under the effect of ionol (3,5) and "diphenyl" (2,4).

(see Fig. 34B) [150]. It can be seen that ionol caused the free radical level to decrease significantly. Kinetic curves of the free radical concentration in the region of drug injection demonstrates that the free radical content decays exponentially as shown by the semilogarithimic plots (Fig. 34D). Values of the kinetic constant K (rate of decrease in free radical concentration) as well as rate of spleen weight increase in relation to control values are given in the tabulation.

Inhibitor	Dose (mg/kg)	Rate of increase of spleen weight (percentage of control)	K x 10^{-3} hr^{-1}
Ionol	100	0.91	3.5
	150	0.73	6.9
	200	0.55	34.0
"Diphenyl"	125	0.67	4.4
	175	0.62	10.4

Similar observations were made after injection of another radical inhibitor, 4-oxy-3,5-di-tert-butyl-4'-methyldiphenyl ("diphenyl"), into leukemic animals. The injected doses were 125 and 175 mg/kg, equivalent to ionol doses of 100 and 150 mg/kg (Fig. 34C). It had previously been shown that equimolar doses of the substituted oxydiphenyls have a profound effect on leukemia. Comparing ionol with diphenyl shows that there is a direct relation between the rate of decrease in free radical levels and the slowdown of spleen weight increase. After termination of drug injections, the shape of the kinetic curves changed. Apparently, the inhibitor action is limited, and after the end of this limited period, the free radical concentration in spleen again begins to increase. In all cases, the free radical content decreased below the normal value shortly before the death of the animals. This phenomenon has been noticed many times and is obviously a unique characteristic of the free radical level in terminal stages of tumor growth.

The model of inhibitor action on the free radical content of animals with leukemia La is one which can probably be used for other types of experimental tumors. In any case, the oral application of ionol results in an inhibition of sarcoma 37 ascites and a decrease in free radical concentration of the ascitic fluid. Thus radical inhibitors lower the free radical content of organs in transplanted leukemia and in ascitic fluid in the case of sarcoma 37. The greatest effect was achieved by injection of the drug at the time of maximal content of free radicals in malignant cells.

The rate constant of the decrease in free radical concentration depends on the dose and correlated with the degree of antitumor activity.

VII. OBSERVATIONS ON CHANGES IN THE FREE RADICAL CONTENT OF HUMAN TUMORS

Even while gathering the first results on kinetic changes in free radical content of experimental tumors, it became apparent that the application of the ESR technique to oncological practice was possible. This has been discussed with respect to early diagnosis of malignant tumors and to possible use for the control of therapeutic efficiency.

An increase in the ESR signal intensity from leukocytes of patients with chronic lymphoid leukemia [121] and from liver of patients with leukemia [151] was reported in early papers. The free radical concentration was found to be twice higher in the peripheral portion of the tumor and 1.5 times higher in the center of the tumor as compared to normal stomach mucosa [152]. Metastases to regional nodes yielded 3.5 times higher signals than unchanged mucosa.

A fivefold increase in the free radical content was observed in cancer of the uterus [153] and of the lung [154], even though lung metastases of a patient with colon carcinoma did not yield ESR signals [155]. Also there was no ESR signals from a glioblastoma and a neurinoma [156].

A detailed investigation on the kinetic changes in the free radical content of leukocytes from patients with chronic myeloleukosis and lympholeukosis was carried out from 1967 through 1970 in the USSR. The free radical concentrations in the leukocytes of 160 patients with chronic myeloleukosis and in leukocytes of 170 patients with chronic lympholeukosis were measured [157,158]. Free radical concentration in the patients' leukocytes were followed from the time of hospitalization throughout the entire course of treatment. Kinetic curves of the change in free radical content of leukemic cells were correlated with the clinical and hematological profile of leukemia during various therapeutic courses. The free radical content in leukocytes of 50 normal donors was used as control values. The radical content was determined in lyophilized and wet samples. In all cases studied, the ESR signal was an asymmetrical singlet, typical of biological tissue, with a g value of 2.003. The ESR spectrum of "leukemic" leukocytes differed neither in shape nor by g factor from controls in either lyophilized or wet samples. Only in one specific case of chronic lymphoid leukemia was there an additional unknown signal with a g factor of 2.09 given by a wet leukocyte preparation. The ESR signal expressed per leukoyte (ca .20 mg dry weight per milliliter of suspension of $2 \cdot 10^8$ cells) appeared to be less intense for wet samples than for lyophilized ones.

From the start of the clinical treatment, the free radical content of leukocytes from patients with chronic myeloleukosis was on the average two times higher; in leukocytes of patients with chronic lymphoid leukemia it was on the average 4 times higher. The increase in free radical concentration of leukocytes from patients with chronic forms of myeloid and lymphoid leukemia was fairly well described by exponential kinetic curves having kinetic constants of 0.19 and 0.20 weeks.

Progress in the disease was characterized by an increase in the free radical content of leukocytes until a rapid accumulation of distrophic changes occurred in the body. At that time, the free radical level in leukocytes began to rapidly decrease (Fig. 35A). The kinetic shape was on the whole similar to the changes of free radical content in organs of animals with experimental leukemia (Fig. 21A).

In cases of successful chemotherapy, the level of free radicals in leukocytes generally decreased with clinical-hematological remission (Fig. 35B). This type of result was obtained while chronic myeloleukemia was being treated with myelosan and 6-mercaptopurine, or chronic lympholeukemia with prednisolone, leukeran, and dicypin. However, in spite of the clinical-hematological remission of patients, the free radical content in leukocytes remained high

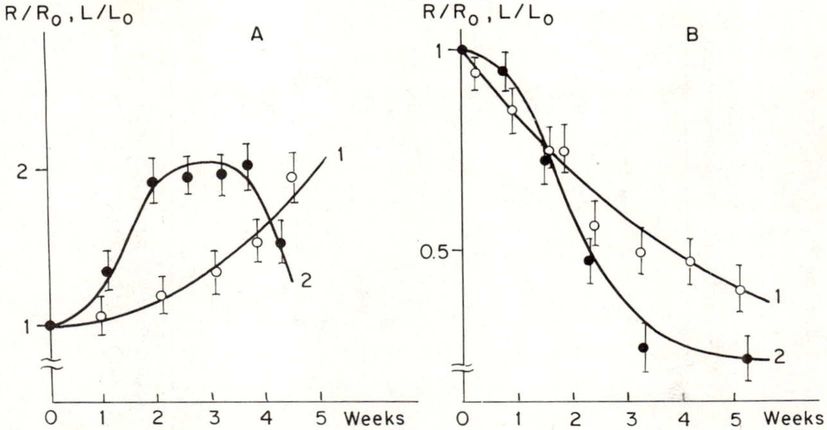

Figure 35 (A) Leukocyte count (1) and relative content of free radicals (2) in blood of patients with chronic lymphoid leukemia. (B) Variations in leukocyte number (2) and in relative content of free radicals in leukocytes (1) of patients with chronic lymphoid leukemia during the course of treatment.

in most cases and did not fall to values characteristic of normal cells. It appears that the higher concentration of free radicals in leukocytes of patients during remission may be evidence of a remaining small percentage of leukemic cells in the peripheral blood.

Observations on six patients discharged from the clinic in a state of remission showed that those with a higher content of free radicals in their leukocytes at the time of discharge relapsed sooner.

Comparing kinetic curves of free radicals in leukocytes of leukemia patients with leukocyte number in the case of successful therapy showed that a decrease in leukocyte free radical level was generally faster than the decrease in leukocyte number. This phenomenon reflects early changes in leukocytes at the molecular level in response to therapy. Table 8 presents data on oxygen consumption, glycolysis (increase in lactic acid content), and ATP content of leukocytes from normal donors compared with those from patients

Table 8 Respiration, Glycolysis, ATP Content, and Free Radical Concentration of Leukocytes as Influenced by Metabolic Poisons[a]

	Consumption of O_2	Increase in lactic acid content	ATP content	Free radical concentration
1. Donors (normal)				
(a) Leukocytes + Gl	1120	7.1	154	100
(b) Leukocytes + Gl + BA	1170	0	49	60
(c) Leukocytes + Gl + SF	1000	0	90	100
(d) Leukocytes + Gl + DNP	1150	12.0	160	80
(e) Leukocytes + Gl + Ol	1130	8.0	150	95
2. Chronic lymphoid leukemia				
(a) Leukocytes + Gl	1104	8.7	112	100
(b) Leukocytes + Gl + BA	1275	0	25	50
(c) Leukocytes + Gl + SF	1300	0	75	95
(d) Leukocytes + Gl + DNP	1320	13.0	100	50
(e) Leukocytes + Gl + Ol	1150	8.0	98	90
3. Chronic myeloid leukemia				
(a) Leukocytes + Gl	112	0	133	100
(b) Leukocytes + Gl + BA	25	0	0	60
(c) Leukocytes + Gl + SF	75	0	80	90
(d) Leukocytes + Gl + DNP	100	9.0	122	40
(e) Leukocytes + Gl + Ol	98	7.0	110	90

[a]Data per 10^9 cells. Average values based on 60 experiments. Error + 10%, Key: Gl, glucose; BA, monobromacetate, 8 x 10^{-4} M; SF, sodium fluoride, 2 x 10^{-2} M; DNP, 2,4-dinitrophenol, 8 x 10^{-4} M; Ol, oligomycin, 2 x 10^{-4} M.

with myeloid and lymphoid leukemia. It can be seen that, during incubation of leukocytes from blood of leukemic patients with various metabolic toxins, the free radical level was reduced considerably (50%) only by 2,4-dinitrophenol, a compound which inhibits oxidative phosphorylation at initial stages. Oligomycin, which inhibits oxidative phosphorylation at later stages, did not markedly affect the free radical level in the "leukemic" leukocytes. It should be noted that incubation of leukocytes from normal donor blood with 2,4-dinitrophenol and oligomycin did not have a pronounced effect on the free radical concentration. The free radical content of leukocytes remained unchanged after suppressing glycolysis by flouride as well as after marked effect on the free radical concentration. However, it may be that in this particular case the change in free radical level results from a damage to the phosphorylation system as well.

These results are evidence of specific shifts in the oxidative phosphorylation system of leukocytes from patients with myeloid and lymphoid leukemia. Additionally, they indicate that leukemic transformation is accompanied by certain changes in the leukemic cell metabolism that seems to involve free radical stages.

Another example of the possible use of the ESR method in clinical practice was provided by Wallace et al. [63]. They reported that of 152 clinical cases of mammary gland tumors examined, 63 were histologically diagnosed as cancer and 89 as various benign tumors. Due to the large experimental number it was possible to show a relation between the g factor of the ESR signal and the probability of finding a malignant tumor, which correlated with higher g factors. The signal width from the pheripheral zone of malignant tumors was 10.6 ± 0.35 G, whereas for normal tissue or benign tumor it was 12.5 ± 0.46 G. The relative free radical concentration in 1 to 4 cm tumors was considerably higher than in adjoining tissue. Such a relation indicates a change in free radical content as one readies more peripheral tumor sections. Wallace et al. [63] stated that tumors of about 4 mm were characterized by the same free radical level as normal tissue. They suggested that this may be a direct consequence of the characteristic changes in free radical content found, for instance, in the development of transplanted tumors.

The changes observed in these experiments (Fig. 36A) were used to characterize the clinical results obtained or the changes in free radical content. The changes in free radical content as a function of tumor doubling time can be plotted assuming that a 1-mm tumor corresponds to 20 doubling times and a 1-cm tumor to 30 doubling times [159]. Wallace et al. [63] considered only the descending section of the free radical curve, i.e., that corresponding to the period of clinical investigation. A satisfactory approximation of the data as well as data on free radical changes in the blood of patients could be obtained (Fig. 36B). When the authors utilized data on the changes of free radical content in the blood for a developing experimental solid tumor [128], a reliable increase in the signal amplitude (g = 2.003) was obtained from the efferent blood of advanced stomach cancer [152].

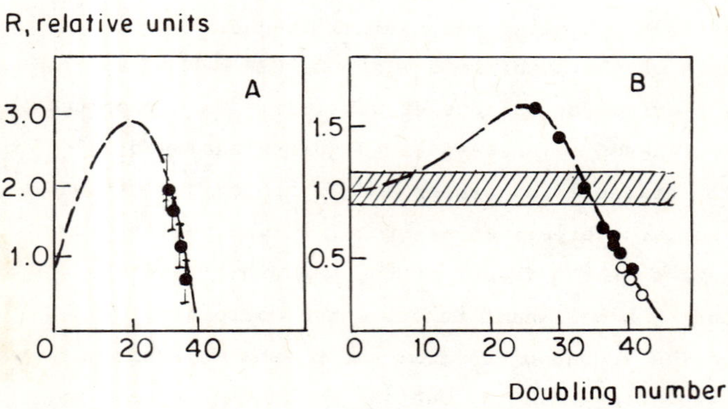

Figure 36 Variations in free radical content of tumors (A) and blood (B) of patients as a function of doubling number. (From Ref. 63.)

Thus the investigation of free radical mechanisms and changes in tissue paramagnetic properties during the course of tumor development has become an important trend in experimental and clinical oncology.

REFERENCES

1. L. Michaelis, *J. Biol. Chem. 84*:777 (1929).
2. L. Michaelis, *Cold Spring Harbor Symposium Quant. Biol. 7*:33 (1939).
3. B. Chance, *Svobodnie radikali v biologicheskikh systemakh*. Izd. IL, Moscow, 1963, p. 11.
4. B. Commoner, J. Townsend, and G. Pake, *Nature 174*:689 (1954).
5. V. N. Varfolomeev, G. N. Bogdanov, V. V. Djyakova, V. M. Pavlova, and N. M. Emanuel, *Biophysica 21*:881 (1976).
6. D. Sh. Burvaev and A. V. Lebanidze, *Biophysica 21*:942 (1976).
7. H. M. Swartz, J. R. Bolton, and D. C. Borg, *Biological Applications of Electron Spin Resonance*. Wiley (Interscience), New York, 1972.
8. L. A. Blumenfeld, *Problemi biologicheskoi fiziki*. Izd. Nauka, Moscow, 1977.
9. W. Pryor (Ed.), *Free Radicals in Biology*, Vols. 1 and 2. Academic Press, New York, 1976.
10. A. N. Saprin, E. V. Klochko, M. V. Chibrikin, K. E. Kruglyakova, and N. M. Emanuel, *Dokl. Akad. Nauk SSSR 169*:222 (1966).
11. N. M. Emanuel, *Ann. N.Y. Acad. Sci. 222*:1010 (1973).
12. N. M. Emanuel, *Kinetika eksperimental'nikh opukholevikh processov*. Izd. Nauka, Moscow, 1977.
13. N. M. Emanuel, R. E. Kavetskii, B. N. Tarusov, and E. P. Sidorik, *Biophysica Raka*. Izd. Naukova dumka, Kiev, 1976.
14. N. M. Emanuel, *Q. Rev. Biophys. 9*:283 (1976).
15. I. N. Naktinis and A. R. Chernyauskene, *Biophysica 19*:1039 (1974).
16. A. F. Vanin, D. Sh. Burbaev, N. L. Voevodskaya, A. V. Lebanidze unpublished data.
17. T. F. Slater, *Free Radicals in Tissue Injury*. Pion, London, 1972.
18. N. Kataoka, A. Imamura, Y. Kawazoe, G. Chihara, and C. Nagata, *Bull. Chem. Soc. Japan 40*:62 (1967).
19. T. Matsushima, I. Kobuna, and T. Sugimura, *Nature 216*:508 (1967).
20. A. Forrester, M. Ogilvy, and R. Thomson, *J. Chem. Soc. (Sect. C)* p. 1081 (1970).
21. N. M. Emanuel and L. P. Lipchina, *Dokl. Akad. Nauk SSSR 121*: 141 (1958).
22. H. Mason, T. Yamano, J. C. North, Y. Hashimoto, and P. Sakagishi, in *Oxidases and Related Redox Systems* (T. S. King, H. S. Mason, and M. Morrison, Eds.), Wiley (Interscience), New York, 1965, p. 879.

23. E. K. Ruuge, V. A. Timoshenkov, and L. A. Blumenfeld, *Biophysica 2*:611 (1966).
24. O. N. Brzhevskaya, V. S. Marinov, O. S. Nedelina, and E. M. Sheksheev, *Biophysica 12*:354 (1967).
25. A. E. Kalmanson and V. M. Chumakov, in *Mekhanismi dikhaniya fotosinteza i fiksatsii azota*. Izd. Nauka, Moscow, 1967, pp. 165-175.
26. I. G. Kharitonenko, V. V. Demin, and A. E. Kalmanson, *Dokl. Akad. Nauk SSSR 170*:29-31 (1966).
27. A. A. Konstantinov and E. K. Ruuge, *Bioorg. Khim. 3*:787 (1977).
28. G. Palmer, F. Müller, and V. Massey, in *Flavins and Flavoproteins*. (H. Kamin, Ed.). University Park Press, Baltimore, 1971, pp. 123-140.
29. E. K. Ruuge and A. A. Konstantinov, *Biophysica 21*:586 (1976).
30. M. K. F. Wikström and J. A. Berden, *Biochim. Biophys. Acta 283*: 403 (1972).
31. P. Mitchell, *FEBS Lett. 56*:1 (1975).
32. B. Commoner and T. Hollocher, *Proc. Nat. Acad. Sci. U.S. 46*: 405 (1960).
33. D. Bäckström, B. Norling, A. Ehrenberg, and L. Ernster, *Biochim. Biophys. Acta 197*:108 (1970).
34. O. N. Brzhevskaya, L. P. Kayushin, M. N. Kondrashova, V. S. Marinov, O. S. Nedelina, and E. M. Sheksheev, *Biophysica 12*: 839-844 (1976).
35. I. N. Nikitina and A. R. Chernyauskene, *Biophysica 19*:1039 (1974).
36. E. K. Ruuge and L. A. Blumenfeld, *Biophysica 10*:689 (1965).
37. R. H. Sands and W. R. Dunham, *Rev. Biophys. 7*:443 (1974).
38. W. Lovenberg (Ed.), *Iron-Sulfur Proteins*. Academic Press, New York, 1973.
39. D. Sh. Burbaev, *Dissertation*, University of Moscow, 1971.
40. N. R. Orme-Johnson, R. E. Hansen, and H. Beinert, *J. Biol. Chem. 246*:1922 (1974).
41. J. Rieske, D. McLennan, and R. Coleman, *Biochem. Biophys. Res. Commun. 15*:338 (1964); *65*:678 (1975).
42. T. Ohnisni, J. A. Hemington, K. F. LaNoue, H. P. Morris, and J. R. Williamson, *Biochem. Biophys. Res. Commun. 55*:272 (1973).
43. T. Ohnishi and J. S. Leigh, *Biochem. Biophys. Res. Commun. 56*: 775 (1974).
44. T. Ohnishi, D. B. Winter, J. Lim, and T. E. King, *Biochem. Biophys. Res. Commun. 61*:1017-1026 (1974).
45. Y. I. Shethna, P. W. Wilson, R. E. Hansen, and H. Beiner, *Proc. Nat. Acad. Sci. U.S. 52*:1263 (1964).
46. R. L. Tsai, I. C. Gunsalus, W. H. Orme-Johnson, R. E. Hansen, and H. Beinert, *Proc. Nat. Acad. Sci. U.S. 59*:959 (1968).
47. T. Ohnichi, D. B. Winter, J. Lim, and T. E. King, *Biochem. Biophys. Res. Commun. 53*:231 (1973).
48. D. Sh. Burbaev and A. V. Lebanidze, *Biophysica 21*:942 (1976).
49. M. Gutman, T. P. Singer, and H. Beinert, *Biochemistry 11*:556 (1972).

50. H. Beinert and W. Lee, *Biochem. Biophys. Res. Commun.* *5*:40 (1961).
51. H. Schleyer, *Ann. N.Y. Acad. Sci.* *212*:57 (1963).
52. G. T. Rikhireva, M. K. Pulatova, L. N. Chekulaeva, A. E. Druzhko, and V. A. Kositsina, *Biophysica* *20*:938 (1975).
53. "*Enzyme Models and Enzyme Structure.*" Brookhaven Symposium in Biology No. 15 (1962).
54. H. Beinert and G. Palmer, *Adv. Enzymol.* *27*:105-198 (1965).
55. G. Palmer, R. C. Bray, and H. Beinert, *J. Biol. Chem.* *239*:2657 (1964).
56. R. C. Bray and T. Vänngard, *Biochem. J.* *114*:725 (1969).
57. R. C. Bray and L. S. Meriwether, *Nature* *212*:467 (1966).
58. R. C. Bray, P. Knowles, and L. Meriwether in *Magnetic Resonance in Biological Systems* (A. Ehrenberg, B. Malmström, and T. Vänngard, Eds.). Pergamon Press, Elmsford, N.Y., 1967, p. 249.
59. L. E. Kozlova, A. N. Saprin, and N. M. Emanuel, *Sb. Aktualnie voprosi sovremennoi onkol*. No. 3, MGU, p. 51 (1975).
60. D. W. Nebert and H. S. Mason, *Cancer Res.* *23*:833 (1963).
61. J. R. Mallard and M. Kent, *Nature* *204*:1192 (1964).
62. N. Brannan, R. Cole, and J. Singley, *Proc. Soc. Exp. Biol. Med.* *123* (1966).
63. J. B. Wallace, B. H. Driscoll, C. G. Kaloniris, and A. Neaves, *Cancer* *25*:1087 (1970).
64. A. J. Vithayathil, J. L. Ternberg, and B. Commoner, *Nature* *207*:1246 (1965).
65. T. S. Shulyakovskaya, S. A. Vlasova, A. N. Saprin, K. E. Kruglyakova, and N. M. Emanuel, in *Tezisi Simpoziuma: Svobodnoradikalnii sostoyaniya i ikh rol pri luchevom porazhenii i slokachestvennom roste*, Moscow p. 99 (1971).
66. N. M. Emanuel, T. S. Shulyakovskaya, and V. A. Kondratyeva, *Dokl. Akad. Nauk SSSR* *209*:1213 (1973).
67. T. A. Shulyakovskaya, A. N. Saprin, and N. M. Emanuel, *Dokl. Akad. Nauk SSSR* *210*:221 (1973).
68. A. N. Saprin and T. S. Shulyakovskaya, *Dokl. Akad. Nauk SSSR* *189*:889 (1969).
69. M. I. Kuzin, V. I. Sachkov, V. L. Sharigin, and A. N. Saprin, *Eksp. Khir. Anesteziol.* *2*:65 (1971).
70. Ya. I. Azhipa, L. P. Kayushin, and E. I. Nikinkin, *Biophysica* *2(4)*:710 (1966).
71. Ya. I. Azhipa, L. P. Kayushin, and E. I. Nikishkin, *Biophysica* *14(5)*:852 (1969).
72. J. C. Woolum and B. Commoner, *Biochim. Biophys. Acta* *201*:131 (1970).
73. A. F. Vanin, L. V. Vakhnina, and A. G. Chetverikov, *Biophysica* *15*:1044 (1970).
74. D. Sh. Burbaev, A. F. Vanin, and L. A. Blumenfeld, *Zh. Strukt. Khimi.* *12*:252 (1971).
75. A. F. Vanin, manuscript in preparation (1978).
76. A. N. Saprin, V. A. Shabalkin, L. E. Kozlova, K. E. Kruglyakova, and N. M. Emanuel, *Dokl. Akad. Nauk SSSR* *181*:1520 (1968).

77. E. M. Mil, L. E. Kozlova, and A. N. Saprin, *Izv. Akad. Nauk SSSR (Biol.)* p. 901 (1971).
78. A. N. Saprin, L. E. Kozlova, V. A. Shabalkin, K. E. Kruglyakova, and N. M. Emanuel, *Izv. Akad. Nauk SSSR (Biol.) 6*:887 (1969).
79. F. T. Matyata, N. Kataoka, S. Nagase, H. Nakada, H. Sato, and H. Sasak, *Cancer Res. 31*:179 (1971).
80. A. F. Vanin, A. A. Kasparov, and E. I. Matkhanov, *Biophysica 16*:472 (1971).
81. H. Kon, *J. Biol. Chem. 243*:4350 (1968).
82. H. Kon, *Biochem. Biophys. Acta 379*:103 (1975).
83. H. Kon and N. Kataoka, *Biochemistry 8*:4757 (1969).
84. A. Szabo and M. F. Perutz, *Biochemistry 15*:4427 (1976).
85. J. C. Maxwell and W. S. Caughey, *Biochemistry 15*:338 (1975).
86. M. R. Borukaeva, L. M. Raikhman, V. A. Shabalkin, and A. N. Saprin, *Dokl. Akad. Nauk SSSR 189* (1969).
87. L. P. Kayushin, M. K. Pulatova, and B. G. Krivenko, *Svobodnie radikali v ikh prevrashchenii v obluchennikh belkakh*. Atomizdat., Moscow, 1976.
88. V. A. Sharpatii, *Pervichnie mekhanismi luchevogo porazheniya biologicheski vazhnikh makromolekul: DNK, belka, DNP, policakharidov*. Doctoral dissertation, Chernogolovka, USSR, 1972.
89. A. F. Usatii, *Biophysica 18*:439 (1973).
90. B. B. Singh and M. G. Ormerod, *Nature 206*:1314 (1965).
91. T. Henriksen, "ESR and the effects of radiation on biological systems." *Nat. Acad. Sci.—Nat. Res. Council, Nucl. Sci. Ser. Rep. 43*:81-100 (1966).
92. F. Patten and W. Gordy, *Proc. Nat. Acad. Sci. U.S. 46*:1137 (1960).
93. G. T. Rikhireva, M. K. Pulatova, M. M. Turganov, N. P. Palmina, and E. B. Burlakova, *Biophysica 23(2)*:395 (1978).
94. M. K. Pulatova, G. T. Rikhireva, M. M. Turganov, E. B. Burlakova, and N. P. Palmina, *Biophysica*, in press.
95. O. X. Avdeeva, V. Ya. Gotlib, G. T. Rikhireva, and M. K. Pulatova, *Biophysica*, in press.
96. S. Ya. Pshezhetskii, A. G. Kotov, V. K. Milinchuk, V. A. Roginskii, and V. I. Tupikov, *ESR svobodnikh radikalov v radiatsionnoi khimii*. Khimiya, Moscow, 1972.
97. A. K. Pikaev, *Solvatirovannii elektron v radiatsionnoi khimii*. Izd. Nauka, Moscow, 1969.
98. J. Duchesne (Ed.), *Physico-chemical Properties of Nucleic Acids* Vol 1. Academic Press, New York, 1973.
99. T. A. Nikolskaya, L. P. Lipchina, G. G. Poroshenko, and M. S. Aksyutina, *Sytologia 2*:486 (1969).
100. E. B. Burlakova, V. D. Gaintseva, and N. P. Palmina, *Dokl. Akad. Nauk SSSR 213:No. 6* (1974).
101. N. P. Palmina, V. D. Gaintseva, and E. B. Burlakova, *Sb. Aktualnie Vopr. Sovremennoi Onkol.* (Moscow) (1975).
102. D. M. Abdurasulov, A. I. Nikolaev, and A. I. Gaznev, *Vopr. Onkol. 13*:50 (1967).

103. T. S. Shulyakovskaya, V. A. Kondratyeva, S. A. Vlasova, and A. N. Saprin, *Dokl. Akad. Nauk SSSR 207*:461 (1972).
104. E. P. Sidorik, Doctoral dissertation, University of Kiev (1969).
105. A. Kh. Kogan, A. S. Sizikh, and A. N. Saprin, in *Tezici Dokl. IY Mezhdunarodnogo Biophysicheskogo Kongressa Moscow* Vol. 4, p. 207 (1972).
106. E. P. Sidorik, N. N. Beregovskaya, E. V. Oleinik, L. M. Korchevaya, and T. N. Yurkovskaya in *Tezici Dokladov I Vsesoyuznogo Simposiuma po Bioneorganicheskoi Khimii* Izd. Naukova dumka, Moscow, 1972, p. 27.
107. E. P. Sidorik, R. E. Kavetskii, E. A. Baglei, T. N. Yurkovskaya, O. P. Melnikov, M. I. Danko, and I. B. Kaganovich, in *Materiali IY Mezhdunarodnogo Biophysicheskogo Kongressa, Moscow* Vol. 4, p. 197 (1972).
108. Ya. I. Azhipa, L. P. Kayushin, and N. A. Bavlova, *Biophysica 14*:1040 (1969).
109. Ya. I. Azhipa, N. A. Pavlova, and L. P. Kayushin, *Biophysica 15*:1056 (1970).
110. E. P. Sidorik, *Kantserogenez*. Izd. Naukova Dumka, Kiev, 1973, pp. 135-148.
111. N. M. Emanuel, A. N. Saprin, T. S. Shulyakovskaya, L. E. Kozlova, and T. V. Bunto, *Dokl. Akad. Nauk SSSR 209*:1449 (1973).
112. O. S. Frankfurt, L. P. Lipchina, T. V. Bunto, and N. M. Emanuel, *Byul. Eksp. Biol. Med. 8*:86 (1967).
113. L. W. Wattenberg, *J. Nat. Cancer Inst. 48*:1425 (1972).
114. L. W. Wattenberg, *J. Nat. Cancer Inst. 50*:1541 (1973).
115. E. B. Burlakova and E. M. Molochkina, *Vopr. Onkol. 20(4)*:62 (1974).
116. T. S. Shulyakovskaya, B. A. Rikova, A. N. Saprin, and M. F. Rozhkov, *Dokl. Akad. Nauk SSSR*, in press.
117. N. M. Emanuel, *Dokl. na Simposiume: Pervichnie mekhanismi biologicheskogo deistviya ioniziruyushchikh izlucheniya, Moscow*. Trudi MOIP, Vol. 7, p. 73 (1960, 1963).
118. N. M. Emanuel and L. P. Lipchina, *Dokl. Akad. Nauk SSSR 125*: 1448 (1959).
119. N. M. Emanuel, *Sb. Puti sinteza i iziskaniya protivoopukholevikh preparatov*. *Medgiz*, Moscow, 1960.
120. M. M. Petyaev, *Biophysicheskie podkhodi k diagnostike zlokachestvennikh opukholei*. Meditsina, Moscow, 1972.
121. N. I. Pavlova and A. R. Livenson, *Biophysica 10*:169(1965).
122. B. Commoner, J. Townsend, and G. E. Pake, *Nature 174*:689 (1954).
123. F. K. Truby and J. W. Goldziecher, *Nature 182*:1371 (1958).
124. I. K. Kolomiitseva, K. L. L'vova, and L. P. Kayushin, *Biophysica 10(5)*:636 (1965).
125. A. N. Saprin, E. V. Klochko, M. V. Chibrikin, K. E. Kruglyakova, and N. M. Emanuel, *Dokl. Akad. Nauk SSSR 169* (1966).

126. A. N. Saprin, E. A. Minenkova, E. G. Nagler, E. V. Koperina, S. A. Kruglyak, K. E. Kruglyakova, E. M. Vermel, and N. M. Emanuel, *Biophysica 2(4)*:616 (1966).
127. A. N. Saprin, E. A. Minenkova, E. G. Nagler, S. A. Kruglyak, K. E. Kruglyakova, and N. M. Emanuel, *Biophysica 12*:No. 6 (1967).
128. A. N. Saprin, Yu. S. Kaznatcheev, E. A. Minenkova, L. G. Nagler, S. A. Kruglyak, K. E. Kruglyakova, and N. M. Emanuel, *Biophysica 12(6)*:1099 (1967).
129. H. M. Swartz and P. L. Guttierrez, *Science 198*:936 (1977).
130. D. H. Driscoll, C. M. Dettmer, J. D. Wallace, and A. Neaves, *Current Mod. Biol. 1*:275 (1967).
131. Mc. Nicolas, D. H. Driscoll, J. D. Wallace, and A. Neaves, *ESR Biol. Symposium, Washington* (1968).
132. Yu. P. Kozlov, *Svobodnie radikali i ikh rol v normal'nikh i patologicheskikh protessakh.* MGU, Moscow, 1973.
133. E. B. Burlakova, S. K. Dobrina, Yu.P. Kozlov, and V. A. Pukhov, *Vestn. MGU (Ser. Biol. i Pochvoved)* No. 2 (1966).
134. A. N. Vodolazovskii, S. K. Dobrina, Yu.P. Kozlev, and B. N. Tarusov, *Dokl. Akad. Nauk SSSR 172*:713 (1967).
135. A. N. Goloschapov, A. N. Alesenko, G. N. Bogdanov, and E. B. Burlakova, *Biophysica 18(6)*:1047 (1973).
136. V. N. Varfolomeev, G. N. Bogdanov, V. V. D'yakova, V. M. Pavlova, and N. M. Emanuel, *Biophysica 21*:881 (1976).
137. N. J. F. Dodd and J. M. Giron-Conland, *Br. J. Cancer 32*:451 (1975).
138. H. M. Swartz, C. Mailer, S. Ambegaoukar, W. E. Antholine, D. McNellis, and S. Schneller, *Cancer Res. 33*:2588 (1973).
139. D. S. Khristianovich, L. M. Dronova, V. N. Varfolomeev, G. N. Bogdanov, and N. M. Emanuel, *Dokl. Akad. Nauk SSSR 236*:1011 (1977).
140. N. M. Emanuel, *Dokl. Akad. Nauk SSSR 217*:No. 1 (1974).
141. G. N. Bogdanov, V. N. Varfolomeev, L. S. Vasil'eva, I. V. Pavlovskii, and N. M. Emanuel, *Dokl. Akad. Nauk SSSR 233*:494 (1977).
142. N. J. F. Dodd, *Br. J. Cancer 34*:556 (1976).
143. G. N. Bogdanov, V. N. Varfolomeev, V. M. Pavlova, and N. M. Emanuel, *Dokl. Akad. Nauk SSSR 226(1)*:207 (1976).
144. H. A. Pohl, P. R. C. Gascoyne, A. Szent-Györgyi, *Proc. Nat. Acad. Sci. U.S. 74*:1558 (1977).
145. T. Ohwishi, J. Hemington, K. Lanoue, H. Morris, and J. Williamson, *Biochem. Biophys. Res. Commun. 55(2)*:372 (1973).
146. C. Mailer, H. M. Swartz, M. Konieczny, S. Ambegaonkar, and V. L. Moore, *Cancer Res. 34*:637 (1974).
147. N. M. Emanuel and T. E. Lipatova, *Dokl. Akad. Nauk SSSR 130*:221 (1960).
148. A. N. Saprin, K. E. Kruglyakova, and N. M. Emanuel, *Trudy MOIP: Fizikokhimicheskie mekhanizmi zlokachestvennogo rosta 32*:197 (1970).
149. E. S. Elkhovskaya, A. E. Kalmanson, L. P. Lipchina, and V. N. Tveritinov, *Dokl. Akad. Nauk SSSR 139*:4 (1961).

150. A. N. Saprin, E. V. Klochko, V. M. Chibrikin, K. E. Kruglyakova, and N. M. Emanuel, *Dokl. Akad. Nauk SSSR 166*(3):447 (1966); *Biophysica 2*(3):443 (1966).
151. L. Mulay and J. Mulay, *Med. Tribune 6*:2 (1965).
152. M. I. Kuzin, N. M. Emanuel, O. C. Shkrob, A. N. Saprin, and M. V. Zaitsev, *Tezisi doklada: Svobodno-radikalnii sostoyaniya i ikh rol pri luchevom porazhenii i zlokachestvennom roste*, Moscow, p. 53 (1973).
153. C. P. Hodgkinson and T. Cole, *Obstet. Gynecol. 25*:411 (1965).
154. Y. Mallard, *Med. Tribune 3* (1965).
155. B. Commoner and J. L. Ternberg, *Proc. Nat. Acad. Sci. U.S. 47*: 1374 (1961).
156. J. Duchesme and A. Van de Vorst, *Bull. Sci. Acad. Roy. Belg. 56*:433 (1970).
157. I. A. Kassirskii, N. M. Emanuel, E. V. Klochko, O. A. Kovalenko, V. I. Pinzur, K. E. Kruglyakova, and V. I. Chibrikin, *Vopr. Gematol. i Perelivaniya Krovi 2*:No. 8 (1976).
158. E. V. Klochko, L. V. Kovalchuk, K. E. Kruglyakova, L. A. Seits, M. C. Luganova, M. N. Blinov, and N. M. Emanuel, *Dokl. Akad. Nauk SSSR 190*:No. 2 (1970).
159. E. Gerstenberg, *Radiobiol. Radiother. 6*:325 (1965).

9

CHEMICAL RADIATION SENSITIZERS IN CANCER THERAPY

Rajagopalan Sridhar

Oklahoma Medical Research Foundation
Oklahoma City, Oklahoma

I. INTRODUCTION

Ionizing radiation such as x- or γ-rays is known to be mutagenic, carcinogenic, and cytotoxic. These effects of radiation reflect the damage sustained by DNA which is the critical target in the cell. A large part of the radiation-induced damage to DNA occurs via free radical reactions. The cytotoxicity of x- or γ-radiation forms the basis for radiotherapy of cancer. Thus, free radicals are involved not only in the triggering of events leading to cancer but also during radiation therapy of cancer.

This chapter is concerned with chemical radiation sensitizers which have been developed with a view to improving the radiocurability of tumors. Some background information is presented in order to acquaint the reader with the problems encountered in radiation therapy of cancer. This chapter will deal with the mechanisms of radiosensitization.

II. RADIOBIOLOGICAL CONCEPTS, THERAPEUTIC INDEX, AND RADIATION SURVIVAL CURVES

The use of x- and γ-rays in cancer therapy has been known for some time. The idea of using chemical agents to improve the effectiveness of radiation therapy is not new [1-5] and many compounds have been tried in combination with radiation for the treatment of malignant disease [6,7]. Until recently, attempts to improve the radiation response of tumors by concomitant drug therapy were unsuccessful mainly because the additive or synergistic effect of the combined treatment was exerted equally on both the normal and malignant tissues without any gain in therapeutic index. Therapeutic index is a measure of the damage to malignant or neoplastic tissue relative to that sustained by normal tissue as a result of the therapeutic treatment. Thus,

$$\text{therapeutic index} = \frac{\text{extent of damage to neoplastic tissue}}{\text{extent of damage to normal tissue}}$$
$$= \frac{\text{number of neoplastic cells killed}}{\text{number of normal cells killed}}$$

In this context it is important to understand what cell killing and cell survival mean in radiobiological parlance. In the case of proliferating cells such as those growing in tissue culture or hematopoietic stem cells, death is defined as the loss of the capacity for sustained proliferation. This loss of reproductive integrity is known as reproductive death. A surviving cell, often referred to as being *clonogenic*, is one which has not lost its reproductive capacity and is able to proliferate indefinitely to form a sizeable clone or colony of cells. As regards differentiated cells, such as those of nerve and muscle tissue, which do not proliferate, cell death has to be defined as the loss of some biological function specific to the particular cell system under study.

When mammalian cells such as Hela or V79 Chinese hamster lung cells are irradiated with x- or γ-rays, only a few cells are killed in the beginning at the lower doses and the extent of killing increases progressively with increasing dose. This increase becomes more pronounced at a "threshold" dose beyond which the survival decreases exponentially with increasing dose. Survival curves are usually obtained by plotting the extent of survival (surviving fraction) on a logarithmic scale as a function of radiation dose on a linear scale. Typically, the survival curves for cells exposed to sparsely ionizing radiations display an initial bending shoulder (or threshold) region at the low doses followed by a linear (or straight) portion at the higher doses (Fig. 1). Survival curves are characterized by D_o and n, which are two parameters that vary with cell line and the experimental conditions. The reciprocal of D_o, is a measure of the slope of the linear portion of the survival curve. D_o is defined as the increment in dose which decreases survival to 37% in the linear portion of the survival curve. The extrapolation number, n, is a measure of the extent of the threshold (shoulder) region of the curve; this number n is the value found on the ordinate by extrapolating the linear portion of the curve such that it intercepts the ordinate. The appearance of the shoulder in the survival curve is attributed to the ability

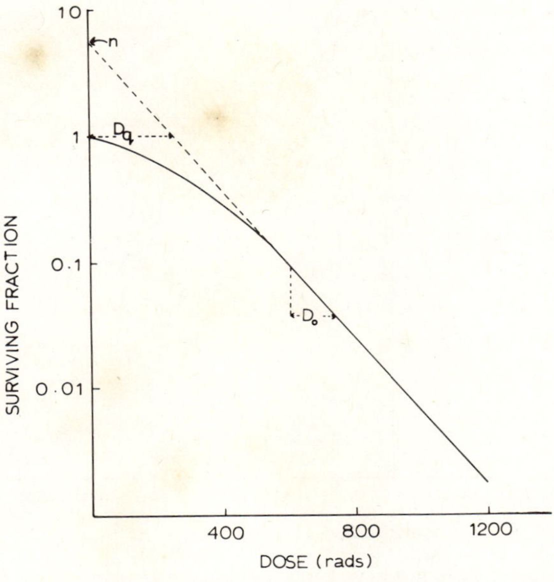

Figure 1 A typical survival curve for mammalian cells exposed to low LET (linear energy transfer) radiation. D_o is a measure of the slope of the linear portion of the survival curve. The extrapolation number, n, is a measure of the extent of the initial "threshold" or shoulder. The quasi-threshold dose, D_q, is the dose at which the linear portion of the survival curve extrapolates to a point corresponding to a surviving fraction of unity; D_q is related to n and D_o by the expression $\log_e n = D_q/D_o$.

of cells to accumulate sublethal damage which can subsequently be repaired. For a constant D_o^{-1}, a decrease in extrapolation number indicates a decrease in the extent of repair. The nature of the sublethal lesions and the mechanisms involved in their repair are not well understood. Repair of sublethal damage is associated with a phenomenon known as Elkind-Sutton repair or split-dose recovery. In a typical split-dose experiment a population of cells is subjected to a dose of radiation sufficient to reduce the survival to a point in the linear region of the survival curve and then the

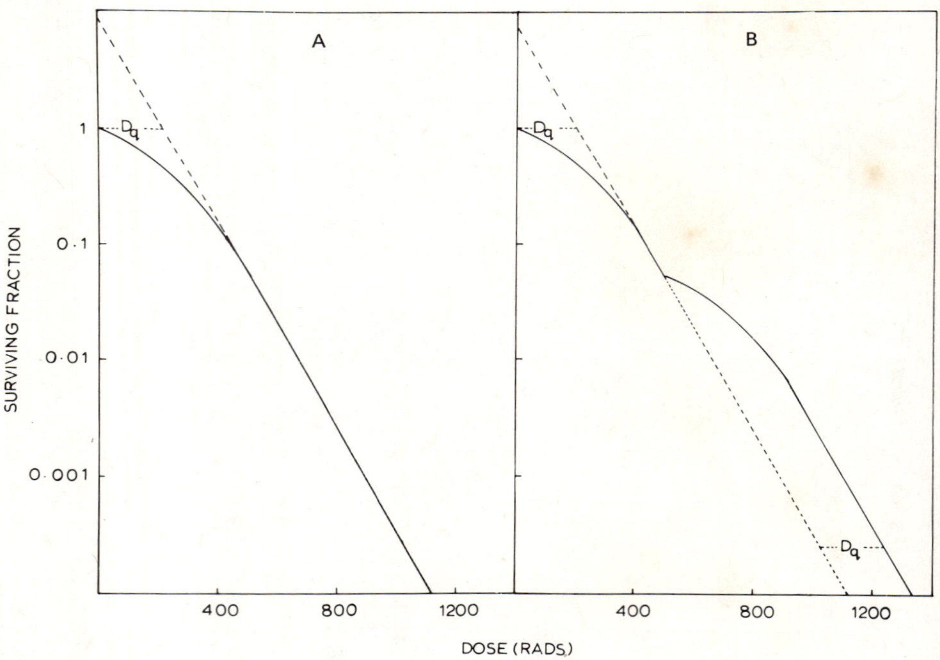

Figure 2 (A) A typical radiation survival curve for cultured mammalian cells which were continuously irradiated. (B) Survival curve obtained when the cells were continously irradiated with 0.5 krad and further irradiation was commenced after a 12-hr interruption during which the cells were incubated to allow repair of sublethal damage to occur (Elkind-Sutton repair). The quasi-threshold dose, D_q, is the amount of dose that is wasted as a result of fractionation. (Based on a figure in Ref. 8.) With permission from *Biochemical Mechanisms of Liver Injury* (T. F. Slater, Ed.), Academic Press, London, 1978.

radiation is stopped for some time, whereupon the population regains its original resistance to low doses of radiation. Thus upon resumption of radiation after the interruption, an additional shoulder appears in the survival curve which eventually becomes linear at higher doses such that the survival is displaced parallel to the one expected for uninterrupted delivery of the total radiation dose [8] (Fig. 2).

III. RADIOSENSITIZERS

Radiosensitizers are chemical agents which possess the potential to increase the lethal effect of ionizing radiation when administered in combination with irradiation. Radiosensitizers may exert their influence by acting prior to, during, or after irradiation. Radiosensitizers which serve to synchronize cells in a radiation-sensitive stage of the cell division cycle will be effective even if present only prior to irradiation. Sensitizers which inhibit cellular repair processes can be expected to be effective even when administered after irradiation. Sensitizers which interact with radiation-induced radicals (e.g., DNA free radicals) exert their effect when present during the administration of radiation.

A. Dose-Modifying and Non-Dose-Modifying Radiosensitizers

Radiosensitizers (and radioprotectors) may or may not be dose-modifying [8]. Dose-modifying radiosensitizers are those chemicals which act in such a way that the ratio of the respective doses with and without the chemical for any level of survival is constant, i.e., the extrapolation number, n, is the same for the semilogarithmic radiation survival curves obtained with and without the chemical. Conversely, non-dose-modifying sensitizers are those which reduce the shoulder in the survival curve without changing the slope by a corresponding amount, i.e., n is different for the sensitized and unsensitized radiation survival curves [8] (Fig. 3).

Due to the efforts of many workers, our current knowledge of tumor biology and radiation effects has advanced to such a level that several chemical radiosensitizers are being used effectively

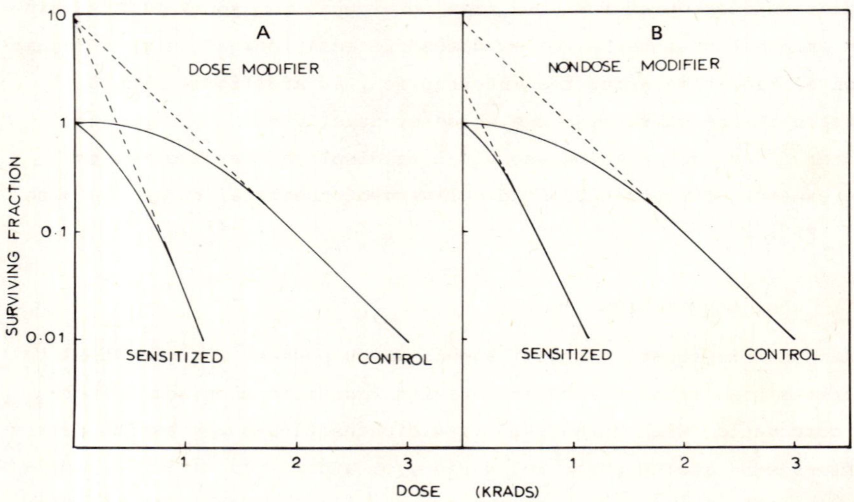

Figure 3 Survival curves obtained on irradiation of mammalian cells with and without a radiosensitizer. (A) The effect of a dose-modifying sensitizer. (B) The effect of a non-dose-modifying radiosensitizer.

in combined modality treatment of certain types of cancer [9,10]. These radiation sensitizers (radiosensitizers) have been designed to selectively sensitize tumor cells while sparing the normal cells as far as possible. The process of chemical radiosensitization is very complex, and extensive research remains to be done in order to provide answers for the many unresolved problems related to tumor radiobiology.

This review will discuss some aspects of using chemicals in combination with radiation for the treatment of malignant tumors. Particular emphasis will be placed on chemicals which specifically sensitize hypoxic cells to radiation, since the presence of radiation-resistant hypoxic cells in solid tumors may be a reason for failure of radiation therapy [11,12]. As mentioned previously, the cytotoxic action of x- and γ-rays forms the basis for radiation therapy of cancer. Since both normal and malignant cells can be killed by radiation, it is very important that the beam of therapeutic radiation be very accurately aimed at the tumor mass and that the dose be carefully controlled. The dose of radiation that can be administered for the purpose of tumor sterilization is limited to a large extent by the amount of damage sustained by normal tissue surrounding or adjoining the tumor.

IV. DIRECT AND INDIRECT ACTION OF IONIZING RADIATION

Absorption of x- and γ-rays by biological material may lead to direct interaction with biologically important molecules such as DNA, RNA, and proteins. During such interactions, energy is deposited directly on a critical target molecule or a part of it which undergoes excitation or ionization within a nanosecond of absorption of radiation. This excitation or ionization may lead to alterations in the structure of the critical target molecule which may cause biological changes. These alterations which result from direct deposition of energy on biologically important molecules are termed *direct effects*. The changes brought about in this manner are said to be due to the direct action of radiation. Since a cell contains a large proportion of water, much of the energy absorbed by the cell is deposited on water molecules, and this causes the generation of free radical species such as H• and e^-_{aq} (solvated electron). These reactive free radicals diffuse to and react with biologically important molecules and alter their chemical structures. Such structural alterations are the result of indirect action of radiation. Chapman et al. [13] have estimated that in irradiated aerobic cell cultures, some 30-40% of cell death is due to direct effects and about 60% is due to HO• produced by indirect effects; <10% may be caused by e^-_{aq}, which is also produced by indirect effects.

Cellular DNA is the prime target molecule responsible for radiation-induced cell death. Numerous radiation-induced lesions or alterations have been recognized in the DNA of irradiated cells [14]. The extent of radiation damage to cells is dependent on the amount of oxygen available to the cell [15]. For example, the number of single-strand breaks produced in DNA of irradiated cells is three to four times higher for air-saturated cells than for cells irradiated under anoxic conditions [16,17]. It has also been demonstrated that many DNA single-strand breaks produced by radiation in aerobic cells are due to the indirect effect mediated by HO• [18]. Reactions involving HO• are to a large extent responsible for the production of neutral free radical centers in the DNA of irradiated cells. Reducing species such as sulfhydryl compounds (which accumulate in cells under anaerobic conditions) can act as free radical scavengers or hydrogen donors. Sulfhydryl compounds react with either HO• or neutral DNA free radicals to prevent damage to DNA. Oxygen, on the other hand, oxidizes the DNA radicals before repair processes involving hydrogen donation can occur [19].

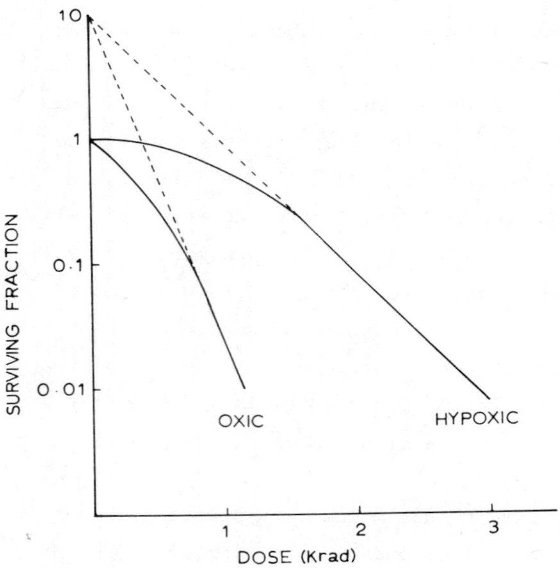

Figure 4 Survival curves obtained on irradiation of cultured mammalian cells under hypoxic (anoxic) and aerobic (oxic) conditions. These curves are typical of several examples in the literature where oxygen appears to be dose modifying. However, there are some instances where oxygen acts as a non-dose-modifying radiosensitizer.

From the foregoing description of the free radical mechanism for the inactivation of cells by ionizing radiation, it is easy to appreciate why oxygen acts as a radiation sensitizer. The importance of this, i.e., the *oxygen effect* in therapeutic radiology, can be appreciated by comparing the survival curves for aerobic and anoxic V79 cell cultures exposed to γ-rays (Fig. 4). The aerobic cultures are nearly three times as sensitive toward radiation damage as anoxic cultures.

V. HYPOXIC CELLS IN TUMORS

In 1934 Crabtree and Cramer [20] made the astute observation that patients with severe anemia or poor circulation did not respond well to radiation therapy because of the lowered levels of oxygen in certain regions of the tumors receiving radiation. These workers were the first to recognize the importance of the aforementioned oxygen effect in radiation therapy. Subsequently, there have been several other reports relating anemia and poor response to radiation therapy of tumors [12,21,22]. Bush et al. [12] have shown that anemia influences the proportion of hypoxic cells in tumors (see also Hill et al. [23]) and have reported that the correction of anemia at the time of radiation treatment leads to an increased cure rate. The radiation resistance of cells under low oxygen levels becomes more important in view of the tendency of solid tumors to develop regions of hypoxia. This is because cellular crowding and congestion cause interruption of blood flow in tumors which have outgrown their vascular supply. Oxygen diffusing from capillaries is rapidly consumed by the nearby cells, and as a result the distant cells become hypoxic. Methods employing microelectrodes for measuring oxygen levels in solid tumors and normal tissues confirm the presence of hypoxic regions in poorly vascularized solid tumors [24]. There is strong evidence for the existence of hypoxic cells in many animal tumors [25], and there is indirect clinical evidence for the occurrence of hypoxic cells in tumors in humans [26-30]. The presence of radiation-resistant hypoxic cells may explain the failure to eradicate tumors by x- or γ-rays [11,12, 24,31].

Several approaches have been suggested to overcome the problem associated with the presence of radioresistant hypoxic cells in tumors. Attempts have been made to increase oxygenation of tumors by: (a) inhalation of hyperbaric oxygen [32,33]; (b) inhalation of pure oxygen [32,34]; (c) inhalation of carbogen, a mixture of 95% oxygen and 5% carbon dioxide [32,35]; and (d) arterial infusion of hydrogen peroxide in the expectation that oxygen will be liberated by the action of catalase-like enzymes in blood [32,36]. These methods have not been encouraging, and their value in the treatment of cancer has not been proven [32].

In an alternative approach, the sensitivity of normal tissues to radiation is reduced by artificially creating regional hypoxia through the application of a tourniquet. This allows the use of a higher dose of radiation for tumor control, since risk to the normal tissue has been diminished [32,37,38].

As we have already seen, the DNA may escape permanent structural alteration if the neutral radicals derived from it are able to interact with reducing species such as sulfhydryl compounds usually present in cells. However, permanent structural alteration could be achieved and the lesion made long lasting if oxidizing agents such as oxygen and strongly electron-attracting chemicals are available for reaction with the neutral DNA radicals formed during radiation.

VI. SOME ASPECTS OF COMBINING RADIATION WITH CHEMICAL AGENTS

The combination of chemical agents and radiation for therapeutic purposes may produce a variety of effects depending on the nature of the chemical. The combination of a cytotoxic drug and radiation may increase the killing of both normal and neoplastic cells in a simple additive manner. This would be of little use in view of the lack of selectivity toward malignant cells; moreover, it may be possible to achieve the same increased level of cytotoxicity by increasing the radiation dose in the absence of the drug. Even a combination of a chemotherapeutic agent and radiation which exerts an additive and nonselective cytotoxic action may be of some value in controlling distant metastases lying in the regions outside the field of radiation treatment. Another situation may exist when the cytotoxic chemical potentiates the radiation effect and the overall cell killing may be more than additive. Again, for such a combination to be useful, the cytotoxic effect should be exerted preferentially on neoplastic cells. Even combinations which do not show any selectivity may be useful if the chemical used in the combination therapy has a tendency to accumulate in tumors. In such a case differential cytotoxicity might be achieved by administering radiation after the chemical has concentrated preferentially in neoplastic tissue. A third possibility involves the use of a chemical which is nontoxic by itself but increases the cytotoxic action of radiation, i.e., the chemical acts exclusively as a radiosensitizer.

It should be possible in principle to achieve preferential killing of tumor cells by the use of chemicals which afford selective protection for normal cells against radiation damage without affecting the radiation response of neoplastic cells.

One of the first attempts at chemical radiosensitization was reported by Werner, who used a preparation of choline borate (Enzytol) in combination with radiation for tumor treatment (see Mitchell [2] and Becker [3]). Since then, dyes such as fluorescein and trypan blue [4,5], cations from metals such as copper [39], vitamin analogs such as menadione and Synkayvite which inhibit mitosis and cellular ATP production [1], respiratory poisons such as iodoacetic acid derivatives [40], conjugated carbonyls such as diacetyl and indanetrione [41], and sulfhydryl-binding agents such as N-ethylmaleimide (NEM) [42] and diamide [43] have all been tested for use as radiosensitizers. Most of these chemicals are of little use in radiation therapy of cancer; some are too toxic to normal tissue and lack selectivity for tumor cells, whereas others are degraded to inactive metabolites under pharmacological conditions.

VII. HYPOXIC CELL RADIOSENSITIZERS

A very promising approach for achieving selective radiosensitization of tumors involves the use of chemicals which selectively radiosensitize hypoxic cells. Since hypoxic cells in tumors constitute a limiting factor in the radiocurability of some tumors, it would be advantageous to use a chemical agent that would sensitize hypoxic cells exclusively without influencing the radiation response of normal well-oxygenated cells. A clinically useful drug for this purpose should be (a) nontoxic to oxic cells and normal tissues, (b) readily absorbed but slowly metabolized by tissues, and (c) capable of rapidly diffusing to the poorly vascularized regions of tumors and selectively sensitizing the hypoxic cells [13,44,45]. One of the most promising sensitizers which possesses most of these favorable properties is metronidazole [Flagyl; or 1-(2-hydroxyethyl)-2-methyl-5-nitroimidazole] [45,46].

Most of the early experiments with radiosensitizers were carried out on bacterial systems [42,46-50] and *Vicia faba* roots [46,49]. Knowledge gained from these studies paved the way for later developments which led to the use of radiosensitizers that were effective in mammalian systems [13,45,51-55]. At present a few chemicals are undergoing clinical trials as radiosensitizers of hypoxic cells in tumors [9,10,56-65], and a considerable part of this review will be devoted to the discussion of hypoxic cell radiosensitizers.

The different sensitizers vary in their selectivity toward hypoxic cells and act by a variety of mechanisms. It is possible to classify the various radiosensitizers according to their mechanism of action. Since many sensitizers act through more than one mechanism of sensitization, they can belong to more than one classification.

A. Sulfhydryl-Binding Agents as Radiosensitizers

Suflhydryl compounds (thiols) are radioprotective agents by virtue of their tendency to act as hydrogen donors and depletion of intracellular pools of sulfhydryl compounds should therefore increase the radiosensitivity. Bridges [42] demonstrated that the sulfhydryl-binding agent NEM sensitized irradiated bacteria, the extent of sensitization being much greater for hypoxic compared to well-oxygenated suspensions of bacteria. Neoarsphenamine [66], p-chloromercuribenzoate [44], iodoacetamide [40], and diamide [diazenedicarboxylic acid bis-(N,N'-dimethylamide)] [43] are other examples of sulfhydryl-binding compounds with radiosensitizing properties. These sensitizers are far too toxic to have clinical potential. Research on these compounds has continued, however, in order to elucidate the mechanistic aspects of radiosensitization. Although these compounds were tested as radiosensitizers because of their ability to depress the sulfhydryl content of cells, there is evidence at least for NEM that the radiosensitizing action may be due to the formation of radiochemically generated free radical derivatives of NEM. Rapid-mixing studies where bacteria were irradiated only a few milliseconds after exposure to the sensitizer have shown that sensitization can occur even when the contact time is insufficient for any chemical reaction to occur between sulfhydryl groups and NEM and the sensitization occurs via a fast radiochemical event leading to free radical species [67]. Radiosensitization of hypoxic mammalian cells has not been observed in experiments employing nontoxic concentrations of NEM [68-72]. The same is true in the case of diamide, which is far too toxic to be of any clinical use [43,73].

B. Inhibitors of Postirradiation Repair Processes as Radiosensitizers

Some of the damage initially inflicted on irradiated mammalian cells can be repaired by cellular enzyme systems, and this accounts for the increased cell-survival seen in split-dose experiments [74]. Split-dose recovery of clonogenic potential known as Elkind-Sutton repair (see Fig. 2) can reduce the radiocurability of tumors [75]. One may expect radiosensitization by chemicals which can either inhibit the enzymes of the repair system or modify the initial radiation-induced lesion in such a way as to preclude its repair.

In this connection it is important to note that there is no evidence for repair processes which are specific for tumors and different from those in normal tissue. Drugs such as actinomycin D [75,76] and adriamycin [75], which are known to bind to cellular DNA, inhibit Elkind-Sutton repair and act as radiosensitizers [75].

C. Radiosensitizers Which Form Toxic Products as a Result of Radiation Action

Certain chemicals such as iodoacetamide give rise to cytotoxic products upon irradiation, and this accounts for their sensitizing property. Cupric salts represent another example of this mode of apparent radiosensitization. Cupric ion undergoes radiation-induced reduction under anaerobic conditions to yield highly cytotoxic cuprous ion, which undergoes rapid oxidation to the less toxic cupric ion in the presence of oxygen. Therefore, cupric ion produces an apparent radiosensitizing effect in hypoxic bacteria [39,77,78] and hypoxic mammalian cells [79] without any effect on the corresponding oxygenated systems.

D. Halogenated Pyrimidine Analogs as Radiosensitizers

Halogenated pyrimidines such as 5-chlorodeoxyuridine, 5-bromodeoxyuridine, 5-iododeoxyuridine, and 5-fluorodeoxyuridine are synthetic analogs of thymidine (5-methyldeoxyuridine). The resemblance of thymidine to the chloro or bromo or iodo pyrimidine derivative is much closer than that between thymidine and 5-fluorodeoxyuridine. The reason for this becomes clear when one considers that the van der Waals radius of the methyl group (2.00 Å) is nearly the same as that of an atom of chlorine (1.80 Å) or bromine (1.95 Å) or iodine (2.15 Å) but much larger than that of atomic fluorine (1.35 Å). Thymidine analogs such as 5-chlorodeoxyuridine, 5-bromodeoxyuridine, and 5-iododeoxyuridine can be substituted for thymidine in the structure of DNA, and they can act as radiosensitizers when incorporated into the DNA chain [80,81]. It is believed that radiosensitization occurs because incorporation of bromodeoxyuridine into the DNA leads to an increase in the number of radiation and radiochemical lesions which may cause permanent cellular damage [82]. Adams [83] has suggested a mechanism involving long-range intramolecular electron migration, and fast luminescence studies by Fielden and Lillicrap [84] using DNA containing structurally incorporated 5-bromodeoxyuridine have provided evidence for long-range migration of some form of energy along the DNA chain. The thymidine analog, 5-fluorodeoxyuridine is not incorporated into mammalian DNA; however, this fluorinated analog is used to enhance the incorporation of other halogenated thymidine analogues into DNA because de novo synthesis of thymidylic acid is inhibited by the fluorinated analog. The enzyme thymidylate kinase, a key enzyme for de novo synthesis of thymidylic acid is inhibited by 5-fluorodeoxyuridine. Many other halogenated pyridines such as 5-bromodeoxyuridine, 5-chlorodeoxyuridine, 5-iododeoxyuridine, 5-bromodeoxycytidine, 5-iododeoxyuridine, and trifluoromethyldeoxyuridine can replace thymidine in mammalian DNA. More than 50% of thymidine in cellular DNA can be replaced by the halogenated thymidine analogs [81]. Incorporation of the analogs is possible only in the case of dividing cells, and this seriously limits their use since a considerable proportion of cells in tumors are usually in a noncycling state. Moreover, selective incorporation of the analogs into the tumor is made difficult due to the uptake by normal dividing cells. Other factors which curtail the clinical usefulness of thymidine analogs include enzymatic dehalogenation, enzymatic removal, rapid excretion, and toxicity. Gain in therapeutic ratio has, however, been claimed in clinical trials utilizing intra-arterial infusion directed toward brain tumors [81,85,86]. Although thymidine analogs have been of very limited use in radiotherapy of cancer, they have proved valuable in radiobiological research.

E. Stable Free Radicals as Radiosensitizers

Since it was recognized that oxygen is an excellent radiosensitizer of hypoxic cells, experiments were initiated with nitric oxide which is isoelectronic with oxygen). Nitric oxide further resembles oxygen in being a paramagnetic gas. High toxicity of nitric oxide precludes its clinical use, although its effectiveness as a sensitizer has been shown in test systems in vitro. More encouraging results have been obtained with derivatives of nitric oxide, viz., the nitroxyls. Nitroxyls are stable free radicals which can modify the radiation response of cells because of their tendency to react with radiation induced DNA-free radicals [87,88] and by their ability to inactivate repair processes by reacting with essential sulfhydryl groups of some important cellular molecules. Emmerson [89] was the first to demonstrate the effectiveness of triacetoneamine-N-oxyl (TAN), a stable nitroxyl compound, as a sensitizer. Since then other workers have also investigated the radiosensitizing properties of TAN, and the metabolic inactivation of this chemical by high cell densities (such as those encountered in tissues) has been demonstrated [90]. Norpseudopelletierine-N-oxyl (NPPN) and 2,2,6,6-tetramethyl-4-piperidinol-N-oxyl (TMPN) are better radiosensitizers of hypoxic cells than TAN [91]. There is no evidence to suggest that these compounds may be clinically useful [92,93]. Furthermore, they are prone to inactivation by whole blood [93]. Cooke et al. [94] have carried out extensive studies on the use of the bifunctional nitroxyl compound bis(2,2,6,6-tetramethyl-1-oxyl-4-piperidinyl)succinate (Ro-03-6061) as a radiosensitizer of hypoxic cells. Cooke and co-workers have demonstrated that the binitroxyl, Ro-03-6061 is a very efficient radiosensitizer of hypoxic cells which, in addition to reducing the D_0 value of the hypoxic survival curve, reduces the shoulder region of the curve; for radiation doses less than 1000 rad, it appears to be a more powerful sensitizer than oxygen [94]. Cooke et al. [94] have pointed out that the biradical (Ro-03-6061) is more effective than the corresponding monofunctional compounds because of its potential to combine with two radiation-induced DNA-free radical centers in extreme proximity and produce double lesions which may be very toxic [95]. Calculations have indicated that the free radical centers in Ro-03-6061 are separated by approximately 14 Å at room temperature [96], and a cellular concentration of 600 mM would be required in the case of mononitroxyls to furnish two nitroxyl groups in such close intermolecular spacing, i.e., 14 Å apart [94]. The bifunctional sensitizer Ro-03-6061 is not very soluble in aqueous systems (solubility limit is around 100 μM), and therefore other more soluble biradical derivatives, such as N,N-bis(2,2,6,6-tetramethyl-1-oxyl-4-piperidinyl)succinamide (Ro-03-9199) and bis(2,2,6,6-tetramethyl-1-oxyl-4-piperidinyl)cis-1,2-cyclopropanedicarboxylate (Ro-03-8767) have been studied. The solubility limit of Ro-0-8767 is 200 μM, whereas Ro-03-9199 is more soluble [97]. The radiosensitizing action of biradicals was shown to be dependent on the cell-drug contact time at both 20° and 37° [97]. (Temperatures throughout this chapter are given in degrees Celsius.) The three biradicals were tested at equimolar concentrations at 20° for their ability to decrease the D_o value. On the basis of the drug-cell contact time required to decrease the D_o value, the order was Ro-03-8767 > Ro-03-9199 > Ro-03-6061. However, when they were compared on the basis of contact time required to reduce the extrapolation number n (i.e., to reduce the shoulder region), the order was Ro-03-9199 > Ro-03-8767 > Ro-03-6061 [97]. From these results Millar et al. [97] concluded that D_o and n are modified by different mechanisms. Further support for this conclusion had been provided by Millar et al. [98] in an earlier study employing combinations of drugs to sensitize hypoxic mammalian cells. None of the mononitroxyls tested had any effect on the shoulder region of the hypoxic cell radiation survival curve, i.e., they did not alter the extrapolation number n. All three biradicals tested thus far have the ability to reduce the shoulder region of the hypoxic survival curve, and in this respect they resemble the sulfhydryl-binding compound diamide (see Fig. 5) [43,73]. Millar et al. [97] have stated that the modification of the shoulder region of the hypoxic cell radiation survival curves by these chemicals may be a consequence of reduction in cellular repair capacity brought about by interference with cellular biochemistry. Although this may be applicable in this particular instance, it is important to realize that the presence of a shoulder in the survival curve indicates that

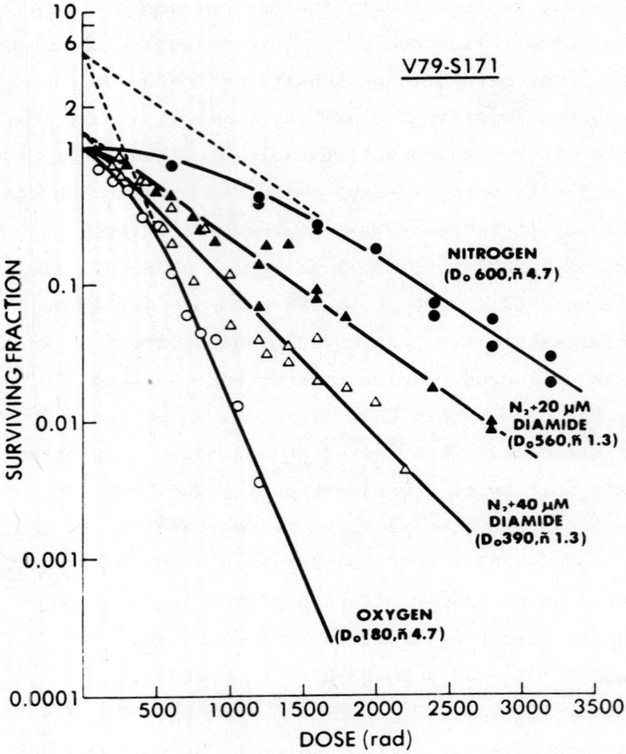

Figure 5 Survival of V79-S171 Chinese hamster cells after x-irradiation in the presence or absence of diamide. Cells were irradiated while suspended at 2 x 16^6 cells/ml in phosphate buffered saline (0°). Gassing with oxygen or nitrogen was for 20 min before irradiation. In these experiments, 8 μM diamide completely oxidized the cellular reduced glutathione (GSH). The D_q (quasi-threshold dose) for the four curves are O_2, 278; N_2, 929; N_2 + 20 μM diamide, 147; N_2 + 40 μM diamide, 102. (From Ref. 73.)

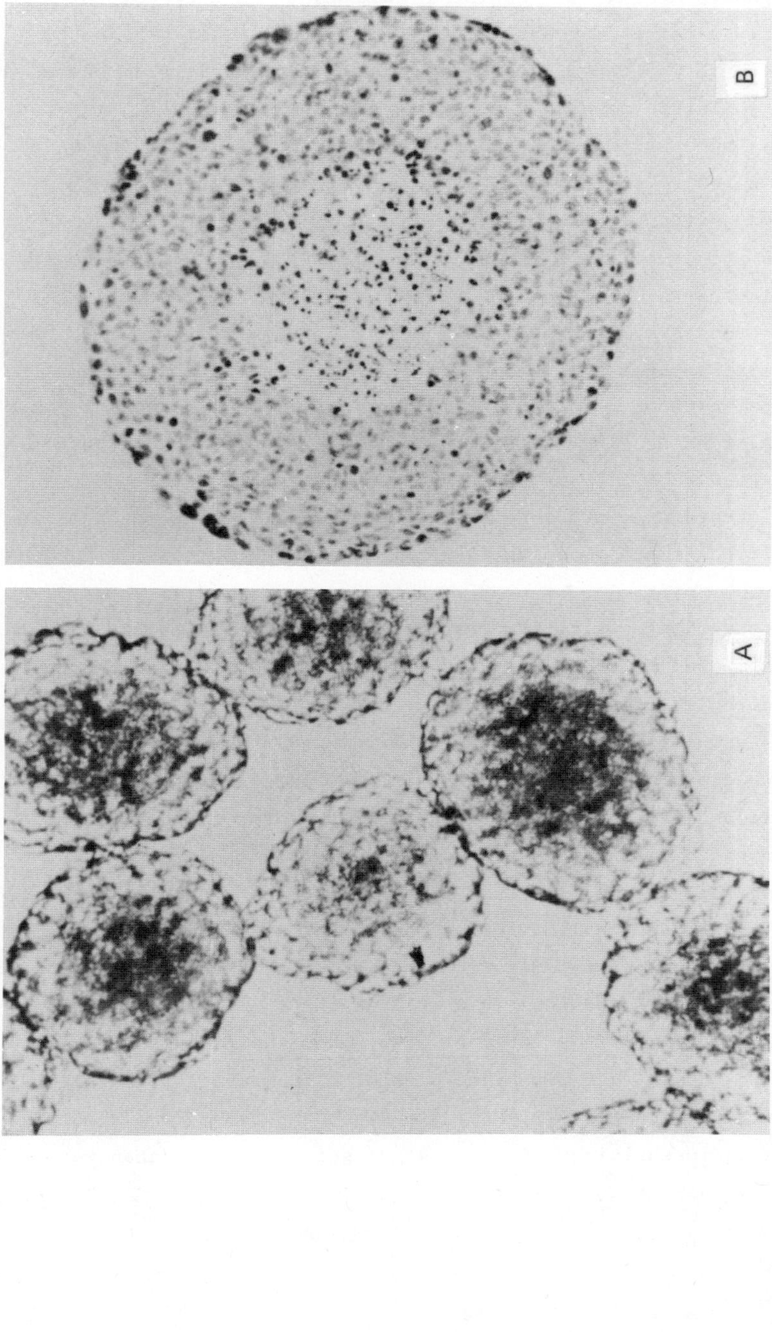

Figure 6 (A) Phase contrast wet-mount of multicell spheroids x 120. (B) Section through center of multicell spheroid showing necrotic center. Diameter = 450 µM. (From Ref. 45.)

the cells have accumulated sublethal damage, which they may not necessarily repair. Conversely, the absence of a shoulder in the survival curve indicates lack of accumulation of sublethal damage by cells while the repair processes may not necessarily be affected. Ro-03-6061 appears to be the most efficient sensitizer of the three biradicals tested so far [97]. The shoulder modifying property of Ro-03-6061, evident in experiments conducted at 20° and 37°, was lost when the temperature during irradiation was 4° [97]. However, the ability of Ro-03-6061 to decrease D_o was not lost at 4° [97].

Electron spin resonance (ESR) studies on aqueous solutions of the three biradicals show that the degree of interaction between the two radical centers is different in each case (personal communication from T. A. Jenkins cited in Ref. 97). The trans isomer of the cis-compound Ro-03-8767, which might well be interesting from the standpoint of mechanistic studies, was not effective as a radiosensitizer of hypoxic cells at its solubility limit (≃10 μM) in aqueous systems [97].

The superior sensitizing properties of the biradical Ro-03-6061 as compared to the monoradicals TAN and TMPN has been demonstrated using multicellular spheroids of V79 cells [99,100]. These spheroids are composed of tightly adhering cells which can contain internal regions of chronic hypoxia and necrosis similar to poorly vascularized solid tumors (see Fig. 6). Spheroids further resemble solid tumors in having a mixture of noncycling and cycling cells. They provide a good model system for studying the effect of metabolic activity and diffusion limitations due to cellular congestion and necrotic zones on the radiosensitizing action of chemicals [101,102]. It was found that Ro-03-6061 (0.1 mM) present for 1 hr before and during irradiation of V79 intracellular spheroids (≃600-μM diameter) at 37° with ^{60}Co γ-rays withstood metabolic inactivation and eliminated the hypoxic cells [99,100]. Ro-03-6061 (0.1 mM) was a more effective radiosensitizer of hypoxic cells in spheroids than was TAN or TMPN (0.6 mM) [99,100].

F. Electron-Affinic Compounds as Radiosensitizers

Adams and Dewey [41] were the first to state that the radiosensitizing properties of a chemical may be related to its electron affinity or oxidizing power. The significance of redox characteristics in the radiosensitization of hypoxic cells has been amply demonstrated [103-106]. As mentioned earlier in this chapter, the electron-affinic compounds act as radiosensitizers by removing an electron from neutral DNA radicals (produced as a result of either the direct or indirect effect of radiation) before repair through the action of hydrogen donors can occur [13]. It is worth mentioning that many of the electron-affinic radiosensitizers of hypoxic cells also have the property of preferential or selective toxicity directed against hypoxic cells [45,102,107-111] even in the absence of radiation. This is indeed an added bonus in the radiation therapy of poorly vascularized solid tumors containing radiation-resistant hypoxic cells [45,102]. The importance of redox properties in the radiation-independent cytotoxic action of nitroaromatic compounds has been well recognized [13,106,112,113]. The cytotoxic properties of electron-affinic compounds in the absence of radiation will not be discussed here because it is discussed in detail in Chapter 13 of this book.

Strongly electron-affinic chemicals such as quinones, other dicarbonyl derivatives, indanetrione, unsaturated esters, pyruvic acid derivatives, and nitroaromatic compounds have all been tested for possible use as sensitizers and found effective in the case of hypoxic cells in bacterial and/or mammalian systems [44]. Over the years nitroaromatic compounds including the nitroheterocyclics have shown more promise than any other class of electron-affinic sensitizers. In this respect it is important to mention that two nitroheterocyclic compounds belonging to the nitroimidazole class, namely, metronidazole and misonidazole, are currently undergoing clinical trials as radiosensitizers in Britain and Canada [9,10,56-63].

The radiosensitization of hypoxic Chinese hamster cells in tissue culture by p-nitroacetophenone (PNAP) [114,115] set the stage for many fruitful studies in radiation biology. The testing of PNAP as a sensitizer of hypoxic cells in tumor-bearing animals was hampered by the low solubility of the chemical. A more soluble derivative of PNAP, viz., p-nitro-3-dimethylpropiophenone hydrochloride (NDPP), was synthesized and tested as a radiosensitizer in hypoxic bacterial and mammalian cell systems [44]. NDPP was effective at very low concentrations as a radiosensitizer of hypoxic cells, although not without a significant sensitizing effect on oxygenated cells [44]. After considerable research on various substituted acetophenones, the importance of the nitro group—relative to the carbonyl group—in the radiosensitization of hypoxic cells was recognized [106,114].

Since the electron affinity of a nitrofuran derivative is greater than that of the corresponding nitrobenzene derivative [13,116], Chapman and co-workers screened a variety of nitrofurans for their radiosensitizing ability using tissue culture and animal tumor model systems [13]. It was found that NF-167, a water-soluble derivative of nifuraldezone, at a concentration of 250 μM sensitized hypoxic cells in tissue culture to radiation killing to the same extent as oxygen present in air-saturated medium [13]. Air-saturated medium contains 280 μM oxygen, which is a nearly 10 times higher concentration than is necessary for complete radiosensitization by oxygen [15]. Had the comparison been made at a lower concentration, i.e., between NF-167 (30 μM) and oxygen (30 μM), then the superior radiosensitizing power of oxygen would have been evident. Many of the chemically reactive sensitizers which were effective in conventional tissue culture systems such as monolayer or single cell suspension cultures were ineffective when tested as hypoxic cell sensitizers in irradiated animal tumor systems [13]. This was attributed to the inability of the chemical to reach the hypoxic regions of tumors and also due to the difficulty in maintaining a high enough concentration of the chemical (perhaps due to metabolic inactivation) in the tumor during the period of radiation delivery [13]. Chemicals which were effective as sensitizers in animal tumors were characterized by a fairly long half-life in blood

($T_{1/2}$ > 80 min), high lipid solubility, and high water solubility [13]. It soon became clear that chemicals which were very reactive were also prone to metabolic inactivation and thus would be unlikely to reach the target cells in poorly vascularized tumors. There was obviously a need for radiosensitizers that were less reactive than compounds such as NF-167, if they were to be pharmacologically and toxicologically acceptable.

As stated earlier, a clinically useful radiosensitizing chemical must (a) be nontoxic to oxic cells and normal tissue, (b) be readily absorbed but slowly metabolized, and (c) diffuse to poorly vascularized central regions of tumors and selectively sensitize the hypoxic cells [13,45,46]. One of the most promising sensitizers which meets most of these requirements is 1-(2-hydroxyethyl)-2-methyl-5-nitroinidazole (metronidazole; Flagyl) [49]. Metronidazole will be discussed with another promising analog, misonidazole, in detail later in connection with their use as radiosensitizers in clinical trials [9,10,56-63]. Raleigh and coworkers [117] have studied structure activity relationships in the nitropyrrole series and have found that, in general, the radiosensitizing ability increases with the electron affinity of the nitropyrrole derivative [117]. Of the several radiosensitizers tested in vivo, N-hydroxy-ethyl-2-cyano-5-nitropyrrole (NP-1) was the most effective [117]. In this series of compounds N-hydroxy-ethyl derivatives were less toxic than the N-methyl, N-ethyl, and N-n-propyl derivatives [117].

Since the ultimate goal is to use radiosensitizing agents in a clinical situation, it is extremely important to search for agents which are nonmutagenic and noncarcinogenic. Details about the mutagenicity and carcinogenicity of radiosensitizing chemicals will not be given here because they have been dealt with in Chapter 13 of this book. Rupp and colleagues [118] have examined a variety of radiosensitizers and found that 2,5-dinitroimidazole (KA 121) is a very promising compound for future study because of its low mutagenic and toxic activity and good radiosensitizing potential. Rockwell [119] has recently compared the radiosensitizing effectiveness of KA 121 with that of misonidazole, metronidazole, 2-methylthio-3-nitropyrrole (3-NPR) and 2-amino-5-nitrothiazole (ANT) using EMT6 mouse mammary tumor cells in vivo and in vitro. All the chemicals tested were toxic to hypoxic but not aerobic cells after a 24-hr exposure to the drug at a concentration of 10^{-3} M [119]. When tested as radiosensitizers at 10^{-4} concentration, the dose-modifying factors (DMFs) were 1.8, 1.6, 1.5, and 1.5 for KA 121, misonidazole, 3-NPR, and ANT, respectively, and the oxygen-enhancement ratio (OER) of the test system was 3.2 [119]. Brown et al. [120] have stated that the lipophilicity of radiosensitizers may be very important for effectiveness in vivo while being relatively unimportant in vitro. Based on this premise, a number of 2-nitroimidazole analogs differing in their octanol-water partition coefficients were tested for their radiosensitizing and cytotoxic potential using the EMT6 mouse tumor model. Some interesting compounds have already been recognized in this study [120] (see Figs. 7 and 8). In a comparative study of several nitroimidazoles, it became apparent that the radiosensitizing potential, as determined by in vitro experiments, was predominantly dependent on electron affinity whereas the influence of lipophilicity was negligible [103]. The drug has only the cell membrane as a barrier to entry into the cell in conventional monolayer and single cell suspension tissue culture systems. However, in the case of animal tumors, drug uptake by the tumor cells is in competition with other drug-metabolizing systems, organ compartments, and whole blood, and it is in this context that lipophilicity of radiosensitizers become important [120]. Attempts have been made to develop Hansch correlations [121-123] for radiosensitizers, placing due emphasis on electron affinity and lipophilicity. Adams et al. [103,104,113] and Wardman et al. [122] have convincingly demonstrated that sufficient knowledge and experimental results exist to permit selection of chemical structures with predictably good radiosensitizing or cytotoxic properties purely on the basis of structure-activity relationships developed by using the Hansch correlation approach [121-123].

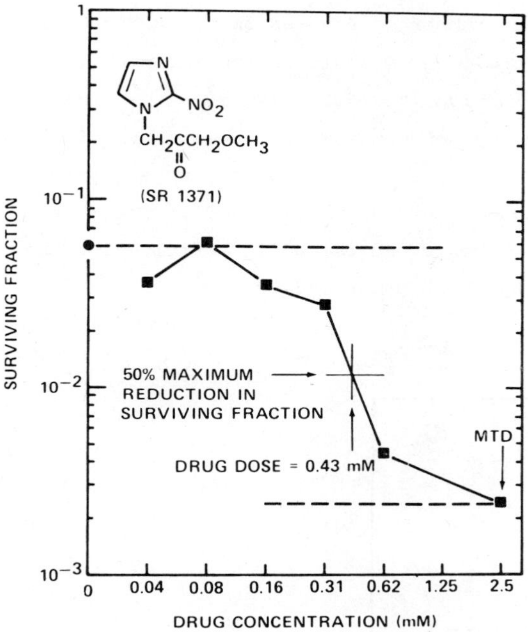

Figure 7 Survival of EMT6 tumor cells irradiated with 1,750 rad in vivo at 30 min after various doses of the 2-nitroimidazole SR 1371. Key: ● saline; ■ drug. (From Ref. 120.)

It must be emphasized that predictions about the radiosensitizing potential of a compound based on its electron affinity alone would be possible only when information is available on a series of compounds which are very closely related structurally. Watts and Jacobs [124] have examined the radiosensitization by several nitro compounds and have come to the conclusion that several compounds, especially those containing a halogen atom or an ionizable group, may behave anomalously and hence that sensitizing potential in vitro cannot be predicted on the basis of the measured electron affinity of these chemicals.

Of the many nitroaromatic and nitroheterocyclic compounds that have been tested for their radiosensitizing ability, it must be emphasized that only two, viz., metronidazole and misonidazole, have shown sufficient promise to merit clinical trial [9,10,56-63].

There have been an impressive number of studies on the radiosensitizing features of metronidazole and misonidazole using both in vitro [45,110,126,127] and in vivo [127-132] systems. Both these chemicals exhibit radiation-independent toxicity toward hypoxic cells at 37° but not at 0°, and no toxicity is observed at either temperature under aerobic conditions [45,102,107,109]. These results indicate the need for hypoxia and metabolic activity for the expression of cytotoxicity by these drugs [45,102].

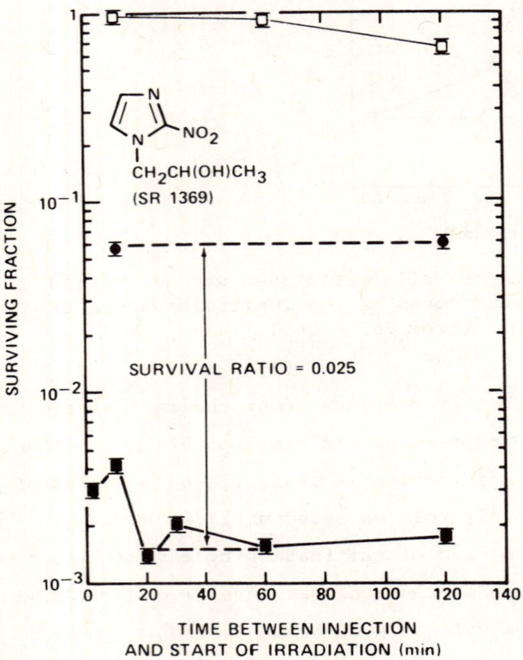

Figure 8 Survival of the cells of EMT6 tumors irradiated in air-breathing mice with 1,750 rad at various times after an intraperitoneal (i.p.) injection of saline or the 2-nitroimidazole SR 1369. Key: ■ drug, 0 rad; ● saline, 1,750 rad; drug 5 mM, 1,750 rad. (From Ref. 120.)

G. Mechanism of Hypoxic Cell Radiosensitization by Metronidazole and Misonidazole

Neutral free radical sites develop in the DNA of irradiated cells either as a result of the direct or the indirect action of radiation [13]. Hydrogen donors such as sulfhydryl derivatives have the capacity to effect "repair" by converting the free radical site back to the original and undamaged structure. Metronidazole and misonidazole resemble oxygen in being electron affinic, and as such they can accept an electron from neutral DNA-free radical centers to generate a reactive DNA cation which binds to an anion such as hydroxyl (OH-) so that DNA undergoes a structural alteration which constitutes a permanent lesion. The concentration of oxygen in air-saturated medium is about 10 times higher than necessary for maximum radiosensitization [15]. Therefore, under aerobic conditions there is enough oxygen to react with all the free radicals, and addition of metronidazole or any other sensitizer yields no further benefit. Thus metronidazole and misonidazole do not show any radiosensitizing ability when tested under aerobic conditions. Nitroradical anions formed by the addition of electrons to these compounds in a one-electron transfer process interact rapidly with oxygen to yield superoxide anion with concomitant regeneration of the parent nitrocompound. This may be another reason why no sensitization of aerobic cells is observed with these chemicals. The phenomenon of radiosensitization of hypoxic cells by metronidazole and misonidazole is a complex one. Willson [46] has stated that radiosensitization by metronidazole may involve several factors. This may very well also be true for misonidazole, which is very similar in action to metronidazole [46]. If we thus extend Willson's observations on metronidazole to misondiazole, we can list the following as prominent factors involved in the sensitizing action of these two drugs: (a) the tendency of these chemicals to form nitro radical anions whose reactivity with oxygen is such that the presence of nitroradical anions becomes unlikely under aerobic conditions; (b) the capacity of these drugs to react with radiation-induced free radical; (c) the tendency of these chemicals to undergo biochemical reduction to yield free radical derivatives; (d) the ability of these chemicals to react rapidly with sulfhydryl groups in the presence of iron and slowly if at all in its absence; and (e) the possibility of toxic product through complex formation between iron and free radials derived from these drugs.

H. Interaction Between Electron-Affinic Sensitizers

Studies by Shenoy et al. [133] on the kinetics of preirradiation sensitization of mammalian cells by oxygen showed that when oxygen was admitted 3 to 5 msec prior to delivery of radiation pulse an enhancement ratio (ER) of only 1.7 was observed and the full sensitizing effect of oxygen was seen only when oxygen was present for at least 40 msec and an (ER) of 2.8 could be achieved. Thus oxygen effect can be separated into two components, viz., the "fast" component yielding an enhancement ratio (ER) of 1.7 and the "slow" component necessary for complete sensitization by oxygen. Adams et al. [134] found that the electron-affinic sensitizer, PNAP gives a maximum sensitizer enhancement ratio (ER) of 1.7 at a concentration of 0.4 mM and increasing the concentration beyond this threshold does not increase the sensitization. Adams et al. [134] have suggested that PNAP modifies the intracellular lesions which contribute to the fast component of oxygen sensitization without influencing lesions responsible for the slower component. McNally and DeRonde [135] have studied the effect of pairs of the sensitizers PNAP, misonidazole, and oxygen. They suggest that their studies indicate that at low concentrations sensitizers interact only with those lesions which are responsible for the fast component of oxygen sensitization.

An ultrafast mixing and irradiation technique known as the *gas* explosion method has been used extensively by Michael et al. [136,137] for studying the mechanistic features of the radiosensitization process. A brief description of the technique is as follows: The test system such as *Serratia marcescens*, preincubated with or without the nongaseous chemical under investigation, is deposited as a monolayer on a millipore membrane filter. This filter is placed in a chamber which is flushed with a gas of known composition, usually nitrogen with very low oxygen content. The cells are exposed to a single 5-nsec pulse of electrons from a Febetron 706 electron pulse generator. A fast shot of gas (usually oxygen) is admitted into the chamber through a high-speed solenoid valve. The concentration of the gas passing across the membrane filter is monitored continuously by a fast photometric device. For this purpose, a narrow beam of ultraviolet (UV) light is passed across the surface to which the membrane is attached and the optical absorption is measured at the appropriate wave length as the gas flows across the path of the UV beam. Electronic devices allow the arrival time of the gas front at the millipore filter to be preset to occur at the required instant prior to, during, or after the electron pulse. The chamber is flushed with the appropriate humidified gas such as nitrogen or nitrogen containing a small amount of oxygen for about 2.25 min and the bacteria are then exposed to the oxygen shot and electron pulse. The time resolution of this technique is estimated to be about 100 μsec. After irradiation, the bacteria are removed from the filters and assayed for colony forming ability. It has become possible to apply the gas explosion method to the study of mammalian tissue culture cells grown as monolayers attached to plastic surfaces which do not have a strong tendency to absorb oxygen (Dr. Howard Michaels, personal communication).

Using the gas explosion method, Michael et al. [136,137] have studied the interaction between radiosensitizers and the lesions responsible for the fast and slow components of the oxygen effect. At low concentrations oxygen, TAN and misonidazole preferentially sensitize the damage responsible for the slow component. At higher concentrations oxygen and TAN interact with lesions responsible for the fast component via a reaction which competes with the repair process. Misonidazole, on the other hand, operates mainly by reaction with an earlier even shorter-lived form of oxygen-dependent damage. Misonidazole does not appear to compete directly with the repair reaction which is in competition with the TAN- or oxygen-mediated reactions involved in causing permanent damage [137]:

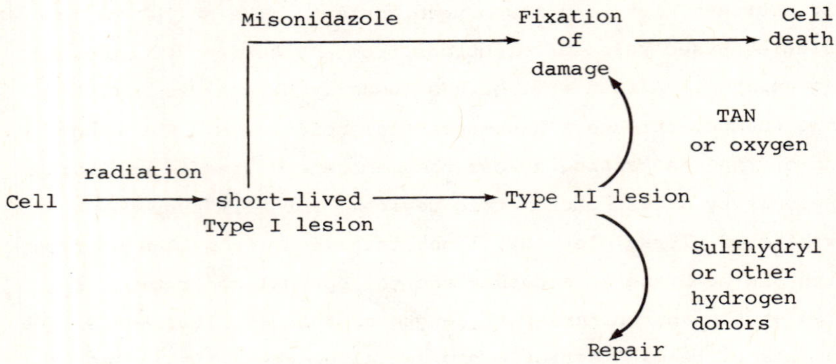

The reactivity of misonidazole with Type I lesion is greater than TAN or O_2. However, at high concentrations, both TAN and O_2 can react with Type I lesion. The reaction of misonidazole with Type II lesion does not appear to contribute significantly to the radiosensitization process [137]. This attractive scheme of events must be treated as conjecture until the nature of the Type I and type II lesions are better understood in chemical terms. The interaction of misonidazole, oxygen, and TAN with thiols merits further scrutiny in relation to this scheme.

Bakale and Gregg [138] have proposed a novel mechanism for the formation of radiation-induced cellular lesions which involves dry electrons e^-_{dry} (unsolvated electrons) [139] and dry or unsolvated holes $H_2O^+_{dry}$ [140]. The model proposed by Bakale and Gregg [138] is consistent with the oxygen effect and hypoxic cell radiosensitizing action of radiosensitizers. In this model, cell damage is supposed to occur via physicochemical interactions of $H_2O^+_{dry}$ with biomolecules such as DNA and damage is averted if an excess electron is delocalized on the DNA. Oxygen and other electron-affinic radiosensitizers act as traps for e^-_{dry} and promote $H_2O^+_{dry}$-mediated damage to DNA [138]:

$$H_2O \xrightarrow{\text{Radiolysis}} e^-_{dry} + H_2O^+_{dry}$$

The solvation times for these unsolvated species is around 600 picoseconds and the pre-solvation migration distances in structured water surrounding polar biomolecules are 350 Å and 85 Å for e^-_{dry} and $H_2O^+_{dry}$ respectively.

$$\text{DNA} + H_2O^+_{dry} \longrightarrow \text{Damage}$$

$$\text{DNA} + e^-_{dry} \longrightarrow \begin{array}{c}\text{DNA with}\\ \text{excess delocalized}\\ \text{negative charge}\end{array} \xrightarrow{H_2O^+_{dry}} \text{no damage}$$

$$\left.\begin{array}{l} O_2 + e^-_{dry} \longrightarrow O_2^{\dot{-}} \\ ArNO_2 + e^-_{dry} \longrightarrow ArNO_2^{\dot{-}} \end{array}\right\} \begin{array}{l}\text{Sequestration of } e^-_{dry} \text{ by oxygen}\\ \text{and other electron-affinic}\\ \text{radiosensitizers promotes}\\ \text{damage caused by } H_2O^+_{dry}\end{array}$$

I. Effects of Sensitizers on Cellular Respiration

Since oxygen is an efficient radiosensitizer, it is important to consider the effect of other chemical radiosensitizers on oxygen consumption by tumor cells. Some radiosensitizers alter cellular respiration rates and this could influence both oxygen availability and the extent of hypoxia in poorly vascularized solid tumors. (This important aspect is dealt with in detail in Chapter 13 of this book.)

J. Platinum Complexes as Sensitizers

The antitumor activity of platinum complexes has been known for some time [141,142]. Several platinum complexes have been used in combination with radiation in cancer therapy and recently Douple and Richmond have investigated the radiosensitization of hypoxic cells by cis- and trans-dichlorodiammineplatinum (II) [143,144]. The probable mechanisms by which cis-dichlorodiammineplatinum sensitizes cells to radiation have been discussed by Richmond and Simic in a recent paper [145].

K. Further Considerations Regarding Hypoxic Cell Sensitization

It is well known that the relative protection due to hypoxia can be diminished considerably by irradiation with high LET (linear energy transfer) particles, e.g., heavy charged-particle beams [146]. The radiosensitizing agents have been found effective in combination with both low and high LET radiation [147-152].

Considerable research effort is currently being directed toward the search for and development of newer and better hypoxic cell sensitizers for use in combined modality treatment of cancer and already promising compounds are beginning to emerge [120].

Although this chapter has dealt mainly with radiosensitization during single doses of radiation, the sensitization accompanying multifraction radiotherapy has more clinical relevance. Since radiation is generally administered in fractionated doses during cancer therapy, it is important to investigate radiosensitization with

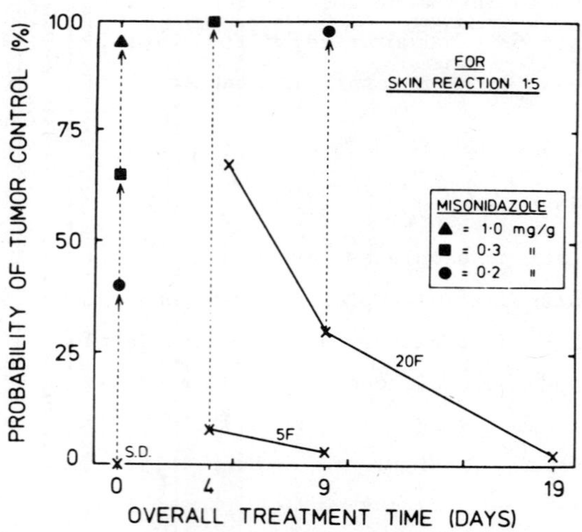

Figure 9 Optimum therapy of the MT tumor (i.e., probability of tumor control at x-ray doses which are double those required to give an average acute skin reaction of 1-5). The solid lines represent the level of effectiveness for x-ray only schedules, whereas the improvement due to the use of misonidazole is shown by vertical arrows. (From Ref. 156)

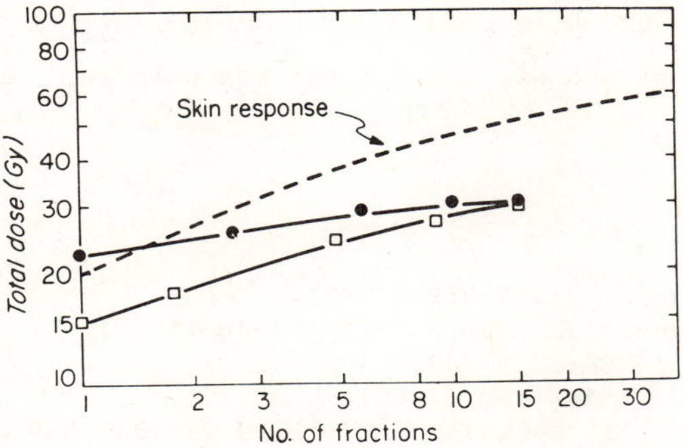

Figure 10 Isoeffect curves for the KHT Sarcoma irradiated in the presence (squares) or absence (circles) of misonidazole (0.5 mg/g), derived at a survival level of 3×10^{-4} from the results in Figures 1, 2, and 3. The broken line representing skin response is taken from Fowler [157]. (From Ref. 153.)

fractionated irradiation. Recently data have become available on the effect of radiosensitizers used in combination with fractionated irradiation (Fig. 9) [153-156]. Hill and Bush [153] have found that the degree of sensitization by misonidazole declines with fraction size. Even though misonidazole gives a higher degree of sensitization when administered in conjunction with larger dose fractions, the smaller dose fractions probably give a better therapeutic index where normal tissue tolerance has to be taken into account (Fig. 10) [153,157]. The effectiveness of metronidazole in clinical trials (see Fig. 11) should encourage researchers in radiation biology.

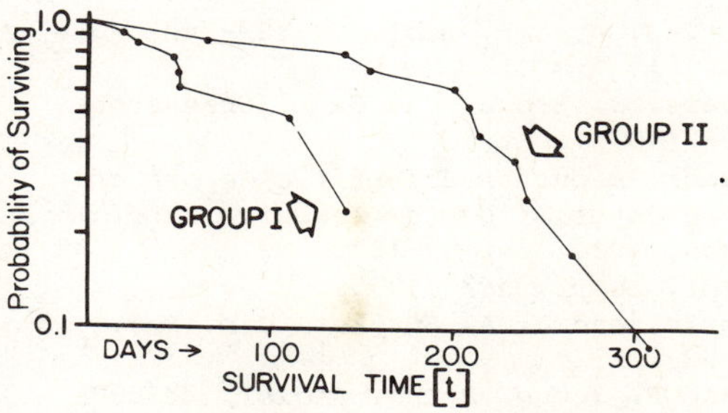

Figure 11 Kaplan-Meier survival plots showing the probability of surviving past time t for patients in Group 1 (radiation alone, 15 patients) and in Group II (radiation combined with metronidazole, 16 patients). The superiority of the Group II survival is significant at the 0.02 level (Wilcoxon-Gehan test statistic). (From Ref. 63.) Reprinted, by permission of the *New England Journal of Medicine* 294:1364 (1976).

ACKNOWLEDGMENT

Thanks are due to Ms. Lisa Gropf and Ms. Ellie Miser for help in preparation of this chapter.

REFERENCES

1. J. S. Mitchell, *Br. J. Cancer 2*:351-359 (1948).
2. J. S. Mitchell, *Br. J. Cancer 37 (Suppl. III)*:11-16 (1978).
3. J. Becker, *Strahlentherapie 80*:85 (1949).
4. J. C. Mottram, *Br. Med. J. 1*:149-150 (1929).
5. J. Russ and M. Scott, *Proc. R. Soc. London [Biol.] 118*:316-320 (1935).
6. G. E. Adams, *Adv. Radiat. Chem. 3*:125-208 (1972).
7. N. M. Bleehan, *Br. Med. Bull. 29*:54-58 (1973).
8. R. L. Willson, in *Biochemical Mechanisms of Liver Injury* (T. F. Slater, Ed.), Chap. 4. Academic Press, London, 1978, pp. 124-224.
9. R. C. Urtasun, P. R. Band, J. D. Chapman, A. F. Wilson, B. Marynowski, and E. Starreveld, *N. Engl. J. Med. 295*:901 (1976).
10. R. H. Thomlinson, S. Dische, A. J. Gray, and L. M. Errington, *Clin. Radiol. 27*:167 (1976).
11. J. F. Fowler, *Clin Radiol. 23*:257 (1972).
12. R. S. Bush, R. D. T. Jenkin, W. E. C. Allt, F. A. Beale, H. Bean, A. J. Dembo, and J. F. Pringle, *Br. J. Cancer 37 (Suppl. III)*:302-306 (1978).
13. J. D. Chapman, A. P. Reuvers, J. Borsa, J. A. Henderson, and R. D. Migliore, *Cancer Chemother. Rep. 58*:559-570 (1974).
14. C. D. Town, K. C. Smith, and H. S. Kaplan, *Curr. Top. Radiat. Res. 8*:351-399 (1973).
15. P. Howard-Flanders and T. Alper, *Radiat. Res. 7*:518-540 (1957).
16. M. C. Paterson, K. J. Kroozen, and R. B. Setlow, *Int. J. Radiat. Biol. 23*:495-508 (1973).
17. B. Palcic and L. D. Skarsgard, *Int. J. Radiat. Biol. 21*:417-433 (1972).
18. R. Roots and S. Okada, *Int. J. Radiat. Biol. 21*:329-342 (1972).
19. J. D. Chapman, A. P. Reuvers, J. Brosa, and C. L. Greenstock, *Radiat. Res. 56*:291-306 (1973).
20. H. B. Crabtree and W. Cramer, *Eleventh Scientific Report on the Investigation of the Imperial Cancer Research Fund*. Taylor & Francis, London, 1934, pp. 89-101.
21. P. J. Taskinen, *Acta Radiol. Suppl. 287* (1979).
22. P. Hierlihy, R. D. T. Jenkin, and J. A. Stryker, *Can. Med. Assoc. J. 100*:1100 (1969).
23. R. P. Hill, R. S. Bush, and P. Yeung, *Br. J. Radiol. 44*:299 (1971).

24. P. Kolstad, *Vascularization, Oxygen Tension and Radiocurability in Cancer of the Cervix: A Colpophotographic Polarographic and Clinical Study*. Universitetsforlaget, Oslo, 1964.
25. L. H. Gray, A. O. Conger, M. Ebert, S. Hornsey, and O. C. A. Scott, *Br. J. Radiol. 26*:638-648 (1953).
26. H. A. S. Van den Brenk in *Frontiers of Radiation Therapy and Oncology* (J. M. Vaeth, Ed.). Karger, Basel, 1968, pp. 141-148.
27. J. M. Henk, P. B. Kunkler, and N. K. Shah, *Clin. Radiol. 21*: 223-231 (1970).
28. R. H. Thomlinson and L. H. Gray, *Br. J. Cancer 9*:539-549 (1955).
29. H. A. Van den Brenk, *Am. J. Roentgenol Radium Ther. Nucl. Med. 102*:8 (1968).
30. J. M. Henk and C. W. Smith, *Lancet ii*:104 (1977).
31. A. O. Badib and J. H. Webster, *Acta Radiol. Ther. Phys. Biol. 8*:247-257 (1969).
32. W. Duncan, *Br. Med. Bull. 29*:33-38 (1973).
33. I. Churchill-Davidson, C. Sanger, and R. H. Thomlinson, *Lancet i*:1091-1095 (1955).
34. N. T. S. Evans and P. F. D. Naylor, *Br. J. Radiol. 36*:418-425 (1963).
35. L. A. Du Sault, *Br. J. Radiol. 36*:759-754 (1963).
36. J. T. Mallams, J. W. Finney, and G. A. Balla, *South. Med. J. 55*:230-232 (1962).
37. H. A. S. Van Den Brenk, R. C. Kerr, J. P. Madigan, N. M. Cass, and W. Richter, *Am. J. Roentgenol. Radium Ther. Nucl. Med. 96*:760-776 (1966).
38. H. Suit and R. Lindberg, *Am. J. Roentgenol. Radium Ther. Nucl. Med. 102*:27-37 ((1968).
39. W. A. Cramp, *Nature 206*:636-637 (1965).
40. W. R. Franks, M. M. Shaw, and W. H. Dickson, *Am. J. Cancer 22*: 601-605 (1934).
41. G. E. Adams and D. L. Dewey, *Biochem. Biophys. Res. Commun. 12*: 473-477 (1963).
42. B. A. Bridges, *Nature 188*:415 (1960).
43. J. W. Harris, W. M. Wara, and L. J. Kane, *Int. J. Radiat. Biol. 26*:227 (1974).
44. G. E. Adams, *Br. Med. Bull. 29*:48-53 (1973).
45. R. Sutherland, J. Richardson, C. Koch, and R. Sridhar, in *Metronidazole*. Excerpta Medica, Amsterdam, 1977, pp. 193-203.
46. R. L. Willson, in *Metronidazole*. Excerpta Medica, Amsterdam, 1977, pp. 147-175.
47. P. T. Emerson and P. Howard-Flanders, *Nature 204*:1005-1006 (1964).
48. P. T. Emerson and P. Howard-Flanders, *Radiat. Res. 26*:54-63 (1965).
49. J. L. Foster and R. L. Willson, *Br. J. Radiol. 46*:234-235 (1973).

50. R. L. Willson, W. A. Cramp, and R. M. J. Ings, *Int. J. Radiat. Biol. 26*:557-569 (1974).
51. J. C. Asquith, J. L. Foster, and R. L. Willson, *Br. J. Radiol. 46*:648 (1973).
52. E. J. Hall and J. D. Chapman, *Br. J. Radiol. 47*:513-514 (1974).
53. J. E. Thomson and A. M. Rauth, *Radiat. Res. 60*:489-500 (1974).
54. J. D. Chapman, A. P. Reuvers, and J. Borsa, *Br. J. Radiol. 46*:623-630 (1973).
55. J. C. Asquith, J. L. Foster, R. L. Willson, R. Ings, and J. A. McFadzean, *Br. J. Radiol. 47*:474-481 (1974).
56. S. Dische, M. I. Saunders, M. E. Lee, G. E. Adams, and I. R. Flockhart, *Br. J. Cancer 35*:456 (1977).
57. A. J. Gray, S. Dische, G. E. Adams, I. R. Flockhart, and J. L. Foster, *Clin. Radiol. 27*:151 (1976).
58. T. L. Phillips, T. H. Wasserman, R. J. Johnson, C. J. Gomer, G. A. Lawrence, M. L. Levine, W. Sadee, J. S. Penta, and D. J. Rubin, *Br. J. Cancer 37 (Suppl. III)*:276-280 (1978).
59. H. D. Kogelnik, H. J. Meyer, K. Jentsch, T. Szepesi, K. H. Karcher, E. Maida, B. Mamoli, P. Wessely, and F. Zaunbauer, *Br. J. Cancer 37 (Suppl. III)*:281-285 (1978).
60. C. R. Wiltshire, P. Workman, J. V. Watson, and N. M. Bleehan, *Br. J. Cancer 37 (Suppl. III)*:286-289 (1978).
61. P. J. D. K. Davies, M. J. Peckham, and G. G. Stell, *Br. J. Cancer 37 (Suppl. III)*:290-296 (1978).
62. A. B. M. F. Karim, *Br. J. Cancer 37 (Suppl. III)*:299-301 (1978).
63. R. Urtasun, P. Band, J. D. Chapman, M. L. Feldstein, B. Mielke, and C. Fryer, *N. Engl. J. Med. 294*:1364 (1976).
64. R. Catane, M. Rosencweig, J. G. Schwade, B. F. Issel, and F. M. Muggia in *Proc. C.R.O.S. Conference on Combined Modalities: Chemotherapy/Radiotherapy*, Hilton Head Island, S.C., Nov. 15-18, 1978, Book of Abstr., p. 61.
65. M. Amer, A. Weaver, G. Vaishampayan, and M. Al-Sarraf, in *Proc. C.R.O. Conference on Combined Modalities: Chemotherapy/Radiotherapy*, Hilton Head Island, S.C., Nov. 15-18, 1978, Book of Abstr., p. 62.
66. M. M. Klingerman and L. Schulhof, *Radiat. Res. 39*:571-579 (1969).
67. G. E. Adams, M. S. Cooke, and B. C. Michael, *Nature 219*:1368-1369 (1968).
68. H. Moroson and M. Furlan, *Int. J. Radiat. Biol. 13*:585-589 (1978).
69. M. Klimek, *Neoplasma 13*:31-33 (1966).
70. M. Klimek and L. Zemanova, *Neoplasma 17*:209-212 (1970).
71. A. G. Tarasenko, I. V. Nekrasova, and E. Ya Graevskii, *Dokl. Akad. Nauk SSSR 193*:1406-1408 (1970).
72. B. F. Kimler, W. K. Sinclair, and M. M. Elkind, *Radiat. Res. 71*:204-213 (1977).
73. J. W. Harris and J. A. Power, *Br. J. Cancer 37 (Suppl. III)*: 84-89 (1978).
74. M. M. Elkind and H. Sutton, *Radiat. Res. 13*:556-593 (1960).

75. J. E. Byfield, *Cancer Chemother. Rep. [Pt. 1] 54(4)*:527-538 (1974).
76. M. M. Elkind, G. F. Whitmore, and T. Alescio, *Science 143*: 1454-1457 (1964).
77. W. A. Cramp, *Radiat. Res. 30*:221-236 (1967).
78. W. A. Cramp, *Radiat. Res. 33*:10 (1968).
79. I. P. Hesslewood, W. A. Cramp, D. C. H. McBrien, P. Williamson, and K. A. K. Lott, *Br. J. Cancer 37 (Suppl III)*:95-97 (1978).
80. B. Djordjevic and W. Szybalski, *J. Exp. Med. 112*:509-531 (1960).
81. W. Szybalski, *Cancer Chemother. Rep. (Pt. 1) 58(4)*:539-557 (1974).
82. H. S. Kaplan, in *Radiation Protection and Sensitization* (H. Moroson and M. Quintiliani, Eds.). Taylor & Francis, London, 1970, pp. 35-42.
83. G. E. Adams, *Curr. Top. Radiat. Res. 3*:35-93 (1967).
84. E. M. Fielden and S. C. Lillicrap, in *Radiation Protection and Sensitization* (H. Moroson and M. Quintiliani, Eds.). Taylor & Francis, London, 1970, pp. 81-86.
85. K. Sano, T. Hoshino, and M. Nagai, *J. Neurosurg. 28*:530-538 (1968).
86. F. Mundinger, P. Vogt, C. Jobske et al., *Strahlentherapie 143*: 318-328 (1972).
87. F. F. Nakken, T. Silleland, and T. Brustad, *FEBS Lett. 8*:33-36 (1970).
88. T. Brustad, H. Bugge, W. B. G. Jones, and E. Wold, *Int. J. Radiat. Biol. 22*:115-129 (1972).
89. P. T. Emmerson, *Radiat. Res. 30*:841-849 (1967).
90. D. A. Agnew and L. D. Skarsgard, *Radiat. Res. 51*:97-109 (1972).
91. P. T. Emmerson, E. M. Fielden, and I. Johansen, *Int. J. Radiat. Biol. 19*:229-235 (1971).
92. R. P. Hill, E. M. Fielden, S. C. Lillicrap, and J. A. Stanley, *Int. J. Radiat. Biol. 27*:499-501 (1975).
93. N. M. Blackett, W. E. Wolliscroft, E. M. Fielden, and S. C. Lillicrap, *Radiat. Res. 58*:361-372 (1974).
94. B. C. Cooke, E. M. Fielden, M. Johnson, and C. E. Smithen, *Radiat. Res. 65*:152-162 (1976).
95. G. W. Barendsen, in *Theoretical and Experimental Biophysics* (A. Cole, Ed.), Vol. 1. Arnold, London, 1967, pp. 167-231.
96. S. H. Glarum and J. H. Marshall, *J. Chem. Phys. 47*:1374-1378 (1967).
97. B. C. Millar, E. M. Fielden, and C. E. Smithen, *Br. J. Cancer 37 (Suppl. III)*:73-79 (1978).
98. B. C. Millar, E. M. Fielden, and C. E. Smithen, *Radiat. Res. 69*:489 (1977).
99. R. Sridhar, E. C. Stroude, and W. R. Inch, *Proc. 11th Int. Congr. Biochem.* Toronto, July 8-13, 1979, Book of Abstr.
100. R. Sridhar, E. C. Stroude, and W. R. Inch, *Proc. 6th Int. Cong. Radiat. Res.*, Tokyo, May 13-19, 1979.
101. R. M. Sutherland and R. E. Durand, *Curr. Top. Radiat. Res. 2*: 87-139 (1976).

102. R. Sridhar, C. Koch, and R. Sutherland, *Int. J. Radiat. Oncol. Biol. Phys. 1*:1149-1157 (1976).
103. G. E. Adams and M. S. Cooke, *Int. J. Radiat. Biol. 15*:457 (1969).
104. G. E. Adams, I. R. Flockhart, C. E. Smithen, I. J. Stratford, P. Wardman, and M. E. Watts, *Radiat. Res. 67*:9 (1976).
105. M. Simic and E. L. Powers, *Int. J. Radiat. Biol. 26*:87 (1974).
106. J. A. Raleigh, J. D. Chapman, J. Borsa, J. Kremers, and A. P. Reuvers, *Int. J. Radiat. Biol. 23*:377 (1973).
107. J. K. Mohindra and A. M. Rauth, *Cancer Res. 36*:930-936 (1976).
108. J. M. Brown, *Radiat. Res. 72*:46 (1977).
109. E. J. Hall and J. E. Biaglow, *Int. J. Radiat. Oncol. Biol. Phys. 2*:521-530 (1977).
110. B. A. Moore, B. Palcic, and L. D. Skarsgard, *Radiat. Res. 67*: 459-573 (1977).
111. E. O. Petterson, *Br. J. Cancer 37 (Suppl. III)*:107-110 (1978).
112. J. D. Chapman, J. A. Raleigh, J. Borsa, R. G. Webb, and R. Whitehouse, *Int. J. Radiat. Biol. 21*:476 (1972).
113. G. E. Adams, E. D. Clarke, R. S. Jacobs, I. J. Stratford, R. G. Wallace, P. Wardman, and M. E. Watts, *Biochem. Biophys. Res. Commun. 72*:824 (1976).
114. J. D. Chapman, R. G. Webb, and J. Borsa, *Int. J. Radiat. Biol. 19*:561-573 (1971).
115. G. E. Adams, J. C. Asquith, D. L. Dewey, J. L. Foster, B. D. Michael, and R. L. Willson, *Int. J. Radiat. Biol. 19*:575-585 (1971).
116. J. Sasaki, *Pharm. Bull. (Tokyo) 2*:104-108 (1954).
117. J. A. Raleigh, J. D. Chapman, A. P. Reuvers, J. E. Biaglow, R. E. Durand, and A. M. Rauth, *Br. J. Cancer 37 (Suppl III)*:6-10 (1978).
118. W. D. Rupp, Z. Mroczkowski, and K. C. Agrawal, *Br. J. Cancer 37 (Suppl. III)*:60-63 (1978).
119. S. Rockwell, *Br. J. Cancer 37 (Suppl. III)*:212-215 (1978).
120. J. M. Brown, N. Y. Yu, M. J. Cory, R. B. Bicknell, and D. L. Taylor, *Br. J. Cancer 37 (Suppl. III)*:206-211 (1978).
121. P. Wardman, E. D. Clarke, I. R. Flockhart, and R. G. Wallace, *Br. J. Cancer 37 (Suppl. III)*:1-5 (1978).
122. C. Hansch, in *Drug Design* (E. J. Ariens, Ed.), Vol. 1., Academic Press, New York, 1971.
123. C. Hansch and J. M. Clayton, *J. Pharm. Sci. 62*:1 (1973).
124. M. E. Watts and R. S. Jacobs, *Br. J. Cancer 37 (Suppl. III)*: 80-83 (1978).
125. E. J. Hall and L. Roizin-Towle, *Radiology 117*:453-457 (1975).
126. G. F. Whitmore, S. Gulyas, and A. J. Varghese, *Br. J. Cancer 37*:115-119 (1978).
127. H. B. Stone and H. R. Withers, *Radiology 113*:441-444 (1974).
128. A. C. Begg, P. W. Sheldon, and J. L. Foster, *Br. J. Radiol. 47*:399-404 (1974).
129. J. Denekamp, B. D. Michael, and S. R. Harris, *Radiat. Res. 60*:119-132 (1974).
130. A. M. Rauth and K. Kaufman, *Br. J. Radiol. 48*:209-220 (1975).

131. J. M. Brown, *Radiat. Res. 64*:633-647 (1975).
132. M. J. Haynes and W. R. Inch, *Radiat. Res. 68*:184-187 (1976).
133. M. A. Shenoy, J. C. Asquith, G. E. Adams, B. D. Michael, and M. E. Watts, *Radiat. Res. 62*:498 (1975).
134. G. E. Adams, B. D. Michael, J. C. Asquith, M. A. Shenoy, M. E. Watts, and D. W. Whillians, in *Radiation Research: Proc. 5th Int. Congr. Radiat. Res.* (O. F. Nygaard, H. F. Adler, and W. K. Sinclair, Eds.). Academic Press, New York, 1975, p. 478.
135. N. J. McNally and J. DeRonde, *Br. J. Cancer 37 (Suppl. III)*: 90-94 (1978).
136. B. D. Michael, G. E. Adams, H. B. Hewitt, W. B. G. Jones, and M. E. Watts, *Radiat. Res. 54*:239 (1973).
137. B. D. Michael, H. A. Harrop, R. L. Maughan, and K. G. Patel, *Br. J. Cancer 37 (Suppl. III)*:29-33 (1978).
138. G. Bakale and E. C. Gregg, *Br. J. Cancer 37 (Suppl. III)*:24-28 (1978).
139. J. W. Hunt, *Adv. Radiat. Chem. 5* (1976).
140. W. H. Hamill, *J. Phys. Chem. 73*:1431 (1969).
141. M. J. Cleare, *Inorg. Persp. Biol. Med. 1*:19 (1977).
142. B. Rosenberg, *Naturwissenschaften 60*:399 (1973).
143. E. B. Douple and R. C. Richmond, *Proc. C.R.O.S. Conference on Combined Modalities: Chemotherapy/Radiotherapy*, Hilton Head Island, S.C., Nov. 15-18, 1978.
144. E. B. Douple and R. C. Richmond, *Br. J. Cancer 37 (Suppl. III)*: 98-101 (1978).
145. R. C. Richmond and M. G. Simic, *Br. J. Cancer 37 (Suppl. III)*: 20-23 (1978).
146. C. A. Tobias, *Radiology 108*:145 (1973).
147. M. R. Raju, J. P. Frank, E. Bain, and T. T. Trujillo, *Radiat. Res. 71*:233 (1977).
148. E. J. Hall, L. Roizin-Towle, R. B. Theus, and L. S. August, *Radiology 117*:173 (1975).
149. M. R. Raju, H. I. Amols, and S. G. Carpenter, *Br. J. Cancer 37 (Suppl. III)*:189-193 (1978).
150. N. M. McNally, *Int. J. Radiat. Biol. 29*:191 (1976).
151. J. D. Chapman, R. C. Urtasun, E. A. Blakely, K. C. Smith, and C. A. Tobias, *Br. J. Cancer 37 (Suppl. III)*:184-188 (1978).
152. J. D. Chapman, E. A. Blakeley, K. C. Smith, and R. C. Urtasun, *Int. J. Radiat. Oncol. Biol. Phys. 2 (Suppl. 2)*:97 (1978).
153. R. P. Hill and R. S. Bush, *Br. J. Cancer 37 (Suppl. III)*:255-258 (1978).
154. J. Denekamp and F. A. Stewart, *Br. J. Cancer 37 (Suppl. III)*: 259-263 (1978).
155. J. F. Fowler, P. W. Sheldon, J. Denekamp, and S. B. Field, *Int. J. Radiat. Oncol. Biol. Phys. 1*:579 (1976).
156. P. W. Sheldon and J. F. Fowler, *Br. J. Cancer 37 (Suppl. III)*: 242-245 (1978).
157. J. F. Fowler, G. E. Adams, and J. Denekamp, *Cancer Treatment Rev. 3*:227-257 (1976).

10

THE ROLE OF FREE RADICALS IN ARYLAMINE CARCINOGENESIS

Robert A. Floyd

Oklahoma Medical Research Foundation
Biomembrane Research Laboratory
Oklahoma City, Oklahoma

I. INTRODUCTION

The concept that free radicals in some unspecified manner play an important role in cancer has been prevalent for many years. It appears to have originated with and was strengthened by the many investigations which showed that x-irradiation induced cancer and that the action of x-irradiation involved primary free radical events. In recent years, the demonstrations that antioxidants in the diet help in general to protect against cancer development and the general idea that antioxidants quench free radical events has forced greater in-depth examination of the relationship of free radicals and cancer. The subject of the present chapter is to review the metabolism of the arylamine carcinogens with particular emphasis on the free radical aspects.

II. ARYLAMINE CARCINOGENESIS: HISTORICAL PERSPECTIVE

Although the general public is now becoming aware that cancer is a resultant of contact with environmental agents, this has been concluded as being true by a few medical research investigators for more than 200 years. John Hill, an English physician, first noted in 1761 the association of tobacco usage with increased tumor incidence. Percival Potts published a treatise in 1776 noting the high incidence of skin cancer in chimney sweeps and warning that soot in contact with the skin was the causative agent. It was late in the nineteenth century that cancers caused by the arylamine carcinogens first became recognized. Clothing dye manufacturing plants which produced aniline-based dyes were prevalent in Europe during that time. It was noted that aniline dye workers developed a higher amount of bladder tumors than the normal population, and hence it became customary to call these aniline cancers. But from the present perspective it is now possible to state that these tumors were not caused by aniline per se but by other chemicals, most likely biphenyl or other arylamine compounds, which were by-products in the manufacture of aniline.

The most widely studied arylamine carcinogen is 2-acetylaminofluorene (AAF), which was synthesized to be used as a pesticide in 1940. Fortunately, it was tested and found to produce tumors in rats and therefore was not released into the market place. AAF, when fed usually as 0.03% of the diet, produces tumors of the liver in male rats and tumors of the mammary gland in female rats along with some small-intestinal tumors, as well as earduct gland tumors in both sexes. The chemistry as to how AAF eventually produces tumors is not known with certainty at the present time, but this has been studied extensively [1,2].

III. METABOLIC ACTIVATION OF 2-ACETYLAMINOFLUORENE

The chemical AAF per se does not cause tumors but first must be acted upon by cellular initiated reactions and thus converted to activated forms of this chemical. This point has been shown by many research workers and discussed at length [1,2]. The first necessary alteration that must occur in the activation of AAF is N-hydroxylation to form N-hydroxy-2-acetylaminofluorene (N-OH-AAF). This was first demonstrated in 1960 by Cramer et al. [3] and has subsequently been repeatedly proven in work conducted by the Miller group [4-6], Gutman et al. [7-9], and many others. The reaction is shown in Eq. (1). N-OH-AAF is not considered the ultimate reactive form of AAF, one reason being that it will react only very slowly, if at all, with proteins and nucleic acids. It is usually considered that these macromolecules must be modified in some fashion to bring about genetic alteration leading to cancer.

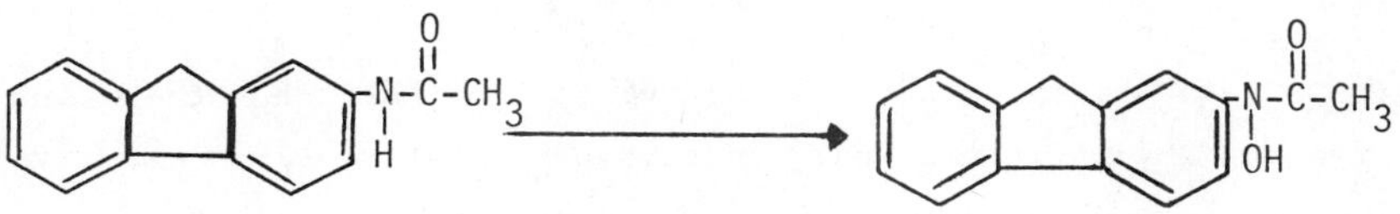

(1)

IV. ACTIVATION OF N-HYDROXY-2-ACETYLAMINOFLUORENE

A. Sulfate Ester of N-Hydroxy-2-acetylaminofluorene

The reaction demonstrating formation of the sulfate ester of N-OH-AAF (denoted as N-OSO_3-AAF) is shown in Eq. (2).

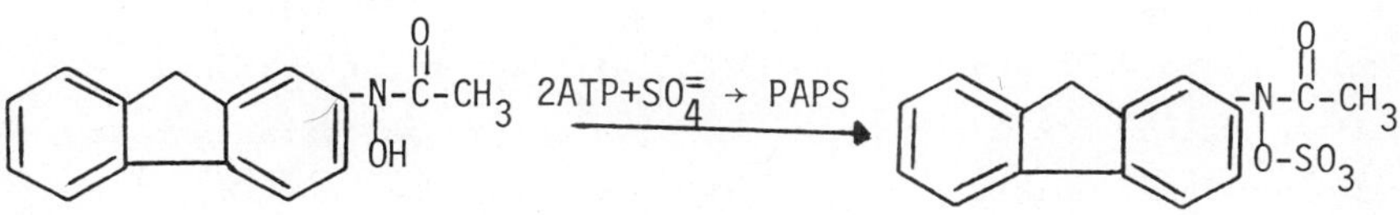

(2)

There has been much research conducted which strongly implicates N-OSO_3-AAF as the ultimate reactive form of AAF, especially in liver carcinogenesis [1,2]. The most convincing arguments center upon animal experiments demonstrating the additive effect of a low-sulfur plus p-hydroxyacetaniline-supplemented diet, which produces a decrease in AAF-induced liver carcinogenesis [10,11]. According to the logic of the experimental design, p-hydroxyacetaniline competes with AAF metabolism to the ultimate reactive form in two essential steps: one is in the N-hydroxylation step, and the second is in the sulfate ester formation step. The competing reactions are shown in Eq. (3).

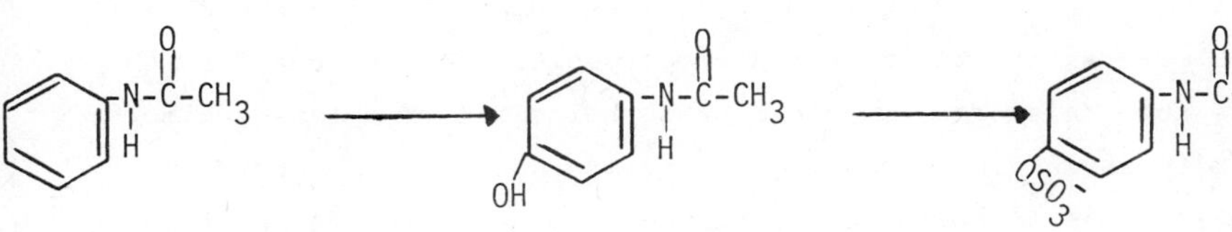

(3)

The results obtained reinforced the logic in the sense that p-hydroxyacetaniline itself lowered AAF carcinogenesis; but when the low-sulfur diet was also imposed, this additionally suppressed tumor formation.

The aforementioned experiments as well as others [1,2] implicate N-OSO_3-AAF as the ultimate reactive form of AAF. But there are many reasons why it is necessary to consider other possible reactive forms of AAF. Several of the reasons are as follows: (a) The enzyme responsible for the formation of N-OSO_3-AAF, sulfotransferase, is a cytosolic enzyme [12]. In order for N-OSO_3-AAF to reach the nucleus from the site of its formation in the cytosol, it must transmit a large amount of intracellular space. Therefore, since the lifetime of this chemical is very short due to decomposition, and its reactivity with many compounds is so great, it seems highly unlikely that any of the formed N-OSO_3-AAF reaches the critical target molecules in the nucleus. (b) The enzyme sulfotransferase is very strongly inhibited by Cu^{2+} [12], yet the cupric chelate of N-OH-AAF is very carcinogenic when implanted in the rat [6]. This is due largely to the continued presence of N-OH-AAF at the site of application, presumably as the cupric chelate [6]. (c) The mammary gland and earduct gland tissues are highly susceptible to AAF and N-OH-AAF, yet these tissues do not have detectable amounts of sulfotransferase [13]. (d) It has been shown that with different species of rats the sulfotransferase activity of the livers of the various strains correlated with the toxicity of AAF rather than the carcinogenicity of the compound [14]. (e) Repeated injections of N-OSO_3-AAF elicited very few tumors at the site of injection [15], whereas N-OH-AAF elicited many more tumors. (f) The experiments involving acetaniline quoted earlier as supporting the concept of N-OSO_3-AAF as the ultimate reactive species are confounded in the sense that this drug, in addition to being N-hydroxylated, also inhibits arylhydroxamic acid acyltransferase [16], an enzyme which King [16] has found to be effective in transferring AAF residues to nucleic acids.

The formation of the sulfate ester of N-OH-AAF is the most widely touted mechanism to account for AAF carcinogenesis. However, the arguments against this scheme, especially in nonhepatic tissue, as just developed, forces continued search for a mechanism which would encompass all the observations. Many other N-OH-AAF activation schemes have been proposed to account for AAF carcinogenesis. I will not review these in this chapter for they have been reviewed previously [1,2]. It is sufficient to state that AAF activation is still an open question, but free radical aspects have been considered only to a limited extent thus far. I will review these aspects in detail.

B. Peroxidase-Free Radical Activation Route of N-Hydroxy-2-acetyl-aminofluorene

1. *Historical perspective*

The first experiments implicating the possibility of a free radical activation route of N-OH-AAF was conducted by Bartsch et al. [17] and almost simultaneously by Forrester et al. [18]. The first experiments were conducted in an organic solvent system. It was demonstrated using one-electron oxidants such as ferricyanide that N-OH-AAF was oxidized to the nitroxyl free radical (N-$\dot{O}$-AAF) and that two of these free radicals dismutate to form 2-nitrosofluorene (NOF) and N-acetoxy-2-acetylaminofluorene (N-OAc-AAF), both of which are considered more potent forms of the carcinogen. Therefore, according to this scheme, the nitroxyl free radical of N-OH-AAF is an intermediate in the activation of this carcinogen.

Bartsch and Hecker [19] realized the potential implications of their findings and attempted to extrapolate and determine if they were biologically relevant. Utilizing horseradish peroxidase (HRP), H_2O_2, and N-OH-AAF, Bartsch and Hecker [19] demonstrated that N-OH-AAF was converted into NOF and N-OAc-AAF during the reaction; the nitroxyl free radical of N-OH-AAF was observed by means of electron spin resonance (ESR) spectroscopy. However, they [19] were unable to demonstrate that peroxidizing rat liver microsomes would activate N-OH-AAF via the peroxidase mechanism. It was at this point that we began work on this problem in our laboratory. We have been essentially the only group that has concentrated on this problem since then and therefore a major portion of this review will concentrate on our work.

2. *Reexamination of the peroxidase activation of N-hydroxy-2-acetylaminofluorene*

I became convinced that in view of the many observations [20-39] demonstrating that dietary antioxidants in general slowed down or prevented tumor formation induced by chemical carcinogens it was necessary to examine chemical carcinogen metabolism with particular emphasis on free radical aspects. Utilizing rat liver microsomes as the catalytic system, we demonstrated [40] that N-OH-AAF was oxidized in this system when cumene hydroperoxide was used as the substrate. An unknown compound was formed from N-OH-AAF having an absorbance shoulder at 320 nm. Later we were able to show that the unknown was 2-nitrofluorene, which was formed by cumene hydroperoxide oxidation of 2-nitrosofluorene—itself formed either by dismutation via the peroxidase route or via deacylation of N-OH-AAF. We found that the K_m for N-OH-AAF in this system was 2.6 μM and that the reaction did not occur after heating the microsomes [40].

The mechanism proposed by Bartsch and Hecker [19] to account for HRP-H_2O_2 activation of N-OH-AAF is shown in Figure 1. HRP is converted to compound I (HRP-I) by H_2O_2 and in this state, a one-electron donor, in the present case N-OH-AAF, reduces the enzyme to the compound II (HRP-II) state forming the nitroxyl free radical of N-OH-AAF (N-$\dot{O}$-AAF) in the process. Compound II is then reduced to the original state of the enzyme by a fresh molecule of N-OH-AAF, thus forming another N-$\dot{O}$-AAF molecule. Two molecules of N-$\dot{O}$-AAF then dismutate to form one molecule of 2-nitrosofluorene (NOF) and one molecule of N-acetoxy-2-acetylaminofluorene (N-OAc-AAF). In the experiments conducted by Bartsch and Hecker [19], the H_2O_2/HRP ratio was 50,000 based upon our duplication of their experiments [41]. If the H_2O_2/HRP ratio exceeds 20, then HRP is converted into the catalytically inactive compound III state [42]. For these reasons, we repeated the experiments of Bartsch and Hecker utilizing small pulses of H_2O_2 such that it was possible to determine on a molecular basis the mechanism of N-OH-AAF activation in this system.

Our results [41] demonstrated unequivocally that in the HRP-H_2O_2 system, for every two molecules of N-OH-AAF oxidized, one was converted to NOF and one to N-OAc-AAF. We discovered that ascorbate prevents the oxidative activation of N-OH-AAF in this system and that ascorbate is oxidized preferentially over N-OH-AAF. This observation may have some important implications in terms of explaining the observations that antioxidants tend to inhibit carcinogenesis. We made some other interesting observations, namely that HRP, after being in contact with the active carcinogens NOF and N-OAc-AAF for several minutes, was still very active as a peroxidase. We also noted that HRP when in the compound II state would oxidize N-OH-AAF. All of these experiments and many others do not prove conclusively that the peroxidase-free radical route shown in Figure 1 is the correct mechanism to explain N-OH-AAF oxidative activation, but all of our results are consistent with this mechanism.

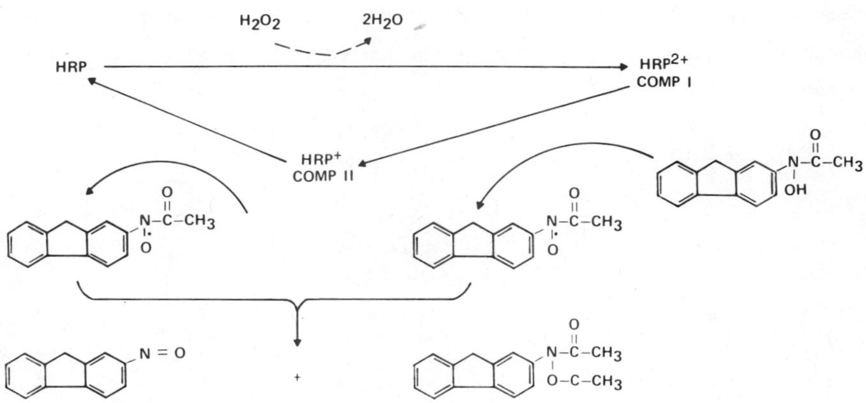

Figure 1 Scheme to explain the horseradish peroxidase (HRP) activation of N-hydroxy-2-acetylaminofluorene via a nitroxyl free radical intermediate: HRP represents the native ferric-iron oxidation state of the enzyme; HRP^{2+} represents compound I of HRP; HRP^{+} represents compound II, which is the one-electron reduction species of HRP^{2+}.

3. *Lipid hydroperoxide-heme oxidative activation of N-hydroxy-2-acetylaminofluorene*

The activation of N-OH-AAF in the HRP-H_2O_2 system forced us to examine whether there was a biologically relevant system that might be operating. Because the association of free radicals in the lipid peroxidation process, the yield of a product, i.e., lipid hydroperoxide, that is similar to H_2O_2 and the known inhibition of this process by antioxidants enforced us to examine this process in greater detail to determine the possible relevance to carcinogenesis. We decided to investigate whether a lipid hydroperoxide would act as a substrate in the oxidative activation of N-OH-AAF. The lipid hydroperoxide, actually the 13-L-hydroperoxy-cis-9, trans-11-octadecadienoic acid isomer, was synthesized using linoleic acid and soybean lipoxygenase [43]. We used the synthesized linoleic acid hydroperoxide (LAHP) as the substrate and hematin as well as methemoglobin as the heme compound and demonstrated that N-OH-AAF was oxidatively activated to NOF and N-OAc-AAF in this system.

Careful optical monitoring of the reaction substrates (LAHP and N-OH-AAF) and one product (NOF) demonstrated that the rate of LAHP and also N-OH-AAF consumption exactly matched kinetically NOF formation [43]. Utilizing absorbance changes of the substrates and products it was possible to write the stoichiometry of the reaction as shown in Eq. (4).

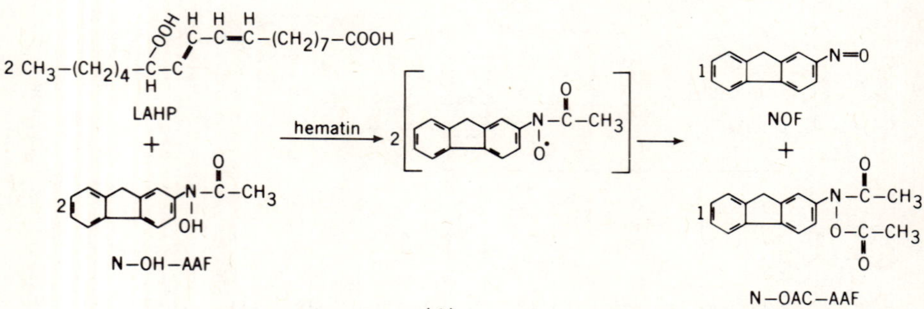

(4)

Thus, two molecules of LAHP plus two molecules of N-OH-AAF reacted in the presence of hematin or methemoglobin as a catalyst to produce two molecules of the nitroxyl free radical of N-OH-AAF, which then dismutate to form one molecule of NOF and one molecule of N-OAc-AAF. Thin-layer chromatography was also used to confirm our conclusions. Actually, in this study, it was impossible to state unambiguously that the nitroxyl free radical was an obligatory intermediate even though the free radical was observed by means of ESR spectroscopy during the reaction.

If the nitroxyl free radical of N-OH-AAF is an obligatory intermediate in the oxidative activation of this carcinogen, then the equations shown in Figure 2 can be written [44]. The final equation shows that according to this model the rate of N-OH-AAF oxidation can be equated to the rate of N-$\dot{O}$-AAF change as well as to the square of the amount of N-$\dot{O}$-AAF present at any time. This equation predicts that it will be possible to compute one variable, i.e., either the rate of N-OH-AAF oxidation or the amount of N-$\dot{O}$-AAF at any time, if the other variable is known. It also indicates that the dismutation constant k_2 can be calculated from this equation if indeed a time in the reaction is chosen such that the rate of N-$\dot{O}$-AAF change is zero.

$$\text{N–OH–AAF} \underset{k_{-1}}{\overset{k_1}{\rightleftharpoons}} \text{N–}\dot{\text{O}}\text{–AAF}$$

$$\text{N–}\dot{\text{O}}\text{–AAF} + \text{N–}\dot{\text{O}}\text{–AAF} \xrightarrow{k_2} \text{NOF} + \text{N–OAc–AAF}$$

$$\frac{d[\text{N–OH–AAF}]}{dt} = -k_1[\text{N–OH–AAF}] + k_{-1}[\text{N–}\dot{\text{O}}\text{–AAF}]$$

$$\frac{d[\text{N–}\dot{\text{O}}\text{–AAF}]}{dt} = -k_2[\text{N–}\dot{\text{O}}\text{–AAF}]^2 + k_1[\text{N–OH–AAF}] - k_{-1}[\text{N–}\dot{\text{O}}\text{–AAF}]$$

$$-\frac{d[\text{N–OH–AAF}]}{dt} = \frac{d[\text{N–}\dot{\text{O}}\text{–AAF}]}{dt} + k_2[\text{N–}\dot{\text{O}}\text{–AAF}]^2$$

Figure 2 Equations to describe the one-electron oxidation of N-hydroxy-2-acetylaminofluorene (N-OH-AAF) to form the nitroxyl free radical and its subsequent dismutation to 2-nitrosofluorene (NOF) and N-acetoxy-2-acetylaminofluorene (N-OAC-AAF). (Redrawn from Ref. 44.)

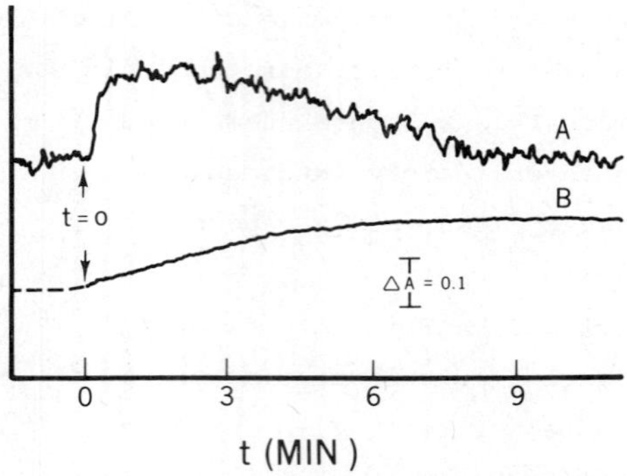

Figure 3 Traces showing monitoring of the amount of the nitroxyl free radical of N-OH-AAF and the amount of N-OH-AAF oxidized during the hematin-cumene hydroperoxide catalyzed oxidation of the carcinogen. The top trace represents a monitor of the amount of free radical present, and the bottom trace is an optical measure of the amount of N-OH-AAF present. (From Ref. 44.)

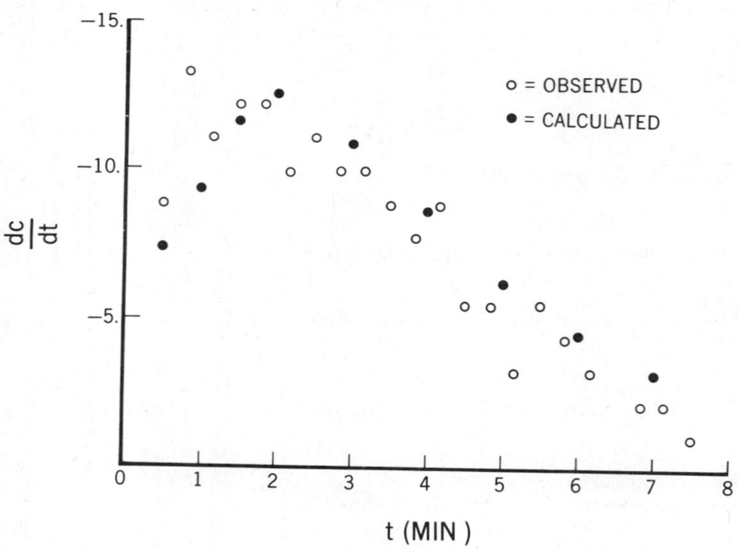

Figure 4 Calculated and observed rate of N-OH-AAF oxidation in the hematin-cumene hydroxperoxide system, the data of which are shown in Figure 3. The data were calculated using the last equation shown in Figure 2. (From Ref. 44.)

Figure 3 shows the results of monitoring both the amount of N-Ȯ-AAF present using ESR spectroscopy as well as the amount of N-OH-AAF by optical absorbance during the cumene hydroperoxide-hematin-induced oxidative activation of N-OH-AAF. LAHP gave the same results as cumene hydroperoxide. It is clear that the amount of N-Ȯ-AAF becomes zero when the rate of N-OH-AAF oxidation becomes zero. Figure 4 shows that using the amount of N-Ȯ-AAF at any time in the reaction and the bottom equation of Figure 2 it is possible to calculate the rate of N-OH-AAF oxidation. It is clear that the computed value at any time in the reaction is equivalent to the observed value. This is the best demonstration thus far that N-Ȯ-AAF is an obligatory intermediate in the oxidative activation of N-OH-AAF. The calculated dismutation constant of N-Ȯ-AAF is $2.7 \times 10^5\ M^{-1}\ sec^{-1}$ in the phosphate buffer system [45]. Thus, the free radical N-Ȯ-AAF is a fairly long-lived species in a physiological media. If one reckons that at a specific time 1 nmol of N-Ȯ-AAF is present, then 1 sec after that time only 0.1% of this radical will have dismutated to form NOF and N-OAc-AAF. The reactivity of N-Ȯ-AAF per se with biological macromolecules is unknown, but viewed from the present perspective it may be an important consideration.

4. *Spin-trapping of free radicals in the hydroperoxide-heme system*

It is possible that electron removal from N-OH-AAF to form N-Ȯ-AAF in the hematin-lipid hydroperoxide system could occur via a primary free radical as is shown in Eq. (5).

$$(\mathrm{ROOH} \xrightarrow{\text{hematin}} \mathrm{R\dot{O}}) + \text{N-OH-AAF} \longrightarrow \text{N-}\dot{\mathrm{O}}\text{-AAF} + \mathrm{ROH} \qquad (5)$$

We have tested this idea using the spin-trapping technique [45]. Spin-trapping is a technique developed in recent years that is very useful for clarifying free radical events in certain chemical reactions [46]. The principles of the technique are illustrated in Eq. (6).

(6)

The spin-trap, in this case 5,5-dimethylpyrolline-1-oxide (DMPO), is reacting with a free radical to form the spin adduct. The spin adduct formed is in general much more stable than the primary free radical and has magnetic parameters such that an analysis of the ESR spectrum will allow one to determine in principle the free radical which has added to the spin-trap.

In the case of the cumene hydroperoxide-hematin or the lipid hydroperoxide-hematin system, we have found that primary free radicals could not be spin-trapped using DMPO. In fact, we have found that this system actually oxidized the spin-trap into the ketone of the α-carbon [45]. This reaction is shown in Eq. (7). Molecular

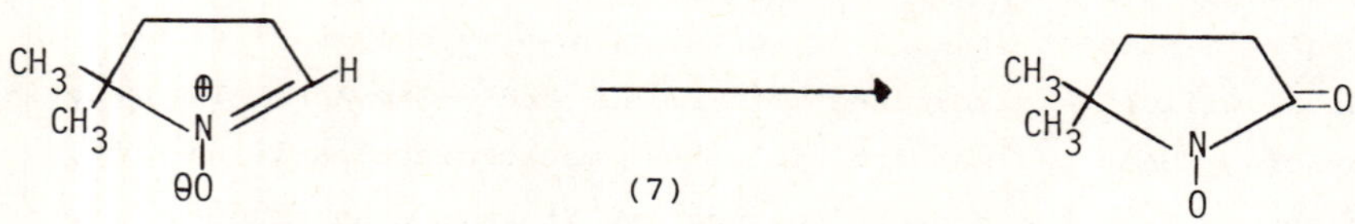

(7)

oxygen enhanced the yield of this reaction, and we found that oxidizable carcinogens such as N-OH-AAF and NOF competed for the site of oxidation of DMPO [45]. This reaction has some interesting into an organic compound and that this is also what occurs in the P_{450} system. We are now using oxygen-17, which has a nuclear magnetic moment of 5/2, to test if the origin of the oxygen is from molecular oxygen, water, or cumene hydroperoxide.

5. *Peroxidase activation of N-OH-AAF in rat mammary gland parenchymal cells*

We have found that a homogenate of rat mammary gland parenchymal cells activates N-OH-AAF via the peroxidase route and that antioxidants inhibit this process [47]. Rat mammary gland parenchymal cells isolated by the method of Topper et al. [48] were sonicated and then N-OH-AAF was added. If to this mixture linoleic acid

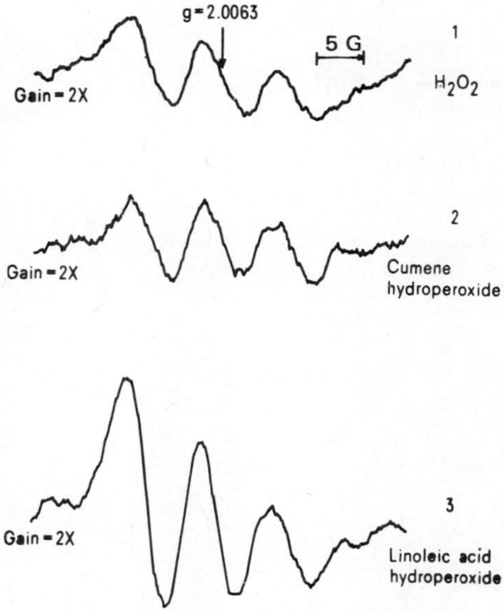

Figure 5 ESR spectra of rat mammary parenchymal cell homogenate to which have been added N-OH-AAF and H_2O_2 (5 mM), cumene hydroperoxide (5 mM), or linoleic acid hydroperoxide (0.4 mM). (From Ref. 47.)

hydroperoxide was added, a distinctive ESR spectrum of the nitroxyl free radical of N-OH-AAF appeared (Fig. 5). Thin-layer chromatography of the samples revealed that NOF and N-OAc-AAF were formed, indicating that the peroxidase route of N-OH-AAF activation was operating in this cell homogenate. Very interestingly, we have found that linoleic acid hydroperoxide was 10 to 20 times more effective than H_2O_2 or cumene hydroperoxide in initiating N-OH-AAF oxidation in this system. Also we found that the antioxidants ascorbate, propyl gallate, and glutathione prevented N-OH-AAF oxidation, producing in the case of the propyl gallate and ascorbate a free radical if these antioxidants were used in addition to the carcinogen (Fig. 6).

6. *P_{420} Activation of N-OH-AAF in a peroxidase mechanism*

Recently, we have shown that if P_{450} or P_{450}-enriched particles are converted into P_{420}, then this heme protein is active in the peroxi-

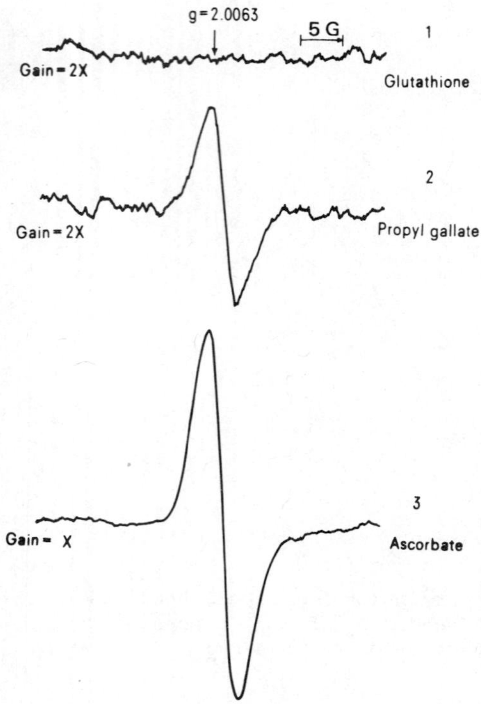

Figure 6 ESR spectra of rat mammary parenchymal cell homogenate to which have been added N-OH-AAF, 5 mM H_2O_2, and either reduced glutathione, propylgallate, or ascorbate. (From Ref. 47.)

dative type activation of N-OH-AAF. P_{450}, in its native electronic state, is not as active as P_{420}. The high-spin form of P_{420}, is two times as active in this regard as the low-spin form of the heme protein. Interestingly, in terms of hydroperoxide specificity, we have found, as with the mammary gland parenchymal cell preparation, that linoleic acid hydroperoxide is about 20 times more active than H_2O_2 or cumene hydroperoxide as a substrate for N-OH-AAF oxidation [49]. These results were obtained with P_{450} particles (trypsin-treated phenobarbital-induced rat liver microsomes) and are shown in Figures 7 and 8. Purified P_{450} (kindly supplied by Dr. M. J. Coon) yielded the same type of results as P_{450} particles [49]. The antioxidants butylated hydroxytoluene (BHT) and butylated hydroxy-

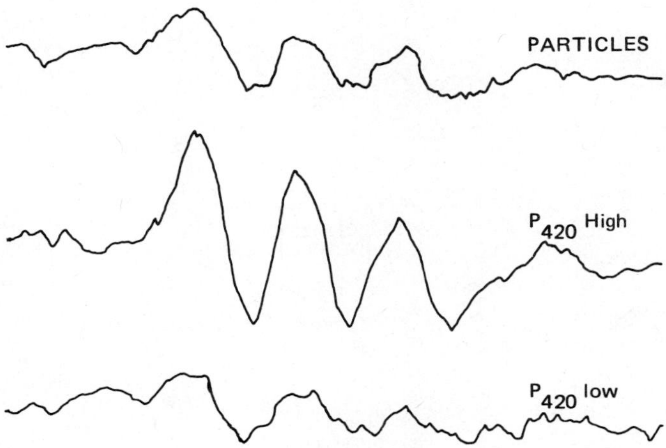

Figure 7 P_{450} particle-catalyzed oxidation of N-hydroxy-2-acetylaminofluorene into the nitroxyl free radical intermediate. Linoleic acid hydroperoxide was used at a level of 0.2 mM and N-OH-AAF at 0.14 mM. P_{450} was converted into high- or low-spin forms [49]. An ESR scan was started 20 sec after the peroxide was added as the final component to the mixture. (From Ref. 49.)

anisole (BHA) prevented N-OH-AAF oxidation (Table 1). Carbon monoxide and SKF-525A are not effective as inhibitors of the P_{420} system. Pyridine stimulates the system. Another very interesting observation that we have made is that rat liver microsomes prepared from phenobarbital-induced rats catalyze the oxidation of N-OH-AAF as observed by ESR spectroscopy of N-$\dot{O}$-AAF when linoleic acid hydroperoxide is used as a substrate (Fig. 9) [49]. This is not the case with microsomes from normal rat livers.

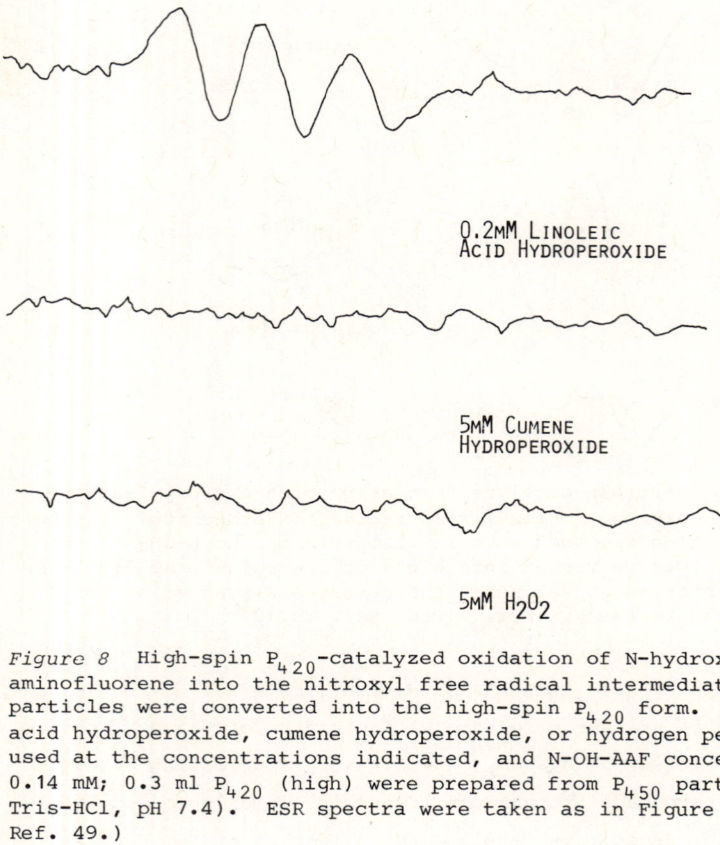

Figure 8 High-spin P_{420}-catalyzed oxidation of N-hydroxy-2-acetylaminofluorene into the nitroxyl free radical intermediate. P_{450} particles were converted into the high-spin P_{420} form. Linoleic acid hydroperoxide, cumene hydroperoxide, or hydrogen peroxide were used at the concentrations indicated, and N-OH-AAF concentration at 0.14 mM; 0.3 ml P_{420} (high) were prepared from P_{450} particles (65 mM Tris-HCl, pH 7.4). ESR spectra were taken as in Figure 7. (From Ref. 49.)

Table 1 Effectors of P_{420}-Spin Form Catalyzed Oxidation of N-OH-AAF[a]

	Control peak height (%)	
Effector	P_{420} (high)	P_{420} (low)
Pyridine (100 mM)	150	140
NaCN (20 mM)	0	0
BHA (1 mM)	13	60
BHT (1 mM)	0	0
Ascorbate (0.1 mM)	0	0
SKF 525A (0.1 mM)	125	110
LAHP (0.2 mM)	0	0

[a]0.14 mM N-OH-AAF, 0.2 mM LAHP. Effectors as shown; spin forms from P_{450} "particles" in 65 mM Tris-HCl, pH 7.4.

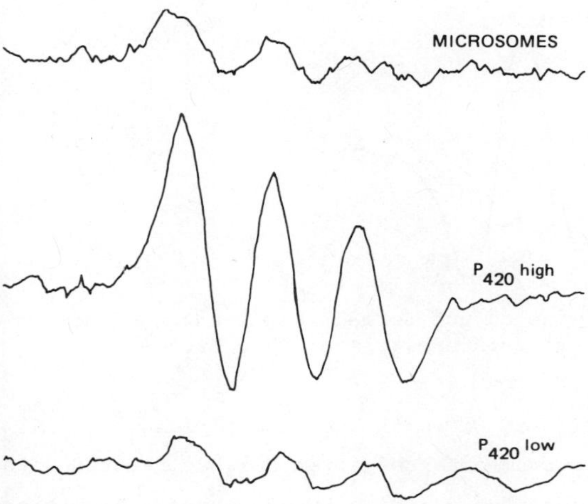

Figure 9 Rat liver microsomes catalyzed oxidation of N-hydroxy-2-acetylaminofluorene into the nitroxyl free radical intermediate. Microsomes were prepared from phenobarbital-induced rats. Microsomes similar to those used to obtain the top spectrum were treated to convert P_{450} into either the high-spin P_{420} form (second spectrum) or low-spin P_{420} form (lowest spectrum 49). Linoleic acid hydroperoxide (0.2 mM) and N-OH-AAF (0.14 mM) were used; equivalent amounts of microsomes in 65 mM Tris-HCl, pH 7.4. The spectra were recorded as in Figure 7. (From Ref. 49.)

C. Addition of 2-Nitrosofluorene to Membrane Lipids

1. *Historical perspective*

We have discovered that NOF adds to the double bonds of membrane lipid components in an "Alder-ene"-type mechanism to form an adduct which either exists as a nitroxyl free radical or the reduced form, N-hydroxyamine [50]. The first experiments were conducted using linoleic acid as well as oleic acid plus NOF [51]. We were testing another hypothesis at the time, but repeatedly obtained a signal of nitroxyl free radical type when only two components were present—oleic acid and NOF [51]. Repeated experimentation to eliminate a catalyst such as iron ions did not alter our results, and hence we became convinced that NOF was reacting directly with oleic acid. Long-chain carbon compounds having at least one double bond all reacted to yield the nitroxyl free radical adduct, but those not having a double bond would not react with NOF [51].

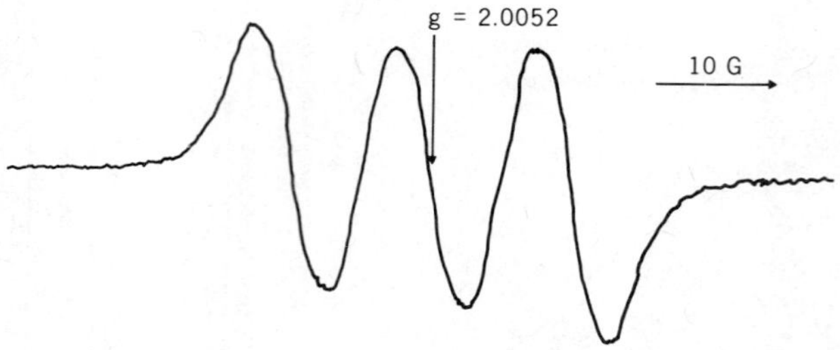

Figure 10 ESR spectrum of the adduct formed when 2-nitrosofluorene reacts with 2,3-dimethyl-2-butene. (From Ref. 50.)

2. Mechanism of the reaction of NOF with membrane lipids

Figure 10 shows the ESR spectrum obtained after NOF has reacted with the simple double bond hydrocarbon 2,3-dimethyl-2-butene. The three-line nitroxyl free radical spectrum is apparent. No hyperfine structure is present because we have not degassed the system. In 1966, Sullivan [52] noted that nitrosobenzene reacted with 2,3-dimethyl-2-butene to yield a nitroxyl free radical. On the basis of ESR hyperfine analysis and also after synthesis of the expected product, he predicted the scheme shown in Figure 11 to account for the results. Nitrosobenzene adds to the double bond of 2,2-dimethyl-2-butene

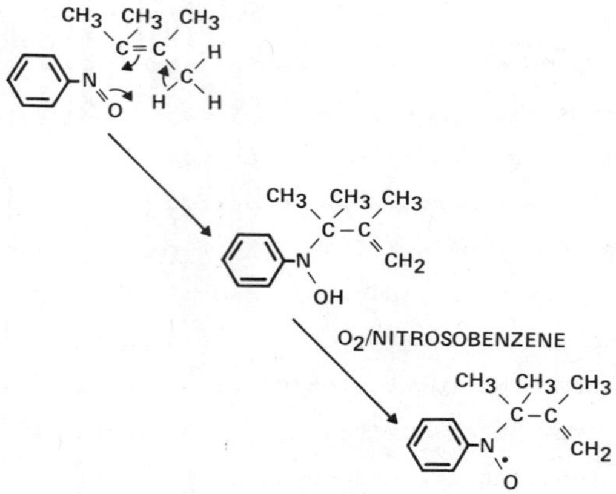

Figure 11 Scheme redrawn from Sullivan [52] to explain nitrosobenzene addition to 2,3-dimethyl-2-butene to form a nitroxyl free radical adduct.

to yield the hydroxyamine intermediate, which is then oxidized to the nitroxyl free radical. Sullivan was unable to detect the intermediate, but 4 years later Knight [53] was successful in isolating the intermediate in good yield. All of our results thus far suggest that NOF is adding to the unsaturated double bonds of fatty acids in the "Alder-ene"-type mechanism proposed by Sullivan. Figure 12 shows NOF adding to oleic acid via the "Alder-ene" mechanism.

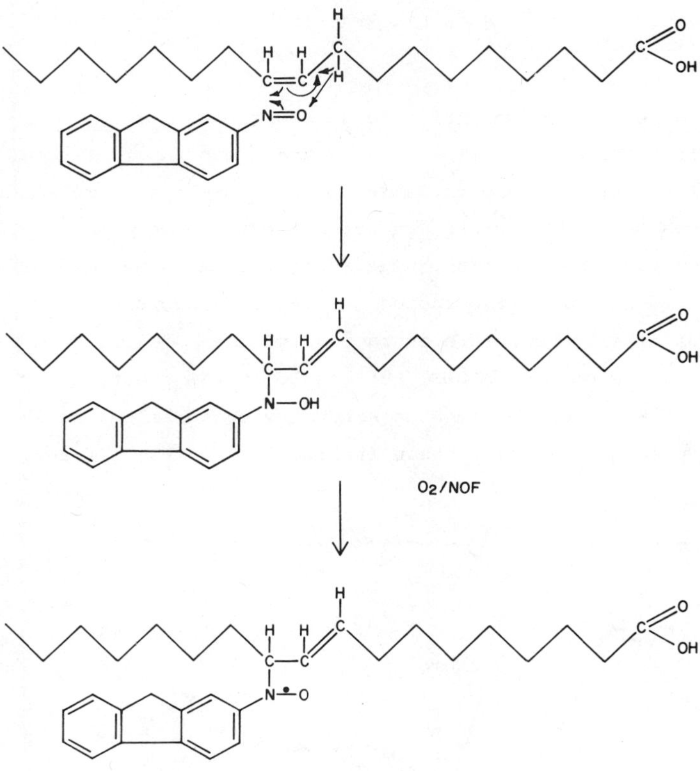

Figure 12 Scheme showing the addition of 2-nitrosofluorene to oleic acid to form a nitroxyl free radical adduct. (From Ref. 50.)

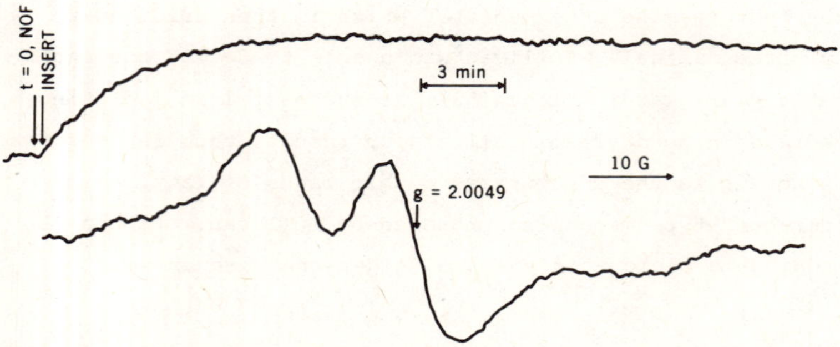

Figure 13 Time course of the formation of and resultant ESR spectrum of the 2-nitrosofluorene adduct to rat liver microsomal membranes. (From Ref. 50.)

3. 2-Nitrosofluorene reaction with natural membranes

We have found that NOF adds to lipids in natural membranes. Figure 13 shows the time course of the formation of the NOF-lipid complex, which we will designate as N-$\dot{O}$-LAF, in rat liver microsomes as well as the resultant ESR spectrum. The ESR spectrum is typical of a highly immobilized nitroxyl free radical. Figure 14 shows that if the total lipids of microsomes are extracted and then sonicated to form liposomes and reacted with NOF, the resultant ESR spectrum is very similar to that obtained with the natural microsomes per se. Therefore, it appears that the protein fraction has little, if any,

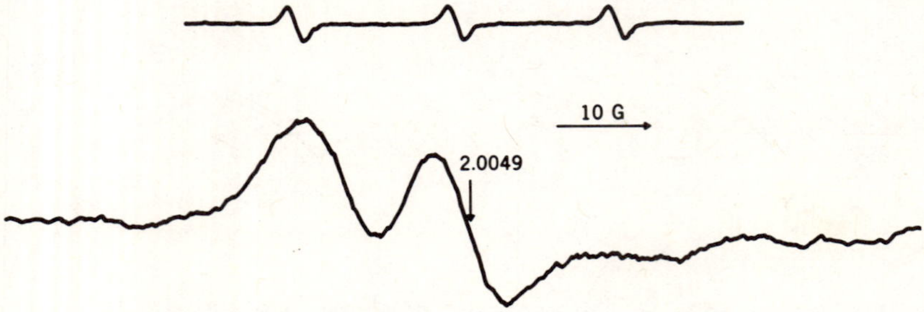

Figure 14 ESR spectrum of the nitroxyl free radical adduct produced after 2-nitrosofluorene has reacted with liposomes formed from the total lipids of rat liver microsomes. The top spectrum is that of the stable free radical 4-hydroxy-2,2,6,6-tetramethylpiperidinooxy used as a standard. (From Ref. 50.)

influence on the formation of N-Ȯ-LAF. If the total amount of N-Ȯ-LAF is determined by hand integration of the derivative spectra using the Reimann sum method [54], it is possible to determine the amount of N-Ȯ-LAF formed as a function of NOF added to the liposomes. The results for the microsomes as well as for the liposomes derived from the microsomes are shown in Figure 15. It can be seen that the profiles are similar for the two preparations, yet the half-maxima occur at about 100 μM NOF in the microsomes and at about 50 μM in the liposomes—at which point the percentage N-Ȯ-LAF formed from the added NOF is about 2% and 3%, respectively. Therefore, the yield of N-Ȯ-LAF is low compared to the amount of NOF added but increases as the NOF/lipid ratio decreases. We are now conducting experiments in an attempt to determine what components of the membrane influence the amount of N-Ȯ-LAF formed.

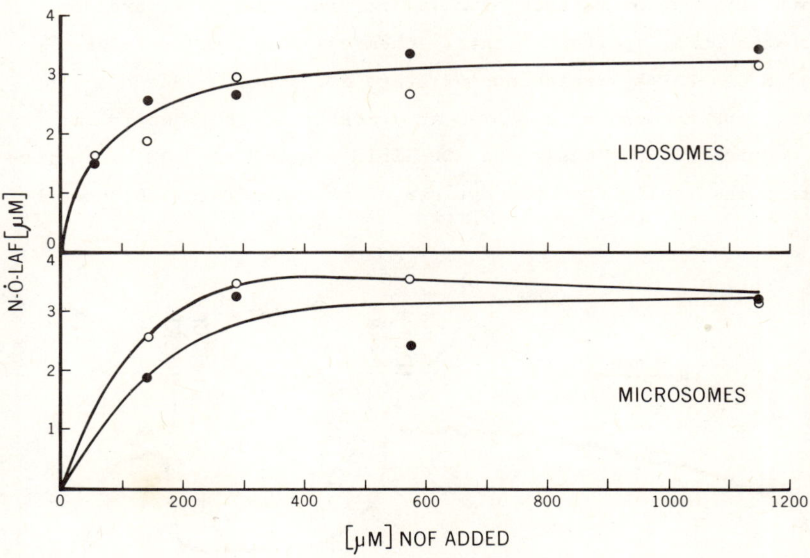

Figure 15 The concentration of the nitroxyl free radical adduct (N-Ȯ-LAF) formed in liposomes as described in Figure 14 and rat liver microsomes (Fig. 13) as a function of 2-nitrosofluorene concentration. (From Ref. 50.)

4. Redox changes of N-Ȯ-LAF in natural membranes

We have found that N-Ȯ-LAF formed in rat liver microsomal membranes can be reduced to the hydroxyamine by addition of reduced pyridine nucleotides and then subsequently oxidized back to the nitroxyl free radical form by addition of ferricyanide. This is shown in Figure 16. A significant portion of the pyridine nucleotide induced reduction of N-Ȯ-LAF is mediated through membrane enzyme components. This is illustrated by the results shown in Figure 17. After the microsomes were heated at 80° for 15 min (top scan), the NADPH-induced reduction of N-Ȯ-LAF was much slower than that observed in the unheated microsomes.

5. Significance of 2-nitrosofluorene-lipid adduct

Since NOF reacts rapidly with natural membranes, we must now ask: What is the relevance of this observation to AAF carcinogenesis? This question may be of considerable importance. Perusal of some of the data obtained by Bartsch et al. [15] presented in Figure 18 illustrates some important points. Chemical modification of N-OH-AAF and N-OAc-AAF by replacing myristate for acetate caused a considerable enhancement of sarcomas at 7 months after injection of these chemicals. Obviously the NOF-lipid adduct (Fig. 12) is quite different chemically from the myristate compounds (Fig. 18) used by

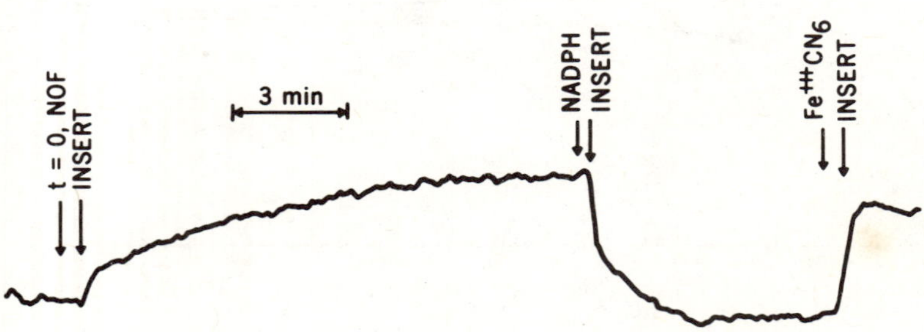

Figure 16 Reduced pyridine nucleotide induced reduction of the 2-nitrosofluorene-lipid adduct and subsequent oxidation by ferricyanide. The smaller amplitude after oxidation is due to dilution of the sample. (From Ref. 50.)

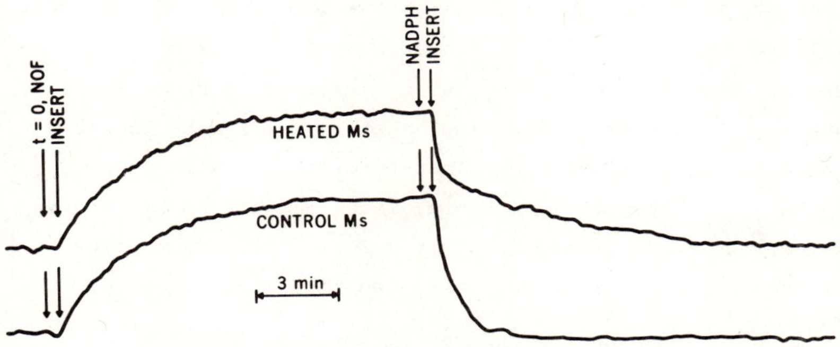

Figure 17 Formation of the 2-nitrosofluorene-lipid adduct and subsequent reduced pyridine reduction of the free radical in control and heated (80° for 15 min) rat liver microsomes. (From Ref. 50.)

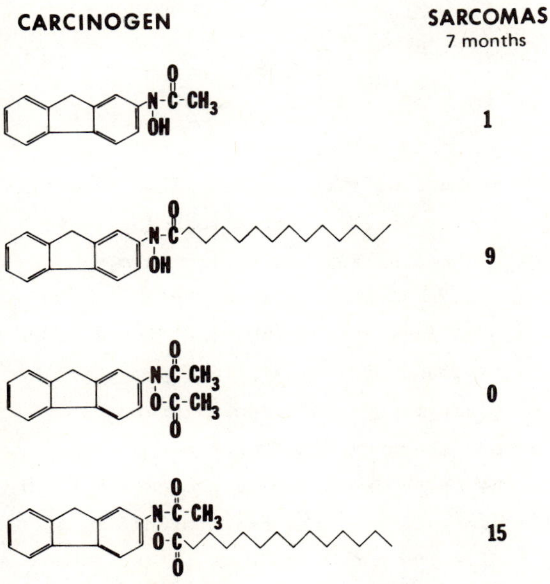

Figure 18 Summary of some of the data presented by Bartsch et al. [15], showing the number of sarcomas formed after 7 months in rats receiving the various carcinogens.

Bartsch et al. [15], but there appear to be some similarities — especially in view of the fact that Bartsch et al. claim the reason the myristate analogs are more potent carcinogens is because these compounds reside at the place of implantation for a much longer time period that the acetate analogs. Therefore, it is quite possible that the NOF-lipid adduct, in a sense, localizes NOF such that its mechanism of action is spread out over a longer time span. The points raised by these questions need much more investigation.

6. *Formation of 2-nitrosofluorene by deacylation of N-hydroxy-2-acetylaminofluorene*

It had been expected that the deacylation of N-OH-AAF would form N-OH-AF [55], which then would be rapidly oxidized to NOF, as shown in Eq. (8).

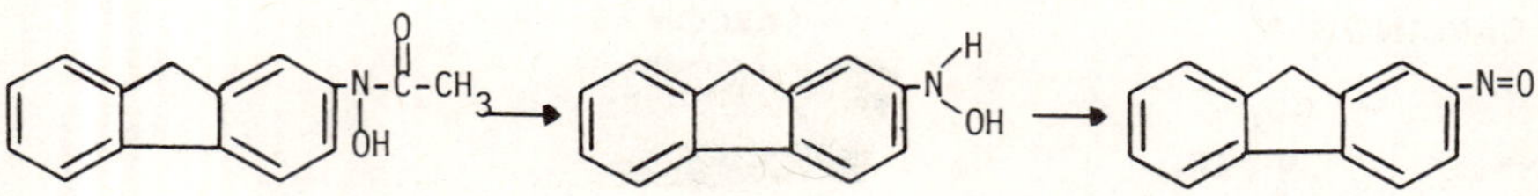

(8)

Nonetheless, we investigated the notion that the NOF-lipid adduct is formed from N-OH-AAF in rat liver subcellular fractions [56]. We discovered that rat liver microsomes as well as nuclei catalyze the formation of the NOF-lipid adduct (N-$\dot{O}$-LAF) from (N-OH-AAF). The results shown in Figure 19 demonstrate that incubation of microsomes with N-OH-AAF produces a small amount of N-$\dot{O}$-LAF (top spectrum); but if the sulfhydryl reagent p-hydroxymercuribenzoate is present, then more N-$\dot{O}$-LAF is apparent (second spectrum from the top). It has been shown in bacteria that the sulfhydryl reagent prevents nitroxide reduction [57], and therefore we interpret the small signal in the absence of this reagent as indicating the N-$\dot{O}$-LAF is in the reduced state, i.e., the hydroxyamine form. We have demonstrated N-$\dot{O}$-LAF reduction and subsequent oxidation in microsomes [50]. The results shown in Figure 19 demonstrate that P_{450} particles also are capable of N-$\dot{O}$-LAF formation from N-OH-AAF.

We have imposed various treatments in order to elucidate the nature of the microsomal catalyzed conversion of N-OH-AAF to N-$\dot{O}$-LAF

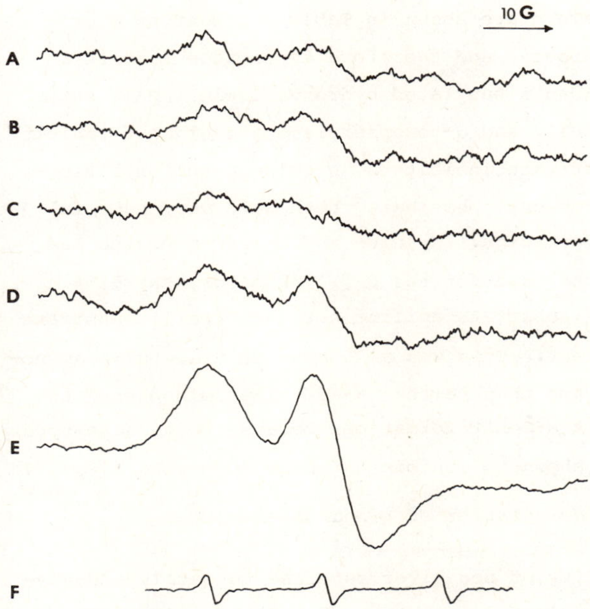

Figure 19 Formation of the nitrosofluorene-lipid adduct during incubation of rat liver microsomes with N-hydroxy-2-acetylaminofluorene. The top trace (A) is the ESR spectrum obtained after 30-min incubation (37°) of rat liver microsomes with N-OH-AAF. The second spectrum (B) was obtained in the same manner as the top spectrum except p-hydroxymercuribenzoate was present to inhibit N-Ȯ-LAF reduction. The third spectrum (C) was obtained after incubating N-OH-AAF with P_{450}-particles, and the fourth spectrum (D) was obtained with similar P_{450}-particles except p-hydroxymercuribenzoate was present. The fifth spectrum down (E) was obtained after incubating rat liver microsomes with 2-nitrosofluorene. The bottom spectrum is that of the ESR standard.

[56]. The results obtained are shown in Table 2. Heating microsomes inactivates the process and therefore implicates that it is enzymatic. The antioxidants butylated hydroxyanisole (BHA), butylated hydroxylotuene (BHT), and α-tocopherylacetate did not inhibit the process; but two esterase inhibitors, α-toluene sulfonyl fluoride and diethyl p-nitrophenyl phosphate, both were very powerful inhibitors of N-Ȯ-LAF formation. Cyanide and carbon monoxide had little effect, as was the case for the P_{450} inhibitor, SKF-525A. Neither the type I P_{450} substrate aniline nor the type II substrate acetaniline had any effect. This was also true of N-acetyl-p-aminophenol. It is significant that neither N-acetyl-p-aminophenol nor the antioxidants inhibit N-Ȯ-LAF formation, because both these types of chemicals have been shown to inhibit AAF carcinogenesis [10,11].

7. *Hog liver esterase deacylation of N-hydroxy-2-acetylaminofluorene*

We have tested the ability of hog liver esterase to catalyze deacylation of N-OH-AAF and eventual formation of N-Ȯ-LAF [56]. Figure 20 shows that hog liver estrase when incubated with N-OH-AAF and

Table 2 Effects of Various Treatments on N-OH-AAF —— N-Ȯ-LAF by Rat Liver Microsomes

Additive	[Additive]/[N-OH-AAF]	Function	Control (%)
Heat at 85° (15 min)	——	Enzyme denatured	0
p-Hydroxymercuribenzoate	0.75	Electron transport inhibition	100
Butylated hydroxytoluene	4.5	Antioxidant	87
Butylated hydroxyanisole	4.5	Antioxidant	91
α,α-Tocopheryl acetate	15.4	Antioxidant	96
α-Toluene sulfonyl fluoride	1.5	Esterase inhibitor	0
Diethyl p-nitrophenyl phosphate	1.5	Esterase inhibitor	0
NaCN	0.75	Peroxidase inhibitor	61
N-Acetyl-p-aminophenol	1.5	Deacylase competitor	108
Carbon monoxide	Bubbled in 2 min	P_{450} inhibitor	80
SKF-525A	0.75	P_{450} inhibitor	108
Aniline	1.0	P_{450} inhibitor	111
Acetanilide	2.0	P_{450} inhibitor	111

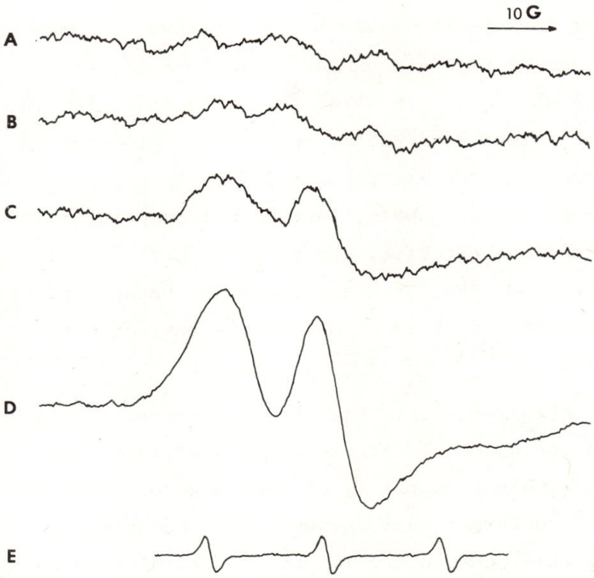

Figure 20 The ability of hog liver esterase to deacylate N-hydroxy-2-acetylaminofluorene leading to nitrosofluorene-lipid adduct formation. The top trace (A) shows the ESR spectrum obtained after incubating N-OH-AAF in the presence of the esterase and liposomes formed from egg phosphatidyl choline. The second trace (B) shows the spectrum obtained after incubating N-OH-AAF with esterase and liposomes formed from the total lipids of rat liver microsomes. The third trace (C) was obtained as in trace B except that the liposomes were added after incubation of N-OH-AAF with the esterase only. The fourth trace shows the spectrum obtained after 2-nitrosofluorene had reacted with liposomes formed from egg phosphatidyl choline. The bottom spectrum is that of the ESR standard.

either egg phosphatidyl choline liposomes or liposomes formed from the total lipid extract of rat liver microsomes demonstrated only a small amount of N-$\dot{O}$-LAF formation. But when N-OH-AAF first was incubated with the esterase and then liposomes were added later, substantial amount of N-$\dot{O}$-LAF was formed. We have demonstrated by optical spectroscopy as well as by thin-layer chromatography that NOF is formed from N-OH-AAF in the presence of hog liver esterase. Therefore, we [56] interpret the experiments as indicating that when liposomes were present during esterase action, the liposomes solubilized N-OH-AAF and thus in effect competed with the enzyme.

An investigation of the kinetic behavior of hog liver esterase in relation to deacylation of N-OH-AAF demonstrated that the kinetic constants were K_m = 45.5 μM and V_{max} = 25.8 nmol min^{-1} mg^{-1} protein [56]. In addition to hog liver esterase many different protease-type enzymes have been shown to be ineffective [56]. Thus pepsin, chymotrypsin, cathepsin D, ficin, papain, leucine aminopeptidase, phospholipase B, acetylcholine esterase, and trypsin (both type I and V) did not deacylate N-OH-AAF. It was also noted that the polymeric form of hog liver esterase (type I) was active but not the monomeric form (type II).

8. Behavior of 2-nitrosofluorene-lipid adduct in membranes

The molecular configuration of the 2-nitrosofluorene-lipid adduct in membranes is not known. The accessibility of the carcinogen-lipid adduct and the molecular conformational changes which occur in the adduct as membrane properties change are important questions to be answered. We have begun an attempt at answering these questions [56]. We have found that as the temperature changes the rotational mobility of N-$\dot{O}$-LAF changes. Figure 21 illustrates the differences in rotational mobility at two different temperatures. At 10° the spectrum is that of a highly immobilized nitroxide; the characteristic wide and shallow high-field transition is apparent. But at 50° there is a sharp and narrow high field transition positioned nearer the middle transition. The 50° spectrum is indicative of a much more mobile nitroxide radical.

The ratio of the mobile high field transition to the low field transition, as defined in Figure 21, is plotted vs. temperature for N-$\dot{O}$-LAF in egg phosphatidyl choline liposomes and rat liver microsomes in Figure 22. The carcinogen-lipid adduct begins to exhibit enhanced rotational mobility at about 25° in the phosphatidyl choline liposomes but only at 45° in the microsomes. In independent studies using nitroxyl-labeled stearic acid spin probes, we found that a phase transition near the polar region of the membrane occurred at about 25-30° in phosphatidyl choline liposomes. The reduced rotational mobility of N-$\dot{O}$-LAF in natural membranes as

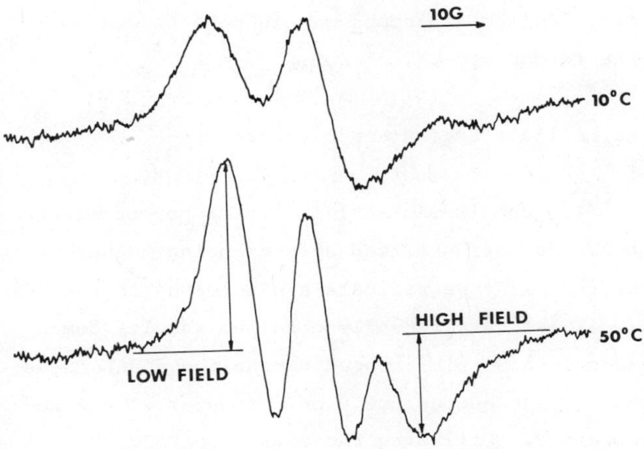

Figure 21 ESR spectra of the 2-nitrosofluorene-lipid adduct in egg phosphatidyl choline liposomes at 10° and 50°.

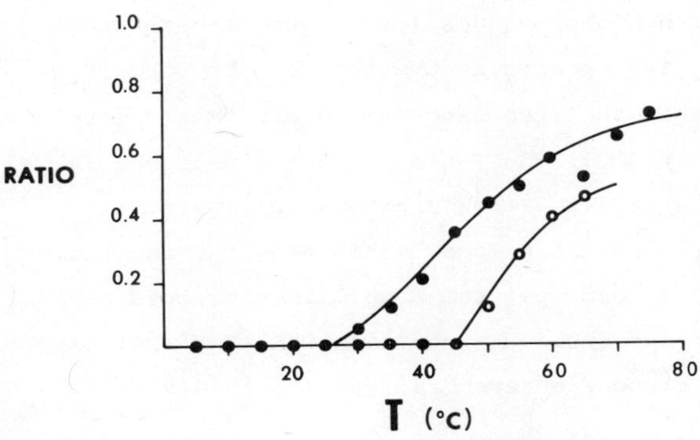

Figure 22 Rotational mobility of nitrosofluorene-lipid adduct in egg phosphatidyl choline liposomes and rat liver microsomes as a function of temperature. The ratio of the high-field to low-field peak as defined in Figure 21 is plotted against temperature (°C). The closed circles represent data obtained from liposomes; the open circles, from microsomes.

compared to phosphatidyl choline liposomes may in part be due to proteins present in the former case.

D. Spin-Trapping Free Radicals from N-Acetoxy-2-acetylaminofluorene Decomposition

N-Acetoxy-2-acetylaminoflourene (N-OAc-AAF) is a very potent mutagen and ultimate carcinogen. We have utilized spin-trapping techniques in an attempt to understand if free radicals are involved in the action of N-OAc-AAF [58]. We have recently obtained results demonstrating that free radicals are spin-trapped during the spontaneous decomposition of N-OAc-AAF in aqueous solution. Figures 23 and 24 present the results obtained. Utilizing the spin-trap DMPO (5,5-dimethyl-1-pyrroline-1-oxide) and PBN (phenyl-t-butylnitrone), we have obtained spectra indicating that carbon-centered free radicals are produced in the spontaneous decomposition of N-OAc-AAF. Even though the solutions in these cases were 40% methanol, very similiar results were obtained with aqueous-ethanol and aqueous-hexamethylphophoramide solutions. The spectra in the the case of DMPO (Fig. 23) can be interpreted that the free electron interacts with the nitroxyl nitrogen ($A_n \simeq 14.3$ G), the β-proton on the α-carbon ($A_\beta^H \simeq 9.8$ G), and a γ-proton ($A_\gamma^H \simeq 1.4$ G). The arrows in Figure 23 point to transitions of a radical trapped other than the one indicated above. With time (second spectrum down), this trapped radical tends to disappear. The spectrum obtained with PBN (Fig. 24) also indicated trapping of a carbon-centered radical ($A_N \simeq 14.9$ G; $A_\beta^H \simeq 3.3$ G).

On the basis of these results, it is very difficult to predict the molecular structure of the spin-trapped radical. Janzen and Liu [59] have obtained γ-proton splittings with spin-trapped alkoxy radicals having the general formula as shown for the trapped methoxy radical.

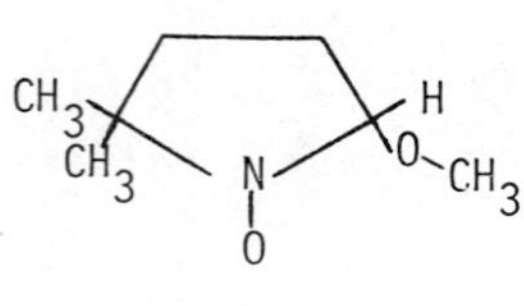

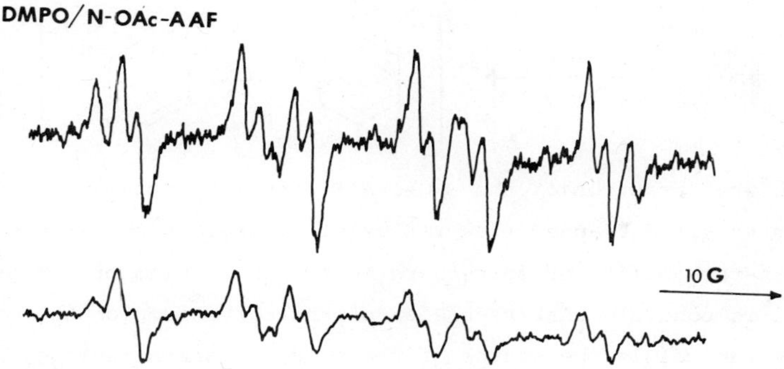

Figure 23 ESR spectra obtained after incubation of N-acetoxy-2-acetylaminofluorene in the presence of the spin trap, viz., 5,5-dimethyl-1-pyrroline-1-oxide. The top spectrum was taken within 45 min after the carcinogen aqueous decomposition was initiated, and the lowest spectrum was taken after the top spectrum was obtained and shows the change of the spectrum with time.

The coupling constants obtained in benzene range as such: $A_N = 12.24 - 13.58$ G; $A_\beta^H = 6.83 - 9.63$ G; and $A_\gamma^H = 1.85 - 2.06$ G. The values we obtained in the spontaneous decomposition of N-OAc-AAF are in this range. It should be pointed out that they [59] did not obtain γ-proton coupling with trapped acetyl radicals.

Scribner et al. [60] postulated that N-OAc-AAF decomposes in aqueous solution as shown in Eq (9). Thus a nitrenium ion is in equilibrium with a carbonium ion. It is difficult to reconcile our results with the proposed mechanism shown in Eq. (9).

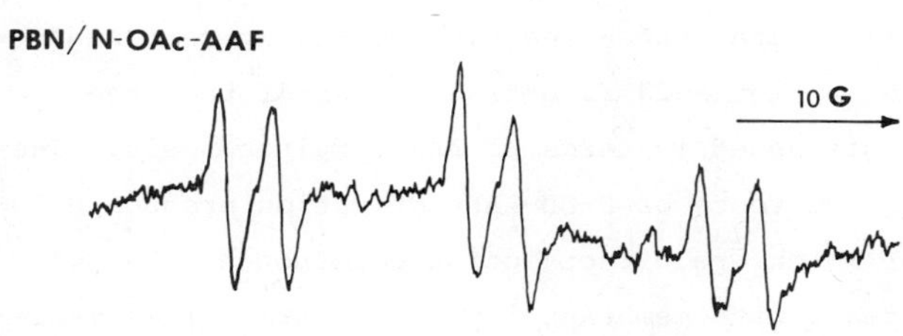

Figure 24 ESR spectrum obtained after incubating N-acetoxy-2-acetylaminofluorene with the spin-trap phenyl-t-butylnitrone.

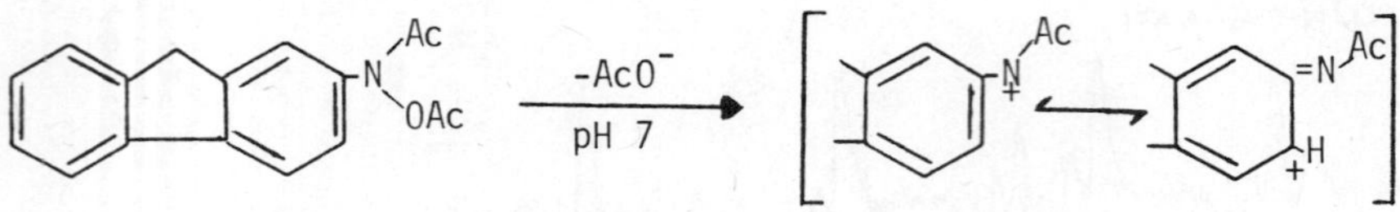

(9)

We do not know what fraction of the N-OAc-AAF actually decays in a manner such as to yield trapped free radicals. There are no reported observations on the ability of spin-traps to trap nitrinium or carbonium ions. It is possible that reduction of the carbonium ion to produce a free radical is the source of the trapped carbon radical, but we have data that argue against this point. Thus, we have observed that reducing agents such as ferrocyanide and ascorbate do not increase the amount of trapped radicals. In addition, it would be difficult to explain the γ-proton coupling even if the reduced carbonium ion radical is trapped. If the nitrenium ion is reduced to produce a free radical, the resultant nitrogen-centered radical if spin-trapped would yield a spectrum indicative of a nitrogen-centered radical, but this is contrary to what we have obtained with both DMPO and PBN (Fig. 23 and 24).

V. THE IMPORTANCE OF FREE RADICALS IN ARYLAMINE CARCINOGENESIS

It is not possible at the present time to assess the true importance of free radicals in arylamine carcinogenesis. The work we have described in this chapter certainly has shown that these carcinogens will undergo reactions involving free radicals, and there is a very high degree of probability that these reactions occur in carcinogenesis. The scheme shown in Figure 25 illustrates most of the free radical routes I have discussed in terms of AAF carcinogenesis. The free radical or peroxidase route of N-OH-AAF activation produce NOF and N-OAc-AAF, both of which are direct-acting carcinogens. As we have shown, NOF will react with membrane lipids to form a free radical adduct. It is known that deacylation of N-OH-AAF does occur and this will lead to NOF, which then would be expected to combine with membranes. We have also shown that deacylation of N-OH-AAF to form NOF and, subsequently, N-$\dot{O}$-LAF formation does occur in microsomal as well as nuclear membranes.

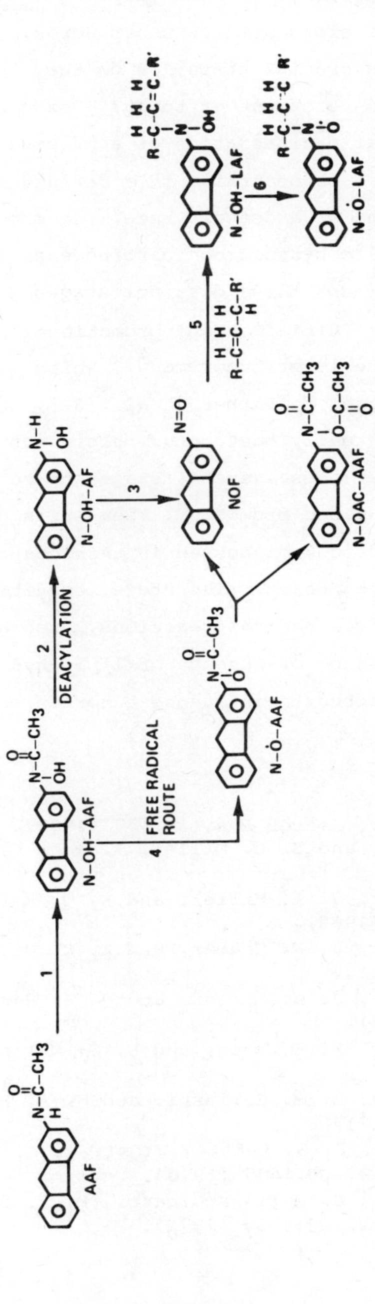

Figure 25 Scheme showing some of the free radical aspects discussed in this chapter. (From Ref. 50.)

We must then ask: What is the probability that any of the free radical events discussed in this chapter will have any effect on eventual tumor development? In other words, are any of the free radical reactions crucial steps, or do they influence crucial steps in the eventual development of tumor? One can consider a palpable tumor as the final manifestation of a process involving several essential steps, each occurring in a defined chronological sequence and each occurring at a certain rate. In mice, in skin tumors induced by polycyclic hydrocarbon carcinogens, there appears to be at least two and perhaps three distinct stages in tumor development [61]. The initiation and promotional stages are the two most clear-cut. In the liver tumor model, which involves hyperplastic nodules, as proposed by Farber et al. [62], there are probably many stages which eventually must occur before actual tumors are present. But if we consider these several stages to be grouped into two broad phases, initiation and promotion, then it is quite possible that the free radical reactions discussed in this chapter may influence either one or both phases. Therefore, conditions or agents which influence these free radical reactions, such as perhaps antioxidant level or the presence or absence of lipid hydroperoxides, would certainly be expected to influence tumor formation.

REFERENCES

1. J. A. Miller, *Cancer Res.* *30*:559 (1970).
2. J. A. Miller and E. C. Miller, *J. Nat. Cancer Inst.* *47*: v (1971).
3. J. W. Cramer, J. A. Miller, and E. C. Miller, *J. Biol. Chem.* *235*:885 (1960).
4. J. A. Miller, J. W. Cramer, and E. C. Miller, *Cancer Res.* *20*: 950 (1960).
5. E. C. Miller, J. A. Miller, and H. A. Hartmann, *Cancer Res.* *21*: 815 (1961).
6. J. A. Miller, M. Enomoto, and E. C. Miller, *Cancer Res.* *22*:1381 (1962).
7. H. R. Gutman, S. B. Galitski, and W. A. Foley, Cancer Res. 27: 1443 (1967).
8. H. R. Gutman, D. S. Leaf, Y. Yost, R. E. Rydell, and C. C. Chen, *Cancer Res.* *39*:1485 (1970).
9. H. R. Gutman, D. Malijka-Giganti, E. J. Berry, and R. E. Rydell, *Cancer Res.* *32*:1554 (1972).

10. J. H. Weisburger, R. S. Yamamoto, G. M. Williams, P. H. Grantham, T. Matsushima, and E. K. Weisburger, *Cancer Res. 32*: 491 (1972).
11. L. C. Mohan, P. H. Grantham, E. K. Weisburger, J. H. Weisburger, and J. B. Idoine, *J. Nat. Cancer Inst. 56*:763 (1976).
12. S.-C. G. Wu and K. D. Straub, *J. Biol. Chem. 251*:6529 (1976).
13. C. C. Irving, D. H. Janss, and L. T. Russell, *Cancer Res. 31*: 387 (1971).
14. C. C. Irving, *Cancer Res. 35*:2959 (1975).
15. H. Bartsch, C. Malaveille, H. F. Stich, E. C. Miller, and J. A. Miller, *Cancer Res. 37*:1461 (1977).
16. C. M. King, *Cancer Res. 34*:1503 (1974).
17. H. Bartsch, M. Traut, and E. Hecker, *Biochim. Biophys. Acta 237*:556 (1971).
18. A. R. Forrester, M. M. Ogilvy, and R. H. Thomson, *J. Chem. Soc. (Org.)* p. 1081 (1970).
19. H. Bartsch and E. Hecker, *Biochem. Biophys. Acta. 237*:567 (1971).
20. R. J. Shamberger and G. Rudolph, *Experientia 22*:116 (1966).
21. R. J. Shamberger, *J. Nat. Cancer Inst. 44*:931 (1970).
22. R. J. Shamberger, *J. Nat. Cancer Inst. 48*:1491 (1972)
23. R. J. Shamberger, E. Rukovena, A. K. Longfield, S. A. Tytko, S. Deodhar, and C. E. Willis, *J. Nat. Cancer Inst. 50*:863 (1973).
24. R. J. Shamberger, F. F. Baughman, S. L. Kalchert, C. E. Willis, and G. C. Hoffman, *Proc. Nat. Acad. Sci. U.S. 70*: 1461 (1973).
25. R. J. Shamberger, T. L. Andreaone, and C. E. Willis, *J. Nat. Cancer Inst. 53*:1771 (1974).
26. R. J. Shamberger, S. A. Tytko, and C. E. Willis, *Arch. Environ. Med.* Sept.-Oct. p. 231 (1976).
27. L. W. Wattenberg, *Fed. Proc. 31*:633 (1972).
28. L. W. Wattenberg, *J. Nat. Cancer Inst. 48*:1425 (1972).
29. L. W. Wattenberg, *Proc. Am. Assoc. Cancer Res. 14*:7 (1973).
30. L. W. Wattenberg, *J. Nat. Cancer Inst. 50*:1541 (1973).
31. L. W. Wattenberg, *J. Nat. Cancer Inst. 52*:1583 (1973).
32. P. Borchert and L. W. Wattenberg, *J. Nat. Cancer Inst. 57*:173 (1976).
33. S. L. Haber and R. W. Wissler, *Proc. Soc. Exp. Biol. Med. 111*: 774 (1962).
34. S. S. Epstein, S. Joshi, J. Andrea, J. Forsyth, and N. Mantel, *Life Sci. 6*:225 (1967).
35. B. M. Ulland, J. H. Weisburger, R. S. Yamamoto, and E. K. Weisburger, *Food Cosmet. Toxicol. 11*:199 (1973).
36. P. H. Grantham, J. H. Weisburger, and E. K. Weisburger, *Food Cosmet. Toxicol. 11*:209 (1973).
37. J. I. Goodman, J. E. Trosko, and J. D. Yager, Jr., *Chem.-Biol. Interactions 12*:171 (1976).
38. T. J. Slaga and W. M. Bracken, *Cancer Res. 37*:1631 (1977).
39. J. U. Skaare, I. Nafstaf, and H. K. Dahle, *Toxicol. Applied Pharmacol. 42*:19 (1977).

40. R. A. Floyd, *Life Sci.* *18*:189 (1976).
41. R. A. Floyd, L. M. Soong, and P. L. Culver, *Cancer Res.* *36*:1510 (1976).
42. D. Keilin and E. F. Hartree, *Biochem. J.* *49*:88 (1951).
43. R. A. Floyd, L. M. Soong, R. N. Walker, and M. Stuart, *Cancer Res.* *36*:2767 (1976).
44. R. A. Floyd and L. M. Soong, *Biochim. Biophys. Acta.* *498*:244 (1977).
45. R. A. Floyd and L. M. Soong, *Biochem. Biophys. Res. Commun.* *74*:79 (1977).
46. E. G. Janzen, *Acc. Chem. Res.* *4*:31 (1971).
47. D. L. Reigh, M. Stuart, and R. A. Floyd, *Experientia* *34*:107 (1978).
48. Y. J. Topper, I. Oka, and B. K. Vanderhaar, *Methods Enzymol.* *39*:443 (1975).
49. D. L. Reigh and R. A. Floyd, *Cancer Biochem. Biophys.* 5:213 (1981).
50. R. A. Floyd, L. M. Soong, M. A. Stuart, and D. L. Reigh, *Arch. Biochem. Biophys.* *185*:450 (1978).
51. R. A. Floyd, *Experientia* *33*:197 (1977).
52. A. B. Sullivan, *J. Org. Chem.* *31*:2811 (1966).
53. G. T. Knight, *Chem. Commun.* p. 1016 (1970).
54. P. B. Ayscough, *Electron Spin Resonance in Chemistry.* Methuen, London, 1967, p. 442.
55. R. A. Cardona and C. M. King, *Biochem. Pharmacol.* *25*:1051 (1976).
56. J. E. Steward and R. A. Floyd, *Cancer Biochem. Biophys.* 5:47 (1980), and unpublished results (1980).
57. J. J. Baldassare, D. E. Robertson, A. G. McAfee, and C. Ho, *Biochemistry* *13*:5210 (1974).
58. R. N. Walker and R. A. Floyd, unpublished results (1980).
59. E. G. Janzen and J. I.-P. Liu, *J. Magnetic Resonance* *9*:510 (1973).
60. J. D. Scribner, J. A. Miller, and E. C. Miller, *Cancer Res.* *30*:1570 (1970).
61. R. K. Boutwell, *CRC Critical Rev. Toxicol.* Jan. p. 419 (1974).
62. E. Farber, S. Parker, and M. Gruenstein, *Cancer Res.* *36*:3879 (1976).

11

DETECTION OF FREE RADICALS BY SPIN-TRAPPING USING THE NITROXYL METHOD

Edward G. Janzen and Edward R. Davis*

Guelph-Waterloo Centre for Graduate Work in Chemistry
University of Guelph
Guelph, Ontario, Canada

I. INTRODUCTION

Electron spin resonance (ESR) spectroscopy provides an excellent tool for the detection of atoms, free radicals, and paramagnetic transition metal centers. Detection by ESR is nondestructive, sensitive, and specific. Many applications of ESR spectroscopy are discussed in the literature. The theory is developed in numerous monographs and reference texts [1]. Excellent instrumentation is available at a reasonable price.

However the ESR method has some limitations. Many free radical reactions are rapid under conditions where the steady state concentration of free radicals is below the detection threshold of ESR spectrometers. Also certain atoms and free radicals do not give ESR spectra in the condensed phase. For example, hydroxy radicals have not been detected by ESR in solution even under the best of conditions. These limitations are discussed in more detail elsewhere [2]. A method was needed to extend the limits of ESR spectroscopy so that lower concentrations of free radicals of any structure could be detected in a condensed phase.

The spin-trapping method was developed to answer this need. In this technique, typical reactive short-lived free radicals are chemically "converted" into relatively unreactive persistent free

---

*Present affiliation: University of Delaware, Newark, Delaware

radicals whose ESR spectra are distinctive for the radical trapped. If R• is the radical to be detected and X=Y an unsaturated function capable of trapping R•, X and Y can be selected so that the addition products R-X-Y• are relatively long-lived under the reaction conditions of the experiment and ESR spectra are clearly dependent on the nature of R•:

$$R\cdot + X{=}Y \rightarrow R{-}X{-}Y\cdot$$

When X=Y is the nitroso or nitrone function the addition products are nitroxyl radicals:

(nitroso)

(nitrone)

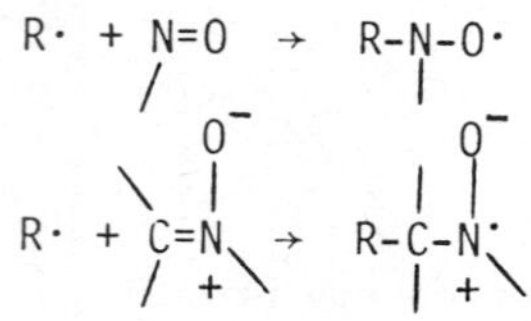

Nitroxyl radicals are relatively stable and long-lived free radicals readily detected by ESR methods. The unpaired electron is approximately equally distributed over both atoms of the nitroxyl-function:

Dimerization of nitroxyl radicals is not significant. Mechanisms of decay usually involve disproportionation, α- or β-bond cleavage or reduction by hydrogen atom transfer. The chemistry of nitroxyl radicals is reviewed in a number of books [3].

The unsaturated compounds which give nitroxyl radicals upon addition of radicals are called spin traps [4]. The addition products are called spin adducts. The most popular spin traps are NtB, PBN and DMPO.

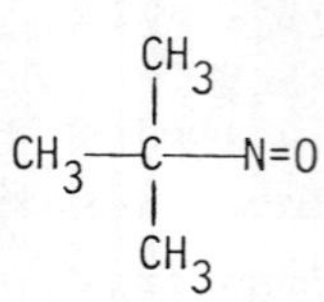

NtB

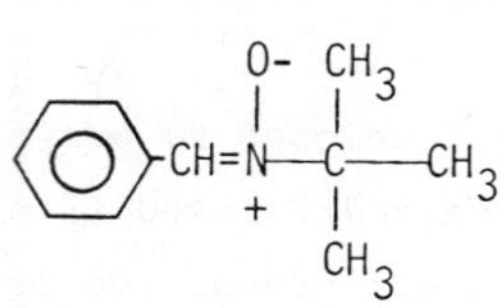

PBN

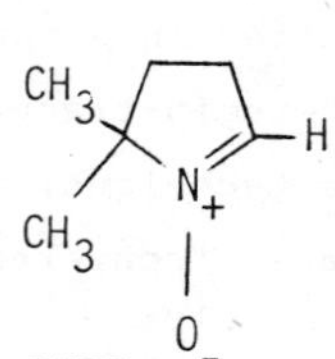

DMPO

The correct name for NtB is 2-methyl-2-nitrosopropane, but since this name does not point to the presence of the very important tert-butyl group the names and abbreviations nitroso-tert-butane (NtB) or tert-nitrosobutane (tNB) have been proposed. NtB will be used in this chapter. PBN stands for α-phenyl-N-tert-butylnitrone. The correct name for DMPO is 5,5-dimethyl-1-Δ-pyrroline-N-oxide.

Other nitroso compounds have also been used for spin-trapping, e.g., 3-methyl-3-nitrosobutanone [5], 2-methyl-2-nitrosopropanol [6], nitrosobenzene [7], 2,4,6-tri-tert-butylnitrosobenzene [8], and nitrosodurene [9]:

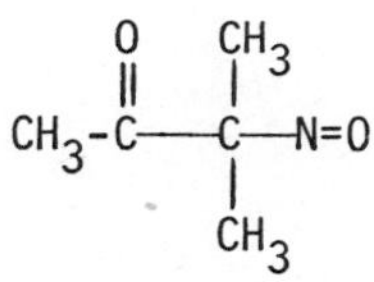

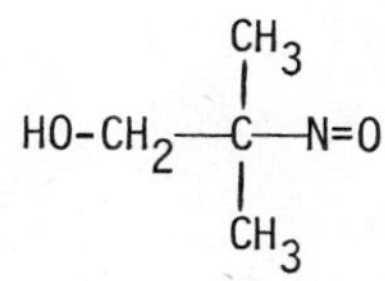

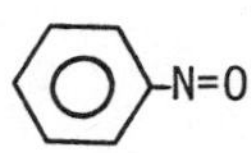

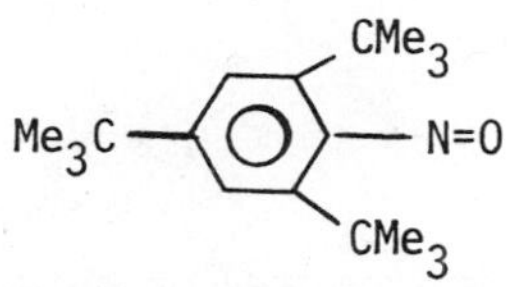

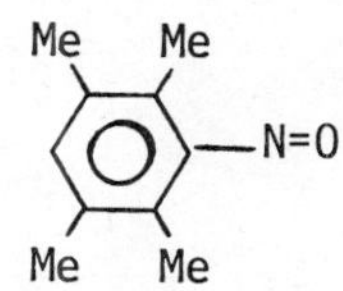

Me_3C = *tert*-butyl Me = Methyl

A variety of nitrone spin-traps have recently been described, e.g., p-nitro, methoxy, chloro, and cyanophenyl tert-butyl nitrones [10], 4-pyridyl-N-oxide tert-butyl nitrones [12], 2-, 3-, 4-methyl pyridinium tert-butyl nitrones [11], 2-sodium sulfonatophenyl tert-butyl nitrone [13], and phenyl-5-methyl-2,2-dialkyl-1,3-dioxanyl nitrones [11]:

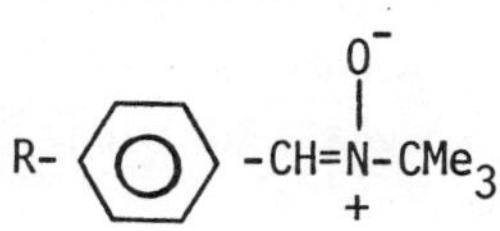

substituted PBNs

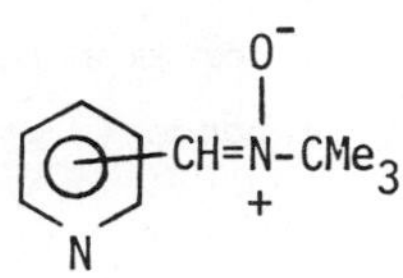

2-, 3-, 4-PyBNs

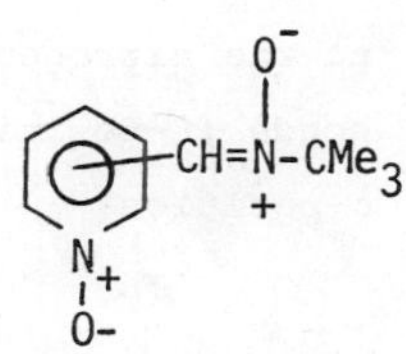

2-, 3-, 4-PyOBNs

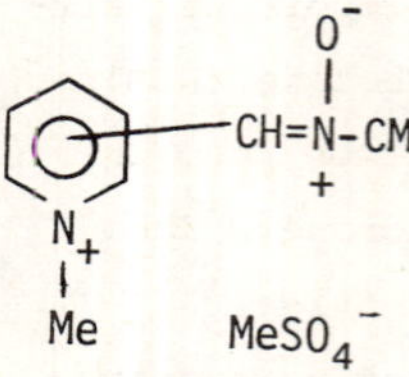

2-, 3-, 4-MePyBNs

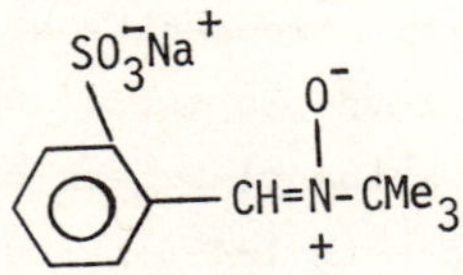

2-SSPBN

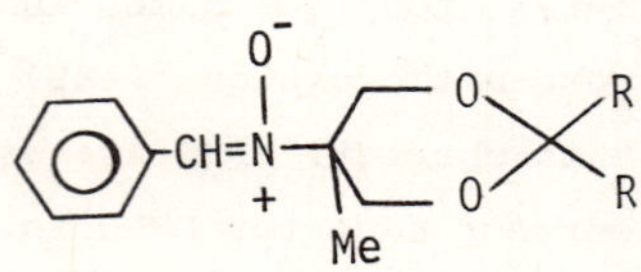

PDNs

Some substituted DMPOs have been studied, e.g., 2-methyl [14], 2-phenyl [15], and 4-phenyl [15]:

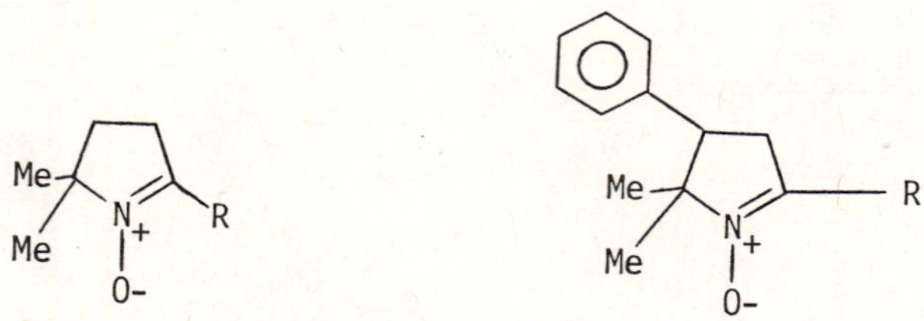

R-methyl, phenyl

II. ELECTRON SPIN RESONANCE SPECTRA OF SPIN ADDUCTS

A. Vocabulary and Classification

The ESR spectrum of most nitroxyl spin adducts is dominated by the triplet splitting of the nitrogen nucleus (^{14}N has a spin of 1; the number of lines in the hyperfine splitting pattern equals $2nI + 1$, where I = nuclear spin and n = number of magnetically equivalent nuclei interacting with the unpaired electron). Since for oxygen-16 $I = 0$, further splitting might come from nuclei directly attached to the nitrogen (α-position), or from nuclei attached through two bonds (β-position), or three bonds (γ-position), or even four bonds, δ-position:

```
 O·
 |
-N-W-X-Y-Z
   α β γ δ
```

Spin adducts of NtB serve as good examples to illustrate the kind of spectra to be expected from various trapping situations. Thus when a radical R• is trapped by NtB, the spin adduct produced always has at least one carbon in the α-position, three carbons in the β-position, and nine hydrogens in the γ-position.

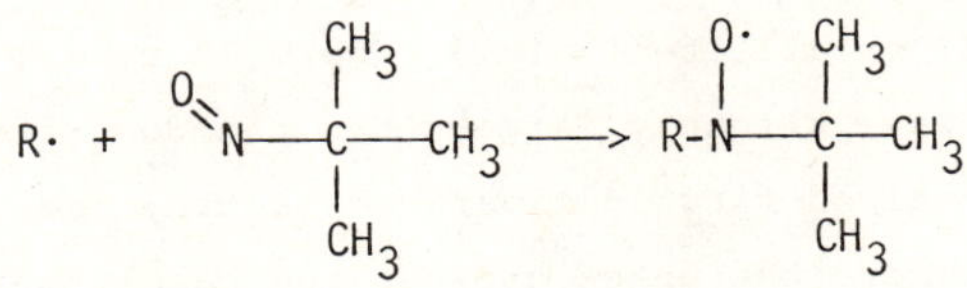

Since for carbon-12, I = 0 the nearest nuclei with spin are the hydrogens in the γ-position (for hydrogen-1, I = 1/2). For a freely rotating tert-butyl group the γ-hydrogen splitting is usually too small to detect, although this splitting does contribute to the line width of the spectrum.

Thus the ESR spectrum will consist of splitting from ^{14}N and from R. The ^{14}N splitting consists of three equally spaced lines or groups of lines of equal intensity. For example, when R = tert-butyl, a triplet spectrum is obtained (see Fig. 1).

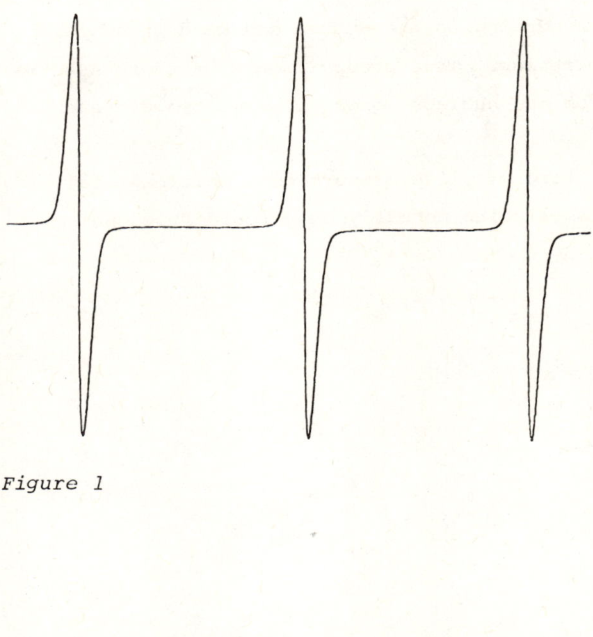

Figure 1

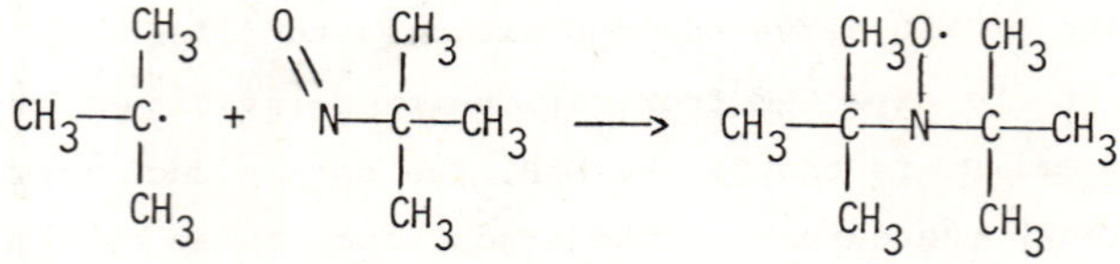

The hyperfine splitting constant (hfsc) is the spacing between the lines measured in guass (G) or millitesla (mT), where 10 mT = 1 G. The up-field spacing is always slightly larger than the down-field spacing. For perfectly resolved first derivative spectra, the spaces between the points where the trace crosses the baseline is measured and averaged. For less well-resolved spectra the spaces between the maxima and the minima are averaged; note that care must be taken to include the same number of up-field spaces as down-field spaces into the average. The repeatability is typically ±0.05 G, although for very small splittings sometimes better reproducibility can be realized. The N-hfsc can vary from 14.5 G in nonpolar solvents to 17 G in water. The larger hfsc in polar solvents reflects the larger contribution to the resonance hybrid of the dipolar structure and thus the larger spin density on the nitrogen atom of the nitroxyl function.

The types of splitting patterns obtained from different spin adducts can be grouped according to the number and kind of nuclei which give hyperfine splitting. Thus group I would be those spectra which show splitting from one nucleus, group II from two different kinds of nuclei, etc.

Thus spectra from nitroxyls like di-tert-butyl nitroxyl fall into group I, hyperfine splitting normally only found from one nucleus.

B. Spin Adducts of Methinyl Radicals

When R• is a methinyl radical, $\overset{A}{\underset{G}{}}>\dot{C}H$, additional splitting from the attached hydrogen is expected:

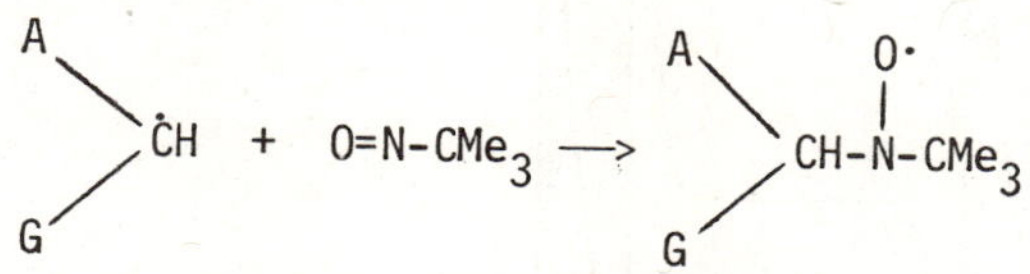

Since the nuclear spin of the hydrogen nucleus is 1/2, the triplet splitting due to ^{14}N will be further split into doublets ($2nI + 1 = 2$). The total number of lines will be 6 (3 x 2). This number is a maximum since fewer lines can exist if exact overlap occurs. The splitting patterns for these nitroxyls can be grouped into the following situations (for tabulation the symbol for N-hfsc is a_N or a^N):

Group II

1. a_N is larger than a^H_β; a^H_β less than line width.
   Note that the spectrum consists of a triplet of lines, although the larger line width clearly contains the possibility of one additional hfsc (see Fig. 2).

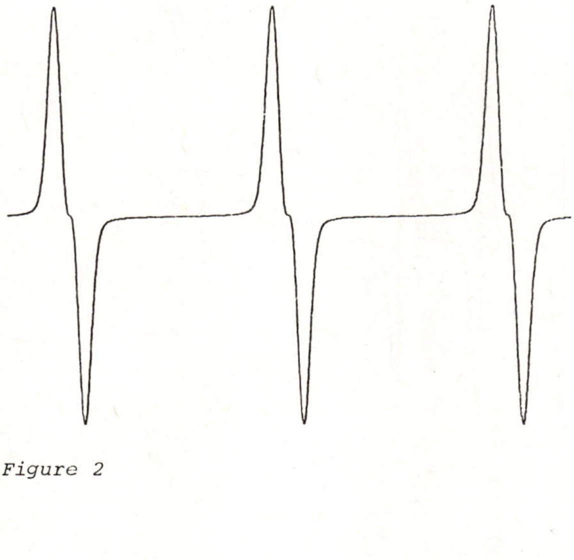

Figure 2

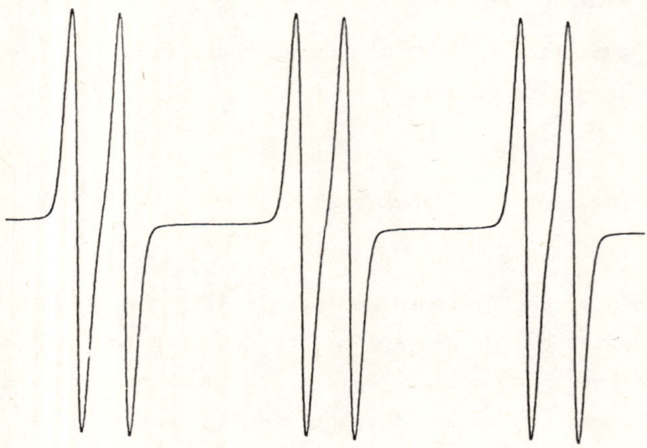

Figure 3

2. a_N is larger than a^H_β but less than $a^H_\beta = 0.5\ a_N$. This spectrum is very common and is distinctively a "triplet of doublets" spectrum (see Fig. 3).
3. $a_N = 2\ a^H_\beta$ or $a^H_\beta = 0.5\ a_N$. Note that this spectrum consists of six equally spaced intense lines (Fig. 4A). As such the spectrum can not be assigned without some knowledge of the history of the sample.

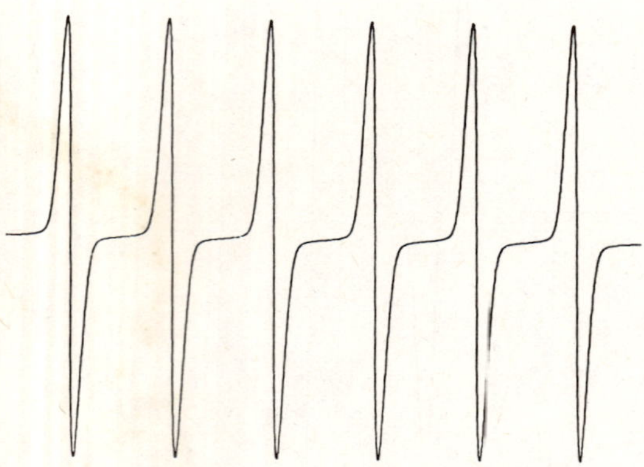

Figure 4A

For example a nuclear spin of 5/2 could give this kind of splitting pattern ($2nI + 1 = 6$, when $n = 1$). In fact solutions of Mn(II) ions in water give such a spectrum ($I = 5/2$ for ^{55}Mn), although the spacing is much larger (90-100 G). Also, the g value of the spectrum should permit assignment to a nitroxyl radical. However, even if clearly due to a nitroxyl radical, the spectrum does not give information as to whether the splitting pattern is due to $a_N = 2a_\beta^H$ or $a_\beta^H = 3a_N$. Here an unequivocal assignment could come only from isotopic substitution. Thus hydrogen-1 could be replaced by hydrogen-2 (deuterium). The spin of ^2H is 1. Thus the ^1H doublet would be replaced by a ^2H triplet with approximately 1/6.5 spacing. Figure 4B,C shows the spectra obtained when ^1H is exchanged with ^2H assuming $a_N = 2a_\beta^H$ or $a_\beta^H = 3a_N$. An alternate method would be to replace ^{14}N ($I = 1$) with ^{15}N ($I = 1/2$). Deuterium exchange would be less costly in most cases.

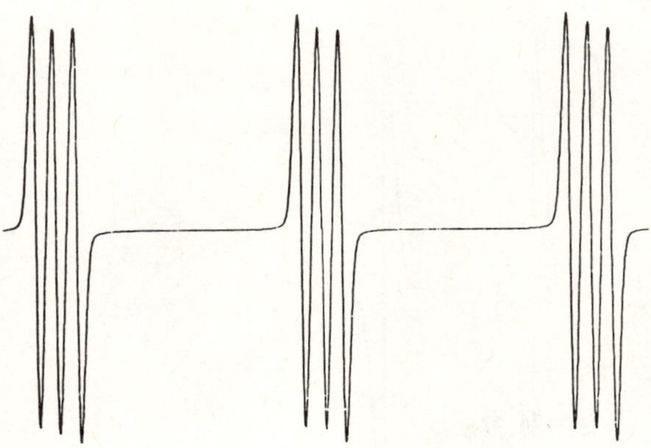

Figure 4B

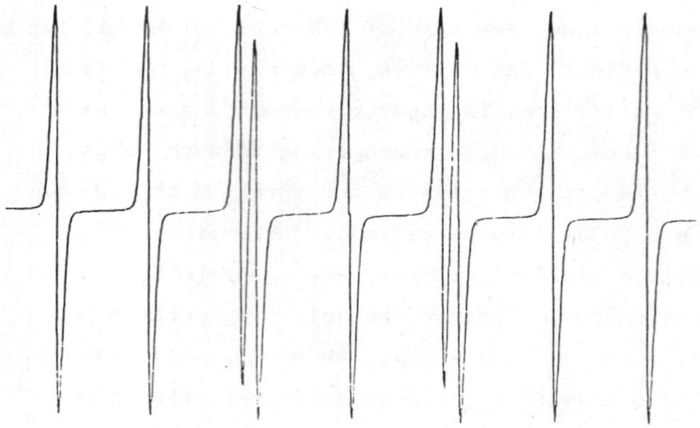

Figure 4C

4. a_N is smaller than $2a_\beta^H$ but larger than a_β^H.
This spectrum still consists of a triplet of doublets (Fig 5). The spacing between peaks 1 and 2, 3 and 4, and 5 and 6 is the β-H-hfsc. This spectrum is of interest in comparison to Figure 7 (discussed later).

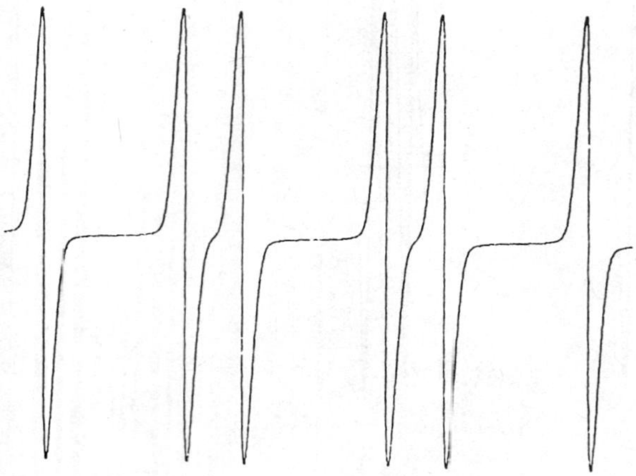

Figure 5

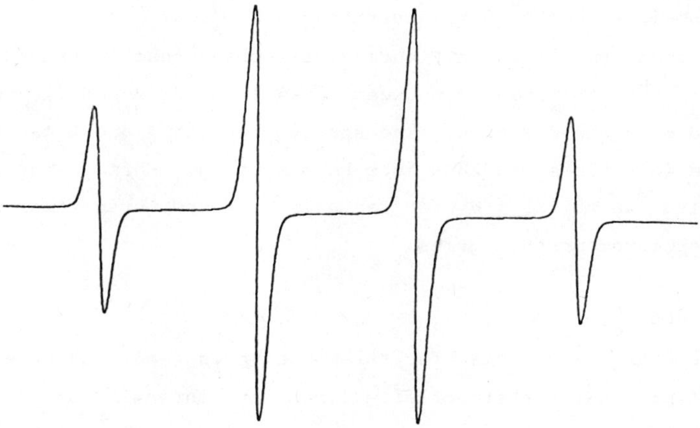

Figure 6

5. $a_N = a^H_\beta$.

Because of identical overlap, this spectrum only consists of four lines (Fig. 6). The intensity sequence is 1:2:2:1. The pattern from three equivalent nuclei with I = 1/2 would also be a quartet, but the intensity sequence would be 1:3:3:1.

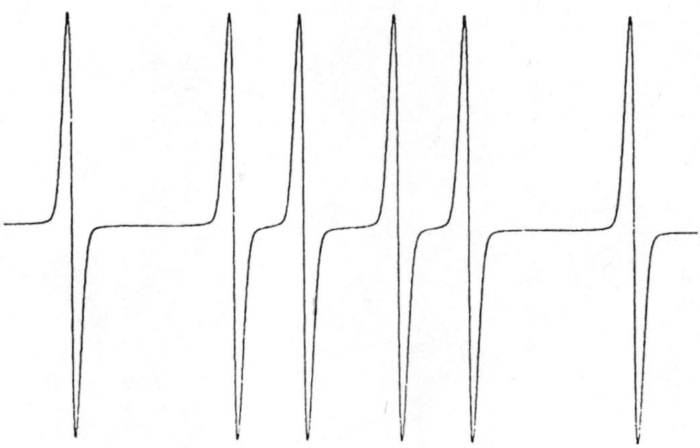

Figure 7

6. a_N, smaller than a_β^H but larger than 0.5 a_β^H.
This spectrum, shown in Figure 7, resembles that in Figure 5 except that the spacing between peaks 1 and 3, 2 and 5, and 4 and 6 is the β-H hfsc. The spacing between 1 and 2 can be found four times and thus must be due to the N-hfsc. The spacing due to the β-H hfsc can only be found three times in any spectrum in this group.

7. a_N = 0.5 a_β^H.
The spectrum with this hfsc relationship is seen in Figure 8. A peculiar pattern of five lines with intensity sequence 1:1:2:1:1 is seen.

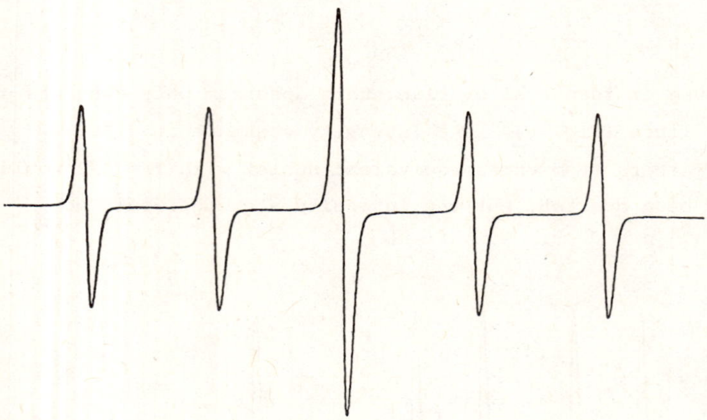

Figure 8

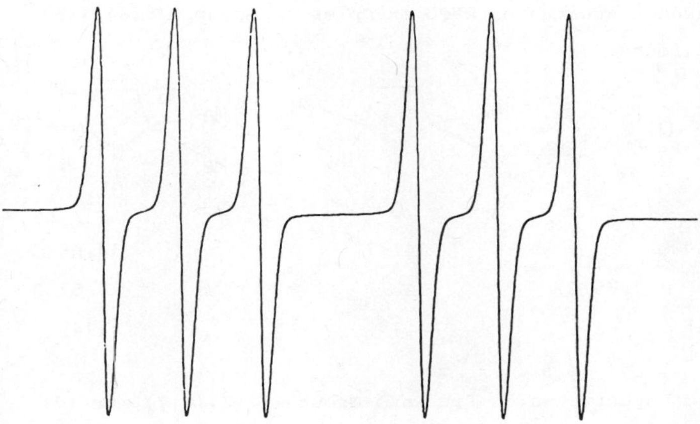

Figure 9

8. a_N is much smaller than a_β^H.

Figure 9 shows the spectrum. This pattern is distinctively a doublet of triplets.

The magnitude of a_β^H is very sensitive to the nature of the groups A and G in AGCHN(O•)CMe_3. If A and G are bulky carbon groups (even when A = G = methyl) the β-H hfsc is much smaller than a_N and a typical triplet of doublets is obtained [Group II (2)]:

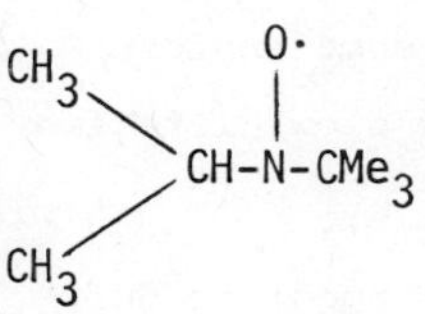

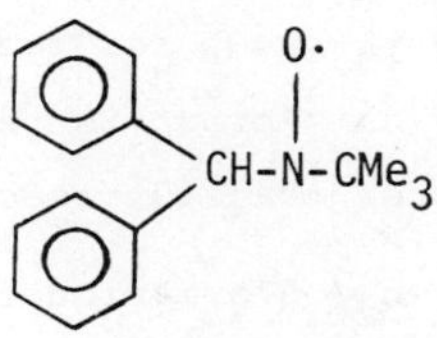

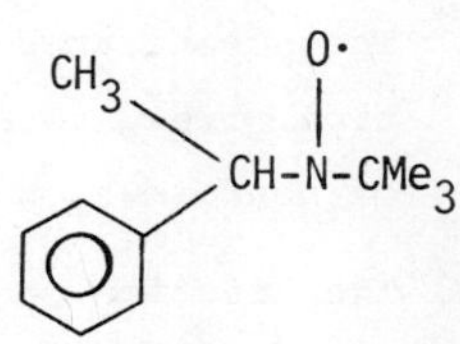

a_N =	16.8 ($CHCl_3$)	15.25 (CH_3OH)	14.82 (C_6H_6)
a_β^H =	1.8	3.14	3.47
Ref.	[16]	[17]	[18]

In some spin adduct spectra of DMPO examples of Group II (4), (5), and (6) can be found.

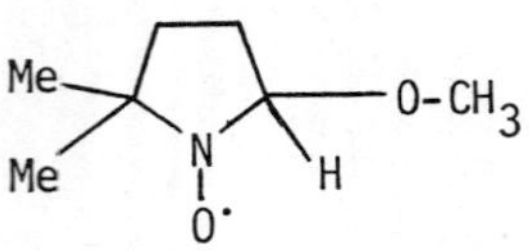

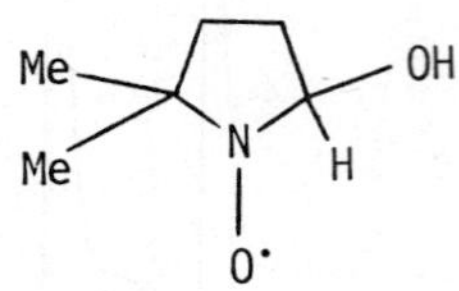

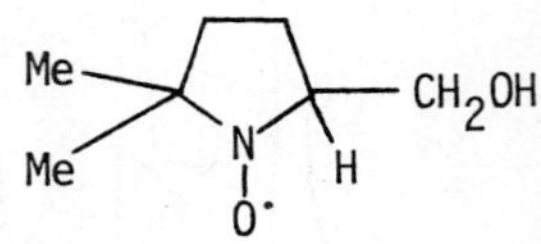

a_N =	13.58 G	14.77	14.66 G
a_β^H =	7.61 G (a_γ^H = 1.85 G)	14.77	20.67 G
Ref.	[14]	[19]	[14]

Examples of spectra which fit into Group II (3), (7), or (8) are rare if known.

C. Spin Adducts of Methylenyl Radicals

When the trapped radical is methylenyl, K-CH_2·, the ESR spectrum of the spin adduct can be relatively simple or quite complicated depending on the nature of the group attached, the temperature, and the solvent system used:

$$\text{K-CH}_2^{\cdot} + \text{O} = \text{N-CMe}_3 \longrightarrow \text{K-CH}_2\text{-}\overset{\text{O}\cdot}{\overset{|}{\text{N}}}\text{-CMe}_3$$

Three possibilities will be considered:

1. The substituted methylenyl group is "locked" in some conformation, i.e., rotation of this group is relatively slow: lifetime of conformation is greater than 10^{-5} sec.

2. The lifetime of any given conformation is of intermediate duration: 10^{-5} to -10^{-7} sec.

3. The substituted methylenyl group is rotating "freely": lifetime of conformation is less than 10^{-7} sec.

The last possibility is most readily understood in that if the methylenyl group is rotating freely the two β-hydrogens are magnetically equivalent. This produces a triplet pattern with 1:2:1 intensity sequence (actually area ratio) so that a spectrum consisting of three groups of 1:2:1 triplets will be observed.

If this possibility is taken as a subset of Group III where the spectrum is due to the hyperfine splitting from three different nuclei except that two nuclei are magnetically equivalent, the following situations can be visualized:

Group III (Two Magnetically Equivalent Hydrogens)

1. a_N is larger than a_β^H; a_β^H is less than line width.
   The spectrum in Figure 10 illustrates this situation, where the magnitude of a_β^H is not quite large enough to break out from the line width.

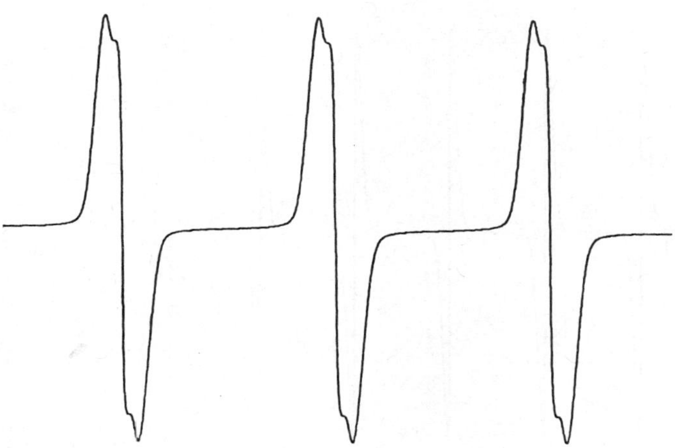

Figure 10

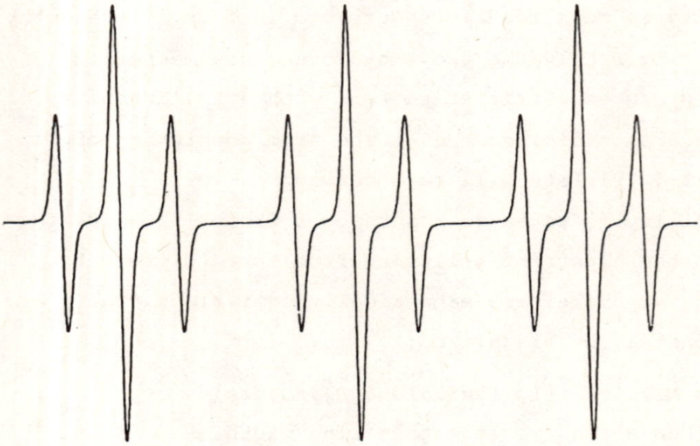

Figure 11

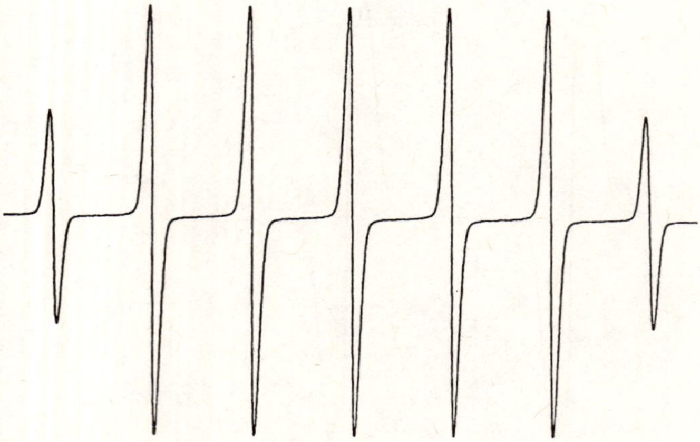

Figure 12

2. a_N is larger than a^H_β and larger than $2a^H_\beta$.

   Figure 11 is most easily recognized as the case where the splitting pattern consists of a triplet of 1:2:1 triplets. Nine lines are found in the full spectrum (3 x 3). The spacing for both the β-H and N-hfsc's can be measured six times.

3. $a_N = 2a^H_\beta$.

   In this case identical overlap produces only seven lines (Figure 12). Two lines are lost when the intensity 1 lines of the center triplet overlap with the inner intensity 1 line of the other two triplets.

4. a_N is larger than a^H_β but smaller than $2a^H_\beta$.

   In Figure 13 the spacing between peaks 1 and 2, 2 and 4, 3 and 5, 5 and 7, 6 and 8, and 8 and 9 are due to the β-H hfsc. The N-hfsc can also be measured six times: peaks 1 and 3, 2 and 5, 3 and 6, 4 and 7, 5 and 8, and 7 and 9.

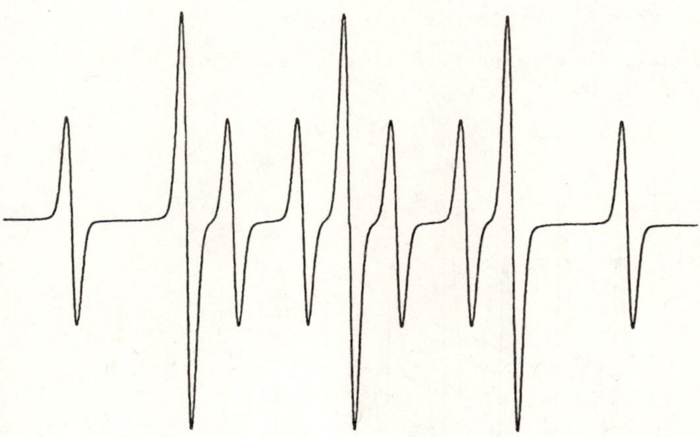

Figure 13

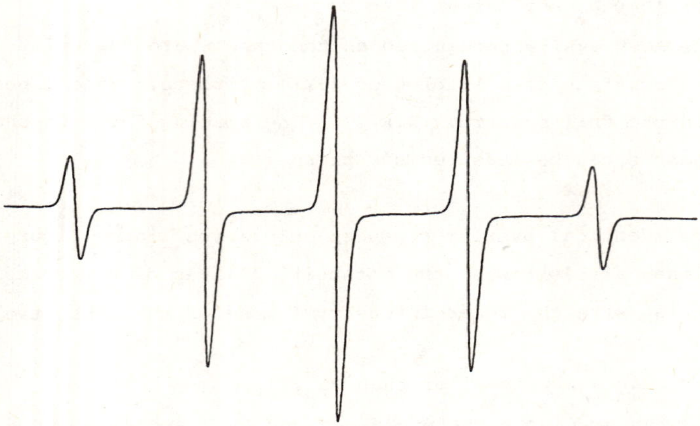

Figure 14

5. $a_N = a_\beta^H$.

   When $a_N = a_\beta^H$ four peaks are lost (Fig. 14). The intensity sequence is 1:3:4:3:1.

6. a_N is smaller than a_β^H but larger than 0.5 a_β^H.

   Figure 15 clearly shows a pattern of triplets where a_β^H is larger than a_N.

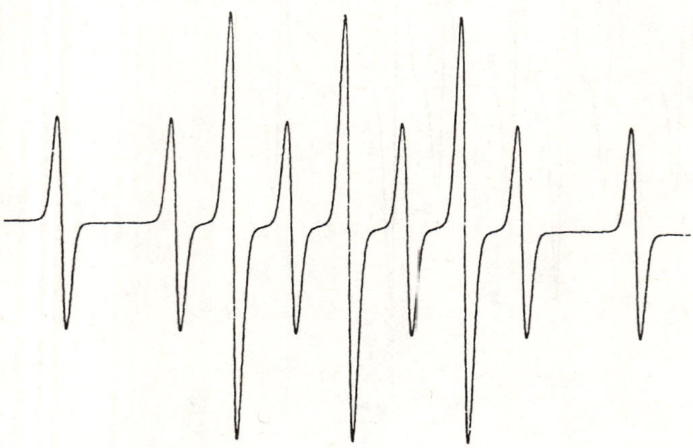

Figure 15

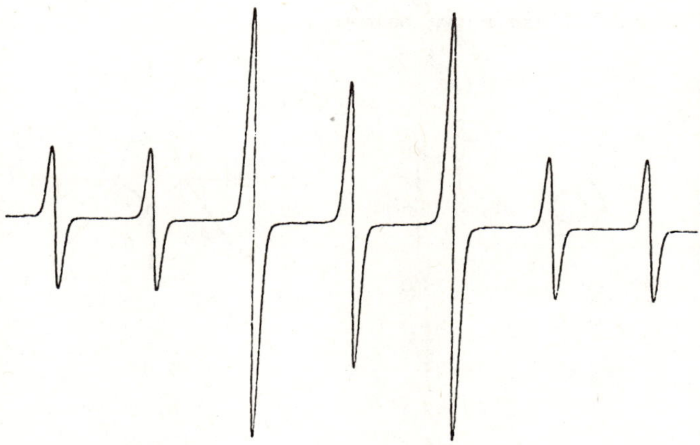

Figure 16

7. $a_N = 0.5\ a^H_\beta$, or $2a_N = a^H_\beta$.
   This situation produces a peculiar spectrum where the intensity sequence is 1:1:3:2:3:1:1 (Fig. 16).

8. a_N is much smaller than a^H_β.
   In this spectrum the large 1:2:1 triplet is split into the smaller 1:1:1 triplets (Fig. 17).

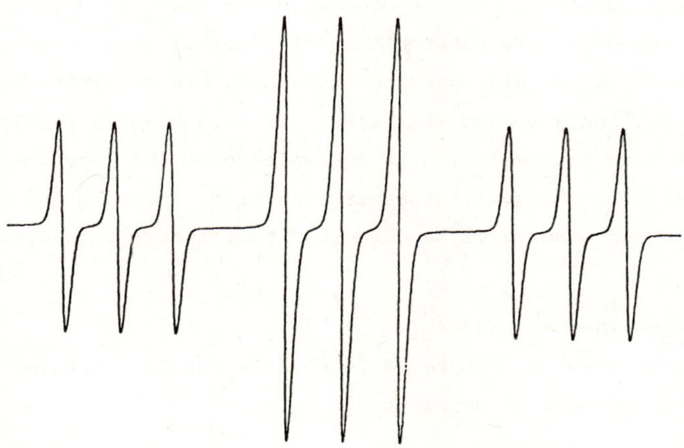

Figure 17

Some examples of Group III are shown below:

GROUP III (2)	III (3)	III (6)
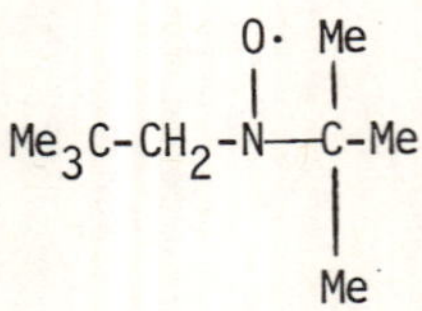	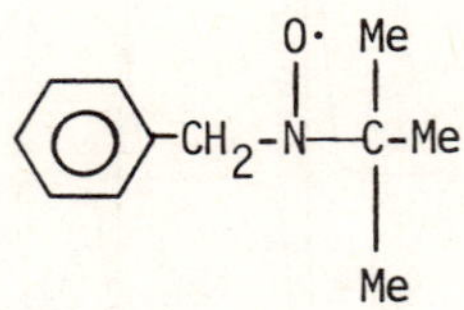	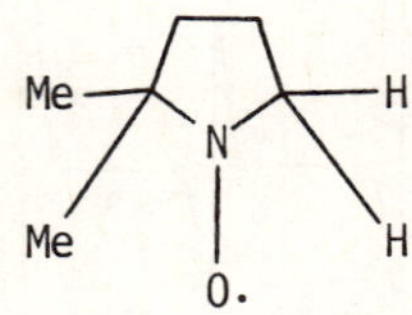
a_N = 13.44	14.84	14.43
a_β^H = 4.33	7.42	18.89
Ref. [20]	[21]	[14]

2. *Group III (Two Nonequivalent Hydrogens)*

When the methylene group is locked in some conformation (lifetime longer than 10^{-5} sec), two different β-H hfsc's are expected. The spectrum consists of a pattern of 1:1:1 triplets split into doublets of doublets. Obviously many possibilities exist if every line in the group II spectra shown earlier were to be split into an additional doublet. However, if the angular relationship between the methylene hydrogens is considered, it becomes clear that the values for the β-H hfsc can only vary between certain limits. Thus, whatever the value for the β-H hfsc when the hydrogens are magnetically equivalent (6-20 G), one β-H hfsc will always be larger and one will always be smaller than this value. The equivalence of the β-hydrogens can be removed by changes in substitution or by lowering the temperature. Only two cases will be discussed to illustrate these points.

1. a_N is larger than $a_\beta^{H^1}$ and $a_\beta^{H^2}$.

   The spectrum shown in Figure 18 is clearly due to a triplet split into doublets of doublets.

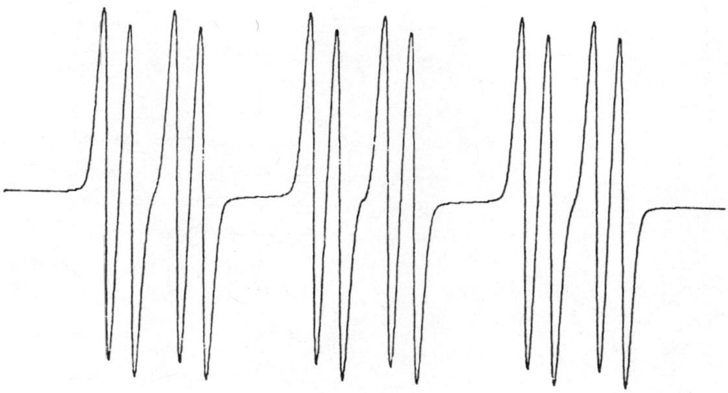

Figure 18

2. a_N is larger than $a_{\beta}^{H^1}$ but smaller than $a_{\beta}^{H^2}$.
This case is shown in Figure 19. This spectrum derives from the situation illustrated in Figure 7, where the β-H hfsc is much larger than the N-hfsc. Clearly these are complicated spectra. A detailed discussion of all possible combinations is beyond the scope of this chapter. For some examples see Ref. [22].

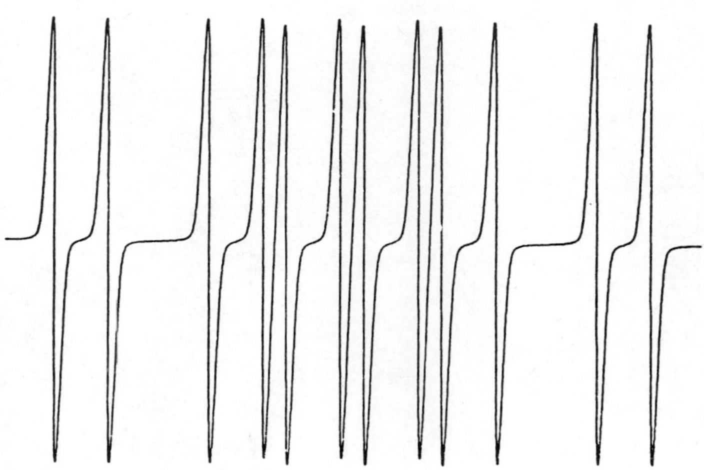

Figure 19

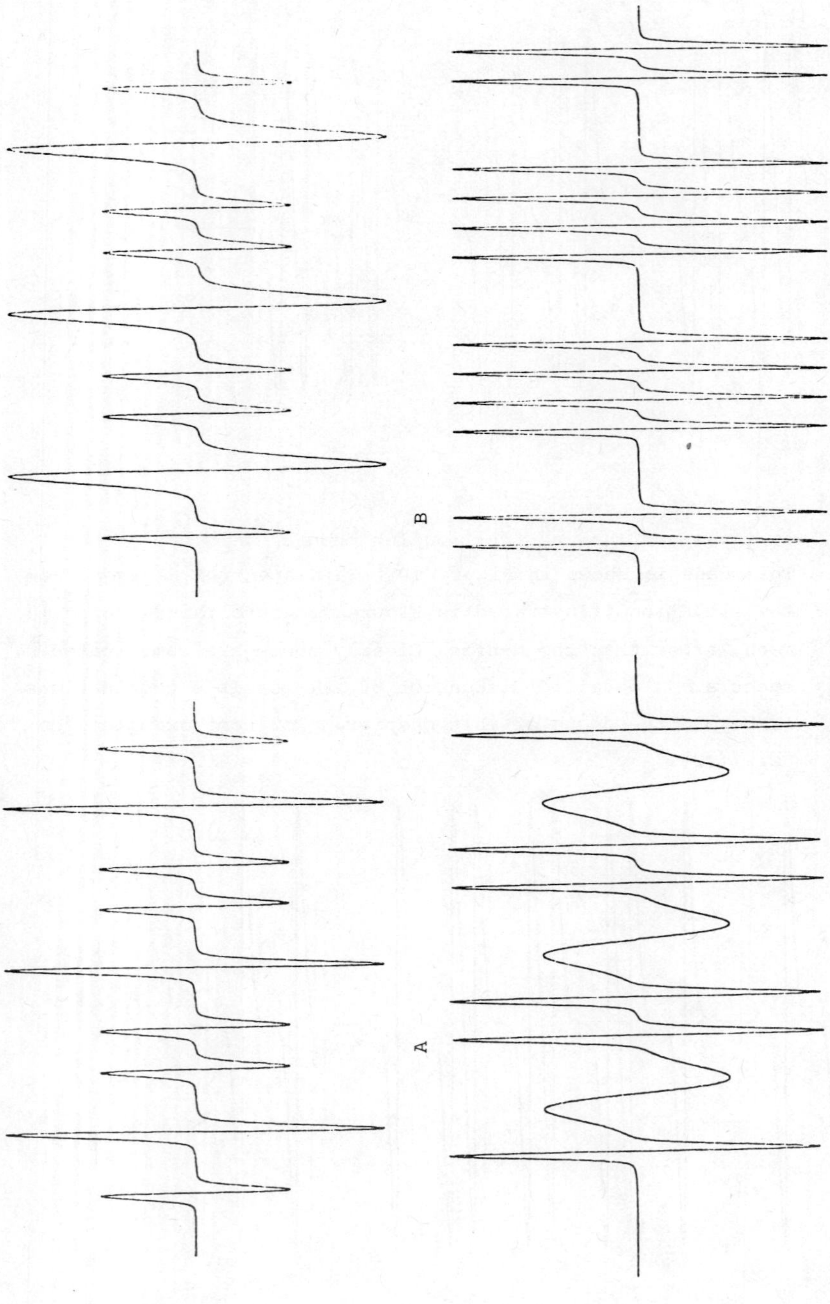

Figure 20

3. Group III (Two Almost Equivalent Hydrogens)

An intermediate case can be visualized where the lifetime of a given conformation is between 10^{-5} and 10^{-7} sec. Whereas in the fast limit a simple set of three 1:2:1 triplets might be found (e.g. see Fig. 11), when the temperature is reduced the intensity 2 lines broaden and separate into new sets of doublets (see Fig. 20 for a simulation of this process). Spectra which have sharp-broad-sharp lines are said to exhibit "line-width alternation" (LWA) which is indicative of time-dependent hfsc's.

D. Spin Adducts of Methyl Radicals

When the trapped radical is methyl, the β-hydrogens are magnetically equivalent—even at quite low temperatures:

$$CH_3\cdot + O{=}N\text{-}CMe_3 \longrightarrow CH_3\text{-}\underset{}{\overset{\displaystyle O\cdot}{\overset{|}{N}}}\text{-}CMe_3$$

The splitting patterns consist of a triplet of 1:3:3:1 quartets. Some typical situations follow:

Group IV (Three Magnetically Equivalent Hydrogens)

1. a_N is much larger than a_β^H.

   Figure 21 shows the typical triplet of quartets.

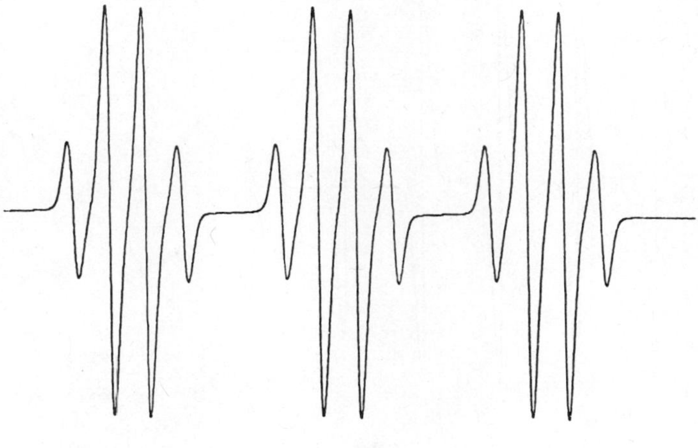

Figure 21

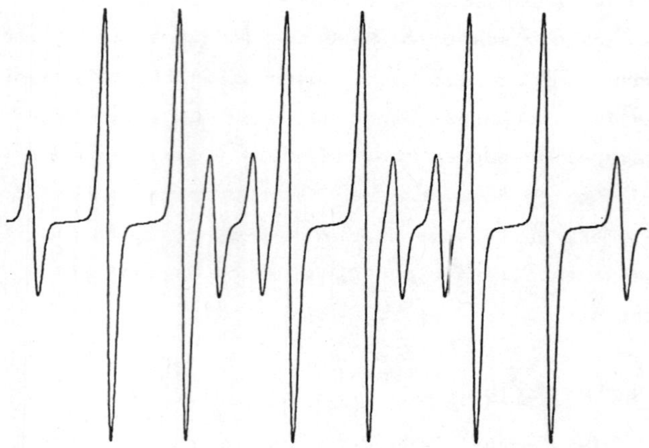

Figure 22

2. a_N is somewhat larger than a^H_β.
   A number of possibilities fall into this category. Figure 22 shows one of these.
3. $a_N = 2a^H_\beta$.
   An interesting intensity sequence is produced by this equivalence (Fig. 23). The eight peaks are in the ratio 1:3:4:4:-4:3:1.

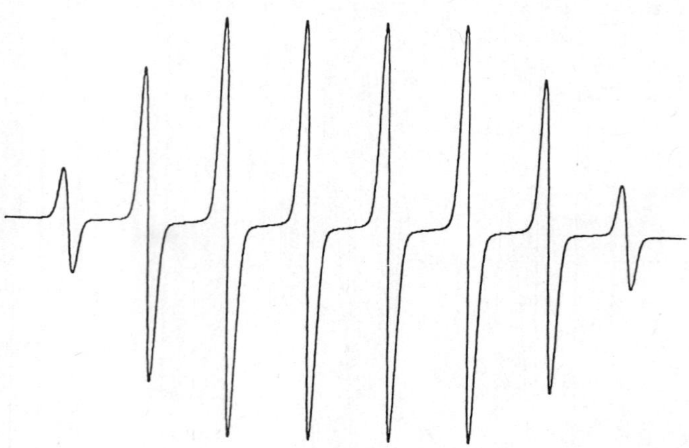

Figure 23

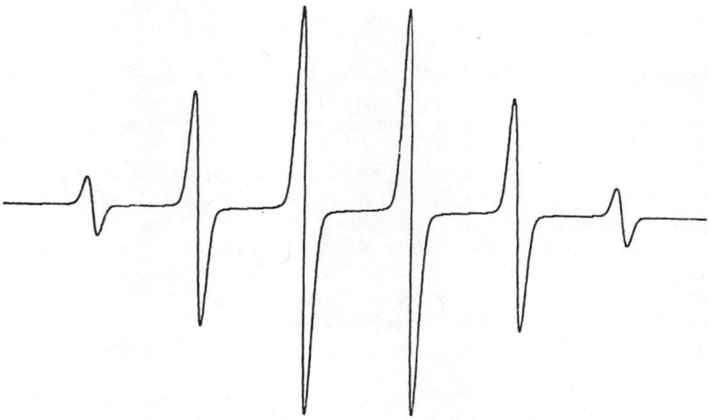

Figure 24

4. $a_N = a_\beta^H$.

Figure 24 shows the six peaks which result from this situation. The intensity ratio is 1:4:7:7:4:1.

Methyl and trifluoromethyl tert-butyl nitroxide are examples of Group IV (2) and (4) (for fluorine-19, I = 1/2):

	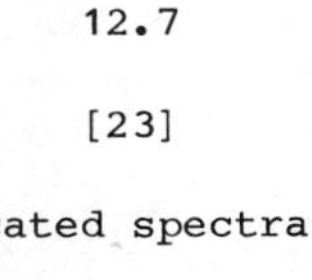	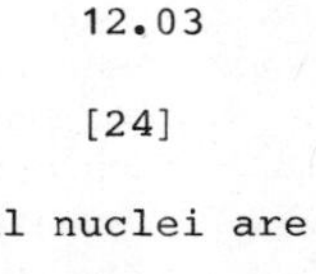
a_N	16.4 (CH_3OH)	12.03 (C_6H_6)
a_β^H	12.7	12.03
Ref.	[23]	[24]

More complicated spectra are obtained when additional nuclei are introduced, e.g., those of nitrogen and chlorine. However, even these can be derived from the basic patterns shown in Groups I, II, III, or IV.

REFERENCES

1. J. E. Wertz and J. R. Bolton, *Electron Spin Resonance*, McGraw-Hill, New York (1972).
2. E. G. Janzen, in *Free Radicals in Biology* (W. A. Pryor, Ed.), Vol. IV, 1980, p. 115
3. L. J. Berliner (Ed.), *Spin Labelling Theory and Applications*. Academic Press, New York, 1976; A. R. Forrester, J. M. Hay, and R. H. Thomson, *Organic Chemistry of Stable Free Radicals*, Academic Press, New York, 1968; E. G. Rozantsev, *Free Nitroxyl Radicals* (transl. from Russian by B. J. Hazzard). Plenum Press, New York, 1970; J. F. Keana, *Chem. Rev. 78*:37 (1978).
4. E. G. Janzen, *Accounts of Chem. Res. 4*:31 (1971).
5. C. Lagercrantz and K. Torssell, *Acta Chem. Scand. 22*:1935 (1968).
6. J. J. M. C. De Groot, G. J. Garssen, J. F. G. Vliegenthart, and J. Boldingh, *Biochim. Biophys. Acta 326*:279 (1973).
7. S. Terabe and R. Konaka, *J. Am. Chem. Soc. 91*:5655 (1969).
8. T. Hosogai, N. Inamoto, and R. Okazaki, *J. Chem. Soc. [Org.]* p. 3399 (1971); S. Terabe and R. Konaka, *J. Am. Chem. Soc. 93*:4306 (1971); *J. Chem. Soc.*, Perkin II p. 369 (1973).
9. S. Terabe, K. Kurama, and R. Konaka, *J. Chem. Soc. Perkin II* p. 1252 (1973).
10. E. G. Janzen, C. A. Evans, and Y. Nishi, *J. Am. Chem. Soc. 94*:8236 (1972).
11. E. G. Janzen, R. L. Dudley, and R. V. Shetty, *J. Am. Chem. Soc. 101*:243 (1979).
12. E. G. Janzen, Y. Y. Wang, and R. V. Shetty, *J. Am. Chem. Soc. 100*:2923 (1978).
13. E. G. Janzen and R. V. Shetty, *Tetrahedron Lett.* p. 3229 (1979).
14. E. G. Janzen and J. I.-P. Liu, *J. Mag. Res. 9*:510 (1973).
15. R. L. Eggers and W. R. Goldman, unpublished work.
16. G. Chapelet-Letourneux, H. Lemaire, R. Lenk, M. A. Marechal, and A. Rassat, *Bull. Soc. Chim. Fr.* p. 3963 (1968).
17. E. G. Janzen, D. E. Nutter, Jr., and C. A. Evans, *J. Phys. Chem. 79*:1983 (1975).
18. E. G. Janzen and I. G. Lopp, *J. Phys. Chem. 76*:2056 (1972).
19. E. G. Janzen, D. E. Nutter, Jr., E. R. Davis, B. J. Blackburn, J. L. Poyer, and P. B. McCay, *Can. J. Chem. 56*:2237 (1978).
20. E. G. Janzen and I. G. Lopp, *J. Mag. Res. 7*:107 (1972).
21. E. G. Janzen, T. Kasai, and K. Kuwata, *Bull. Chem. Soc. Japan 46*:2061 (1973).
22. C. Lagercrantz and M. Setaka, *Acta Chem. Scand. B28*:619 (1974).
23. G. Chapelet-Letourneux, H. Lemaire, and A. Rassat, *Bull. Soc. Chim. Fr.* p. 3283 (1965).
24. K. J. Klabunde, *J. Am. Chem. Soc. 92*:2427 (1970).

12

A POSSIBLE ROLE OF FREE RADICALS IN THE MUTAGENICITY OF N-HYDROXY-2-ACETYLAMINOFLUORENE

Jack A. Hinson,* Larry S. Andrews,† and James R. Gillette

National Heart, Lung, and Blood Institute
National Institutes of Health
Bethesda, Maryland

I. INTRODUCTION

Many of the present concepts of the role of metabolic activation in chemical carcinogenesis and mutagenesis were derived from studies on the aromatic amide 2-acetylaminofluorene (2-AAF) [1-3]. In rats this compound produces tumors in the liver, mammary gland, earduct, and small intestine [1]. Its carcinogenic activity, however, is apparently mediated through the formation of a metabolite, N-hydroxy-2-acetylaminofluorene (N-hydroxy-2-AAF) [4,5], which is formed by a liver microsomal cytochrome P_{450}-mixed function oxidase [6]. Even though N-hydroxy-2-AAF may be a key intermediate in the carcinogenesis of 2-AAF, there is evidence that it is metabolized further to the ultimate carcinogenic form or forms. The mechanisms of the conversion to the ultimate carcinogens, however, is not completely understood. Work by E. C. Miller, J. A. Miller, and several co-workers [7-11], as well as other investigators [12,13], has suggested that the formation of N-O-esters of N-hydroxy-2-AAF may be important. They showed that the N-O-acetate ester of N-hydroxy-2-AAF (N-acetoxy-2-AAF) as well as other synthetic esters are very potent in producing tumors at the site of injection in rats [7]. Moreover, they showed

*Present affiliation: National Research Center for Toxicological Research, Jefferson, Arkansas
†Present affiliation: ARCO Chemical, New Town Square, Pennsylvania

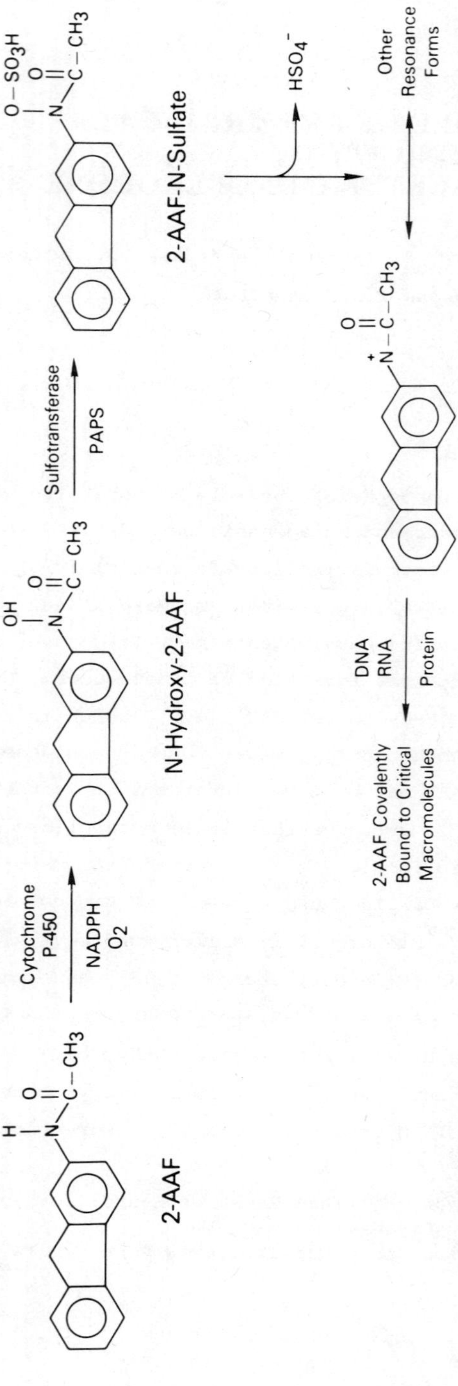

Figure 1 Postulated mechanism for conversion of 2-AAF to a reactive metabolite.

that the N-acetoxy-2-AAF breaks down to a reactive metabolite which covalently binds to guanine, methionine, and other nucleophilic sites of macromolecules. The reactive metabolite was postulated to be a nitrenium or amidonium ion formed by heterolytic cleavage of the N-O-acetate bond. It was also postulated that the reaction of this metabolite of N-acetoxy-2-AAF with target macromolecules resulted in altered nucleic acids and carcinogenesis [7].

Although N-acetoxy-2-AAF is very carcinogenic, its role in the carcinogenesis of N-hydroxy-2-AAF is questionable since it has not been shown to be a significant metabolite of N-hydroxy-2-AAF. Evidence has been presented, however, that the N-O-sulfate ester of N-hydroxy-2-AAF (2-AAF-N-sulfate) may be an important metabolite in the hepatocarcinogenesis of N-hydroxy-2-AAF and covalent binding of the 2-AAF moiety to nucleophilic sites on protein, RNA, or DNA [1-3,7,9-12]. For example, the hepatocarcinogensis of N-hydroxy-2-AAF in a number of different species correlated directly with in vitro liver sulfotransferase activity when N-hydroxy-2-AAF was used as a substrate, and the conversion of N-hydroxy-2-AAF to a reactive metabolite in vivo as measured by protein covalent binding (Table 1) [11]. It was thus postulated that 2-AAF-N-sulfate spontaneously decomposes to a nitrenium ion which reacts with critical sites in

Table 1 Comparison of Hepatic N-Hydroxy-2-AAF Sulfotransferase Activity and In Vivo Formation of Hepatic Protein-Bound Methionyl-2-AAF Derivatives in Species Susceptible and Resistant to N-Hydroxy-2-AAF Hepatocarcinogenesis

Species and sex	Liver sulfotransferase in vitro (μg 0.04 ml supernatant/30 min)		Liver protein in vitro (μg/5 g liver)		Hepatocarcinogenicity of N-hydroxy-2-AAF
Rat (M)	23 ± 2.0	(13)	38 ± 9	(17)	+++++
Rat (F)	4 ± 0.5	(3)	4 ± 1	(5)	+
Mouse (M)	<0.5	(3)	<1.5	(3)	+
Hamster (M)	<0.5	(3)	<0.3	(3)	+
Guinea pig (M)	<0.3	(3)	<0.3	(3)	-
Rabbit (M)	2 ± 0.5	(3)	<0.5	(3)	-

Source: Data from Ref. 11. The authors would like to thank Dr. E. C. Miller for allowing us to present these data.

hepatic macromolecules, including certain loci on DNA [1-3,7,11]. This postulated mechanism for conversion of 2-AAF to this reactive metabolite is presented in Figure 1.

Mechanisms other than sulfation of N-hydroxy-2-AAF, however, are apparently operative in 2-AAF-induced tumors in the mammary gland and Zymbal's gland (sebaceous gland of the auditory canal) since the levels of sulfotransferase present in both of these tissues are negligible [14]. A number of other mechanisms for activation of N-hydroxy-2-AAF to reactive metabolites have thus been suggested. One mechanism is the metabolism of N-hydroxy-2-AAF by the enzyme acyltransferase. This enzyme, which is present in several different tissues, catalyzes the transfer of the acetyl group from nitrogen to oxygen to form N-acetoxy-2-aminofluorene by an intermolecular mechanism. This leads to covalent binding of aminofluorene residues to nucleophilic sites on RNA and DNA—presumably by a heterolytic cleavage of the N-O-bond, forming the nitrenium ion of 2-aminofluorene [12,15-17]. Also glucuronide and phosphate conjugates of N-hydroxy-2-AAF have been suggested as being potentially important intermediates in the carcinogenesis of N-hydroxy-2-AAF [12,18,19].

II. CORRELATION BETWEEN CARCINOGENS AND MUTAGENS

Recently, Ames and co-workers developed a simple and sensitive assay to study the mutagenic effects of carcinogens in specially derived strains of *Salmonella typhimurium* that require histidine for growth [20-22]. In this assay bacteria and mutagen are added to top agar, mixed, and poured onto minimal agar plates. When the test compound is not a mutagen per se, enzyme preparations, usually 9,000 x *g* of rat liver supernatant (S-9), are frequently included to convert the chemical to an ultimate mutagen. The mutagenicity of the compound is then determined by counting the number of bacteria converted to prototrophy (a nonhistidine requirement) within 24-48 hr. This assay has yielded an excellent correlation between mutagenicity and carcinogenicity. Thus far it has been shown that approximately 90% of the known carcinogens are mutagens and 90% of the mutagens are known carcinogens [22].

The aforementioned bacterial mutagenesis assay has been used extensively by a number of laboratories to determine the mutagenic and presumably carcinogenic potential of metabolites of various compounds [23,24], including those of 2-AAF [21,25].

III. MUTAGENICITY OF DEACETYLATED METABOLITES OF 2-ACETYLAMINOFLUORENE

Durston and Ames found that 2-AAF was mutagenic in this assay in the presence but not in the absence of S-9. Moreover, N-hydroxy-2-AAF alone was weakly mutagenic, but the mutagenicity was greatly increased by the addition of S-9. By contrast both N-hydroxy-2-aminofluorene and 2-nitrosofluorene were very mutagenic either in the presence or absence of S-9 [25]. Thus, the bacterial assay suggested that the deacetylated 2-AAF metabolites, 2-nitrosofluorene and N-hydroxy-2-aminofluorene might be more important in the carcinogenicity of 2-AAF than had been previously believed.

It seemed possible to us that the mutagenicity of 2-nitrosofluorene and N-hydroxy-2-aminofluorene is mediated by the same intermediate since these two substances are readily interconverted in the bacteria by reduction and oxidation, respectively. Also a one-electron reduction of 2-nitrosofluorene or a one-electron oxidation of N-hydroxy-2-aminofluorene would yield a N-hydroxy-2-aminofluorene radical (a nitroxide radical). Thus any one or all of these three species could be mutagenic.

The mutagenicity of 2-nitrosofluorene and N-hydroxy-2-aminofluorene was examined in a bacterial tester strain (TA 100) which contains nitro reductase and a modification of the same strain which lacks nitro reductase (TA 100 FR). In strain TA 100 both 2-nitrosofluorene and N-hydroxy-2-aminofluorene were mutagenic as expected. In strain TA 100 FR, however, 2-nitrosofluorene was not mutagenic but N-hydroxy-2-aminofluorene was as mutagenic as it was in strain TA 100 [26]. Thus, the mutagenesis of 2-nitrosofluorene is apparently mediated by reduction by bacterial nitro reductases to N-hydroxy-2-aminofluorene, a known arylating agent. It is still possible, however, that a N-hydroxy-2-aminofluorene radical may be formed during autoxidation of N-hydroxy-2-aminofluorene and that this radical may be mutagenic. At present we cannot distinguish between these mechanisms.

IV. MUTAGENICITY OF 2-ACETYLAMINOFLUORENE-N-SULFATE

A. Role in Sulfation in Covlaent Binding and Mutagenesis of N-Hydroxy-2-acetylaminofluorene

Since De Baun et al. [11] showed that a reactive metabolite was formed after sulfation of N-hydroxy-2-AAF and its formation directly correlated with hepatocarcinogenicity of N-hydroxy-2-AAF, our laboratory examined the role of sulfation and deacetylation in the mutagenicity of N-hydroxy-2-AAF. In this study the 100,000 x *g* supernatant fraction of rat liver was chromatographed on Sephadex G-25 to remove low-molecular-weight molecules and used as a source of the enzyme required for the synthesis of 3'-phosphoadenosine-5'-phosphosulfate (PAPS) and the transfer of the sulfate to N-hydroxy-2-AAF. On addition of p-nitrophenylsulfate (pNPS) and a catalytic amount of adenosine-3',5'-diphosphate (PAP), the sulfate from pNPS is transferred enzymatically to PAP producing PAPS and p-nitrophenol [27,28]. In the presence of N-hydroxy-2-AAF, the sulfate of PAPS is then enzymatically transferred to form 2-AAF-N-sulfate and PAP as shown in Figure 2. With this assay the rate of sulfation of N-hydroxy-2-AAF can be followed by measuring the rate of formation of p-nitrophenol spectrophotometrically at 405 nm. Moreover, the covalent binding between protein and radiolabel originating from N-hydroxy-2-AAF was measured as an indirect assay of the formation of the reactive metabolite of N-hydroxy-2-AAF [28,29].

1. p - Nitrophenylsulfate + PAP ⇌ PAPS + p - Nitrophenol
2. N - Hydroxy - 2 - AAF + PAPS ⟶ PAP + 2 - AAF - N - Sulfate
3. 2 - AAF - N - Sulfate ⟶ Covalent Binding

Figure 2 Reactions occurring under the conditions of the sulfotransferase assay. (From Ref. 28.)

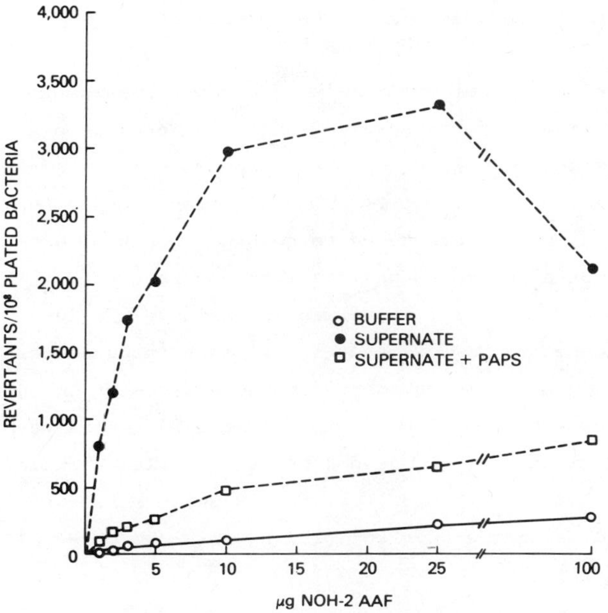

Figure 3 Mutagenesis of N-hydroxy-2-AAF. (Data from Ref. 31.)

The mutagenicity of the enzymatically generated 2-AAF-N-sulfate in the bacterial mutagenesis assay was studied with the tester strain TA 1538. N-hydroxy-2-AAF in buffer alone was only weakly mutagenic, as has been previously reported [25] (Fig. 3). When the Sephadex G-25 treated 100,000 x *g* supernatant of rat liver was added without the cofactors required for PAPS formation, however, a potent mutagen was formed (Fig. 3). Other investigators have attributed the increase in mutagenesis of N-hydroxy-2-AAF in the presence of 100,000 x *g* rat liver supernatant to the formation of N-hydroxy-2-aminofluorene by deacetylases (amidases) and an acyltransferase [30-35]. Surprisingly, the addition of PAP and pNPS greatly decreased the mutagenesis of N-hydroxy-2-AAF (Fig. 3). This decrease in mutagenesis under sulfating conditions was unexpected since it appeared to be inconsistent with the view that this metabolic transformation is important in the hepatocarcinogenicity of N-hydroxy-2-AAF [30,31].

B. Effect of Ascorbate on Mutagenesis and Metabolism of Generated 2-Acetylaminofluorene-N-Sulfate

In an effort to determine whether the remaining mutagenesis was due to the reactive metabolite or to N-hydroxy-2-aminofluorene, we examined the effect of various compounds on the mutagenesis of 2-AAF-N-sulfate in the bacterial assay (Table 2). The addition of DNA, RNA, guanine, and methionine, which were known to combine with the reactive metabolite, did not alter mutagenesis. When ascorbic acid, NADPH, or NADH were added, however, there was approximately a 10-fold increase in mutagenesis (Table 2 and Fig. 4) even though ascorbic acid itself was only slightly mutagenic in the absence of 100,000 x *g* supernatant enzymes under these conditions (Table 2). This was completely unexpected since ascorbic acid decreases covalent binding of the reactive metabolite to protein (Fig. 5) [29,31]. Moreover, ascorbic acid also decreased the covalent binding of N-hydroxy-2-AAF to DNA (peak B in Fig. 6) [36].

Table 2 Effects of Various Agents on the Mutagenesis of N-Hydroxy-2-AAF

	Buffer	Supernatant fraction	Supernatant fraction + PAPS
No addition	65	1750	252
Guanine (50 μM)	91	1740	196
Guanosine (50 μM)	135	1870	298
Guanosine-5'phospate (1.8 mM)	78	1800	285
RNA (0.3 mg/ml)	95	1610	275
DNA (0.2 mg/ml)	130	2040	225
Glutathione (1.0 mM)	85	1700	280
Methionine (10 mM)	92	—	280
Cysteamine (1 mM)	50	—	277
D,L-α-Tocopherol acetate (100 μg/ml)	—	—	260
2-Aminofluorene (20 μM)	73	1870	255
2-Acetylaminofluorene (200 μM)	70	1800	260
NADPH (1 mM)	—	2600	2800
NADH (1 mM)	—	2380	2490
Ascorbic acid (1.9 mM)	92	2896	2910

Source: Data from Ref. 31. Reprinted with permission from *Biochem. Pharmacol. 27*: L. S. Andrews, J. A. Hinson, and J. R. Gillette, 1978, Pergamon Press, Ltd.

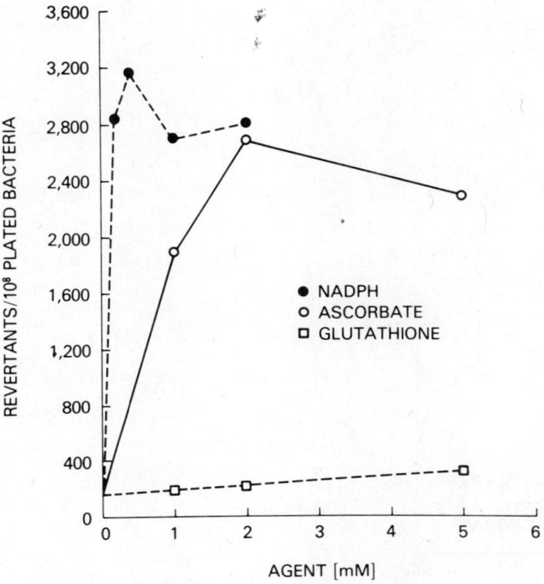

Figure 4 Effect of NADPH, ascorbate, and glutathione on mutagenesis of N-hydroxy-2-AAF under sulfating conditions. (Data from Ref. 31.)

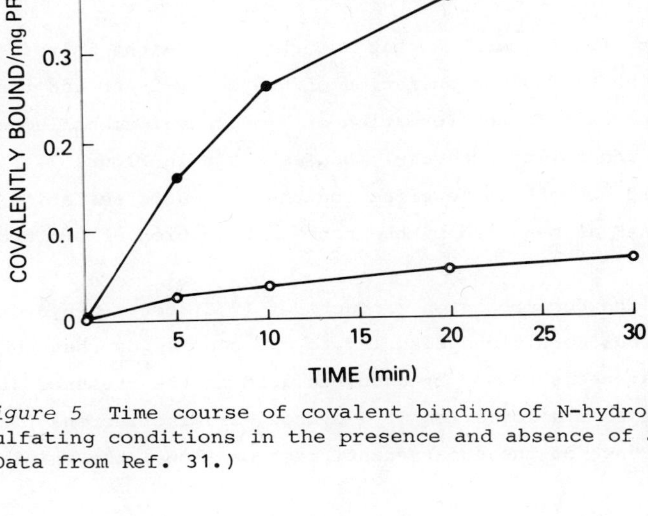

Figure 5 Time course of covalent binding of N-hydroxy-2-AAF under sulfating conditions in the presence and absence of ascorbic acid. (Data from Ref. 31.)

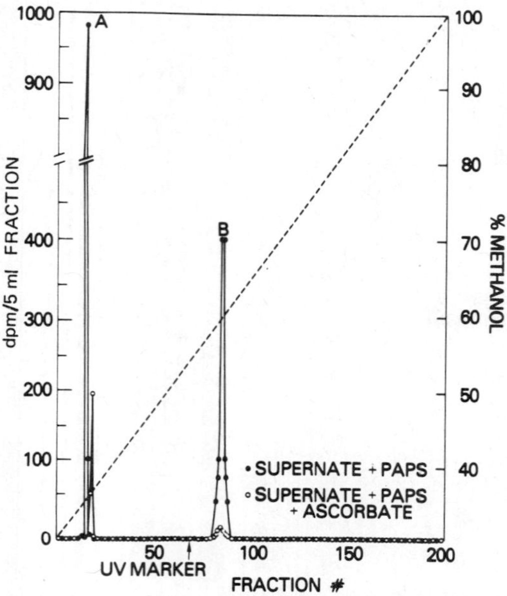

Figure 6 Covalent binding of [^{14}C]acetyl-N-hydroxy-2-AAF to calf thymus DNA. [^{14}C]Acetyl-N-hydroxy-2-AAF was incubated under sulfating conditions with calf thymus DNA plus and minus ascorbate. The DNA was isolated and hydrolyzed. The hydrolyzate was applied to a Sephadex LH-20 column and the radioactivity eluted with methanol. Peak B reflects covalent adduct formation with DNA, whereas peak A may represent radiolabel covalently bound to protein. (Data from Ref. 36.)

At first it seemed possible that ascorbic acid might increase mutagenesis by inhibiting the sulfation of N-hydroxy-2-AAF and thereby lead to an increase in the formation of N-hydroxy-2-aminofluorene by deacetylases and acyltransferase. But as shown in Figure 7, ascorbic acid had no detectable effect on the enzymatic sulfation of N-hydroxy-2-AAF as measured by the rate of formation of p-nitrophenol.

A study of the decomposition products of [^{14}C]acetyl-N-hydroxy-2-AAF under various conditions also led to the conclusion that the increase in mutagenesis caused by ascorbic acid in the presence of PAPS was not due to the formation of N-hydroxy-2-aminofluorene (Table 3). In the presence of the supernatant fraction alone, the primary

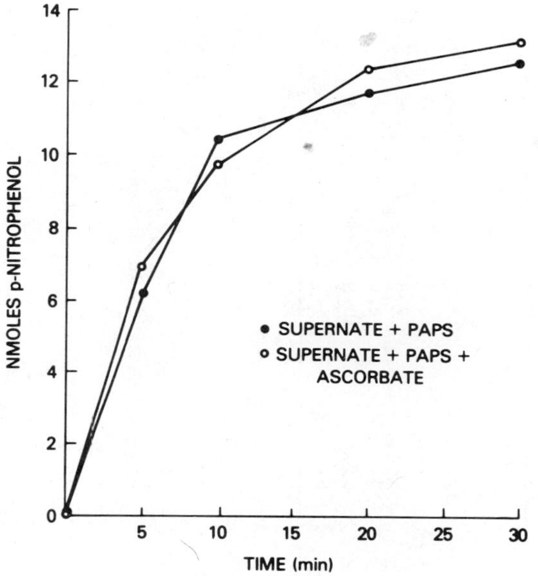

Figure 7 Effect of ascorbic acid on sulfation of N-hydroxy-2-AAF studied by using the pNPS-PAP coupled assay. (Data from Ref. 31.)

radiolabeled metabolites were 2-AAF and acetic acid. Acetic acid presumably was formed by the hydrolysis of N-hydroxy-2-aminofluroene by deacetylases and acyltransferase, and thus its formation also represents the maximum amount of N-hydroxy-2-aminofluorene that could be formed in the system. With supernatant fraction plus a PAPS-generating system a number of different metabolites were isolated but no acetic acid was found, indicating that little, if any, N-hydroxy-2-aminofluorene was formed. Among the metabolites formed were 2-AAF, a radiolabeled product covalently bound to protein, a dimer of 2-AAF, a phosphate conjugate of 7-hydroxy-2-AAF, and a metabolite having molecular weight corresponding to a p-nitrophenol adduct of 2-AAF. When ascorbic acid was incubated with radiolabeled N-hydroxy-2-AAF, the supernatant fraction and a PAPS generating system, there was approximately a sixfold increase in 2-AAF formation at the expense of the formation of the dimer, the p-nitrophenol adduct, the phosphate conjugate, and covalent binding to

Table 3 Metabolism of [^{14}C]Acetyl-N-hydroxy-2-AAF Under Sulfating Conditions[a]

Compound	Buffer	Buffer + ascorbate	Supernatant fraction	Supernatant fraction + ascorbate	Supernatant fraction + PAPS	Supernatant fraction + PAPS + ascorbate
NOH-2-AAF	100.0	96.6	86.0	73.3	24.1	23.7
2-AAF	0.0	3.4	4.5	12.5	11.2	62.4
Dimer	0.0	0.0	0.0	0.0	22.1	1.4
p-Nitrophenol adduct	0.0	0.0	0.0	0.0	7.8	0.8
Acetic acid	0.0	0.0	9.4	14.1	0.0	0.0
Phosphate adducts	0.0	0.0	0.0	0.0	30.8	11.0
Covalently bound material	0.0	0.0	0.1	0.1	4.0	0.7

[a]The data are expressed as percentages.

Source: Data from Ref. 31. Reprinted with permission from *Biochem. Pharmacol. 27*: L. S. Andrews, J. A. Hinson, and J. R. Gillette, 1978, Pergamon Press, Ltd.

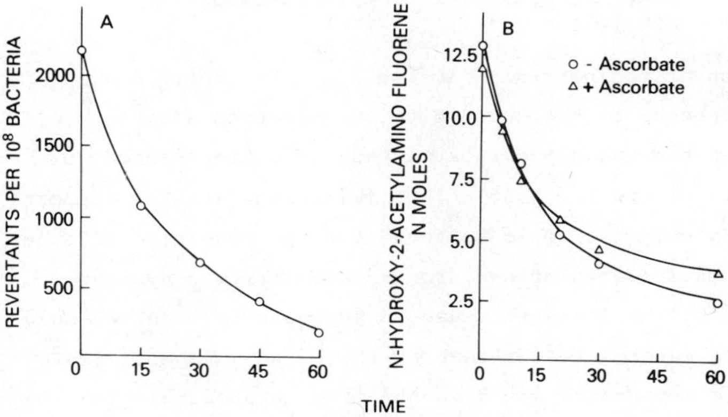

Figure 8 A time course of N-hydroxy-2-AAF mutagenesis under sulfating conditions plus ascorbic acid. Bacteria were added at the indicated time. (Data from Ref. 31.)

protein, but there was no evidence that N-hydroxy-2-aminofluorene was formed, since acetic acid was not found as a metabolite. Thus, ascorbic acid apparently reduced the reactive metabolite to 2-AAF, which is nonmutagenic, and thereby inhibited its reaction with various nucleophiles in the incubation mixture. The effects of NADPH on the reactive metabolite appeared to be similar to that of ascorbic acid since it also blocked covalent binding to protein with a concomitant increase in 2-AAF formation [37].

The potential role of an unstable intermediate that did not lead to covalent binding was subsequently investigated by adding the bacteria at various times to incubation mixtures containing sulfotransferase, the PAPS-generating system, N-hydroxy-2-AAF, and ascorbic acid. These data are presented in Figure 8A with a time course of disappearance of N-hydroxy-2-AAF from the incubation mixtures (Fig. 8B). The similarity of the curves indicate that the mutagenic intermediate did not accumulate in the incubation mixture and thus was short-lived.

C. Postulation of Free Radicals of 2-Acetylaminofluorene as a Mutagenic Metabolite

Since a reduction of the nitrenium ion to 2-AAF by ascorbic acid presumably would occur by two successive one-electron additions, it seems likely that the mutagenic intermediate is a free radical of 2-AAF as shown in Figure 9 [31,36]. The following findings support this view: (a) No evidence could be found for the formation of other known mutagenic metabolites of N-hydroxy-2-AAF, i.e., N-hydroxy-2-aminofluorene. (b) The large increase in mutagenesis occurred only when the reactive metabolite was reduced to the nonmutagen, 2-AAF. (c) The mutagenic metabolite was a short-lived metabolite. (d) The formation of the mutagenic metabolite was directly related to the sulfation of N-hydroxy-2-AAF. A similar mechanism is envisioned for the large increase in mutagenesis when the reactive metabolite is generated in the presence of NADPH (see Table 2 and Fig. 4) since NADPH blocked covalent binding to protein and reduced the reactive metabolite to 2-AAF. Whether the reduction of the nitrenium ion to the mutagenic metabolite by NADPH requires an enzyme, however, remains to be determined. The amount of the 2-AAF free radical formed in liver by either ascorbic acid or NADPH would depend on the rate of formation of the 2-AAF-N-sulfate and thus on the activity of the sulfotransferase in liver. Thus, the finding of De Baun et al. (Table 1) [11] that the species difference in hepatocarcinogenesis of N-hydroxy-2-AAF parallels the species difference in sulfotransferase activity and the covalent binding of N-hydroxy-2-AAF is consistent with the view that the free radical is the ultimate carcinogen. It is noteworthy that the concentrations of ascorbic acid and NADPH used in mutagenesis study shown in Table 2 are similar to those present in liver.

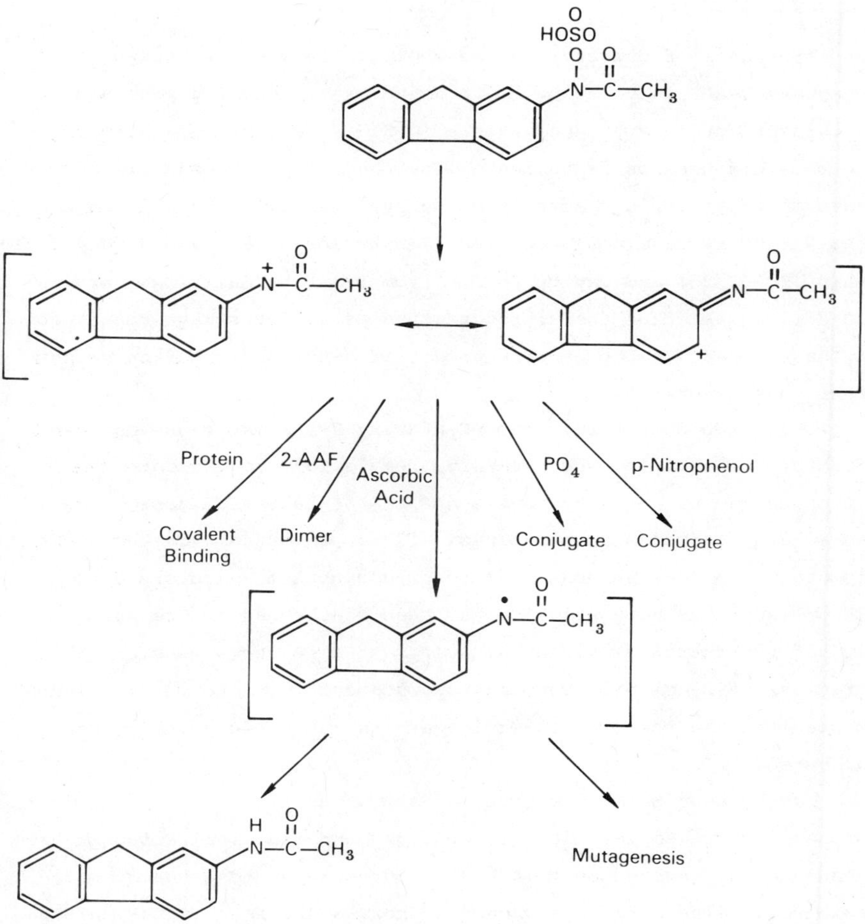

Figure 9 Proposed mechanism for conversion of 2-AAF-N-sulfate to a mutagenic metabolite.

V. CONCLUSION

The importance of deacetylated metabolites in the carcinogenicity N-hydroxy-2-AAF is unknown. Both N-hydroxy-2-aminofluorene and 2-nitrosofluorene are carcinogenic [1-3,7], but the mechanism of their carcinogenesis is unclear since they are both metabolized—at least in part—through acetylated derivatives [1]. The findings that N-hydroxy-2-aminofluorene and 2-nitrosofluorene are highly mutagenic [20-22,25] and may be formed from 2-AAF by mammalian enzymes [30-35] suggest that the aforementioned metabolites may account for at least a part of the carcinogenesis of N-hydroxy-2-AAF in organs other than liver [7].

Other radicals formed from N-hydroxy-2-AAF may cause mutagenic effects. For example, when we used Fremy's salt to generate the nitroxide radical of N-hydroxy-2-AAF in the bacterial assay, there was a large increase in mutagenesis [37]. Unfortunately the radical dismutates to the mutagens 2-nitrosofluorene and N-acetoxy-2-AAF [38,39] and thus no conclusion has been reached as to the mutagenicity of the nitroxide radical itself. It also seems possible that a radical of 2-aminofluorene may be mutagenic, but to our knowledge no studies on either its formation or its mutagenicity have been performed.

The concept that free radicals are mutagenic or carcinogenic is not new. For example, ultraviolet light and ionizing radiation are mutagenic and carcinogenic [40,41]. Our data suggest that a mutagenic metabolite, produced when N-hydroxy-2-AAF is sulfated in the presence of ascorbic acid, is a free radical. If it is, free radicals may play more important roles in mutagenesis and carcinogenesis than previously believed.

NOTE ADDED IN PROOF

After the submission of this paper, Wirth and Thorgeirsson [42] obtained evidence that ascorbic acid increases the mutagenic activity of N-OH-2-FAA in the presence of a purified transacetylase but not in that of a purified sulfotransferase. Thus the stimulatory effects of ascorbic acid do not appear to be due to the proposed free radical of 2-FAA.

REFERENCES

1. J. H. Weisburger and E. C. Weisburger, *Pharmacol. Rev. 25*:1-66 (1973).
2. J. A. Miller, *Cancer Res. 30*:559-576 (1978).
3. E. C. Miller, *Cancer Res. 38*:1479-1496 (1978).
4. J. W. Cramer, J. A. Miller, and E. C. Miller, *J. Biol. Chem. 235*:885-888 (1960).
5. E. C. Miller, J. A. Miller, and H. A. Hartmann, *Cancer Res. 21*: 815-824 (1961).
6. S. S. Thorgeirsson, D. J. Jollow, H. A. Sasame, I. Green, and J. R. Mitchell, *Mol. Pharmacol. 9*:398-404 (1973).
7. E. C. Miller and J. A. Miller, *Ann. N. Y. Acad. Sci. 163*:731-750 (1969).
8. P. D. Lotlikar, J. D. Scribner, J. A. Miller, and E. C. Miller, *Life Sci. 5*:1263-1269 (1966).
9. J. R. DeBaun, J. Y. Rowley, E. C. Miller, and J. A. Miller, *Proc. Soc. Exp. Biol. Med. 129*:268-273 (1968).
10. J. R. DeBaun, J. Y. R. Smith, E. C. Miller, and J. A. Miller, *Science 167*:184-186 (1970).
11. J. R. DeBaun, E. C. Miller, and J. A. Miller, *Cancer Res. 30*: 577-595 (1970).
12. C. M. King and B. Phillips, *Science 159*:1351-1353 (1968).
13. C. C. Irving and R. A. Veazey, *Cancer Res. 29*:1799-1804 (1969).
14. C. C. Irving, D. H. Janss, and L. T. Russell, *Cancer Res. 31*: 387-391 (1971).
15. H. Bartsch, C. Dworkin, E. C. Miller, and J. A. Miller, *Biochem. Biophys. Acta 304*:42-55 (1973).
16. H. Bartsch, C. Dworkin, J. A. Miller, and E. C. Miller, *Biochem. Biophys. Acta 286*:272-298 (1972).
17. C. M. King and W. T. Allaben, in *Conjugation Reactions in Drug Biotransformation* (A. Aitio, Ed.). Elsevier, New York, 1978, pp. 431-441.
18. C. C. Irving, *Cancer Res. 26*:1390-1396 (1966).
19. C. C. Irving, R. A. Veazey, and L. T. Russell, *Chem. Biol. Interactions 1*:19-26 (1969).
20. B. N. Ames, J. McCann, and E. Yamasaki, *Mutation Res. 31*:347-364 (1975).
21. J. McCann, E. Choi, E. Yamasaki, and B. N. Ames, *Proc. Nat. Acad. Sci. U.S. 72*:5135-5139 (1975).
22. J. McCann and B. N. Ames, *Proc. Nat. Acad. Sci. U.S. 73*:950-954 (1976).
23. A. W. Wood, W. Levin, A. Y. H. Lu, H. Yagi, O. Hernandez, D. M. Jerina, and A. H. Conney, *J. Biol. Chem. 251*:4882-4890 (1976).
24. D. R. Thakker, H. Yagi, A. Y. H. Lu, W. Levin, A. H. Conney, and D. M. Jerina, *Proc. Nat. Acad. Sci. U.S. 73*:3381-3385 (1976).
25. W. E. Durston and B. N. Ames, *Proc. Nat. Acad. Sci. U.S. 71*: 737-741 (1974).
26. L. S. Andrews, L. R. Pohl, J. A. Hinson, and J. R. Gillette, *Toxicol. Appl. Pharmacol. 48*:A48 (1979).

27. J. D. Gregory and F. J. Lipmann, *J. Biol. Chem. 229*:1081-1090 (1957).
28. G. J. Mulder, J. A. Hinson, and J. R. Gillette, *Biochem. Pharmacol. 26*:189-196 (1977).
29. G. J. Mulder, J. A. Hinson, and J. R. Gillette, *Biochem. Pharmacol. 27*:1641-1649 (1978).
30. G. J. Mulder, J. A. Hinson, W. L. Nelson, and S. S. Thorgierson, *Biochem. Pharmacol. 26*:1356-1358 (1977).
31. L. S. Andrews, J. A. Hinson, and J. R. Gillette, *Biochem. Pharmacol. 27*:2399-2408 (1978).
32. C. E. Weeks, W. T. Allaben, S. C. Louie, E. J. Lazear, and C. M. King, *Cancer Res. 38*:613-618 (1978).
33. S. Sa Kai, C. E. Reinhold, P. J. Wirth, and S. S. Thorgeirsson, *Cancer Res. 38*:2058-2067 (1978).
34. D. L. Stout, J. N. Baptist, T. S. Matney, and C. R. Shaw, *Cancer Lett. 1*:269-274 (1976).
35. D. McGregor, *Mutation Res. 30*:305-316 (1975).
36. L. S. Andrews, J. M. Fysh, J. A. Hinson, and J. R. Gillette, *Life Sci. 24*:59-64 (1979).
37. L. S. Andrews, J. A. Hinson, and J. R. Gillette, unpublished results.
38. H. Bartsch and E. Hecher, *Biochim. Biophys. Acta 237*:567-578 (1971).
39. R. A. Floyd and L. M. Soong, *Biochim. Biophys. Acta 498*:244-249 (1977).
40. J. H. Robbins, K. H. Kraemer, M. A. Lutzner, B. W. Festoff, and H. G. Coon, *Ann. Intern. Med. 80*:221-248 (1974).
41. L. Stelter, C. von Sonntag, and D. Schulte-Frohlinde, *Int. J. Radiat. Biol. 29*:255-269 (1976).
42. P. J. Wirth and S. S. Thorgeirsson, *Mol. Pharmacol. 19*:337-344 (1981).

13

THE METABOLIC ACTIVATION OF CARCINOGENIC NITRO COMPOUNDS TO OXYGEN-REACTIVE INTERMEDIATES

John E. Biaglow and Marie E. Varnes

Case Western Reserve University
Cleveland, Ohio

Cameron J. Koch

Cross Cancer Foundation
Edmonton, Alberta, Canada

Rajagopalan Sridhar

Oklahoma Medical Research Foundation
Oklahoma City, Oklahoma

I. INTRODUCTION

There is considerable interest in the mutagenic and cytotoxic properties of nitro aromatic compounds such as the nitrofurans, nitrobenzenes, and nitroimidazoles because of the potential use of these drugs in cancer therapy both as radiosensitizers and as cytotoxic agents with selectivity for hypoxic cells [1-10]. The cytotoxicity of these compounds as well as their mutagenic or carcinogenic potential may be related to the metabolic reduction of the nitro group to free radicals or nitroso intermediates [3,7,8]. Mutagenicity and cytotoxicity are greater under anaerobic conditions, suggesting that some of the toxic products are inactivated by reaction with oxygen [3,9-12]. Nevertheless, damage due to the metabolism of the nitro compounds does occur in the presence of oxygen as evidenced by the toxic effects of nitrofurans on primaquine-sensitive individuals [13] and on human lung tissue in vivo and cultured cells in vitro [3]. Damage to other tissue such as the brain may also occur under oxygenated conditions [8]. In this review we will deal with cellular, microsomal, and chemical reduction of various nitroaromatic compounds to oxygen-reactive intermediates and describe the consequences of this reaction.

A. Enzymatic Reduction of Nitroaromatic Compounds

Reduction of nitroaromatic compounds in cells and tissues is due principally to the nitroreductases, which are widely distributed in both prokaryotic and eukaryotic organisms [3,14-18]. The effectiveness of nitroreductases is strongly inhibited by oxygen, and reduction of the nitro group generally occurs only under anaerobic conditions [14-19]. The inhibition by oxygen occurs because the first step in reduction involves the production of an oxygen-reactive nitro radical. The reaction with oxygen results in the production of superoxide and reformation of the parent nitro compound. The rapid reaction of the nitro radical with oxygen prevents the addition of a second electron to the nitro radical anion, which is critical for nitroreductase action [14-18]. Even though the formation of radical anions occurs because of the high electron affinity of the nitro group in the various compounds, the electron affinity of the radical anions themselves is usually much less than that of oxygen. However, certain very strongly electron affinic compounds such as 4-nitroquinoline-1-oxide (4-NQO), 3,5-dinitrobenzonitrile (DNBN), or 1-(2,4-dinitrophenyl)pyridinium chloride (DNPC) are reduced by nitroreductases even under aerobic conditions (our unpublished observations). The nitro radical anions of 4-NQO, DNBN, and DNPC are electron affinic enough to compete with oxygen for abstraction of a second electron from the enzyme [17].

Reduction of various nitro compounds as well as the production of superoxide and nitro radical anions and superoxide has been demonstrated directly and indirectly in experiments with purified enzymes [15,19], with microsomes [15,17,18], and in reactions with ascorbate [20-22]. Biaglow et al. [17,23,24] have shown that similar reactions can also occur in intact cells provided the nitro compounds do not

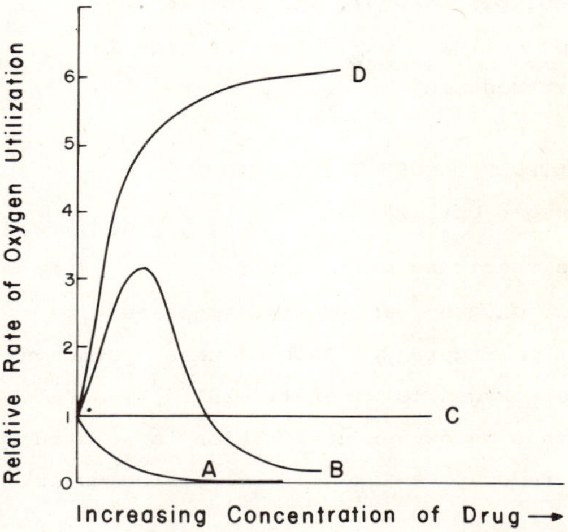

Figure 1 Effects of nitro compounds on cellular oxygen utilization The patterns represented are (A) inhibition, (B) stimulation followed by inhibition, (C) no effect, and (D) stimulation.

bear any charge centers which prevent transport through cell membranes. However, even in the case of charged molecules, a limited amount of reduction occurs [17].

B. Effects of Nitro Compounds on Oxygen Utilization

Because nitro compounds accept electrons from flavoproteins (usually involved in electron transport), they can influence oxygen utilization by cells, microsomes, mitochondria, or enzymes involved in certain redox reactions [25]. Effects on cellular oxygen utilization may follow one of four different patterns as depicted by the four hypothetical curves A, B, C, and D in Fig. 1. Addition of increasing amounts of nitro compound to respiring cells in suspension may cause inhibition of oxygen consumption (A), stimulation followed by inhibition (B), no effect (C), or stimulation of oxygen consumption (D). The pattern obtained is in part dependent upon the redox potential (electron affinity) of the nitro compound, the lineage and growth conditions of the cells under study, as well as cellular pools of reducing substrates such as glutathione (GSH), nonprotein thiols

(NPSH), reduced pyridine nucleotides (NADPH), and glucose [25]. Each of the observed effects on cellular oxygen utilization is explained in detail in the next section.

II. MECHANISMS INVOLVED IN ALTERED OXYGEN UTILIZATION

A. Stimulation of Cellular Oxygen Utilization

Addition of nitro compounds to respiring cells may result in stimulation of oxygen consumption if oxygen-reactive intermediates are produced via reactions A and B in Figure 2. Both of these pathways eventually lead to reduction of oxygen to water [17,25]. As a rule, the effect of the nitro compounds on oxygen utilization is related to their electron affinity or redox potential [17,26]. Biaglow et

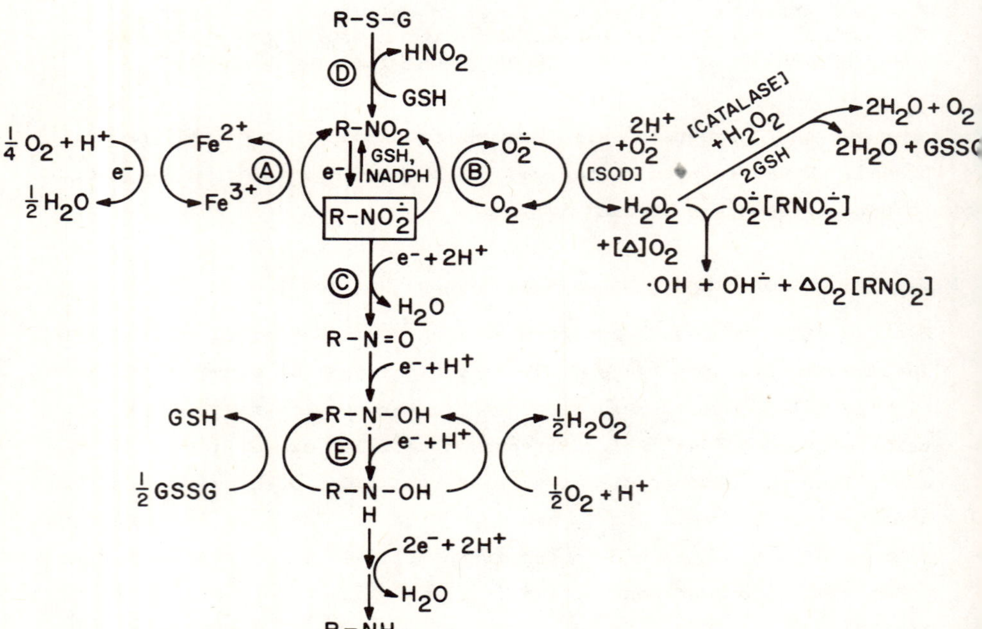

Figure 2 Metabolic scheme indicating various reactions of the nitro radical anion as well as possible sites of reaction of glutathione (GSH) with nitro compounds. Abbreviations: SD, superoxide dismutase; GSSG, oxidized GSH. (From Ref. 24.).

al. [17] have shown that, in the presence of glucose, nitrobenzene derivatives such as p-nitroacetophenone (PNAP), p-nitro-3-dimethylaminopropiophenone hydrochloride (NDPP), and DNBN, which possess oxidation-reduction potentials more positive than -0.35 V with respect to a standard calomel electrode (SCE), stimulate oxygen utilization by cells. DNBN produced the maximum stimulation, as seen in the plot of initial rates of oxygen consumption against the oxidation-reduction potentials of the added nitro compound presented in Figure 3A. It is noteworthy that even though trinitrobenzene sulfonate (TNBS) and DNPC have higher (more positive) oxidation reduction potentials than DNBN, they do not produce correspondingly higher rates of cellular oxygen consumption [17,26,35]. This is perhaps due to the charged nature of TNBS and DNPC, which would prevent their passage through cellular membranes. Similar stimulation of oxygen consumption of V79 Chinese hamster lung cells and Ehrlich ascites cells in culture has been observed for nitrofurans [27], most of which have reduction potentials greater than -0.32 V, and also for many nitroaromatic compounds (Table 1).

Stimulation of cellular oxygen consumption by the nitro compounds is due to the formation of nitro radical anion intermediates via a one-electron transfer step, as previously mentioned. Besides reacting directly with molecular oxygen, with concomitant regeneration of the original drug molecule (reaction B in Fig. 2), the nitro radical anions are also capable of donating an electron to the mitochrondrial cytochrome chain (reaction A in Fig. 2). Such an interaction initiates the shunting of electrons from the microsomes to the mitochrondria, which results in reduction of O_2 to water. Evidence of interaction between nitro radical anions and cytochromes $c + c_1$ has been obtained for 4-NQO and 5-nitro-2-furaldehyde diacetate (NFA) [18]. Reduction of cytochromes $c + c_1$ in Ehrlich cells by 4-NQO- or NFA-mediated electron transfer was monitored spectrophotometrically by observing an increase in absorbance at 550 nm vs. 540 nm (Fig. 4). For maximal drug effects, cells were pretreated with rotenone, which inhibits normal cytochrome reduction by block-

Table 1 The Effect of Various Nitro Compounds on Oxygen Consumption of V79 and Ehrlich Cells[a]

Drug	Relative rate of oxygen utilization					
	Ehrlich tumor cells[b]		V79 lung cells		Ehrlich tumor cells + KCN (nmol O_2/min)	
	- glucose	+ glucose	- glucose	+ glucose	+ glucose	- glucose
1. 1-(2,4-Dinitrophenyl)pyridinium chloride (-0.10)	0.74	1.0	1.0	1.58	—	—
2. 4-Nitroquinoline-1-oxide (-0.15)	1.10	3.3	—	—	—	—
3. Menadione (-0.17)	0.71	2.93	—	—	—	—
4. 3,5-Dinitrobenzonitrile (-0.16)	0.68	2.12	1.5	2.20	—	—
5. Trinitrobenzenesulfonate (-0.20)	0.77	0.94	1.0	1.0	30	24
6. 2-Nitro-5-pyridinylthiazole (-0.20)	0.62	2.15	1.68	2.20	102	38
7. 5-Nitro-2-furaldehyde (-0.22)	0.90	2.80	1.04	1.68	138	90
8. 5-Nitrofurate (-0.25)	0.75	1.3	—	—	—	—
9. 5-Nitro-2-furoic acid (-0.25)	0.75	1.3	—	—	—	—
10. Nifuroxime (-0.27)	0.42	1.16	1.01	1.20	35	20
11. Trinitrobenzene (-0.27)	0.64	3.38	—	2.10	—	—
12. p-Nitroacetophenone (-0.29)	0.37	0.70	1.0	1.0	42	18
13. Nitrofurazone (-0.30)	0.53	1.25	0.70	1.60	72	22

14.	Misonidazole (−0.30)	1.0	1.0	1.0	1.0	12	7
15.	N-(5-Nitro-2-furfurylidene)-1-aminohydantoin (−0.30)	0.45	2.1	—	—	—	—
16.	5-Nitro-3-amino-2-furoxazolidione (Furoxone) (−0.32)	1.02	1.5	—	—	—	—
17.	5-Nitro-2-furaldehyde diacetate (−0.32)	0.47	1.41	1.10	1.45	64	36
18.	2-Bromo-5-nitrothiazole (−0.33)	0.40	0.94	0.90	1.50	30	18
19.	Ro-07-0913 (−0.35)	1.0	1.0	1.0	1.0	12	6
20.	Niridazole (−0.38)	1.0	1.0	0.95	1.0	30	15
21.	p-Chloronitrobenzene (−0.38)	0.5	0.5	0.2	0.2	6	6
22.	2-Nitroimidazole (−0.40)	0.78	0.70	0.90	0.9	10	10
23.	m-Nitroacetophenone (−0.40)	0.68	0.58	1.00	0.78	6	6
24.	o-Nitroacetophenone (−0.46)	0.65	0.52	0.20	0.70	6	6
25.	p-Nitrotoluene (−0.47)	0.20	0.20	0.20	0.20	6	6
26.	2-Nitrothiophene (−0.47)	0.80	0.80	0.80	0.80	6	6
27.	3-Nitrothiophene (−0.52)	0.48	0.47	0.70	0.62	7	7
28.	2-Methyl-5-nitroimidazole (−0.57)	0.79	0.51	0.91	0.62	6	6
29.	Metronidazole (−0.60)	0.52	0.52	0.55	0.79	6	6
30.	2-Nitropyrrole (−0.67)	0.64	0.44	0.40	0.60	6	6

[a]Oxygen utilization was initiated by the addition of cells to the reaction medium consisting of 10 mM potassium phosphate buffer and 10 mM Tris buffer in 0.9% NaCl, pH 7.3, 37° in the case of Ehrlich cells. V79 cells were added to reaction medium containing complete growth medium buffered with 10 mM hepes plus 10 mM β-glycerol phosphate, pH 7.3, 37°. Glucose was 10 mM and KCN was 1 mM when used. The relative rate of oxygen utilization was obtained by dividing the rate of oxygen utilization in the presence of drug by the rate obtained in the absence of drug. With KCN-inhibited cells, the rate of oxygen utilization is expressed as nmoles O_2 consumed/min. The reaction volume was 3 ml in all cases. All oxygen values were obtained as initial slope measurements.

[b]The Ehrlich cell density was 6.7×10^6 per milliliter. The V79 cell density was 3.3×10^6 per milliliter.

Source: From Ref. 27.

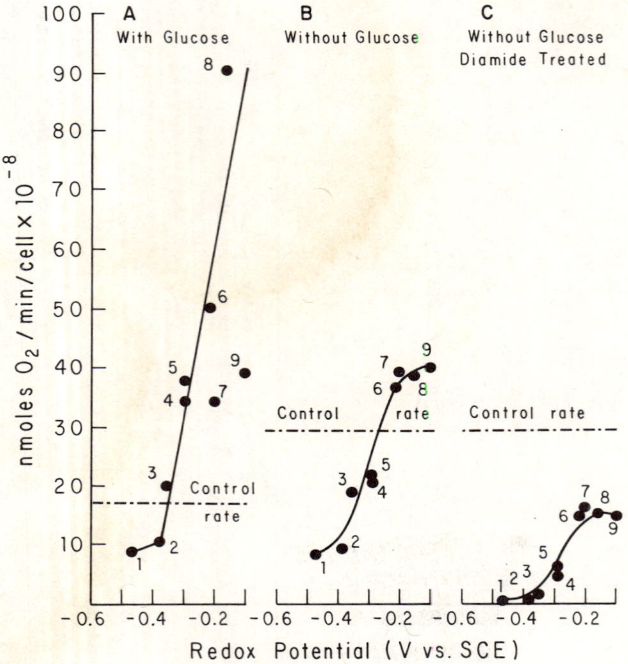

Figure 3 Effects of various nitrobenzene derivatives on oxygen consumption in Ehrlich ascites cells. Control rates of oxidation of the cells in the absence of nitrobenzene. Oxidation was initiated by the addition of cells to the reaction medium (total volume, 3 ml), consisting of 0.05 phosphate buffer in 0.9% NaCl, pH 7.25, at 37°. The amount of each nitrobenzene tested was 6 μmol. Glucose when present was 0.01 M, and where indicated (C), the cells had been treated with 1.5 μmol of diamide. The chemicals tested are listed in order of increasing oxidation-reduction potential: PNT, p-nitrotoluene; PCNB, p-chloronitrobenzene; MNBA, m-nitrobenzamidine; PNAP, p-nitroacetophenone; NDPP, p-nitro-3-dimethylaminopropiophenone hydrochloride; DNBS, 3,5-dinitrobenzamide, TNBS, 2,4,6-trinitrobenzenesulfonate; DNBN, 3,5-dinitrobenzonitrile; DNPC, 1-(2,4-dinitrophenyl)pyridinium chloride. The final cell concentration was 5 x 10^7/3 ml. The rate of oxygen consumption was measured 2 min after addition of the nitrobenzene derivatives. (From Ref. 17.)

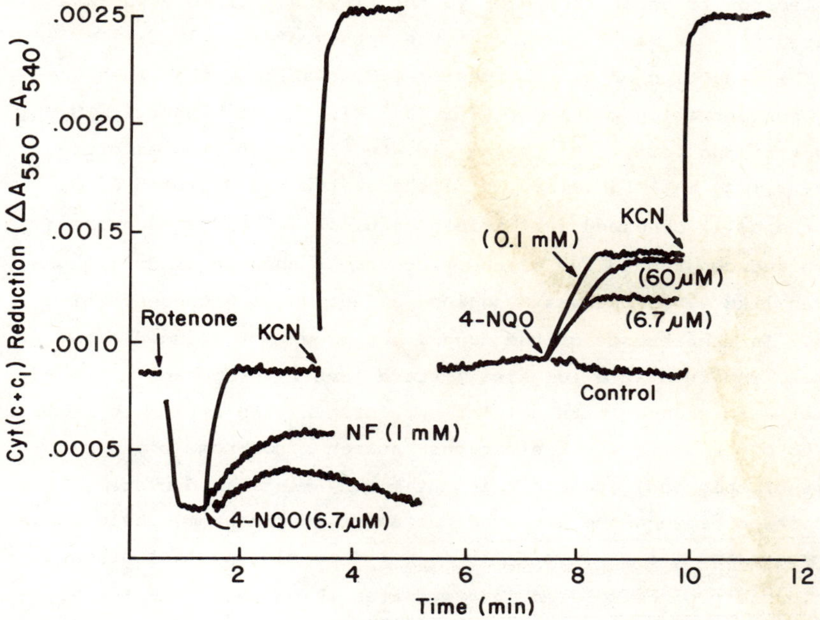

Figure 4 The interaction of nitro radicals with cytochromes (Cyt) c + c_1 in Ehrlich cells. (Left) Reduction of ferricytochrome c + c_1 by 4-NQO and 5-nitro-2-furaldehyde diacetate (NF) in cells pretreated with 1 μM rotenone to inhibit normal electron transport. This particular nitrofuran derivative was chosen because its optical absorption in the 540- to 550-nm range is negligible. (Right) Increase in the steady-state level of reduced cytochrome c + c_1 caused by different additions of 4-NQO in the absence of electron transport inhibitor. KCN (1 mM) was added to each system to indicate the maximum extent of cytochrome c + c_1 reduction. The cells were suspended in 0.9% NaCl solution containing 0.05 M phosphate buffer (pH 7.2) at a density of 9 x 10^6 cells/ml and incubated at 37°. (From Ref. 18.)

ing electron transfer from NADH to the flavins. Glucose was also required for maximal reduction of the cytochromes [18].

The reduction of cytochromes by 4-NQO was transient when the 4-NQO concentration was low (6.7×10^{-6} M), but at higher 4-NQO concentrations (e.g., 10^{-4} M) an elevated level of reduced cytochrome c + c_1 was maintained until the system was depleted of O_2. Once the cells consumed available oxygen, all of the cytochromes became reduced regardless of the presence or absence of drug. NFA behaved like 4-NQO but was much less effective on a concentration basis. In the absence of the drugs, the amount of reduced cytochrome c remained at a low steady-state level. 4-NQO and NFA were also able to reduce intracellular cytochromes when antimycin A was used to block transfer of electrons between cytochrome b and cytochrome c; this indicates that in antimycin A-inhibited systems, the interaction between the drug radical and the cytochrome chain is at the cytochrome c + c_1 level. It is worth mentioning that other nitrofurans produce colored intermediates which absorb in the $A_{550-540}$ region and thus obscure the cytochrome changes [18]. The reduction of cytochrome c + c_1 (Fe^{3+}) by 4-NQO-derived radicals could also be demonstrated in the absence of rotenone (right-hand side of Fig. 4). The addition of 4-NQO (final concentrations 6.7×10^{-6} M, 6×10^{-5} M, and 10^{-4} M) to the cells in the presence of glucose led to new, higher levels of reduced cytochrome c + c_1 (Fe^{3+}) by 4-NQO$^{\cdot-}$. In the absence of glucose, no effect was seen, even in the presence of 10^{-4} M 4-NQO [18].

Oxygen consumption by nitro drugs may also occur via the free radical intermediates produced during further metabolic reduction (reaction C and subsequent steps in Fig. 2) to hydroxylamine-derived radicals, which are also believed by several authors to be oxygen reactive [7,18,21]. Another pathway to hydroxylamine exists via nitroso compounds which arise from disproportionation of nitro radical anions. The pathways leading to oxygen consumption via hydroxylamine-derived radicals occur only with a limited number of nitroaromatic derivatives, such as DNBN and 4-NQO, which are characterized by very strong electron affinity and undergo reduction beyond the nitro radical anion stage under aerobic conditions [17]. It is difficult to assess the relative degree of stimulation of oxygen consumption via the hydroxylamine radical pathway as opposed to the alternate route involving the nitro radical anion.

The dramatic effect of glucose on drug-mediated stimulation of O_2 consumption (Fig. 3 and Table 1) can be attributed to increased cellular pools of the reducing species NADPH and NADH, formed during glucose metabolism. Both nitro and hydroxylamine radicals can react with sulfhydryl groups in proteins and with NPSH such as glutathione [24,28-29]. Therefore, the extent of stimulation of oxygen consumption would be influenced by the cellular pools of thiols and the availability of reducing equivalents in the form of NADPH and NADH, which are cofactors for enzymatic reduction of the nitro compounds.

B. Inhibition of Cellular Oxygen Consumption

Both stimulation and inhibition of cellular oxygen consumption effected by nitro compounds can be a result of metabolic activation of these compounds. In the previous subsection we described how stimulation is brought about due to metabolically generated oxygen-reactive radicals derived from nitro compounds [25]. Conversely, inhibition of oxygen consumption occurs because metabolic intermediates derived from nitro compounds, i.e., nitro radical anions, nitroso or hydroxylamine derivatives, serve to deplete GSH, other thiols, and glycolytic intermediates which serve as electron sources in cell respiration [17,18,24].

The activation of all the nitro compounds investigated thus far requires the participation of an electron donor such as NADPH, NADH, or ascorbate [15-22] (see Section V). Additional electrons may be supplied from oxidizable substrates such as glucose via the pentose cycle reduction of $NADP^+$. Therefore, the presence or absence of glucose markedly influences the effect of a number of nitroaromatic derivatives on cellular oxygen consumption (Fig. 3B,C and Table 1). However, this appears to be cell-line dependent since, with the exception of 4-NQO, none of the nitro compounds tested with Ehrlich ascites cells under the conditions set for the experiment detailed in Table 1 stimulated oxygen consumption in the absence of glucose, whereas in the case of V79 Chinese hamster lung cells there was a general decrease in stimulatory activity of the less electron-affinic compounds [17,27].

In the absence of glucose, varying degrees of inhibition can be seen (Table 1). Of the various nitrofurans, nitrofurate (5-nitro-2-furoic acid) is the least inhibitory, while nitrofurantoin [N-(5-nitro-2-furfurylidene)-1-aminohydantoin] causes maximum inhibition [23,27]. Time-dependent progressive inhibition has been observed with several other nitroaromatic compounds in the absence of glucose [23,25,27]. Biaglow et al. [17] demonstrated that a range of effects were possible with DNBN on cells incubated with or without glucose

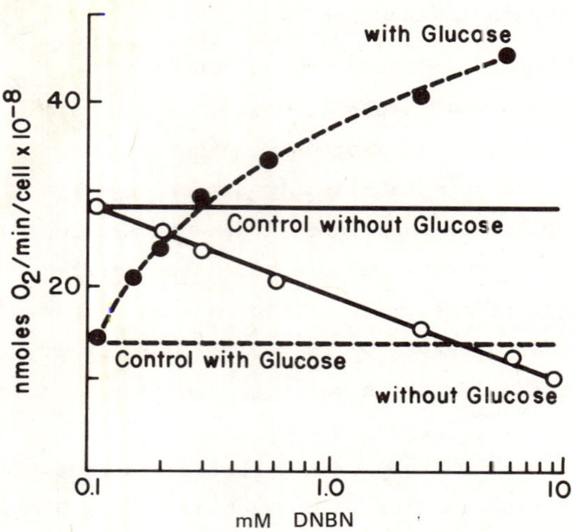

Figure 5 Effect of DNBN on rate of oxygen utilization by Ehrlich ascites cells incubated in the presence and absence of glucose. The experimental conditions were the same as in Figure 3. The cell concentration was 2.5 x 10^7 cells/3 ml. The rate of oxygen consumption in the presence of glucose was measured 2 min after the addition of DNBN and 4 min after the addition of DNBN to the glucose-free reaction medium. The horizontal dashed and solid lines represent the control rates for oxygen utilization with and without glucose. (From Ref. 17.)

(Fig. 5). Increasing concentrations of DNBN stimulated oxygen consumption when glucose was available, but in the absence of glucose all concentrations of DNBN were inhibitory. The degree of inhibition increased with the time of incubation with drug, suggesting depletion of reducing equivalents, i.e., electron sources. Similar effects due to depletion of electron sources have been reported for nitrobenzene and nitrofuran derivatives [17-19,23,24], which are known to stimulate cellular oxygen utilization in the presence of glucose.

The onset of the nitro compound-mediated inhibition of oxygen utilization can be hastened, and the extent of inhibition by the nitro compound can be increased, if the nonmitochondrial GSH and other reducing species are oxidized by the addition of diamide [17, 18,21] at concentrations which do not affect the rate of oxygen consumption in control (nitro compound-free) cultures.

C. Effects on Intracellular Ferricytochrome c + c_1

Evidence for direct inhibition of the mitochrondrial electron transport chain in a manner analogous to that by classical inhibitors such as rotenone and antimycin A was obtained by studying the state of reduction of the cytochromes c + c_1 in the presence of several of the drugs [25,27]. A transfer of electrons to cytochromes c + c_1 (as visualized in Fig. 2, reaction A) would lead to an increased ratio of reduced to oxidized (Fe^{2+}/Fe^{3+}) cytochromes as monitored by an increase in the $A_{550}-A_{540}$ difference spectrum, whereas oxidation of the ferrocytochromes c + c_1 will be seen as a decrease in this ratio. The effects of rotenone and of a number of hypoxic cell radiosensitizers on the red/ox state of cytochrome c + c_1 in Ehrlich cells are shown in Figure 6. Nitrotoluene (1 mM) produced an oxidation that was nearly as great as that observed with 1 μM rotenone. Misonidazole [1-(2-nitro-1-imidazolyl)-3-methoxy-2-propanol] caused a slight decrease in oxidation when tested in either phosphate- or Tris-buffered saline systems. In the Tris-buffered system, metronidazole [1-(β-hydroxethyl)-2-methyl-5-nitroimidazole] produced a slow oxidation of cytochrome c + c_1 (Fig. 6B), whereas no effect was

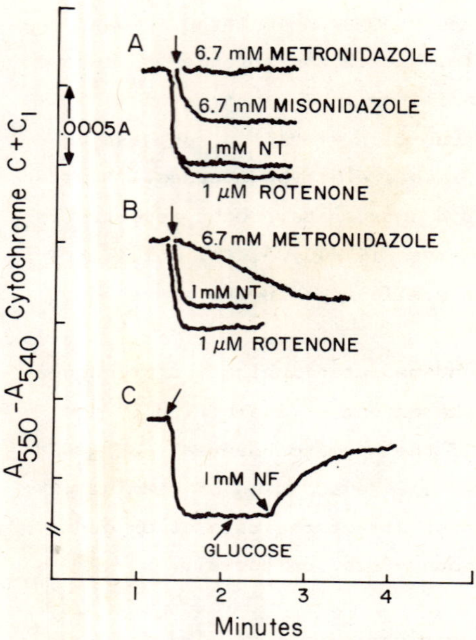

Figure 6 The effect of rotenone, nitrotoluene, metronidazole, and misonidazole on intracellular cytochrome c reduction. The measurements were obtained with the aid of a DW-2 dual beam spectrophotometer. The measurements were initiated by the addition of the drug to 50 mM phosphate or 50 mM tris buffered Ehrlich cells (1.3×10^7 cells/ml), in 0.9% NaCl, pH 7.3, 37°, 3 ml. The solutions were stirred continuously. (A) Phosphate buffered saline systems. (B,C) Tris-buffered systems. (From Ref. 27.)

observed in phosphate-buffered saline (Fig. 6A). By comparison, addition of 1 mM NFA produced a fairly rapid reduction in the cytochomes c + c_1 (Fig. 6C). The site of inhibition is not known; however, the ability of nitro compounds to form charge transfer complexes with riboflavin [32] suggests a flavoprotein site. Similar cytochrome changes have been observed with nitropyrroles and other nitroheterocyclic compounds [27].

There are reasons to believe that all electron-affinic hypoxic cell radiosensitizers, including those stimulating overall oxygen consumption, exhibit an inhibitory component. This idea is supported by the inhibition observed with diamide-treated cells incubated in the absence of glucose and by the gradual, partly irreversible inhibition observed with time in the case of nitrofurans [17, 18,25,27]. On balance, the difference between most radiosensitizing drugs that stimulate and inhibit oxygen consumption may be in the relative efficiencies of the reactions that stimulate oxygen consumption (notably reaction B, etc., in Fig. 2) vs. reactions that lead to the inhibition of respiratory cellular enzymes.

D. Nitrocompounds That Stimulate and Inhibit Cellular Oxygen Utilization

In the case of drugs that exhibit a concentration-dependent change from stimulation of oxygen consumption to inhibition, such as the mitochondrial uncoupler dinitrophenol (DNP) [33,34], their action might be explained by the predominance of the inhibitor activity at higher concentration, rather than by a time-dependent substrate depletion as was observed with the nitrofurans [25]. Notably, DNP does not stimulate oxygen utilization in cyanide-inhibited cells (Fig. 7). In addition to the substrate-dependent change of effect, i.e., from stimulation to inhibition of oxygen consumption, already described for some of the drugs, compounds such as DNP [34], NDPP [35,36], and 2-nitropyrrole (2-NP) [37,38] were found to stimulate oxygen utilization at low drug concentrations and to inhibit oxygen utilization at high concentrations when incubated with V79 cells in complete growth medium. The effects obtained with DNP and 2-NP are independent of the presence of glucose. However, the stimulatory effect obtained with NDPP on Ehrlich cells (Fig. 3A,B,C) [11] is dependent upon the concentration of glucose (Fig. 3B and Table 1) as well as the concentrations of endogenous substrates such as GSH (Fig. 3C). The stimulation of cellular oxygen utilization by DNP is presumably due to its capacity to uncouple oxidative phosphorylation (for a review see Kun [33]). The mechanism of inhibition of cellular oxidation by DNP at concentrations greater than 1 mM is unknown but may be related to its ability to interact with riboflavin, an essential coenzyme involved in mitochrondrial electron transfer [32]. 2-NP may behave like DNP [27,37,38]. The stimulation

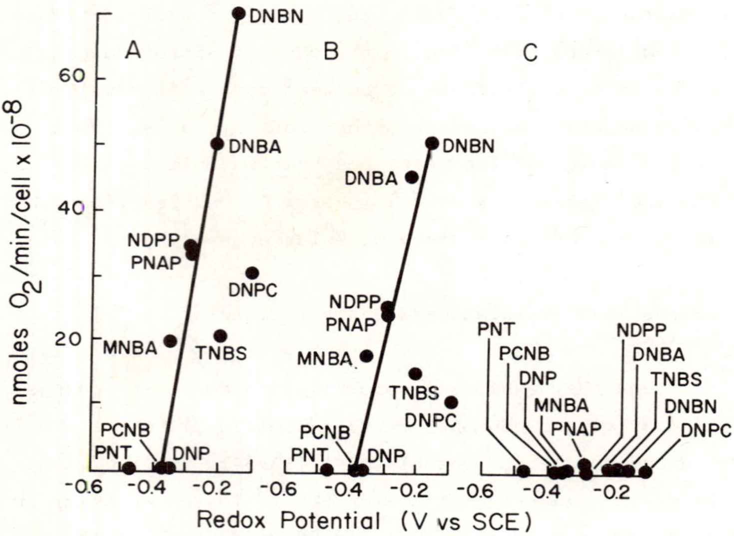

Figure 7 Stimulation of oxygen utilization by nitrobenzene derivatives in Ehrlich ascites cells in the presence of 10 mM glucose, inhibited with either 3 mM KCN (A), 30 μg of antimycin A (B), 30 μg antimycin A + 50 μM diamide (C). The final cell concentration was 5 x 10^7 cells/3 ml. The rates of oxygen consumption were measured 2 min after addition of each nitrobenzene derivative. The derivatives tested are listed in order of increasing oxidation-reduction potential: PNT; PCNB; DNP, 2,4-dinitrophenol; MNBA; PNAP; NDPP; DNBA; TNBS; DNBN; DNPC (cf. Fig. 3). (Data taken from Ref. 17.)

of cellular O_2 consumption by NDPP at low concentrations is presumably due to its capacity to react with intracellular flavin radicals, in a manner similar to the nitrofurans [15,16,23,36] and nitrobenzenes [17,27], to produce an oxygen-reactive species [14-19]. The mechanisms of inhibition of cellular oxidation by NDPP at higher concentrations [35,36] may be due to a reaction with intracellular GSH [39], which inhibits cellular oxidation in a manner similar to the sulfhydryl-reactive agents diamide and NEM [24,31].

E. Conditions Under Which Nitro Compounds Have No Effect on Cellular Oxygen Utilization

All nitro compounds tested so far have been found to influence cellular oxygen consumption in vitro. However, the effects are generally dependent upon drug concentrations and cell density, and it is possible in some cases to select drug concentrations that provide radiosensitization without any effect on cellular oxygen utilization. For example, at concentrations less than 1.0 mM, misonidazole, niridazole [1-(5-nitro-2-thiazolyl)-2-imidazolidinone] and the nitromidazole Ro-07-0913 (Table 1) were found to have no effect on cellular oxygen consumption [25,27,37,40]. In addition, some of the nitropyrroles were found to have only minor effects on cellular oxygen utilization at a concentration of 1.0 mM [37,38]. Even drugs which show little effect on cellular oxygen utilization can stimulate microsomal oxidation of NADPH [14-17]. This will be discussed more thoroughly in Section IV.

III. THE USE OF CYANIDE-INHIBITED CELLS TO STUDY METABOLIC ACTIVATION OF NITRO COMPOUNDS TO OXYGEN-REACTIVE INTERMEDIATES

Potassium cyanide, at concentrations of 0.5-1.0 mM, inhibits mitochondrial respiration by combining with cytochrome a_3 (cytochrome oxidase) [17,18,24,27]. At these concentrations cyanide also inhibits catalase activity [17,18,24,27]. Thus cyanide-inhibited cells are a useful test system for studying the metabolic reduction of nitro drugs to oxygen-reactive intermediates. When cyanide is present, reaction A in Figure 2 is blocked and drug-induced oxygen consumption is due entirely to reaction B and subsequent reactions of $O_2^{\overline{\cdot}}$ and H_2O_2. For compounds with oxidation-reduction potentials more negative than -0.2 V with respect to SCE, nitro radical anions will be the principal oxygen-reactive species, since further reduction (reaction C in Fig. 2) does not occur under aerobic conditions.

With cyanide-inhibited cells, the electron affinity of the nitro compound as well as the concentration of glucose are important factors in determining the degree of stimulation of oxygen consumption (Fig. 7A and Table 1). All nitrobenzenes with redox potentials less negative than -0.35 V vs. SCE stimulate oxygen consumption by cyanide-inhibited Ehrlich cells. Stimulation also occurs with other nitroheterocyclic compounds having redox potentials more positive than -0.4 V vs. SCE (Table 1). The stimulation observed in the presence of the mitochondrial respiratory inhibitor antimycin A (Fig. 7B) was generally less than that in the presence of cyanide. This occurs because antimycin A, unlike cyanide, does not inhibit catalase nor block the entrance of nitro radicals into the mitochondrial cytochrome chain at the level of ferricytochrome c + c_1. In the presence of antimycin A, the degradation of H_2O_2 by catalase leads to the partial regeneration of O_2. When cyanide-inhibited cells were pretreated with diamide and incubated in the absence of glucose, none of the nitrobenzenes tested stimulated oxygen utilization (Fig. 7C) since there are no electron donors under these conditions.

The stimulation of oxygen consumption with cyanide-inhibited cells occurs with a variety of nitro compounds as seen in Table 1 [17,27,35]. Of particular interest were the effects of the carcinogen 4-NQO and the clinically used drugs metronidazole, misonidazole, niridazole, and nitrofurazone (5-nitro-2-furaldehyde semicarbazone). The effects of these drugs on Ehrlich ascites cells (Fig. 8, upper panel) were compared with those obtained with V79 Chinese hamster lung cells grown in vitro (Fig. 8, lower panel). The rate of oxygen consumption for the first minute after addition of drug is plotted against the concentration of the nitro compounds. The stimulation of oxygen utilization for the cyanide-inhibited cells incubated in the presence of glucose increases with the concentration and the electron affinity of the nitroaromatic compound.

4-NQO, with an oxidation-reduction potential of -0.17 vs. SCE [27], can be reduced by cyanide-inhibited cells to 4-HAQO. Both nitro and hydroxylamine intermediates may be produced during this reduction by cells [18,24]. The production of oxygen-reactive intermediates was demonstrated by adding small amounts of 4-NQO to cyanide-inhibited cells, incubating them in medium containing glucose, and monitoring the oxygen consumption until it ceased (Fig. 9). When the oxygen consumption ceases, all of the 4-NQO has been converted to a material with an absorption maximum at 354 nm. The initial absorption peak for 4-NQO is at 373 nm, and the resultant 354 nm peak is characteristic for either 4-HAQO or the glutathione

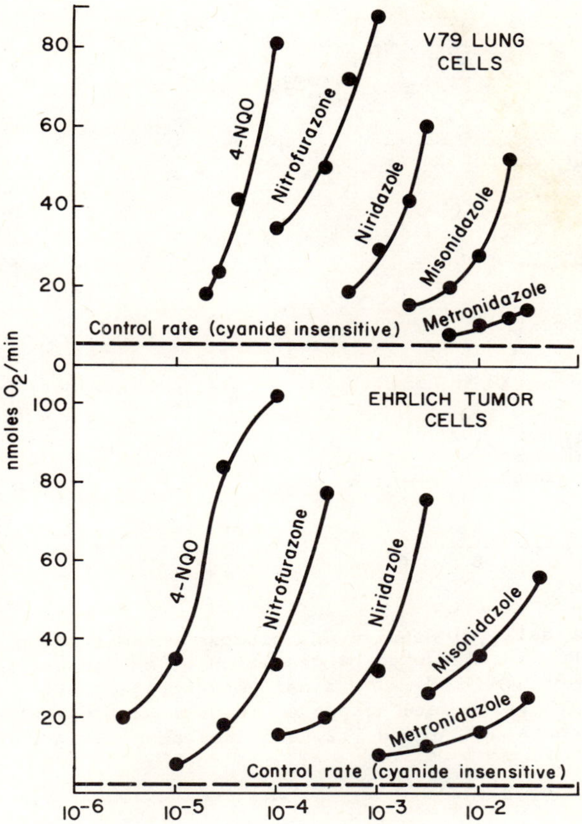

Figure 8 Stimulation of oxygen utilization by nitro compounds with Ehrlich ascites and V79 lung cells in the presence of 10 mM glucose and 1 mM KCN. The rates of oxygen consumption were measured 1 min after addition of each nitro derivative. The reaction medium (total volume 3 ml) consisted of 0.05 M phosphate buffer in 0.9% NaCl, pH 7.25, at 37°. The Ehrlich and V79 cell density, as measured by Coulter counter was 1 x 10^7 cells/ml. Abscissa: molar concentration of the drug.

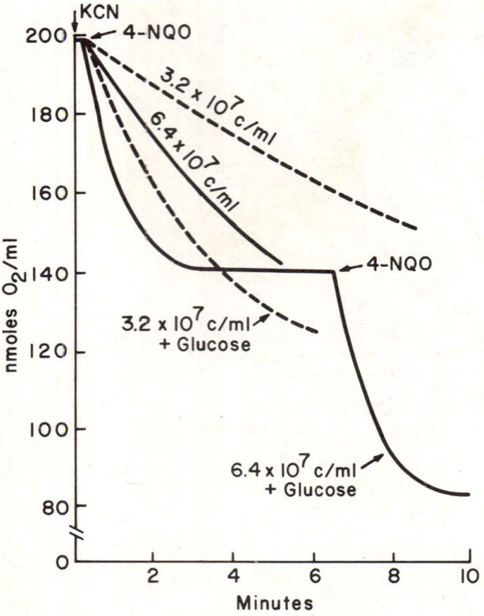

Figure 9 The effect of cell (c) density and glucose concentration on the reduction of 4-NQO was measured by cessation of oxygen utilization in the presence of 1 mM KCN. The final concentration of 4-NQO was 67 μM. Ehrlich cells were suspended in 0.9% NaCl solution containing 0.05 M phosphate buffer (pH 7.2) and incubated at 37°. (From Ref. 18.)

adduct [24]. The cell density had an influence on both the rate of reduction and the amount of oxygen-reactive 4-NQO intermediates produced with cyanide-inhibited cells. The time required for oxygen utilization to cease, after addition of 4-NQO (67 μM), was plotted for different densities of Ehrlich and V79 cells (Fig. 10). It can be seen that the total amount of oxygen consumed during the reduction process decreased with increasing cell density. In addition, the ratio of oxygen consumed to the amount of 4-NQO added (O_2/4-NQO) was also found to decrease with higher cell densities (Fig. 10). Higher cell densities would contain greater concentrations of scavenging species such as GSH as well as catalase and superoxide dismutase, all of which might play a role in decreasing the oxygen utilization [18]. In addition one cannot also exclude the possibility that some

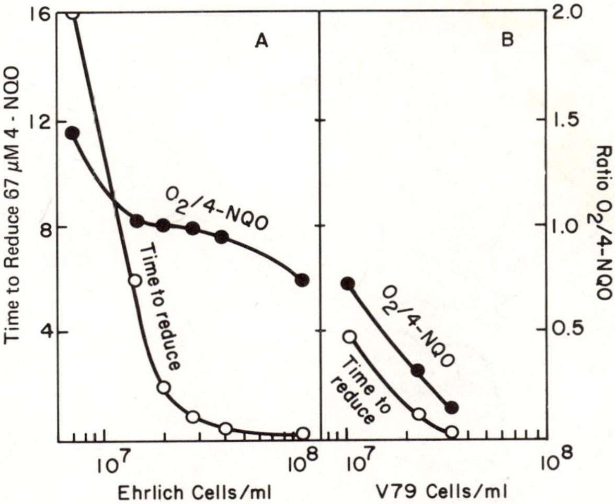

Figure 10 The effects of Ehrlich (A) and V79 (B) cell density on the reduction of 67 μM 4-NQO in the presence of 1 mM KCN, determined by the time for oxygen consumption to cease (left ordinate, open symbols). Also shown are the ratios of oxygen consumed to 4-NQO added (right ordinate, solid symbols) as a function of cell density. The cells were suspended in 0.9% NaCl solution containing 0.05 M phosphate buffer (pH 7.2) and incubated at 37°. (From Ref. 18.)

of the 4-NQO reduction is occurring via a two-electron transfer process which would decrease the production of oxygen-reactive radicals. The time required for reduction of 4-NQO is less with cultured V79 lung cells that may be nutritionally "rich" compared to the Ehrlich ascites tumor cells grown under conditions of limited nutrient availability in vivo. The differences between V79 and Ehrlich cells illustrated another important variable, namely, cell line differences and past history of the cells.

A. Production of Peroxide in Cyanide-Inhibited Cells

Oxygen utilization in cyanide-inhibited cells should result in the net production of peroxide if the oxygen-reactive intermediates transfer electrons to oxygen to produce superoxide, which can dismutate into H_2O_2 (Fig. 2). Peroxide production can be conveniently demonstrated by oxygen evolution upon addition of catalase to the

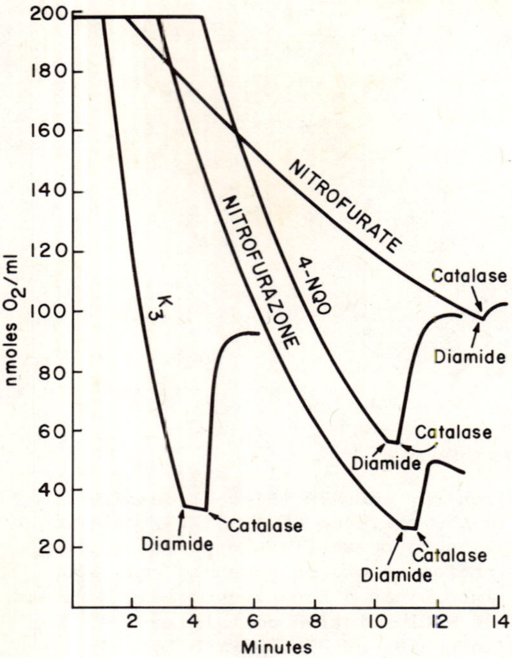

Figure 11 The effect of 4-NQO, nitrofurans, and menadione on the oxygen consumption of Ehrlich cells in the presence of 1 mM KCN. Final concentrations were 0.2 mM for 4-NQO and menadione (K_3) and 1 mM for nitrofurazone and nitrofurate. Diamide (final concentration, 0.17 mM) was added as indicated to inhibit oxygen utilization, and catalase (final concentration, 67 μg/ml) was subsequently added to determine the production of H_2O_2 by the release of oxygen. The cells were incubated at 37° in 0.9% NaCl solution containing 0.05 M phosphate buffer (pH 7.2) at a density of 2 x 10^7 cells/ml. Total volume 3 ml. (From Ref. 18.)

reaction mixture. Figure 11 shows the formation of peroxide with a number of nitrofurans as well as with 4-NQO and menadione (2-methyl 1,4-naphthoquinone [17]). The addition of diamide prior to the addition of catalase ensures that the drug-activating system, which requires reducing equivalents, is depleted and that further oxygen consumption is inhibited. The amount of peroxide produced may vary depending upon the ability of the nitro compound to stimulate oxygen utilization as well as the amount of endogenous GSH available for reaction with superoxide [17,18,24] or nitro radicals [41,42]. The effects of several other nitro compounds on net peroxide production after a 9-min incubation period are listed in Table 2. Only metronidazole, which has the most negative reduction potential of the drugs tested, did not lead to peroxide production (J. E. Biaglow, unpublished observations).

B. Nitroso Effects

Inhibition of cellular oxygen utilization by nitro compounds in the absence of glucose may in part be due to the production of nitroso intermediates. These two-electron reduction products may be formed by any of the following reactions [17,18,24]:

$$R\text{-}NO_2 + 2\bar{e} + 2H^+ \rightarrow R\text{-}N{=}O(\text{nitroso}) + H_2O \qquad (1)$$

$$R\text{-}NO_2^{\bar{\cdot}} + R\text{-}NO_2^{\bar{\cdot}} + 2H^+ \rightarrow R\text{-}N{=}O + H_2O + R\text{-}NO_2 \qquad (2)$$

$$R\text{-}NO_2^{\bar{\cdot}} + O_2^{\bar{\cdot}} + 2H^+ \rightarrow R\text{-}N{=}O + O_2 + H_2O \qquad (3)$$

Further reduction of nitroso intermediates by one-electron addition can produce hydroxylamine-derived radicals, e.g., hydronitroxyl radicals, which may also be reactive with oxygen [17,18,21,24]. In normal cells, nitrosobenzene is a potent inhibitor of oxygen utilization, probably due to its ability to remove endogenous nonprotein

Table 2 The Effect of Nitro Compounds on Oxygen Consumption and Peroxide Production by Cyanide Inhibited Ehrlich Ascites Tumor Cells[a]

Drug	O_2 consumed (nmol)	H_2O_2 produced (nmol)	Ratio H_2O_2/O_2
Metronidazole, 30 mM	222	none	—
Misonidazole, 30 mM	474	126	0.26
Niridazole, 3 mM	420	90	0.21
Nitrofurazone, 0.3 mM	456	150	0.33
4-NQO, 0.05 mM	300	114	0.38

[a]Ehrlich cells were incubated in 0.9% NaCl solution containing 0.05 M phosphate buffer (pH 7.2) at 37° in the presence of 1 mM KCN. Oxygen consumption was measured for 9 min after drug addition, at which time diamide was added (final concentration, 1.33 mM), followed by catalase, 67 μg/ml. Total incubation volume was 3 ml. H_2O_2 production was calculated from the O_2 released.

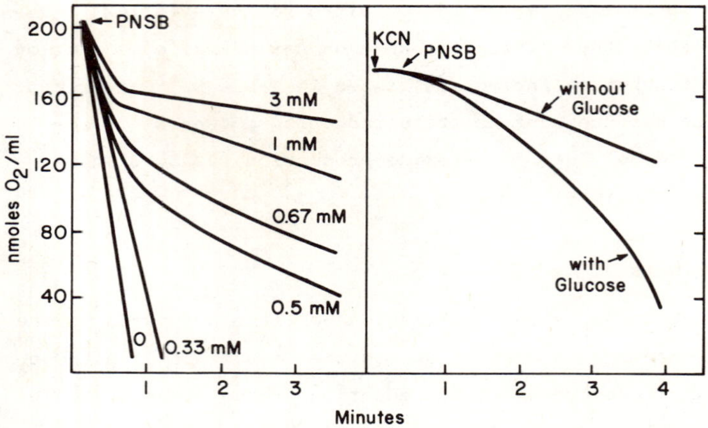

Figure 12 Effect of nitrosobenzene on oxygen utilization by Ehrlich ascites cells. (Left) Effect of drug concentration on inhibition of oxygen utilization. (Right) Stimulation of oxygen consumption by 3 mM KCN-inhibited cells (3 ml 0.05 M phosphate buffer, pH 7.3) in the presence and absence of glucose. All other experimental conditions were the same as in Figure 3. The cell concentration was 5 x 10^7 cells/3 ml. (From Ref. 17.)

thiols, which are necessary for respiration [13,17]. However, nitrosobenzene has a small stimulatory effect on oxygen utilization in the presence of cyanide [17] and this effect is increased by the presence of glucose (Fig. 12).

C. Reactions with Thiols

As seen in Fig. 2, the production of superoxide, peroxide, nitro radicals and other partially reduced intermediates such as nitroso or hydroxylamine radicals may result in an alteration of the non-protein thiol content of mammalian cells. GSH is reactive with $O_2^{\cdot}$, H_2O_2, and nitroso intermediates, and it may be reactive with nitro radicals [17,18,21,24,41,42]. Thus the intracellular pool of non-protein thiols may be depleted either by adduct formation or by oxidation to GSSG [24]. The oxidation of intracellular GSH occurs if the rate of production of $O_2^{\cdot}$ or H_2O_2 is greater than the capacity for endogenous cellular enzymes such as superoxide dismutase or

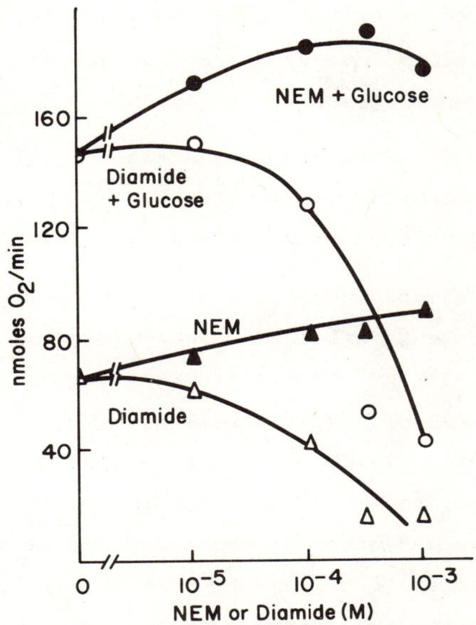

Figure 13 The effects of glucose, NEM, and diamide on 4-NQO-stimulated oxygen consumption of cyanide-inhibited Ehrlich cells. Cells (10^7 cells/ml) were incubated at 37° in 0.9% NaCl solution containing 0.05 M phosphate buffer and 1.0 mM KCN. 4-NQO (5.0×10^{-5} M) was added to 3 ml of cell suspension in a closed compartment, and initial rates of oxygen consumption were determined using a Clark oxygen electrode. Additions of NEM, diamide, or glucose (10 mM) were made immediately prior to addition of 4-NQO. Abscissa: Molar concentration of NEM or diamide. (From Ref. 24.)

catalase as well as glutathione peroxidase to reduce the oxygen radicals or peroxide to water (Eqs. 4 to 7):

$$2R\text{-}NO_2 + 2\bar{e} \rightarrow 2R\text{-}NO_2^{\bar{\cdot}} \text{ (metabolism)} \tag{4}$$

$$2R\text{-}NO_2^{\bar{\cdot}} + 2H^+ + \tfrac{1}{2}O_2 \rightarrow R\text{-}N{=}O + 2H_2O_2 + R\text{-}NO_2 \text{ (spontaneous)} \tag{5}$$

$$2O_2^{\bar{\cdot}} + 2H^+ \rightarrow H_2O_2 + O_2 \text{ (spontaneous or enzyme catalyzed)} \tag{6}$$

$$2H_2O_2 \rightarrow 2H_2O_2 + O_2 \text{ (catalase)} \tag{7}$$

As seen in Eqs. (6) and (7) (Fig. 2), the combined effect of catalase and superoxide dismutase is to reduce oxygen to water. The rates of reaction of various nitro radical anions with oxygen may vary between 10^7 and 10^9 m^{-1} sec^{-1} [17,14,42]. Oxidation of the nonprotein thiols would be expected to occur if insufficient cellular enzyme activities were present to detoxify the $O_2^{\overline{\cdot}}$ and peroxide. Thiols may also be depleted by reactions involving enzymes such as glutathione peroxidase and glutathione-S-transferase [44,45]:

$$NADP^+ + 2GSH \rightarrow NADPH + GSSG + H^+ \text{ (glutathione reductase)} \quad (8)$$

$$H_2O_2 + 2GSH \rightarrow GSSG + 2H_2O \text{ (glutathione-S-peroxidase)} \quad (9)$$

$$R\text{-}NO_2 + GSH \rightarrow RSG + HNO_2 \text{ (glutathione-S-transferase)} \quad (10)$$

The carcinogen 4-NQO produces a rapid depletion of the nonprotein thiols (NPSH), and the NPSH content influences the 4-NQO-catalyzed production of $O_2^{\overline{\cdot}}$ and H_2O_2 as well as the overall reduction of 4-NQO to 4-HAQO; prior removal of NPSH by N-ethylmaleimide (NEM) results in increased oxygen utilization in CN^--inhibited cells (Fig. 13) [24]. As with other nitro drugs, reduction of 4-NQO is dependent on the concentration of glucose and other endogenous substrates. 4-HAQO production is actually increased if glucose is present and prior treatment of CN^--inhibited cells with diamide results in inhibition of radical production because diamide oxidizes both NPSH and NAD(P)H. These results suggest that glucose-mediated NAD(P)H production is more important than thiols in the reduction process. In addition to 4-NQO, other nitro compounds were also found to deplete the NPSH when incubated with aerobic cells. NF was found to be almost as effective as 4-NQO (Fig. 14). Niridazole, after an initial lag period, also showed an ability to oxidize the NPSH. Only a small effect occurred with nitrofurazone [55]. Neither metronidazole nor misonidazole had an effect on the NPSH, even at a concentration of 20 mM [55]. However, Varnes and Biaglow found that misonidazole was able to lower the NPSH of dense suspensions of anaerobic Ehrlich and V79 lung cells [55]. This is perhaps due to the ability of NPSH to react with metabolites of nitro compounds [24, 46,47]. Misonidazole is not conjugated with GSH via glutathione-S-transferase, as determined spectrophotometrically (J. E. Biaglow, unpublished observations). Of additional interest is the finding that nitrosobenzene is extremely reactive with glutathione and also produces a depletion of cellular thiols (Fig. 14). In this connection it is significant that cysteamine protects against the hypoxic

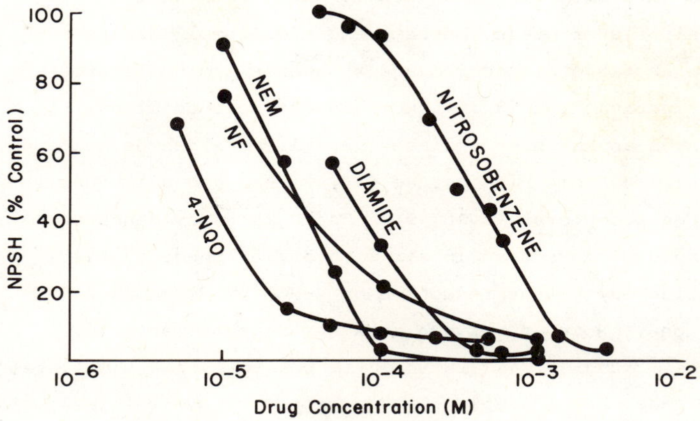

Figure 14 The effects of various drugs on NPSH levels of Ehrlich ascites cells. Cells, at a density of 1×10^7 cells/ml, were incubated with drugs at 37° in 0.9% NaCl solution containing 0.05 M phosphate buffer (pH 7.2). Treatment times were as follows: 4-NQO, 5 min; 5-nitro-2-furaldehyde diacetate (NF), diamide, and NEM, 10 min; nitrosobenzene, 2 min. Control value for NPSH in Ehrlich cells was $3.5 \pm 0.6 \times 10^{-6}$ nmol/cell. (From Ref. 24.)

cytotoxicity of misonidazole [22,28]. Glutathione has also been found to protect against the aerobic cytotoxicity of 4-NQO [24].

IV. EFFECTS OF NITRO COMPOUNDS ON THE OXYGEN CONSUMPTION OF PURIFIED MICROSOMES

The major site for cellular reduction of nitro compounds to oxygen-reactive intermediates is believed to be microsomal since these reactions occur in KCN- and antimycin A-inhibited cells. Results with diamide and with NEM [cf. Figs. 7 and 13 and Refs. 17,18,24, 31,47], both of which are known to inhibit cellular oxygen utilization, also suggest that a major portion of the reactions responsible for the stimulation of oxygen utilization are extramitochondrial, although—as previously mentioned—electron transfer from mitochondrial flavoprotein is possible [48,49]. Under some conditions, as in the case of 4-NQO, electron transfer to cytochrome $c + c_1$ may actually be stimulated (cf. Fig. 18 in Section V and Ref. 24). These results would agree with the general concept that a considerable amount of cellular nitroreductase activity is located on the microsomes [15,16,19].

Microsomes have been reported to have nitro reductase activities due to their flavoprotein content, and microsomal nitroreductase can reduce aromatic nitrocompounds to hydroxylamine intermediates or amine derivatives [3,15] under anaerobic conditions. Other enzymes such as xanthine oxidase [50], liver aldehyde oxidase [51], and NADPH-cytochrome c reductase [19,52] may use nitro derivatives as electron acceptors [15,16,19]. There is no evidence that the nitro compounds are reduced in air with any of these systems. The only exception may be the reduction of 4-NQO by diaphorase in liver preparations. However, oxygen tensions were not actually monitored in these studies, and it is quite possible that the system rapidly became anaerobic [24,25]. In addition to flavoproteins possessing nitroreductase activity, there have been reports that riboflavin mononucleotide plus NADPH, under anaerobic conditions will also reduce nitro derivatives [53] and that ascorbate may reduce 4-NQO under anaerobic conditions [54]. Under certain conditions liver mitochondria can also utilize nitro compounds as electron acceptors, producing oxygen reactive nitro radicals [48,49]. This latter reaction might occur because of the high flavoprotein content in the mitochondria. In general, electron acceptance by nitro compounds is believed to occur at the level of flavoprotein [15,16, 19] and is thought to involve one-electron transfer. Recently it was demonstrated that cellular and microsomal nitroreductase is inhibited by oxygen [15-18]. The mechanism for this aerobic inhibition (lack of reduction of the nitrocompound beyond the radical intermediate) is believed to be due to the reactivity of the nitro radical with oxygen, which precludes further metabolic reduction:

$$R\text{-}NO_2 + e^- \rightarrow R\text{-}NO_2^{\bar{\cdot}} \quad \left[\begin{array}{l} + e^- + 2H^+ \rightarrow R\text{-}N=O + H_2O \\ + O_2 \rightarrow R\text{-}NO_2 + O_2^{\bar{\cdot}} \end{array} \right. \qquad (11)$$

Mason and Holtzman demonstrated this mechanism of inhibition of nitro reductase for purified microsomes [15,16]. Their electron spin resonance (ESR) studies showed that the first step in the reduction process involves the transfer of an electron from the flavin region of the microsomes to produce the radical intermediate [15,16]. The nitro radical, once formed, is extremely reactive with oxygen, as has been demonstrated in chemical systems for many different heterocyclic nitro compounds [14]. The rates for the reaction of these compounds with oxygen are given in Table 3 [17]. Wardman found that the reactivity of the radical anions of monosubstituted nitro derivatives increases as the electron affinity of the nitroaromatic derivatives increases [14]. However, Biaglow et al. [17] did not find this to be true for di- and trisubstituted nitrobenzene derivatives. The nitro radical can react with itself, and the rate constant for the dismutation process is usually larger than that for the reaction with oxygen. This suggests that the nitroso intermediates that are spontaneously formed may react with additional electrons derived from the flavoprotein sites to produce the oxygen-reactive hydroxyamine or hydronitroxyl radical [15,16,26].

Table 3 Rate Constants for Reactions of Nitrobenzene Radical Anion Under Aerobic and Anaerobic Conditions

NB derivative[a]	$k(NB^- + O_2)$ $\times 10^7\ M^{-1}\ sec^{-1}$	$2k(NB^- + NB^-)$ $\times 10^{-1}\ M^{-1}\ sec^{-1}$
PNP	2.0	350
NB	3.0	270
PCNB	2.8	250
PNAP	0.7	150
PNBN	1.5	120
TNBS	0.8	110
DNBN	0.6	130
DNPC	1.2	120

[a]In order of increasing electron affinity.
Source: From Ref. 17.

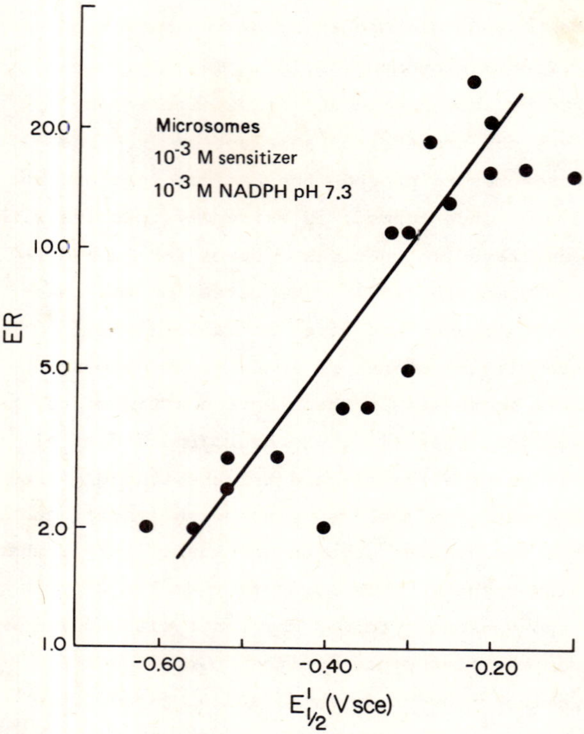

Figure 15 Stimulated oxygen consumption by Ehrlich ascites cell microsomes in the presence of various nitroheterocyclic drugs, showing an exponential increase in rate of oxygen consumption as a function of sensitizer electron affinity. Isolated microsomes (2.0 mg protein) were added to 3 ml of 0.9% NaCl solution containing 0.05 M phosphate buffer (pH 7.25) and 1 mM NADPH at 37°. The reaction was initiated by addition of the respective drug, and the rate of oxygen consumption was determined 1 min later. All of the initial rates were corrected for the microsomal oxidation of NADPH in the absence of drug, which was 1.5 nmol/min. Ordinate: ER refers to enhancement for oxygen consumption. (From Ref. 26.)

Reduction of nitrobenzene derivatives to oxygen-reactive intermediates by microsomes was dependent on electron affinity [17], as was the case in intact cells (cf. Figs. 3 and 7). In the microsomal system, NADH or glucose-6-phosphate and glucose-6-phosphate dehydrogenase could substitute for NADPH as the electron donor, although the rate of oxygen consumption was lower in the presence of NADH. Greenstock et al. [26] extended these studies to additional nitroheterocyclic compounds, as seen in Figure 15; they found that oxygen consumption increases exponentially with nitroheterocyclic electron affinity from a threshold redox potential of -0.65 V vs. SCE, which is consistent with the reactions shown in Figure 16 suggesting the involvement of nitroaromatic compounds ($R-NO_2$) in catalyzing electron transfer from flavin or flavoprotein-containing microsomes (nitroreductase) to oxygen. These results stress the value of using tumor microsomes for studying the relationship between drug electron affinity and oxygen consumption. However, a chief drawback is the fact that the lack of a membrane barrier may give misleading results, so that drugs which stimulate microsomal oxygen consumption by microsomes may have no effect on the oxygen consumption of intact cells [17]. This was found to be true for charged molecules such as TNBS and DNPC (Figs. 3A,B,C, and 7).

Biaglow and Varnes also determined the effect of a number of clinically used nitro compounds and the carcinogen 4NQO on the oxygen utilization of Ehrlich microsomes [30,55]. The most effective

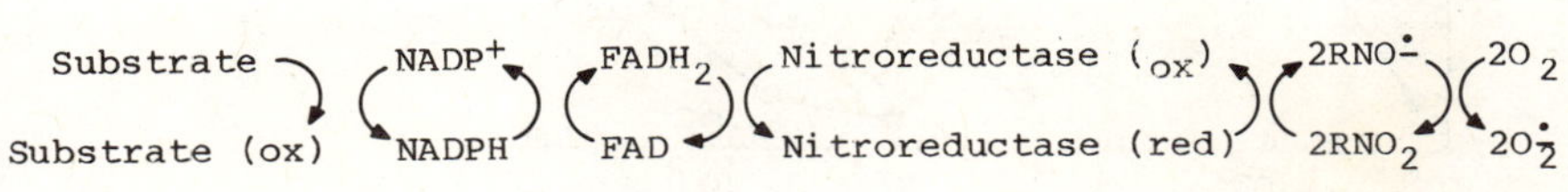

Figure 16 The involvement of nitroheterocyclic compounds (RNO_2) in catalyzing electron transfer from the flavin coenzyme containing microsomal "nitroreductase" to oxygen.

stimulator of oxygen utilization, as well as the most electron-affinic compound, was 4-NQO followed by nitrofurazone, niridazole, misonidazole, and metronidazole in a decreasing order of effectiveness (Fig. 17). These results indicate that all of the clinically used nitrocompounds are capable of being activated by the microsomes to oxygen-reactive intermediates.

A. Production of Superoxide and Peroxide

The stimulation of oxygen utilization as seen in Figs. 1, 7, 8, 15 and 17, may be due to the production of oxygen-reactive intermediates such as nitro radicals and hydroxylamine-derived radicals. Biaglow et al. [17,18] have demonstrated that both superoxide dismutase and catalase influence the rate of oxygen consumption by Ehrlich microsomes in the presence of the nitro compounds. Catalase was found to reduce the rate presumably by recycling the peroxide [17,18,56]. The effects of superoxide dismutase are more compli-

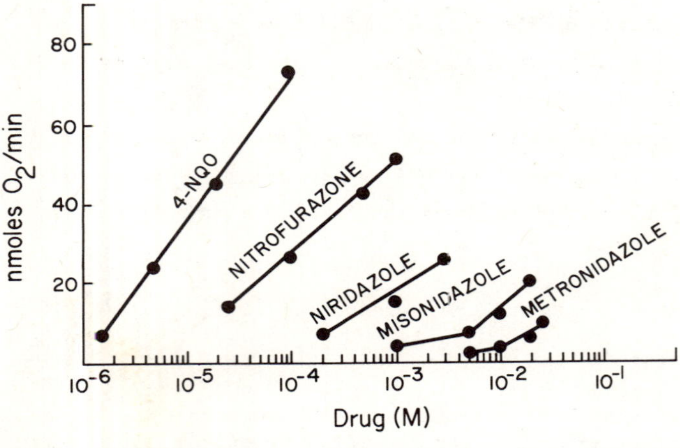

Figure 17 The effect of nitrocompounds on the oxidation of NADPH by Ehrlich cell microsomes. Isolated microsomes (2.0 mg protein) were added to 3 ml 0.9% NaCl solution containing 0.05 M phosphate buffer (pH 7.25) and 3 mM NADPH, at 37°. The reaction was initiated by addition of the respective drug, and the rate of oxygen consumption was determined 1 min later. All of the initial rates were corrected for the microsomal oxidation of NADPH in the absence of drug, which was 1.5 nmol/min.

cated and suggest that oxidation of enzyme-bound NADPH by $O_2^{\bar{\cdot}}$ is prevented [57]. Reduction of $O_2^{\bar{\cdot}}$ by enzyme-bound NADPH would result in the production of peroxide and oxygen consumption. The resultant NADP• radical (also produced by reaction of NADPH with OH• generated by Fenton-type reactions [56]) has been demonstrated to be highly reactive with nitroheterocyclic compounds [17,26]. These results suggest, that in the absence of sufficient catalase and superoxide dismutase, the initiation of chain reactions may occur by the transfer of one electron to produce the oxygen-reactive nitro radical anion [15,16, 17,27]. The combined effects of superoxide dismutase and catalase are greater than for either enzyme alone, indicating that additional reactions are inhibited when both superoxide and peroxide are removed. The ratio of peroxide produced to oxygen consumed is presented in the next-to-last column of Table 4. This ratio is of interest, because as with CN^--inhibited cells, peroxide accumulates in the microsomal system. If all the $O_2^{\bar{\cdot}}$ produced by the superoxide dismutase catalyzed reaction ($2O_2^{\bar{\cdot}} + 2H^+ \rightarrow H_2O_2 + O_2$) resulted in H_2O_2 production, the ratio would have been 1. The ratio was found to be the highest with the menadione (highly electron affinic non-nitro compound), followed by 4-NQO and the nitrofurans. The nitrobenzenes were also found to produce H_2O_2 with the Ehrlich microsomal system [17,18]. A high ratio also suggests the lack of side reactions such as production of OH• or the reaction of superoxide with NADPH to produce additional chain reactions [27].

B. Effects of Clinically Used Nitro Compounds on the Oxygen Consumption of S9 Fraction of Rat Liver

There is considerable interest in the mutagenic and carcinogenic properties of clinically useful nitro compounds such as niridazole, metronidazole, misonidazole, and nitrofurazone [1-8,43,58,59]. Metronidazole, niridazole, nitrofurazone, and 4-NQO have been shown to be mutagenic in the Ames carcinogen assay [1,2,58]. The finding that certain drugs stimulate O_2 utilization with Ehrlich tumor

Table 4 Effects of Superoxide Dismutase and Catalase on Oxygen Consumption by Ehrlich Microsomes in the Presence of Nitrobenzene Derivatives and NADPH[a]

Drug and concentration	Control	Oxygen utilized (nmol/min) + Catalase	+ SD	SD + catalase	H_2O_2/O_2	Reference
1. PNAP, 1 mM	28	12	15	10	0.8	17
2. DNBA, 1 mM	63	30	42	24	0.72	17
3. DNBN, 1 mM	60	34	48	24	0.84	17
4. DNPC, 1 mM	60	36	42	24	0.67	17
5. TNBS, 1 mM	93	66	43	24	0.57	17
6. Nitrofurazone	51	33	35	21	0.7	18
7. Metronidazole, 20 mM	6	3.6	6	3.5	—	55
8. Midonidazole, 20 mM	17	8	11	6.6	—	55
9. Niridazole, 5 mM	33	16	12	8.4	—	55
10. 4-NQO, 0.1 mM	57	29	36	21	0.9	18
11. Menadione, 0.1 mM	80	36	36	19	1.0	18
12. AF-2, 0.1 mM	54	30	51	27	0.63	55

[a]The reaction medium contained 6 mg of microsomal protein and 1 mM NADPH. Catalase and superoxide dismutase, when used, were added at 200 μg/3 ml. Nitro compounds, 3 μmol, were added to the reaction cuvette to initiate the reaction. The ratio of peroxide produced to oxygen consumed was determined during a 5-min incubation period. Rates of oxygen consumption were determined 1 min after addition of drug. The presence of peroxide was determined by oxygen evolution after the addition of catalase. The ratio was obtained by dividing the total amount of peroxide produced by the amount of oxygen consumed.

Source: Data from Refs. 17, 18, and 55.

microsomes (Fig. 17) suggested that similar measurements be performed with the Ames system. The effects of the nitroaromatic drugs on the oxygen consumption of the 9,000 x *g* rat liver supernatant commonly used with the Ames system have been determined [1,2,58]. All of the nitro compounds had the ability to stimulate oxygen utilization of this system in the presence and absence of NADPH (Table 5).

The finding that liver S9 homogenates consume oxygen in the presence of nitro drugs is interesting with respect to the mechanism of action of these drugs in the Ames mutagenicity test system (Table 5). Present evidence suggests that the mutagenic species for many nitro drugs is the nitro radical or a nitroso or hydroxyamine intermediate [3,7,28]. In addition, many of the nitro heterocyclic compounds, especially the nitrofurans, are more mutagenic and cytotoxic under hypoxic conditions, where formation of reduced intermediates is not inhibited [3]. In the previous sections we have shown that reaction of drug intermediates with oxygen can occur. If nitro radical anions or nitroso intermediates are the important mutagenic species in the Ames assay system, then results of mutagenicity testing for drugs such as metronidazole, misonidazole, and 2-(2-furyl)-3-(5-nitro-2-furyl)acrylamide (AF2) may depend on the oxygen tension

Table 5 Effect of Nitro Compounds on Oxygen Consumption of 9,000 x *g* Rat Liver Supernatant[a]

	Oxygen consumed (nmol/min)	
Drug concentration	-NADPH	+NADPH
No drug	18	47
Metronidazole, 10 mM	25	75
Misonidazole, 10 mM	37	144
Niridazole, 1 mM	43	111
Nitrofurazone, 1 mM	60	264
4-NQO, 0.5 mM	50	108

[a]The reaction mixture contained 8 mg biuret protein in 0.02 M hepes buffer, 37°, pH 7.2. The NADPH concentration was 1 mM. The reaction was initiated by the addition of nitro compounds. The values represented in the table are the initial rate of oxygen consumption obtained the first minute after drug addition.

of the test system [55). Factors such as the amounts of S9 homogenate used, cell density, and drug concentration could alter oxygen tension and thus increase or decrease the mutagenic effect. Nagao et al. [60] and others [1,2,58] have shown that the degree of mutagenicity of several chemicals is dependent on the amount of S9 supernatant used in the Ames assay. We suggest, therefore, that oxygen tensions ought to be monitored with the Ames system when testing drugs that are known to stimulate oxygen utilization. This would be true also for many other clinically used drugs, such as adriamycin and bleomycin, which are mutagenic in the Ames system [1,2,58] and which were recently reported to stimulate microsomal oxygen consumption [61].

V. OXYGEN CONSUMPTION BY ASCORBATE-NITRO COMPOUND MIXTURES

Various nitro compounds react with ascorbate in a reaction that consumes oxygen and ultimately results in the oxidation of ascorbate [20-22,25]. The various reactions involved are shown schematically in Figure 18. The study of the reaction with ascorbate is important for the following reasons: (a) ascorbate mimics nitroreductase by donating one electron to the nitro drug; and (b) ascorbate increases the hypoxic cytotoxicity of nitro compounds, presumably by reducing the drugs to cytotoxic radical intermediates. Oxygen consumption in drug ascorbate mixtures may, in part, be due to the transfer of the electron from the nitro radical to oxygen, as seen in Figure 18. This reaction leads to formation of superoxide and results in production of H_2O_2 and possibly even OH•. These three oxygen-derived species are reactive with ascorbate [20-22,25,54].

The oxygen consumption of ascorbate-drug mixtures increases as the polarographically determined electron affinity of the drug increases (Fig. 19). The most reactive drug was DNPC, followed by TNBS and 4-NQO. The least reactive drugs were those possessing the most negative redox potentials, such as 2-NP and p-nitrotoluene [20,21].

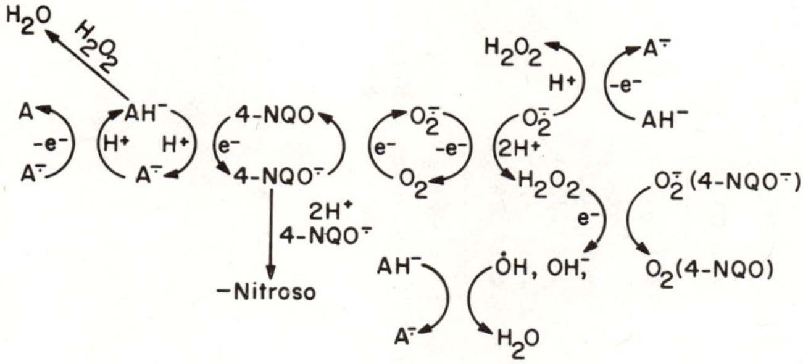

Figure 18 The reaction of ascorbate (AH^-) with 4-NQO to produce the nitro radical ($4\text{-}NQO^{\dot{-}}$). Nitro radical may react with oxygen to produce superoxide or with ferricytochrome c if present. The dismutation of superoxide results in peroxide which may react with either superoxide or ascorbate. Nitro radicals as well as superoxide, in the presence of Fe^{3+} contamination, react with peroxide to produce hydroxyl radical (OH•). (From Ref. 54.)

A simple method used to demonstrate the production of the reactive superoxide radical is to include superoxide dismutase in the reaction mixture. The initial presence of superoxide dismutase reduced the rate of oxygen consumption by 22-38% for the six chemicals tested (Table 6). The largest effect occurred with 4-nitropyridine-1-oxide (4-NPO). The initial presence of catalase also reduced the rate by 27-42%, and the largest effect similarly occurred with 4-NPO. The initial presence of both enzymes reduced the rate by 53-65%. The inhibitory effects of the two enzymes were nearly additive for 4-NQO, TNBS, DNPC, and DNBN, and more than additive for 4-NPO and nitrofurazone (see relative rates in parenthesis). The simultaneous production of H_2O_2 and $O_2^{\dot{-}}$ may result in production of hydroxyl radicals that are known to be extremely reactive with ascorbate [20-22,25,54,63]. However, the production of OH• may also depend upon Fe^{2+} contamination [56]. DMSO has been used previously as a trap for the OH• radical [65], but its oxidation by this radical is complex [65]. We have tested the ability of 10% DMSO in buffer systems to alter the oxygen consumption of ascorbate-drug mixtures. In all cases the presence of DMSO increased the oxygen consumption, presumably by reducing the radicals to water.

Table 6 The Effect of SD and Catalase on the Oxidation of Ascorbate[a]

Drug	Control rate	Oxygen consumed (nmol/min)			
		SD	Catalase	SD + Catalase	10% DMSO
4-NQO	197	150 (0.76)	144 (0.73)	78 (0.40)	262 (1.33)
TNBS	246	191 (0.78)	165 (0.67)	116 (0.47)	264 (1.07)
DNPC[b]	290	209 (0.72)	180 (0.62)	102 (0.35)	399 (1.40)
DNBN	68	45 (0.66)	45 (0.66)	24 (0.35)	72 (1.06)
4-NPO	198	123 (0.62)	114 (0.58)	75 (0.38)	222 (1.12)
NFA[c]	47	30 (0.64)	30 (0.64)	20 (0.43)	54 (1.15)

[a]The reaction mixture contained 1 mM drug in 50 mM potassium phosphate buffered and 0.9% NaCl, pH 7.3, 37°. When used, catalase and SD were 200 μg/3 ml. The values represent the initial rate obtained after the addition of 3 μmol of ascorbate. The values in parenthesis were obtained by dividing the rate obtained in the presence of enzyme or DMSO by the control rate found in column 2.

[b]Dinitropyridinium chloride was tested at 0.5 mM.

[c]Nitrofurazone.

Source: Reprinted with permission from J. E. Biaglow.

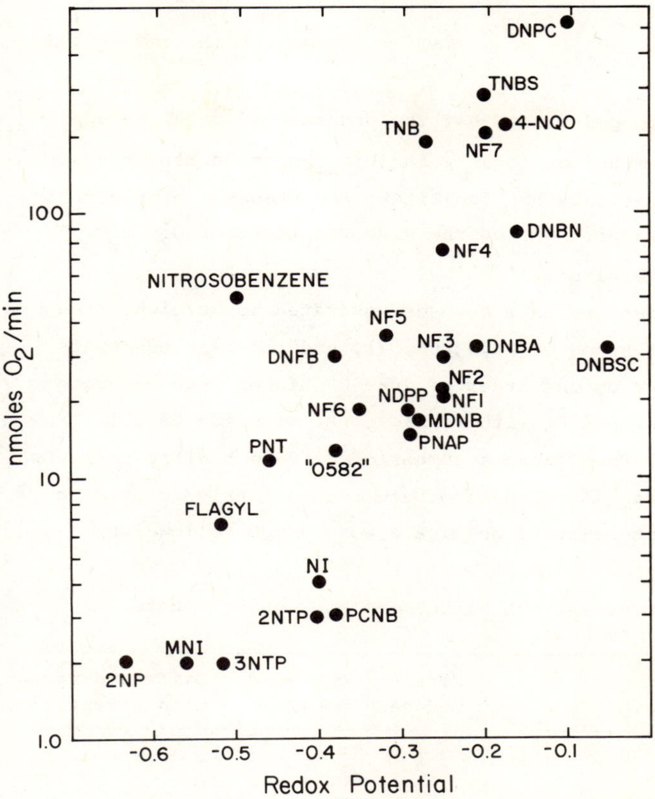

Figure 19 The relationship between drug redox potential (with respect to SCE) and the rate of ascorbate reduction of oxygen. The reaction was initiated by the addition of 3 μmol drug to a reaction mixture containing 1 mM ascorbate, 50 mM potassium phosphate buffer, pH 7.3 in 0.9% NaCl, 37°. The initial rates of oxygen consumption were measured after the addition of drug. The abbreviations and redox potentials of the drugs are as follows: 2-nitropyrrole (2NP, -0.62); 2-methyl-5-nitroimidazole (MNI, -0.56); 3-nitrothiophene (3NTP, -0.52); metronidazole (flagyl, -0.52); nitrosobenzene (-0.50); p-nitrotoluene (PNT, -0.47); 2-nitroimidazole (NI, -0.40); 2-nitrothiophene (2NTP, -0.40); p-chloronitrobenzene (PCNB, -0.38); dinitrofluorobenzene (DNFB, -0.38); nitrofurate (NF6, -0.35); 5-nitrofuraldehydediacetate (NF5, -0.32); misonidazole Ro-07-0582 ("0582", -0.30); p-nitroacetophenone (PNAP, -0.29); p-nitro-3-dimethylaminopropiophenone hydrochloride (NDPP, -0.29); m-dinitrobenzene (MDNB, -0.28); trinitrobenzene (TNB, -0.27); nitrofurazone (NF3, -0.25); nifuroxime (NF3, -0.25); nitrofurantoin (NF2, -0.25); furoxone (NF1, -0.25); 3,5-dinitrobenzamide (DNBA, -0.21); trinitrobenzenesulfonate (TNBS, -0.20); 5-nitrofuraldehyde (NF7, - 0.20); 4-nitroquinoline-N-oxide (4-NQO, -0.17); 3,5-dinitrobenzonitrile (DNBN, -0.16); 1-(2,4-dinitrophenyl)pyridinium chloride (DNPC, -01.); and 3,5-dinitrobenzenesulfonyl chloride (DNBSC, -0.05). (Reprinted with permission from J. E. Biaglow, B. E. Jacobson, M. E. Varnes, and C. Koch, The oxidation of ascorbate by electron affinic drugs and carcinogens, *Photochem Photobiol.* *28*:869-876. Copyright 1978, Pergamon Press, Ltd.)

It is of interest to point out that the presence of catalase and superoxide dismutase do not totally inhibit the oxidation of ascorbate by the nitro radicals but facilitate the transfer of electrons to oxygen to form water; some of the side reactions of $O_2^{\bar{\cdot}}$, H_2O_2, and OH• are also inhibited.

Nitro radical anions have been demonstrated to be highly reactive with ferricytochrome c [21,54,66,67], and the rate constants for the reaction are on the order of 10^3-10^4 higher than the rate constant for reaction of $O_2^{\bar{\cdot}}$ with ferricytochrome c [56,66,67]. Biaglow et al. [21] found that a number of different nitro radicals generated by the reaction with ascorbate reduce ferricytochrome c (Table 7). The most reactive chemicals were 4-NQO followed by

Table 7 The Effect of Various Nitro Compounds on Ascorbate Reduction of Ferricytochrome c[a]

Drug	Ferricytochrome c reduced (nmol/min)	Oxidation-reduction potential
Dehydroascorbate, 100 μM	13	—
Ascorbate, 33 μM	65	—
Alloxan	242	—
2-Nitropyrrole	65	-0.67
Metronidazole	65	-0.60
Nitrosobenzene	770	-0.47
p-Nitrotoluene	65	-0.47
Misonidazole	64	-0.30
5-Nitrofuraldehydediacetate	147	-0.32
Nitrofurazone	268	-0.30
4-Nitropyridine-N-oxide	286	-0.25
Trinitrobenzenesulfonate	230	-0.20
Dinitrobenzonitrile	106	-0.16
4-Nitroquinoline-N-oxide, 1.0 mM	1610	-0.15
0.1 mM	157	
Dinitropyridiniumchloride, 1.0 mM	450	-0.10
0.1 mM	123	

[a]The reaction mixture contained 60 μM ferricytochrome c, buffered with 10 mM potassium phosphate plus 10 mM Tris in 0.9% NaCl, pH 7.4, 37°. The drug concentration was 1 mM except where noted. The values represent the initial rate obtained after the addition of ascorbate to a final concentration of 33 μM.

Source: Reprinted with permission from J. E. Biaglow, B. Jacobson, and M. E. Varnes, The oxidation of ascorbate by electron affinic drugs and carcinogens. *Photochem. Photobiol. 28*:869-876. Copyright 1978, Pergamon Press, Ltd.

nitrosobenzene, DNPC, TNBS, nitrofurazone, NFA, and DNBN [20,21]. Misonidazole (1 mM) and metronidazole (1 mM) were not reactive in this system.

The rate constant for enzymatic dismutation of superoxide (2 x 10^9 mol M^{-1} sec^{-1}) is much larger than the rate constant for reaction of $O_2^{\overline{\cdot}}$ with ferricytochrome c [56,66,67]. Superoxide dismutase, when added to the ascorbate-drug reaction mixture, does not influence the rate of reduction of ferricytochrome c (Table 7). These data, in agreement with the aforementioned rate constants, suggest that $O_2^{\overline{\cdot}}$ is not the reducing species in drug-ascorbate systems. The reduction of ferricytochrome c was greater in the presence of the more electron-affinic chemicals (see Table 7). This is in agreement with their greater ability to accept electrons from ascorbate (see Figs. 2 and 19).

The rate of reduction of ferricytochrome c was greater with 4-NQO than with DNPC, although DNPC is a more effective stimulator of oxygen consumption. On the basis of oxidation-reduction potential, DNPC (-0.1 vs. SCE) should be more reactive toward both oxygen and ferricytochrome c than is 4-NQO (-0.15 vs. SCE). These results suggest that factors such as the positive charge of DNPC may alter the rate of reaction of radicals with ferricytochrome c [54,66,67]. Differences between reactivities of DNPC and 4-NQO toward ferricytochrome c diminished as drug concentration was reduced from 1.0 to 0.1 mM. Also of interest was the finding that nitrosobenzene, the two-electron reduction product of nitrobenzene, was quite effective in stimulating the reduction of ferricytochrome c in the presence of ascorbate (see Fig. 18 and Table 7). The transfer of an electron between nitrosobenzene and ascorbate results in the formation of the hydroxylamine radical [114] which, like the nitro radical, is oxygen-reactive as seen in Eqs. (12) and (13) [7]:

$$\text{R-N=O} + e^- + H^+ \rightarrow \text{R-}\dot{\text{N}}\text{-OH} \tag{12}$$

$$\text{R-}\dot{\text{N}}\text{-OH} + \text{OH}_2 \text{ or ferricytochrome c } (Fe^{3+}) \rightarrow$$
$$\text{R-N=O} + H^+ + O_2 \text{ or ferricytochrome c } (Fe^{2+}) \tag{13}$$

$$2\text{RNO}_2^{\overline{\cdot}} + H^+ \rightarrow \text{R-N=O} + \text{RNO}_2 + H_2O \tag{14}$$

The formation of hydroxylamine radical probably occurs due to the addition of one electron to the nitroso compound (Eq. 12). Our studies [21] were not to distinguish between the two possible pathways (Eqs. 13 and 14) which lead to the formation of nitroso compounds. As mentioned previously, the dismutation reaction (Eq. 14) becomes more important for the polynitro than the mononitro compounds [17]. Ehrlich cells which contain both superoxide dismutase and catalase were effective in slowing down the rate of oxygen consumption for ascorbate-drug mixtures (Table 8), and like the chemical systems the cells could not prevent the formation of the nitro radical. The inhibition of O_2 consumption was not in general as great as was obtained for the purified microsomes (Table 4). This lowered degree of inhibition may reflect the reactivity of nitro radical anions with sulfhydryl groups in proteins and GSH. However, these effects may be cell density dependent. Additional evidence that cellular enzymes were capable of inactivating peroxide and superoxide radicals was obtained by including KCN in the reaction medium. KCN inhibits catalase and superoxide dismutase activities [17,18,68].

As mentioned previously, the ability of ascorbate to increase the hypoxic cytotoxicity of a number of nitro compounds is of potential clinical usefulness [16-18,69]. It remains to be demonstrated whether this activation results in production of mutagenic species known to be generated in similar reactions between microsomes and NADPH in the Ames carcinogen test system [1,2,55]. Clearly more research is needed in this area in order to determine the potential usefulness of the ascorbate-drug interaction [22].

VI. USE OF CYANIDE-INHIBITED CELLS, MICROSOMES, AND ASCORBATE REACTIONS AS SCREENING TESTS FOR POTENTIALLY USEFUL NITRO COMPOUNDS

Cyanide-inhibited cells, purified microsomes, and ascorbate are three different systems which provide a reducing environment required for the activation of nitro compounds to oxygen-reactive intermediates (Figs. 2, 8, 17, and 18). Cyanide-inhibited cells provide a

Table 8 Drug-Catalyzed Ascorbate Oxidation in the Presence of Ehrlich Cells[a]

	Oxygen utilized (nmol/min)				
Drug	Cells	Cells + drug	Cells + drug + ASC	Drugs + ASC	Ratio[b]
DNPC, 0.5 mM	111	111	291	300	0.60
TNBS, 1 mM	90	90	255	248	0.66
4-NQO, 1 mM	114	114	228	192	0.60
DNBN, 1 mM	111	141	204	96	0.66
4-NPO, 1 mM	114	147	270	186	0.66

[a]Ehrlich cells were incubated in 50 mM phosphate buffered 0.9% NaCl, pH 7.3, 37°. The rate of oxygen consumption was determined before and after each sequential addition of drug followed by 1 mM ascorbate. Drug plus ascorbate rates were also determined in the absence of cells. The cell density was 6.7×10^6 Ehrlich cells/ml. Total incubation volume was 3 ml (data taken from a series of similar experiments).

$$^{b}\text{Ratio} = \frac{\text{rate of (cells + drug + ASC) - rate of (cells + drug)}}{\text{rate of (ASC + drug)}}$$

Source: Adapted with permission from J. E. Biaglow, B. Jacobson, and M. E. Varnes, The oxidation of ascorbate by electron affinic drugs and carcinogens. *Photochem. Photobiol. 28*:869-876. Copyright 1978, Pergamon Press, Ltd.

system in which drug penetration and the consequences of nitro radical anion production under aerobic conditions may be studied [17,18, 27-29,55]. The activation of nitro compounds to radical intermediates occurs mainly in the microsomal fraction of cells [14,15,19]. The purified microsomes provide a cell-free system where the production of oxygen-reactive nitro radicals may be studied. Oxygen consumption measurements with either microsomes or cyanide-inhibited cells can be used to identify those nitro compounds which may be too reactive, or mutagenic, for consideration as useful radiosensitizers in vivo [17,20,25].

The fact that ascorbate reacts with various nitro compounds [20-22] provides an additional means for testing drug reactivity by measuring effects on oxygen consumption. As with cyanide-inhibited cells and microsomes [Figs. 2, 15, 16 and 17], the drugs which are too reactive to be useful can be identified and may be excluded from in vitro or in vivo testing (Fig. 19). However, it remains to be determined whether or not the moderate reactivity between ascorbate and misonidazole [22] may be exploited in vivo as a means of enhancing hypoxic cell cytotoxicity (see Section V).

VII. USEFULNESS OF RESPIRATORY INHIBITION BY NITRO COMPOUNDS FOR RADIATION THERAPY

The ability of nitro compounds to sensitize hypoxic cells to ionizing radiation has been related to their Hammett constant or electron affinity, and more recently to their one-electron acceptance capacity [70-72]. These properties alone do not, however, completely account for their radiosensitizing or cytotoxic action [17,25,71-73]. Nitro compounds of increasing electron affinity are also more metabolically active (see Figs. 3, 7, and 8), and one of the more important metabolic effects is alteration of cellular oxygen utilization (Figs. 1 and 3). Such an effect is obviously unimportant under the very hypoxic conditions typically used to determine radiosensitizing ability [17,27,35]. However, effects of drug metabolism on oxygen utilization become important in tumorlike situations since it is the oxygen consumption by cells near the blood vessels that leads to hypoxia for the more distant cells [35,36].

With an in vitro spheroid tumor model [35], manipulation of the rate of oxygen utilization has been found to change the overall radiosensitivity dramatically because of increase or decrease in the number of hypoxic or anoxic cells. This phenomenon was demonstrated with spheroids by Durand and Biaglow, using DNP, nitrotoluene (NT), and DNBN [34,35]. For comparison of the consequences of stimulating oxygen utilization with DNBN or inhibiting the process with NT within a single population of spheroids, a composite curve was constructed by plotting the survival of spheroids when treatment with NT or DNBN immediately preceded a radiation dose of 2100 rad against the measured respiration rate (Fig. 20). A fairly straightforward calculation [35] leads to a prediction of the radiation response as a function of cellular respiration rate (dashed line) and allows a comparison of observed and theoretical responses. The consistent overestimate of survival at the various concentrations of DNBN (a respiratory stimulator) may indicate a small degree of hypoxic cell sensitization by this compound.

The results with spheroids indicate that radiosensitizing chemicals which alter oxygen utilization cannot be adequately tested for potential usefulness in radiotherapy when studied in a monolayer system where hypoxia is uniformly established by the experimental conditions. Since hypoxia may occur in a solid tumor mass due to metabolic reduction of the amount of oxygen available to cells distant from the vascular supply (metabolic barrier), the degree and extent of cellular and tissue hypoxia are critically dependent upon the rate of oxygen utilization. Many electron-affinic drugs, including some of the nitrobenzenes and nitrofurans, have been shown to stimulate cellular oxygen utilization (Fig. 2) or to stimulate oxygen consumption by reaction with ascorbate (Figs. 17, 18, and 19). This property may alter the effectiveness of these drugs as sensitizers in tumors, since they may produce an increased hypoxic region, as they do in the case of spheroids, and thereby reduce any effect that they may have as radiation sensitizers of the hypoxic and anoxic cells [17,18,20,21,23,25,34-38].

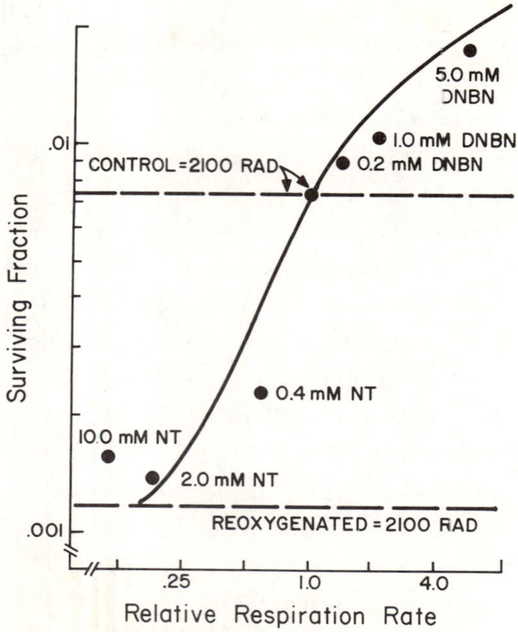

Figure 20 Survival of spheroid cells treated with NT or DNBN immediately prior to receiving 2100 rads, as a function of the relative oxygen utilization rate for each drug treatment. Dashed line represented the predicted survival for spheroids consuming O_2 at the indicated rates. As a first-order approximation, the depth to which oxygen will penetrate the spheroid is proportioned to the inverse square root of the oxygen consumption rate. Histological preparations allow an estimate of the size of the necrotic region. The relative volume of hypoxic cells can easily be estimated if it is assumed that they are located predominantly near the necrotic region of the spheroid. Therefore, for spheroids of radius a (291 μm), necrotic core radius b (72 μm), and "critical" oxygen penetration depth r, the hypoxic fraction F is given by $F = [(a - r)3]/(a^3 - b^3)$ for $a - b \geq r \geq 0$. The spheroids, when totally oxygenated, had a surviving fraction $S_{ox} = 0.0012$ after 2100 rad; if totally anoxic, the survival was $S_{an} = 0.129$. Hence, the net survival is a function of the fraction of cells that are anoxic, i.e., $S = (1-F)S_{ox} + F \cdot S_{an}$. Since the oxygen penetration (r) varies with rate of consumption K so that $r = r_o(K)^{1/2}$, where r_o represents the penetration depth under normal conditions (observed survival = 0.0076), substitution into the equation for S defines r_o, and hence allows calculation of S as a function of K. (From Ref. 35.)

VIII. CYTOTOXIC, RADIOSENSITIZING, AND MUTAGENIC PROPERTIES OF NITRO COMPOUNDS

A. Cytotoxicity and Radiosensitization

Although the powerful bactericidal activity of metronidazole against anaerobic but not aerobic bacteria is well known, the possibility of selective toxicity of this drug towards hypoxic mammalian cells was not explored until it was demonstrated by Sutherland [9], using histological and autoradiographic techniques, that metronidazole preferentially killed noncycling hypoxic cells in the center of multicellular spheroids of V79 Chinese hamster lung cells. Subsequently, several workers [11,12,28,74-77] have shown that both metronidazole and misonidazole are toxic to hypoxic cells at 37° but not at 0°, and no toxicity is observed at either temperature under aerobic conditions (Fig. 21). Also of interest is the finding that small amounts of oxygen, e.g., 1,000-2,000 ppm oxygen, afford significant protection against cytotoxicity. This suggests that both hypoxia and metabolism are necessary for the expression of toxicity [69]. Although aerobic toxicity is not seen with moderate concentrations of these drugs, a pronounced inhibition of growth occurs [74,75].

Detailed growth inhibition and cytotoxicity studies have been carried out with metronidazole and misonidazole using the V79 multicellular spheroid in vitro tumor model system [74,75] (Fig. 22). Spheroids of V79 Chinese hamster lung cells are balls of tightly adhering cells and contain internal regions of chronic hypoxia and necrosis similar to those found in poorly vascularized solid tumors. Both drugs, at concentrations low enough to be physiologically attainable in humans, were preferentially toxic to radioresistant hypoxic cells in spheroids, indicating that pretreatment with these drugs prior to initiation of radiotherapy would be advantageous.

Indeed both metronidazole and misonidazole have been shown to preferentially kill hypoxic cells in animal tumors [12,78,79]. The extent of cell killing often exceeds that expected from killing of the hypoxic cell fraction alone. An explanation for this may be that the hypoxic cells metabolically convert these drugs to deriva-

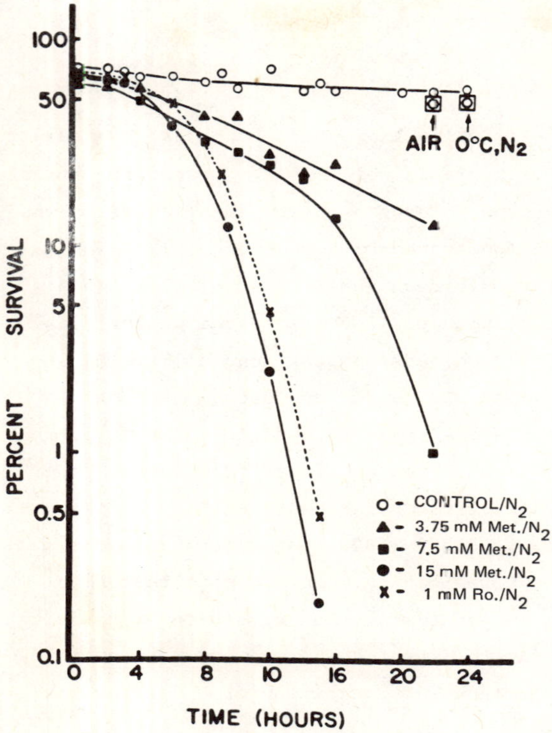

Figure 21 Survival of cells in exponential growth (cycling) in monolayer culture treated in 37° with metronidazole or Ro-07-0582 under hypoxic conditions (<10 ppm O_2). Symbols with squares are for cells treated with 15 mM metronidazole in air at 37° or nitrogen at 0°. Results of the most complete typical experiment where each point represents the average colony count of four separate Petri dishes. The standard error of these counts is within the size of the symbols. (Reprinted with permission from R. Sridhar, C. Koch, and R. Sutherland, Cytoxicity of two nitroimidazole radiosensitizers in an in vitro tumor model. *Int. J. Rad. Oncol. Biol. Phys. 1:* 1149-1157. Copyright 1976, Pergamon Press, Ltd.)

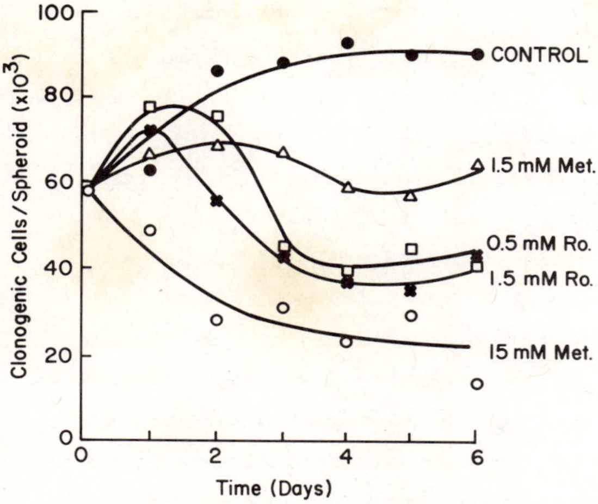

Figure 22 Survival of cells from spheroids treated continuously with metronidazole or Ro-07-0582. Total clonogenic cells per spheroid was determined by multiplying the total cells per spheroid by the fraction of cells capable of forming colonies. Results of the most complete typical experiment. (Reprinted with permission from R. Sridhar, C. Koch, and R. Sutherland, Cytoxicity of two nitroimidazole radiosensitizers in an in vitro tumor model. *Int. J. Rad. Oncol. Biol. Phys. 1*:1149-1157. Copyright 1976, Pergamon Press, Ltd.)

tives which are toxic to both hypoxic and oxygenated cells. Recently, Whitmore et al. [80] have shown that prolonged exposure of misonidazole to hypoxic cells leads to the formation of several derivatives of the compound, at least one of which is toxic to both aerobic and hypoxic cells. This cytotoxic derivative may be a hydroxyamine or a free radical generated from it that is not produced during aerobic treatment of cells. A correlation has been found between toxicity of misonidazole against hypoxic cells and DNA damage [81]. Cytological studies suggest that cytotoxicity of misonidazole is predominantly mediated via an interphase cell death accompanied by a minor effect due to chromosome aberrations and cell death due to unequal distribution of genetic material in progeny cells [82].

As previously stated, there is a positive correlation between metabolic reduction of most nitro drugs and their electron affinity (see Section II and Fig. 15). Electron affinity is readily measured by obtaining the half-wave reduction potential using polarography or pulse radiolysis [66]. The correlation between reduction potential and antibacterial activity of a series of nitroheterocyclics was first observed by Dann and Moller [83] and later by Sasaki [84] and Hirano et al. [85]. With respect to mammalian cells, Adams [71] has correlated the electron affinity of a series of nitro heterocyclics with their ability to act as hypoxic cell radiosensitizers and to cause cell death [73]. Thus, the highly electron-affinic nitrofurans are powerful hypoxic cell radiosensitizers but are also very toxic. Conversely, nitroimidazoles such as metronidazole and misonidazole have moderate capacity to accept electrons and are in general less cytotoxic and hence more useful as radiosensitizers [71-73].

Because of the relevance of hypoxic cell killing to radiation therapy of cancer, it is important to consider the modifying influence of nitro compounds on radiation-mediated killing of hypoxic cells as well as on toxicity in the absence of radiation. The two most important nitro compounds with respect to clinical application are metronidazole and misonidazole. Both drugs sensitize hypoxic cells to radiation damage without affecting the survival of aerobic cells (Fig. 23). The lethality of x-rays towards mammalian cells is approximately three times higher under aerobic conditions than under hypoxic conditions (radiobiological hypoxia ≃1,000 ppm oxygen), and this is implied in the statement that the oxygen enhancement ratio (OER) is 3 for x-rays. In the same way, it is possible to express the degree of hypoxic sensitization by radiosensitizers such as metronidazole and misonidazole in terms of enhancement ratios for cell killing; an enhancement ratio of 1 would indicate lack of sensitization while values greater than 1 would reflect hypoxic cell sensitization. There is abundant literature dealing with sensitization of hypoxic cells to radiation by metronidazole [75,85-91] and misonidazole [76,80,90]. Radiosensitization studies using multicell

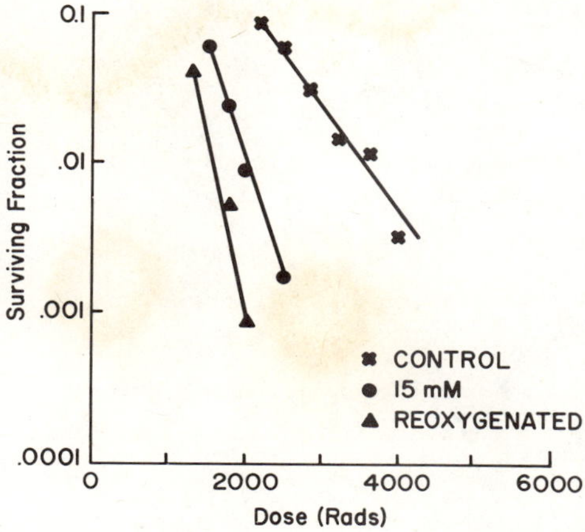

Figure 23 Radiosensitization of spheroids with metronidazole at 37°. The drug was added 1 hr before beginning irradiation at 37°. Reoxygenation of the cells in the center of the spheroids was achieved by lowering the temperature for 20 min before irradiation to 22°. (From Ref. 75.)

spheroids demonstrate the effectiveness of these sensitizers very convincingly [37,74,75]. Radiosensitization studies in vivo also demonstrate that misonidazole (which is more electron affinic) is more effective than metronidazole at equivalent concentrations. For recent summaries of these results see the reviews by Fowler et al. [8] and Adams' group [70-73].

Whitmore et al. [80] have shown that the metabolic activation of misonidazole leads to the formation of toxic products that reduce the shoulder in the radiation survival curve.

The radiosensitizing and cytotoxic properties of metronidazole and misonidazole were important factors in the selection of these drugs for clinical trials in Canada and Britain [96,97]. Although the results of clinical trials have shown some promise, it must be emphasized that these drugs produce undesirable side effects ranging from headache, dizziness, nausea, and vomiting to severe gastrointestinal disturbances and neurotoxicity. Misonidazole is more neurotoxic than metronidazole. The extent and severity of the side effects is dose-dependent and varies from patient to patient.

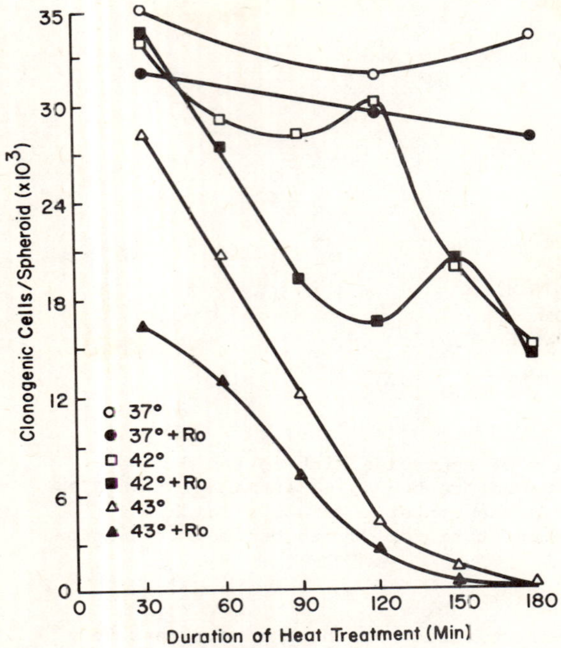

Figure 24 Cytotoxicity of 1.5 mM Ro-07-0582 and hyperthermia in Chinese hamster V79-171b multicell spheroids. A known number of spheroids was treated with hyperthermia for different times with or without Ro-07-0582, then dissociated to single cells and plated for colony formation to determine total clonogenic cells per spheroid. Each point represents cells from 15 to 20 spheroids. (Reprinted with permission from R. Sridhar and R. Sutherland, Hypothermic potentiation of cytotoxicity of Ro-07-0582 in multicell spheroids. *Int. J. Rad. Oncol. Biol. Phys.* 2:531-535. Copyright 1977 by Pergamon Press, Ltd.)

Severe peripheral neuropathy has been known to develop in patients receiving high-dose misonidazole therapy [96-98]. At present it appears that neurotoxicity may be a serious dose-limiting factor in the use of these anticancer agents. Attempts have been made to devise treatment modalities which allow the effective use of these chemicals at low enough doses to curb their mutagenic or carcinogenic potential and, more importantly, to reduce their neurotoxicity [28,99-105].

Recently it has been found that hyperthermia enhances the cytotoxicity of nitroimidazoles [28,35,100-106], e.g., the toxicity of misonidazole toward hypoxic cells in single cell suspension [28, 100,101] is much higher at 42.5° than 37°. Similar effects have been observed in the case of spheroids [102-104] (Fig. 24). Hall and Biaglow [28] have suggested that the cytotoxicity is due to an increased production of toxic metabolites.

As previously indicated, ascorbate increases the cytotoxicity of misonidazole and metronidazole (Fig. 25) whereas sulfhydryl compounds such as cysteamine and glutathione serve as protective agents ([107]; also C. J. Koch, R. L. Howell, and R. Sridhar, unpublished observations).

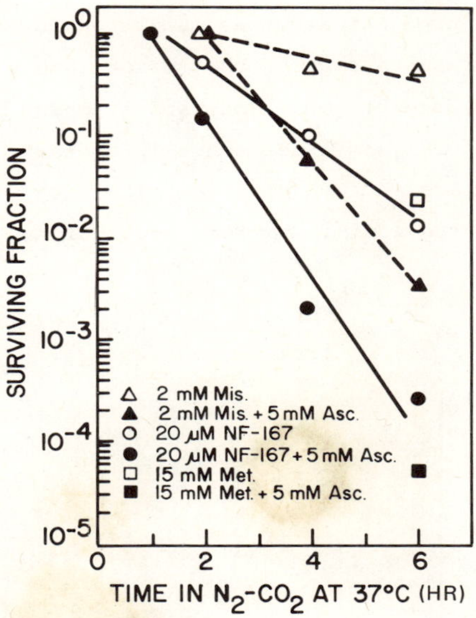

Figure 25 Cytotoxicity of radiosensitizers in the presence or absence of 5 mM ASC under conditions of extreme hypoxia at 37°. There was no cytotoxicity of any drug (combination) in air at 37° or in nitrogen at 0°. Survival at 1 hr for all drugs was not significantly different from controls and points have been omitted for clarity. All cytotoxicity data shown in the figures are for V79 cells. The chambers reach temperature equilibrium about 15 min after immersion in the water bath; the time scale was corrected for this. (From Ref. 22.)

B. Mutagenicity and Carcinogenicity

Another aspect of the use of misonidazole and metronidazole in clinical trials is the possibility that the drugs are carcinogenic [112,113]. The drug is administered orally and is distributed throughout the body at concentrations up to about 1 mM. As a consequence, a large number of normal cells are at risk [6]. To determine the magnitude of this problem, Hall et al. [6] determined the carcinogenic potential of misonidazole using an in vitro transformation assay system, namely the C_3H 10T 1/2 cell line developed by Reznikoff et al. [108]. The number of transformants per surviving cell was determined as a function of x-ray dose and as a function of drug concentration. As seen in Figure 26, a 3-day treatment with misonidazole at a concentration of 1 mM is equivalent to about 125 rad of x-rays; 5 mM drug for 3 days produces about as many transformants as 300 rad of x-rays. These results suggest that misonidazole is a potent mutagen certainly comparable to radiation, and it must be borne in mind that in vivo many more normal cells are in contact with the drug than are irradiated, because in a clinical situation the radiation is localized whereas drug administration results in a total body exposure.

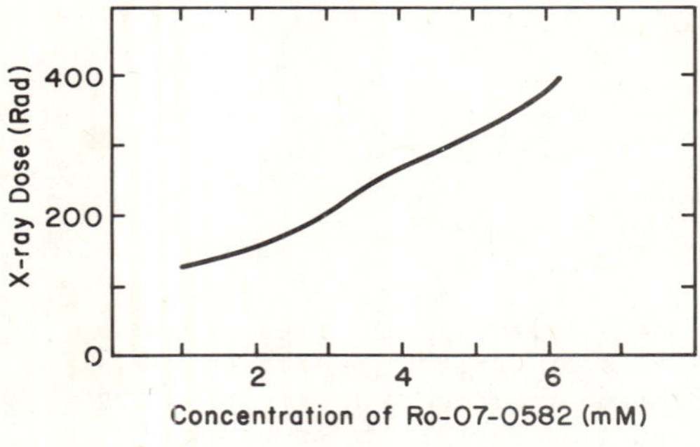

Figure 26 The approximate equivalence between single doses of x-rays and 3-day treatments with misonidazole at various concentrations to produce the same proportion of transformed colonies in C_3H/10T/2 cells. (From Ref. 6.)

In studies with nitroimidazoles, Rupp et al. [113] found that metronidazole, which undergoes metabolic activation by mammalian cells [8,70,71,73], is mutagenic to the tryptophan-requiring strain *Escherichia coli* WP2 uvr A at concentrations as low as 5 μM and causes reversion to tryptophan independence. The mutagenic action of the nitroimidazoles as well as other nitro compounds requires metabolic activation [1,2]. It is of interest that the radiosensitizing capacity, toxicity, and mutagenicity as well as inhibition of β-galactosidase synthesis in *E. coli* strain AB 1157 seem to correlate with the one-electron reduction potentials [113]. The one-electron reduction potentials may be related to metabolic activation if the first step in the reduction process, as described earlier, involves the transfer of a single electron to produce the nitro radical anion.

The nitro group is a required substituent for a number of the related effects of nitroheterocyclics, including toxicity, DNA damage, mutagenicity, carcinogenicity, antitumor activity, and radiosensitizing ability [3,73,109-111]. In addition, experiments with bacterial systems support the importance of metabolic reduction of the nitro group for the DNA-damaging effects of the nitrofurans and related nitro heterocyclics [3,41,109,110]. Absence of the nitro group eliminates the mutagenicity of most furan and other heterocyclic derivatives. The importance of the nitro group for the mutagenicity has led Arcos and Argo [7] to suggest that all nitro compounds have some mutagenic potential. Bacterial mutants lacking one or more of the enzymes involved in nitro reduction are far less susceptible to the toxic and DNA-damaging effects of nitrofurans, including the formation of single-strand breaks in DNA [3,109]. Those strains which do not contain enzymes capable of reducing nitrofurans under aerobic conditions can be made to revert under hypoxic conditions [3]. The requirement for metabolic activation of the nitro heterocyclic chemicals is seen best in the mutagenicity studies of Ames and others (see Table 5). While some of the nitro compounds are mutagenic without the S9 liver preparation, their mutagenicity is increased when incubated in the presence of the liver enzymes [1,2,3,58].

It is interesting to note that no mutants of mammalian cells have been isolated which lack nitroreductase activity, a finding which is consistent with the fact that mammalian cells contain five or more enzymes (e.g., xanthine oxidase, aldehyde oxidase, cytochrome P_{450} reductase, cytochrome c reductase, and diaphorase) which can activate these drugs [15-19]. The food additive AF2, used in Japan, produces azaguanine- and ouabain-resistant mutants to a greater degree when cells are incubated under hypoxic rather than aerobic conditions [3,109,110].

Whether the toxic intermediates resulting from metabolic reduction of nitrofurans, nitroimidazoles, and some of the nitrobenzenes can account for the entire amount of DNA breakage and cell killing by these agents is not known [3,28,109,110]. Although metabolism of nitrofurans and nitroimidazoles is inhibited in the presence of oxygen, partial reduction in air can occur, as evidenced by the formation of the nitrofuran radical anion [14,15,23] and the reduction of the carcinogen 4-NQO in air to yield anionic free radicals [18,55]. The cytotoxicity of 4-NQO in air is greatly reduced in the presence of glutathione [55]. The levels of mutagenicity and cytotoxicity of nitroaromatics under aerobic conditions is lower than those under anaerobic conditions [3]. This latter consideration has important consequences when we are considering the Ames mutagenicity assay system for testing compounds that may stimulate oxygen consumption (see Sections III through VI). The mutagenic species may be the $O_2^{\overline{\cdot}}$ or peroxide formed from it by the spontaneous dismutation process. Moreover, chemical reactions catalyzed by endogenous Fe^{2+} can cause the production of the very toxic hydroxyl radical [56].

IX. CONCLUSION

Nitro compounds continue to be the subject of great interest because of their potential usefulness as chemotherapeutic agents. Of particular importance is the possible use of nitro compounds in combination with radiation and hyperthermia in the multimodality treatment of human cancers. However, as discussed in Section VIII, the mutagenic and carcinogenic properties of several of these compounds limits their usefulness. Even the two most promising drugs, metronidazole and misonidazole, produce severe side effects when administered in large doses to humans.

Both the beneficial and deleterious properties of the nitro compounds are related to their metabolic reduction (Section I), which is influenced by glucose and intracellular substrates (NADPH, thiols), several enzymes (Section IV), and oxygen tension (Section II). In air, reduction of the nitro compound to the nitro radical anion stage can occur. This anion can subsequently react with oxygen, forming superoxide, peroxide, and possibly hydroxyl radical, as shown in Figures 2 and 18. These oxygen-derived species can have mutagenic and cytotoxic effects when not deactivated by cellular enzymes. Thus, it is essential when working with nitro compounds to use reaction conditions under which variables such as cell density and oxygen tension are precisely controlled. Controlled conditions are necessary to avoid artifacts that may lead to erroneous conclusions concerning the effects of nitro compounds on cellular metabolism, cytotoxicity, and mutagenicity.

ACKNOWLEDGMENTS

This work was supported by Grants CA 13747, CA 15901, CA 19283, and CA 21542 awarded by the National Cancer Institute, U. S. Department of Health and Human Resources. We wish to thank Ms. Lisa Gropf for her assistance in the preparation of the final manuscript.

REFERENCES

1. J. McCann, E. Choi, E. Yamasaki, and B. N. Ames, *Proc. Nat. Acad. Sci. U.S. 72*:5135-5139 (1975).
2. J. McCann and B. N. Ames, *Proc. Nat. Acad. Sci. U.S. 73*:950-954 (1976).
3. G. T. Bryan (Ed.), *Carcinogenesis: A Comprehensive Survey, Vol. 4: Nitrofurans: Chemistry, Metabolism, Mutagenesis and Carcinogenesis*. Raven Press, New York, 1978.
4. D. R. McCalla, A. Reuvers, and C. Kaiser, *Biochem. Pharmacol. 20*:3532-3537 (1971).

5. T. Speck, A. B. Stein, and H. S. Rosenkranz, *J. Nat. Cancer Inst. 56*:283-284 (1976).
6. E. J. Hall, R. Miller, M. Astor, and F. Rini, *Br. J. Cancer 37* (Suppl. III):120-124 (1978).
7. J. C. Arcos and M. R. Argo, *Chemical Induction of Cancer*, Vol. 2, Pt. B., Academic Press, New York, 1974.
8. J. F. Fowler, G. E. Adams, and J. Denekamp, *Cancer Treatment Rev. 3*:227-257 (1976).
9. R. M. Sutherland, *Cancer Res. 34*:3501-3503 (1974).
10. R. M. Sutherland, C. J. Koch, J. E. Biaglow, and R. Sridhar, *Proc. Am. Assoc. Cancer Res. 17*:Abstr. No. 883 (1976).
11. J. K. Mohindra and A. M. Rauth, *Cancer Res. 36*:930-936 (1976).
12. J. M. Brown, *Radiat. Res. 72*:469-486 (1977).
13. H. Harris, *The Principles of Human Biochemical Genetics*. Elsevier, New York, 1973, p. 123.
14. P. Wardman and E. D. Clarke, *Biochem. Biophys. Res. Commun. 69*: 942-949 (1976).
15. R. P. Mason and J. L. Holtzman, *Biochem. Biophys. Res. Commun. 67*:1267-1274 (1975).
16. R. P. Mason and J. L. Holtzman, *Biochemistry 14*:1626-1632 (1975).
17. J. E. Biaglow, B. Jacobson, C. Greenstock, and J. Raleigh, *Mol. Pharmacol. 13*:269-282 (1977).
18. J. E. Biaglow, B. E. Jacobson, and O. F. Nygaard, *Cancer Res. 37*: 3306-3113 (1977).
19. J. R. Gillette, *Handbook Exp. Pharmacol. 28*:340-385 (1971).
20. J. E. Biaglow and B. E. Jacobson, *Br. J. Radiol. 50*:844-856 (1977).
21. J. E. Biaglow, B. E. Jacobson, M. E. Varnes, and C. Koch, *Photochem. Photobiol. 28*:869-876 (1978).
22. C. J. Koch, R. L. Howell, and J. E. Biaglow, *Br. J. Cancer 39*: 321-329 (1979).
23. J. E. Biaglow, O. F. Nygaard, and C. L. Greenstock, *Biochem. Pharmacol. 25*:393-398 (1976).
24. M. E. Varnes and J. E. Biaglow, *Cancer Res. 39*:2960-2965 (1979).
25. J. E. Biaglow, The Effects of Anoxic Sensitizers on Cellular Metabolism and Respiration. *J. Pharmacol. Therap. 10*:283-299 (1980).
26. C. L. Greenstock, J. E. Biaglow, and R. E. Durand, *Br. J. Cancer 37* (Suppl. III):11-15 (1978).
27. J. E. Biaglow, C. L. Greenstock, and R. E. Durand, *Br. J. Cancer 37* (Suppl. III):145-149 (1978).
28. E. J. Hall and J. E. Biaglow, *Int. J. Radiat. Oncol. Biol. Phys. 2*:521-530 (1977).
29. J. E. Biaglow, M. E. Varnes, and B. Jacobson, *Radiat. Res. 74*: 499 (1976).
30. J. E. Biaglow, M. E. Varnes, B. Jacobson, and C. B. Schroy, Radiation Sensitizers, Cancer Management, Vol. 5 Chapt. 19 (L. Brady, Ed.) Masson Publishing Co., New York, 127-134 (1980).
31. J. E. Biaglow and O. F. Nygaard, *Biochem. Biophys. Res. Commun. 54*:874-881 (1972).

32. F. J. Bullock, in *Comprehensive Biochemistry*, Vol. 22: *Bioenergetics* (M. Florkin and E. H. Stotz, Eds.), Chap. III. Elsevier, New York, 1967, p. 81.
33. E. Kun, *Biochemical Regulatory Mechanisms in Eukaryotic Cells* (E. Kun and G. Santiago, Eds.). Wiley (Interscience), New York, 1972.
34. R. E. Durand and J. E. Biaglow, *Int. J. Radiat. Biol. 23*:235-239 (1974).
35. J. E. Biaglow and R. E. Durand, *Radiat. Res. 65*:529-539 (1976).
36. R. E. Durand, J. E. Biaglow, and R. M. Sutherland, *Br. J. Radiol. 49*:567-569 (1976).
37. R. E. Durand, J. E. Biaglow, and G. L. Greenstock, *Br. J. Cancer 37* (Suppl. III):150-153 (1978).
38. J. E. Raleigh, J. D. Chapman, A. Reuvers, J. E. Biaglow, R. E. Durand, and A. M. Rauth, *Br. J. Cancer 37* (Suppl. III):6-10 (1978).
39. T. W. Wong and G. Whitmore, *Radiat. Res. 71*:132-142 (1977).
40. J. E. Biaglow, O. F. Nygaard, and C. L. Greenstock, *Radiat. Res. 59*:158 (1974).
41. H. P. Misra, *J. Biol. Chem. 249*:2151-2155 (1974).
42. R. L. Willson, A. J. F. Searle, *Nature 255*:498-500 (1975).
43. P. Wardman, *Radiat. Res. 11*:347-398 (1977).
44. W. B. Jacoby, J. N. Keley, and W. H. Habig, in *Glutathione: Metabolism and Function* (I. M. Aria and W. B. Jacoby, Eds.). New York, Raven Press, 1976, pp. 213-224.
45. J. H. Keen, W. H. Habig, and W. B. Jacoby, *J. Biol. Chem. 251*: 6183-6188 (1976).
46. J. E. Biaglow, O. F. Nygaard, and C. L. Greenstock, *Radiat. Res. 59*:158 (1974).
47. O. F. Nygaard, and J. E. Biaglow, *Radiat. Res. 59*:158-159 (1975).
48. H. I. Hadler and T. M. Cao, *Chem.-Biol. Interactions 20*:219-226 (1978).
49. P. J. Ainsworth, M. Channon, R. Sridhar, B. Gushulak, and E. R. Tustanoff, *Can. J. Biochem. 56*:457-461 (1978).
50. K. Tatsumi, S. Kitamura, and H. Yoshimura, *Arch. Biochem. Biophys. 175*:131-141 (1976).
51. D. R. McCalla, A. Reuvers, and C. Kaiser, *Cancer Res. 31*:2184-2188 (1971).
52. M. Morita, D. R. Fuller, and J. R. Gillete, *Biochem. Pharmacol. 20*:217-226 (1971).
53. D. Voshall and D. O. Carr, *Biochem. Pharmacol. 22*:1521-1523 (1973).
54. J. E. Biaglow, C. Koch, and B. Jacobson, *Biochem. Biophys. Res. Commun. 70*:1313-1323 (1976).
55. J. E. Biaglow, *Radiation Res. 86*:212-242 (1981).
56. I. Fridovich, in *Biochemical and Medical Aspects of Active Oxygen*, (O. Hayaishi and K. Asada, Eds.). University Park Press, Baltimore, 1977.
57. P. C. Chan and B. H. Bielski, *J. Biol. Chem. 250*:7266-7271 (1975).

58. J. McCann and B. N. Ames, *Ann. N.Y. Acad. Sci. 271*:5 (1976).
59. D. I. Edwards and G. E. Mithison, *J. Gen. Microbiol. 63*:297-303 (1970).
60. M. Nagao, T. Yahagi, and T. Sugimura, *Biochem. Biophys. Res. Commun. 83*:373-378 (1978).
61. N. R. Bachur, S. Gordon, and M. V. Gree, *Cancer Res. 38*:1745-1750 (1978).
62. M. Nishikimi, *Biochem. Biophys. Res. Commun. 63*:463-468 (1975).
63. J. L. Redpath and R. L. Willson, *Int. J. Radiat. Biol. 23*:2447-2453 (1974).
64. J. D. Chapman, A. P. Reuvers, J. Borsa, and C. L. Greenstock, *Radiat. Res. 56*:291-301 (1973).
65. J. M. McCord and I. Fridovich, *J. Biol. Chem. 244*:6056-6066 (1969).
66. C. L. Greenstock and I. Dunlop, *Radiat. Res. 56*:428-440 (1973).
67. M. Simic, I. A. Taub, J. Tocci, and U. Hurwitz, *Biochem. Biophys. Res. Commun. 62*:161-166 (1975).
68. S. K. Sahu, L. W. Oberley, R. H. Stevens, and E. F. Riley, *J. Nat. Cancer Inst. 58*:1125-1127 (1977).
69. C. J. Koch and J. E. Biaglow, *J. Cell Physiol. 94*:299-306 (1978).
70. G. E. Adams, *Br. Med. Bull. 29*:48-53 (1973).
71. G. E. Adams, in *Cancer: A Comprehensive Treatise* (F. F. Becker, Ed.), Vol. 6, 1977, Chap. 6, pp. 181-223.
72. G. E. Adams, J. Denekamp, and J. F. Fowler, in *Proceedings of the 9th International Congress of Chemotherapy* (K. Hellman, Ed.). Plenum Press, New York, 1955.
73. G. E. Adams, E. D. Clarke, R. S. Jacobs, I. J. Stratford, R. G. Wallace, P. Wardman, and M. E. Watts, *Biochem. Biophys. Res. Commun. 72*:824-829 (1976).
74. R. Sridhar, C. Koch, and R. Sutherland, *Int. J. Radiat. Oncol. Biol. Phys. 1*:1149-1157 (1976).
75. R. Sutherland, J. Richardson, C. J. Koch, and R. Sridhar, in *Metronidazole: Proceedings, Montreal, May 26-28, 1976*. Excerpta Medica, Amsterdam, 1977, pp. 193-203.
76. B. A. Moore, B. Palcic, and L. D. Skarsgard, *Radiat. Res. 67*:459-473 (1977).
77. E. O. Pettersen, *Br. J. Cancer 37* (Suppl. III):107-110 (1978).
78. C. J. Gomer, R. J. R. Johnson, F. W. Hetzel, and G. Lawrence, *Br. J. Cancer 37* (Suppl. III):228-231 (1978).
79. J. L. Foster, P. J. Conroy, A. J. Searle, and R. L. Willson, *Br. J. Cancer 33*:485-490 (1976).
80. G. F. Whitmore, S. Gulyas, and A. J. Varghese, *Br. J. Cancer 37* (Suppl. III):115-119 (1978).
81. B. Palcic and L. D. Skarsgard, *Br. J. Cancer 37* (Suppl III):54-59 (1978).
82. C. R. Geard, S. F. Povlas, M. B. Astor, and E. J. Hall, *Cancer Res. 38*:644-649 (1978).
83. O. Dann and E. F. Moller, *Chem. Ber. 82*:76-92 (1949).
84. T. Sasaki, *Chem. Pharm. Bull.* (Tokyo) *2*:104-108 (1954).
85. K. Hirano, S. Yoshina, K. Okamura, and I. Suzuka, *Bull. Chem. Soc. Japan 40*:2229-2333 (1967).

86. E. J. Hall and L. Roizin-Towle, *Radiology 117*:453-457 (1975).
87. H. B. Stone and H. R. Withers, *Radiology 113*:1189-1194 (1975).
88. A. C. Begg, P. W. Sheldon, and J. L. Foster, *Br. J. Radiol. 47*:399-404 (1974).
89. M. J. Haynes and W. R. Inch, *Radiat. Res. 68*:184-187 (1976).
90. J. M. Brown, *Radiat. Res. 64*:633-674 (1975).
91. J. Denekamp and S. R. Harris, *Radiat. Res. 61*:191-203 (1975).
92. J. A. Stanley, M. J. Peckham, and G. G. Steel, *Br. J. Cancer 37* (Suppl. III):220-224 (1978).
93. V. D. Courtenay, I. E. Smith, and G. G. Steel, *Br. J. Cancer 37* (Suppl. III):225-227 (1978).
94. P. W. Sheldon and J. F. Fowler, *Br. J. Cancer 37* (Suppl. III): 242-245 (1978).
95. L. M. Van Putten and T. Smink, *Br. J. Cancer 37* (Suppl. III): 246-249 (1978).
96. R. C. Urtasun, J. D. Chapman, M. L. Feldstein, R. P. Band, H. R. Rabin, A. F. Wilson, B. Marynowski, E. Starreveld, and T. Shnitka, *Br. J. Cancer 37* (Suppl. III):271-275 (1978).
97. T. L. Phillips, T. H. Wasserman, R. J. Johnson, C. J. Gomer, G. A. Lawrence, M. L. Levine, W. Sadee, J. S. Penta, and D. J. Rubin, *Br. J. Cancer 37* (Suppl. III):276-280 (1978).
98. M. I. Saunders, S. Dische, P. Anderson, and I. R. Flockhart, *Br. J. Cancer 37* (Suppl. III):268-270 (1978).
99. H. D. Kogelnik, H. J. Meyer, K. Jentsch, T. Szepesi, K. H. Karcher, E. Maida, B. Mamoli, P. Wessely, and F. Zaunbauer, *Br. J. Cancer 37* (Suppl. III):281-285 (1978).
100. I. J. Stratford and G. E. Adams, *Br. J. Cancer 35*:307-314 (1977).
101. N. M. Bleehan, D. Honess, and J. Morgan, *Br. J. Cancer 35*:299-306 (1977).
102. R. Sridhar and R. Sutherland, *Int. J. Radiat. Oncol. Biol. Phys. 2*:531-535 (1977).
103. R. M. Sutherland and R. Sridhar, in *Cancer Therapy by Hyperthermia and Radiation* (C. Streffer et al., Eds.). Urban & Schwarzenberg, Baltimore and Munich, 1978, pp. 271-273.
104. R. Sutherland and W. MacFarlane, *Br. J. Cancer 37* (Suppl. III):168-172 (1978).
105. A. M. Rauth, *Metronidazole: Proceedings, Montreal, May 26-28, 1976*. Excerpta Medica, Amsterdam, 1977, pp. 176-184.
106. R. E. Durand, P. Olive, and J. E. Biaglow, in *Wisconsin Clinical Prospects for Hypoxic Sensitizer Hyperthermia Conference*, (W. L. Caldwell and R. E. Durand, Eds.) Wisconsin Press, 1978, pp. 43-50.
107. E. J. Hall, M. Astor, C. Geard, and J. E. Biaglow, *Br. J. Cancer 35*:810-815 (1977).
108. C. A. Reznikoff, D. W. Brankow, and C. Heidelberger, *Cancer Res. 33*:3231-3238 (1973).
109. D. R. McCalla, P. L. Olive, and Y. Tu, in *Fundamentals in Cancer Prevention* (P. N. Magee, Ed.). University Press, Baltimore, 1976, p. 229.
110. D. R. McCalla and D. Voutsinos, *Mutation Res. 26*:3-16 (1974).
111. J. D. Chapman, A. P. Reuvers, J. Borsa, A. Petkau, and D. R. McCalla, *Cancer Res. 32*:2616-2624 (1972).

112. H. S. Rosenkranz and W. T. Speck, *Metronidazole: Proceedings, Montreal, May 26-28, 1976*. Excerpta Medica, Amsterdam, 1977, pp. 119-125.
113. W. D. Rupp, Z. Mroczkowski, and K. C. Agrawal, *Br. J. Cancer* *37* (Suppl. III):60-65 (1978).
114. M. J. Hampton and R. Sridhar, Abstract of paper presented at the 28th Annual Meeting of the Radiation Research Society, New Orleans, Louisiana, June 1-5, 1980.

AUTHOR INDEX

Numbers in parentheses are reference numbers and indicate that an author's work is referred to although his name is not cited in the text. Italic numbers give the page on which the complete reference is listed.

D

F

G

N

O

S

Y

SUBJECT INDEX

T

U

V

W

X